AF361690

Nanotechnology for Environmental and Biomedical Research

Nanotechnology for Environmental and Biomedical Research

Editor

Giada Frenzilli

MDPI • Basel • Beijing • Wuhan • Barcelona • Belgrade • Manchester • Tokyo • Cluj • Tianjin

Editor
Giada Frenzilli
University of Pisa
Italy

Editorial Office
MDPI
St. Alban-Anlage 66
4052 Basel, Switzerland

This is a reprint of articles from the Special Issue published online in the open access journal *Nanomaterials* (ISSN 2079-4991) (available at: https://www.mdpi.com/journal/nanomaterials/special_issues/nanotech_environ_bio).

For citation purposes, cite each article independently as indicated on the article page online and as indicated below:

LastName, A.A.; LastName, B.B.; LastName, C.C. Article Title. *Journal Name* **Year**, *Volume Number*, Page Range.

ISBN 978-3-0365-0718-7 (Hbk)
ISBN 978-3-0365-0719-4 (PDF)

Cover image courtesy of Mara Palumbo.

Contents

About the Editor . . . . **vii**

Giada Frenzilli
Nanotechnology for Environmental and Biomedical Research
Reprinted from: *Nanomaterials* **2020**, *10*, 2220, doi:10.3390/nano10112220 . . . . **1**

Laura Riva, Nadia Pastori, Alice Panozzo, Manuela Antonelli and Carlo Punta
Nanostructured Cellulose-Based Sorbent Materials for Water Decontamination from
Organic Dyes
Reprinted from: *Nanomaterials* **2020**, *10*, 1570, doi:10.3390/nano10081570 . . . . **5**

Patrizia Guidi, Margherita Bernardeschi, Mara Palumbo, Massimo Genovese,
Vittoria Scarcelli, Andrea Fiorati, Laura Riva, Carlo Punta, Ilaria Corsi and Giada Frenzilli
Suitability of a Cellulose-Based Nanomaterial for the Remediation of Heavy Metal
Contaminated Freshwaters: A Case-Study Showing the Recovery of Cadmium Induced DNA
Integrity Loss, Cell Proliferation Increase, Nuclear Morphology and Chromosomal Alterations
on *Dreissena polymorpha*
Reprinted from: *Nanomaterials* **2020**, *10*, 1837, doi:10.3390/nano10091837 . . . . **23**

Giulia Liberatori, Giacomo Grassi, Patrizia Guidi, Margherita Bernardeschi, Andrea Fiorati,
Vittoria Scarcelli, Massimo Genovese, Claudia Faleri, Giuseppe Protano, Giada Frenzilli,
Carlo Punta and Ilaria Corsi
Effect-Based Approach to Assess Nanostructured Cellulose Sponge Removal Efficacy of Zinc
Ions from Seawater to Prevent Ecological Risks
Reprinted from: *Nanomaterials* **2020**, *10*, 1283, doi:10.3390/nano10071283 . . . . **39**

Stefania Mariano, Elisa Panzarini, Maria D. Inverno, Nick Voulvoulis and Luciana Dini
Toxicity, Bioaccumulation and Biotransformation of Glucose-Capped Silver Nanoparticles in
Green Microalgae *Chlorella vulgaris*
Reprinted from: *Nanomaterials* **2020**, *10*, 1377, doi:10.3390/nano10071377 . . . . **59**

Sujin Park, Jaehyeon Park, Ji Ha Lee, Myong Yong Choi and Jong Hwa Jung
Spectroscopic Study of the Salicyladazine Derivative–UO$_2^{2+}$ Complex and Its Immobilization
to Mesoporous Silica
Reprinted from: *Nanomaterials* **2019**, *9*, 688, doi:10.3390/nano9050688 . . . . **75**

Hye-Min Lee, Byeong-Hoon Lee, Soo-Jin Park, Kay-Hyeok An and Byung-Joo Kim
Pitch-Derived Activated Carbon Fibers for Emission Control of
Low-Concentration Hydrocarbon
Reprinted from: *Nanomaterials* **2019**, *9*, 1313, doi:10.3390/nano9091313 . . . . **83**

Audrey Barranger, Laura M. Langan, Vikram Sharma, Graham A. Rance, Yann Aminot,
Nicola J. Weston, Farida Akcha, Michael N. Moore, Volker M. Arlt, Andrei N. Khlobystov,
James W. Readman and Awadhesh N. Jha
Antagonistic Interactions between Benzo[a]pyrene and Fullerene (C$_{60}$) in Toxicological
Response of Marine Mussels
Reprinted from: *Nanomaterials* **2019**, *9*, 987, doi:10.3390/nano9070987 . . . . **95**

Marianna Santonastaso, Filomena Mottola, Concetta Iovine, Fulvio Cesaroni, Nicola Colacurci and Lucia Rocco
In Vitro Effects of Titanium Dioxide Nanoparticles (TiO$_2$NPs) on Cadmium Chloride (CdCl$_2$) Genotoxicity in Human Sperm Cells
Reprinted from: *Nanomaterials* **2020**, *10*, 1118, doi:10.3390/nano10061118 . . . . **123**

Bianca-Vanesa Boros and Vasile Ostafe
Evaluation of Ecotoxicology Assessment Methods of Nanomaterials and Their Effects
Reprinted from: *Nanomaterials* **2020**, *10*, 610, doi:10.3390/nano10040610 . . . . **139**

Zahra Zahra, Zunaira Habib, Sujin Chung and Mohsin Ali Badshah
Exposure Route of TiO$_2$ NPs from Industrial Applications to Wastewater Treatment and Their Impacts on the Agro-Environment
Reprinted from: *Nanomaterials* **2020**, *10*, 1469, doi:10.3390/nano10081469 . . . . **197**

Ioannis A. Tsakmakidis, Theodoros Samaras, Sofia Anastasiadou, Athina Basioura, Aikaterini Ntemka, Ilias Michos, Konstantinos Simeonidis, Isidoros Karagiannis, Georgios Tsousis, Mavroeidis Angelakeris and Constantin M. Boscos
Iron Oxide Nanoparticles as an Alternative to Antibiotics Additive on Extended Boar Semen
Reprinted from: *Nanomaterials* **2020**, *10*, 1568, doi:10.3390/nano10081568 . . . . **219**

Mohammad Faisal Umar, Faizan Ahmad, Haris Saeed, Saad Ali Usmani, Mohammad Owais and Mohd Rafatullah
Bio-Mediated Synthesis of Reduced Graphene Oxide Nanoparticles from *Chenopodium album*: Their Antimicrobial and Anticancer Activities
Reprinted from: *Nanomaterials* **2020**, *10*, 1096, doi:10.3390/nano10061096 . . . . **235**

Luis Eduardo Genaro, Giovana Anovazzi, Josimeri Hebling and Angela Cristina Cilense Zuanon
Glass Ionomer Cement Modified by Resin with Incorporation of Nanohydroxyapatite: In Vitro Evaluation of Physical-Biological Properties
Reprinted from: *Nanomaterials* **2020**, *10*, 1412, doi:10.3390/nano10071412 . . . . **249**

Tae Hwan Shin, Da Yeon Lee, Abdurazak Aman Ketebo, Seungah Lee, Balachandran Manavalan, Shaherin Basith, Chanyoung Ahn, Seong Ho Kang, Sungsu Park and Gwang Lee
Silica-Coated Magnetic Nanoparticles Decrease Human Bone Marrow-Derived Mesenchymal Stem Cell Migratory Activity by Reducing Membrane Fluidity and Impairing Focal Adhesion
Reprinted from: *Nanomaterials* **2019**, *9*, 1475, doi:10.3390/nano9101475 . . . . **261**

Seira Matsuo, Kenjirou Higashi, Kunikazu Moribe, Shin-ichiro Kimura, Shigeru Itai, Hiromu Kondo and Yasunori Iwao
Combination of Roll Grinding and High-Pressure Homogenization Can Prepare Stable Bicelles for Drug Delivery
Reprinted from: *Nanomaterials* **2018**, *8*, 998, doi:10.3390/nano8120998 . . . . **279**

Elena Sánchez-López, Mariana Guerra, João Dias-Ferreira, Ana Lopez-Machado, Miren Ettcheto, Amanda Cano, Marta Espina, Antoni Camins, Maria Luisa Garcia and Eliana B. Souto
Current Applications of Nanoemulsions in Cancer Therapeutics
Reprinted from: *Nanomaterials* **2019**, *9*, 821, doi:10.3390/nano9060821 . . . . **299**

About the Editor

Giada Frenzilli was born in Pisa in 1968. She is Associate Professor at the Department of Clinical and Experimental Medicine, University of Pisa. She received a degree in Biological Science at Pisa University in the academic year 1991–1992 (110 summa cum laude), and a Ph.D. in Genetics at Ferrara University in 1997, spending part of her doctorate research at the Integrated Laboratory System (NC, USA), pioneering the use of Comet assay in human and environmental monitoring. She received fellowships and contracts from 1997 to 1999 and a Post-Doc fellowship from 1999 to 2002 at Pisa University. Assistant Professor since 2002, she received an award from the Environmental Mutagenesis Society as "best researcher of the year" in 2004, in the environmental mutagenesis field. Giada Frenzilli has more than twenty-five years of experience as a genetic toxicologist, with special emphasis on cytogenetic mechanisms, study design and data analysis. She used a multi-biomarker approach to study the cellular impact of physical and chemical environmental pollutants, like ionizing radiation, loud noise, heavy metals, organic compounds and drug abuse in animal models, human populations and sentinel species. Another field of interest concerns the study of cellular mechanisms underlying the onset of different diseases. At present, one of her main interests is "nano-eco-genotoxicology", aiming to study the cytotoxic and genotoxic effects of nanomaterials by in vitro and in vivo approaches. She has carried out research projects on amorphous silica particles on murine alveolar macrophages (RAW 264.7) and human epithelial lung (A549) cell lines, selected as representatives of occupational and environmental exposure (PRIN 2007 National Program on the "mechanisms responsible for cytotoxicity and genotoxicity of silica NANOPARTICLES and NANOMETRIC fibrous silicates having strictly controlled size, structure and composition" (2007498XRF), Unit Coordinator), and on the evaluation of genotoxic potential of nanosized titanium dioxide, which is one of the most widely used nanomaterials in a range of products/processes, including cosmetics, sun screens, paints, pharmaceuticals, building materials, the paper industry and waste-water treatment in different cell lines (PRIN 2009 National Program on the "Marine ecotoxicology of MANUFACTURED NANOMATERIALS. Biological effects and bioaccumulation of combined exposure to nano-TiO2, metals and dioxin on edible species"). The role of nanomaterials in environmental remediation has been studied in projects based on European funds, i.e., POR CRO FSE 2007–2013, "Use of carbon based-nanoparticles to remediate metal and organic polluted environments. Biological response evaluation in sentinel species", where she served as Principal Investigator and POR-FESR 2014–2020 "Nanomaterials for the remediation of environmental matrices associated with dewatering (NanoBonD)", where she served as Unit Coordinator. As an invited speaker and chair-person at national and international workshops and conferences from 1993 to date, she followed her inclination for multidisciplinarity, dedicating much effort to attracting scientists with different areas of expertise and setting up international collaborations. The quality of Giada Frenzilli scientific activity is supported by 68 publications in peer-reviewed journals and books (H-index= 31), many invited lectures, hundreds of abstracts and participations in conferences. She has tutored of many degree and Ph.D. students, has been teaching at the International Summer School, "From genes to cells: a basic course of molecular, cellular, and ultrastructural biology", from 2013 to date, and currently reviews scientific papers for Mutagenesis, Mutation Research ("Top reviewer" in 2010), Environmental and Molecular Mutagenesis, Aquatic Toxicology, Biomarkers, Chemosphere, Ecotoxicology and Environmental Safety, Comparative Biochemistry and Physiology, Marine Environmental Research, Environmental Pollution, Water Research, PlosOne, Food and

Chemical Toxicology, Nanotoxicology.

 *nanomaterials* 

Editorial

Nanotechnology for Environmental and Biomedical Research

Giada Frenzilli

Department of Clinical and Experimental Medicine, University of Pisa, 56126 Pisa, Italy; giada@biomed.unipi.it;
Tel.: +39 050 2219111

Received: 26 October 2020; Accepted: 5 November 2020; Published: 8 November 2020

Given the high production and broad feasibility of nanomaterials, the application of nanotechnology includes the use of engineered nanomaterials (ENMs) to clean-up polluted media such as soils, water, air, groundwater and wastewaters, and is known as nanoremediation. Contamination by hazardous substances in environmental matrices, including landfills, oil fields, and manufacturing and industrial sites, represents a global concern and needs to be remediated, since it poses a serious risk for the environment and human health. Particular attention should also be focused on the use of medical devices and recent developments in the use of nanoparticles expressed as drug delivery systems designed to treat a wide variety of diseases. This Special Issue, "Nanotechnology for Environmental and Biomedical Research", is characterized by this double role, and is aimed at facing the use of nanotechnology for environmental and human health. It aims at collecting a compilation of articles that strongly demonstrate the continuous efforts in developing advanced and safe nanomaterial-based technologies for nanoremediation, as well as for drug delivery and other biomedical applications. The present Special Issue has covered the most recent advances in the safe nanomaterials synthesis field, as well as in environmental applications, the use of restorative materials, drug delivery, and other clinical applications. In this Special Issue, thirteen selected original research papers and three reviews are collected. More than 80 scientists from universities and research institutions contributed through their research studies and expertise for the success of this Special Issue. The scientific contributions are summarized in the next paragraphs. Ten contributions, including two reviews, are related to environmental applications. Riva and co-authors [1] proposed a nanostructured cellulose-based material as a possible sorbent for the removal of organic dyes from water, demonstrating its sorbent efficiency for four different organic dyes commonly used for fabric printing. The material performance was compared with that of an activated carbon, routinely used for this application, thus highlighting the potentialities and limits of this new material together with the important issue of the regeneration and reuse of the sorbent. The suitability of cellulose-based nanomaterials for the remediation of heavy metal-contaminated environments was then assessed by Guidi and colleagues [2]. They indicated an eco-friendly cellulose-based nanosponge as a safe and effective candidate in the cadmium remediation process, being able to sequester cadmium and restore cellular damage induced by cadmium exposure in the zebra mussel, an animal model typical of freshwater environments, without altering cellular physiological activity. In particular, the authors showed the recovery of cadmium-induced DNA integrity loss, cell proliferation increase, and nuclear morphology and chromosomal alterations in zebra mussel haemocytes. The same ecofriendly nanosponges were demonstrated to be effective in the removal of zinc ions from the seawater environment, through another in vivo study. The contribution by Liberatori et al. [3] confirmed, besides the efficacy of the nanosponge, the recovery of the toxicological responses induced in the marine mussel. The genetic, chromosomal, cellular and tissue alterations induced by zinc ions were actually reported at control levels, supporting the environmentally safe application of cellulose-based nanosponges for heavy metal removal from seawater. Mariano and colleagues [4] demonstrated the possible use of a microalgae as a model microorganism to study silver nanoparticle toxicity, but also to protect against nanoparticle pollution. They showed that silver nanoparticles, internalized in a time- and dose-dependent manner inside large vacuoles, were not released into the medium for almost one week, without undergoing any biotransformation, confirming the role of the microalgae in environmental protection. Still within the field of remediation, Park and co-authors [5] proposed a nanohybrid material able to detect uranyl ions spectroscopically and act as a uranyl ion absorbent in an aqueous system. The contribution has high impact because the uranyl ion, the most soluble toxic uranium species, is considered as an important index for monitoring nuclear wastewater quality. Furthermore, pitch-based activated carbon fibers, prepared by steam activation, were proposed by Lee and colleagues [6] as a solution for unburned hydrocarbon car emission removal. The pitch-based activated carbon fibers actually exhibited enhanced butane working capacity and adsorption velocity when compared to commercial products, with lower concentrations of n-butane due to their characteristic pore structure. Besides nanostructured materials, nanotoxicology and nanoremediation, another keyword was Trojan horse effect. Two contributions faced this research aspect, which concerns the interactions between nanostructured materials and classical pollutants. The assessment of the ecotoxicological effects of the interaction between benzo[a]pyrene and fullerene (C60) was performed by Barranger and co-authors [7] in the marine mussel. They found antagonistic responses at the genotoxic and proteomic level, also showing a complex multi-modal response to environmental stressors in the species used. Another antagonistic interaction was reported by Santonastaso and co-workers [8], who assessed the in vitro effects of titanium dioxide nanoparticles and cadmium interaction in human sperm cells by investigating semen parameters, apoptotic processes, DNA integrity, genomic stability and oxidative stress. They demonstrated that the genotoxicity induced by the co-exposure was lower if compared to single cadmium exposure, suggesting the formation of a sandwich-like structure, with cadmium in the middle, to explain the inhibition of its genotoxicity in human sperms. In order to conclude the section concerning the environmental aspects of nanotechnology research, two valuable reviews are reported here. The one by Boros and Ostafe [9] reviewed the ecotoxicological effects of nanomaterials as well as their testing methods, which are adaptable for nanomaterials. The authors reported a broad spectrum of genetic, molecular, cellular, morphological and behavioral effects, involving a wide range of organisms, such as algae, duckweed, amphipods, daphnids, chironomids, terrestrial plants, nematodes and earthworms. It is interesting to note that, having reported values mainly for aquatic ecotoxicity, the most sensitive test turned out to be the algae assay, and the most toxic nanomaterials were composed of silver, reinforcing the impact of the contribution by Mariano et al. [4]. The other review, by Zahra et al. [10], mainly focused on the aspect of environmental safety, and gave an overview of the potential exposure route of titanium dioxide nanoparticles from industrial applications to wastewater treatment, and the impact of this on the agro-environment. The increasing interest in the role of nanotechnology in nanosafe applications is paralleled in the scientific arena by the extreme need to upgrade our knowledge via the use of nanotechnology in the human health field. In this context, this Special Issue collects five research articles and one review covering the principal aspects of nanotechnology applied to biomedical applications, such as the study of nanoparticles used as antibiotics and restorative material, as well as nanomaterials able to deliver drugs or to modulate stem cell trafficking. The original contribution by the group of Tsakmakidis [11] investigated the effect of Fe_3O_4 nanoparticles on an animal model to assess their ability to substitute antibiotics additives in extending semen storage time. The authors found a significant reduction in the bacterial load in the samples incubated with Fe_3O_4 nanoparticles in comparison with controls after both 24 and 48 hours. Moreover, no adverse effects on sperm characteristics, such as morphology, viability or DNA integrity, were detected, offering important information concerning semen handling in the artificial insemination field. A novel method of preparing reduced graphene oxide from graphene oxide was developed, employing a vegetable extract, by Uman and co-authors [12]. They showed that the "green" modification of graphene oxide leads to an enhancement in antibacterial activity against Gram-positive and Gram-negative bacteria, besides increasing antibiofilm activity on a human breast cancer cell line, thus indicating reduced graphene oxide nanoparticles as a potential anticancer agent. Another biomedical application was proposed by the group of Genaro and colleagues [13], who indicated the incorporation of nanohydroxyapatite in resin-modified glass ionomer cement, a restorative material, as a method to increase the cell viability and biocompatibility performance of odontoblastoid cells. The biophysical effect of nanomaterials within stem cell-based therapies has been investigated by Shin and co-authors [14]. They labeled human bone marrow-derived mesenchymal stem cells with silica-coated magnetic nanoparticles incorporating rhodamine B isothiocyanate, and found decreased cell viability, an increase in intracellular reactive oxygen species, and, most of all, a decrease in stem cell migration activity related with membrane fluidity reduction and focal adhesion impairment. Therefore, the authors highlighted the importance of nanoparticles that are used for stem cell trafficking or clinical applications being labeled using optimal nanoparticle concentrations, so as to preserve human bone marrow-derived mesenchymal stem cells' migratory activity, thus ensuring successful outcomes following stem cell localization. Regarding another key word of the Special Issue, i.e., drug delivery, an interesting contribution came from Matsuo and co-authors [15], who indicated encapsulated lipid-based nanoparticles, prepared from neutral hydrogenated soybean phosphatidylcholine and dipalmitoylphosphatidylglycerol, as an optimal way, after roll grinding and high-pressure homogenization, to prepare stable bicelles for nifedipine delivery. Cryo-transmission electron microscopy and atomic force microscopy were also performed to better understand the structure of such nifedipine-encapsulated lipid-based nanoparticles. So, taking into consideration the result of long term stability, standard nifedipine-encapsulated lipid-based nanoparticle bicelles (5 liposomes/1 micelles) showed the most long-term stabilities, illustrating a useful preparation method for stable bicelles to be employed in the drug delivery field. A review of the role of nanoemulsions in cancer therapeutics [16] effectively concludes this excursus on nanotechnology and biomedical applications. Nanoemulsions are pharmaceutical formulations, made of particles within the nanometric range, which are able to encapsulate drugs that are poorly hydrophilic due to their hydrophobic core. Sánchez-López and co-workers reviewed the characteristics of nanoemulsions that face and overcome problems such as water solubility, targeting specificity and multidrug resistance. Nanoemulsions can actually be modified by the use of ligands of different natures to target specific tumor cell components or to avoid multidrug resistance. A broad spectrum of methodologies through which nanoemulsions can be designed to achieve successful therapeutic outcomes in several types of cancer are reported. Applications of nanoemulsions in colon, ovarian, prostate, breast, lung, leukemia and melanoma cancer therapy, as well as in nanotheragnostics and drug delivery, have been widely discussed. In conclusion, the papers collected in this Special Issue cover the most relevant advanced applications of nanotechnology in the environmental and human health fields, also providing new research directions to be expanded in the near future.

Author Contributions: G.F. solely contributed to the editorial. Author has read and agreed to the published version of the manuscript.

Funding: This research received no external funding.

Acknowledgments: We are grateful to all the authors who contributed to this Special Issue. We also express our acknowledgments to the referees for reviewing the manuscripts.

Conflicts of Interest: The author declares no conflict of interest.

References

1. Riva, L.; Pastori, N.; Panozzo, A.; Antonelli, M.; Punta, C. Nanostructured Cellulose-Based Sorbent Materials for Water Decontamination from Organic Dyes. *Nanomaterials* **2020**, *10*, 1570. [CrossRef] [PubMed]
2. Guidi, P.; Bernardeschi, M.; Palumbo, M.; Genovese, M.; Scarcelli, V.; Fiorati, A.; Riva, L.; Punta, C.; Corsi, I.; Frenzilli, G. Suitability of a Cellulose-Based Nanomaterial for the Remediation of Heavy Metal Contaminated Freshwaters: A Case-Study Showing the Recovery of Cadmium Induced DNA Integrity Loss, Cell Proliferation Increase, Nuclear Morphology and Chromosomal Alterations on Dreissena polymorpha. *Nanomaterials* **2020**, *10*, 1837. [CrossRef]
3. Liberatori, G.; Grassi, G.; Guidi, P.; Bernardeschi, M.; Fiorati, A.; Scarcelli, V.; Genovese, M.; Faleri, C.; Protano, G.; Frenzilli, G.; et al. Effect-Based Approach to Assess Nanostructured Cellulose Sponge Removal

Efficacy of Zinc Ions from Seawater to Prevent Ecological Risks. *Nanomaterials* **2020**, *10*, 1283. [CrossRef] [PubMed]

4.   Mariano, S.; Panzarini, E.; Inverno, M.D.; Voulvoulis, N.; Dini, L. Toxicity, Bioaccumulation and Biotransformation of Glucose-Capped Silver Nanoparticles in Green Microalgae *Chlorella vulgaris*. *Nanomaterials* **2020**, *10*, 1377. [CrossRef] [PubMed]

5.   Park, S.; Park, J.; Lee, J.H.; Choi, M.Y.; Jung, J.H. Spectroscopic Study of the Salicyladazine Derivative⁻UO_2^{2+} Complex and Its Immobilization to Mesoporosorous Silica. *Nanomaterials* **2019**, *9*, 688. [CrossRef] [PubMed]

6.   Lee, H.-M.; Lee, B.-H.; Park, S.-J.; An, K.-H.; Kim, B.-J. Pitch-Derived Activated Carbon Fibers for Emission Control of Low-Concentration Hydrocarbon. *Nanomaterials* **2019**, *9*, 1313. [CrossRef] [PubMed]

7.   Barranger, A.; Langan, L.M.; Sharma, V.; Rance, G.A.; Aminot, Y.; Weston, N.J.; Akcha, F.; Moore, M.N.; Arlt, V.M.; Khlobystov, A.N.; et al. Antagonistic Interactions between Benzo[a]pyrene and Fullerene (C_{60}) in Toxicological Response of Marine Mussels. *Nanomaterials* **2019**, *9*, 987. [CrossRef] [PubMed]

8.   Santonastaso, M.; Mottola, F.; Iovine, C.; Cesaroni, F.; Colacurci, N.; Rocco, L. In Vitro Effects of Titanium Dioxide Nanoparticles (TiO_2NPs) on Cadmium Chloride ($CdCl_2$) Genotoxicity in Human Sperm Cells. *Nanomaterials* **2020**, *10*, 1118. [CrossRef]

9.   Boros, B.-V.; Ostafe, V. Evaluation of Ecotoxicology Assessment Methods of Nanomaterials and Their Effects. *Nanomaterials* **2020**, *10*, 610. [CrossRef] [PubMed]

10.   Zahra, Z.; Habib, Z.; Chung, S.; Badshah, M.A. Exposure Route of TiO_2 NPs from Industrial Applications to Wastewater Treatment and Their Impacts on the Agro-Environment. *Nanomaterials* **2020**, *10*, 1469. [CrossRef] [PubMed]

11.   Tsakmakidis, I.A.; Samaras, T.; Anastasiadou, S.; Basioura, A.; Ntemka, A.; Michos, I.; Simeonidis, K.; Karagiannis, I.; Tsousis, G.; Angelakeris, M.; et al. Iron Oxide Nanoparticles as an Alternative to Antibiotics Additive on Extended Boar Semen. *Nanomaterials* **2020**, *10*, 1568. [CrossRef] [PubMed]

12.   Umar, M.F.; Ahmad, F.; Saeed, H.; Usmani, S.A.; Owais, M.; Rafatullah, M. Bio-Mediated Synthesis of Reduced Graphene Oxide Nanoparticles from Chenopodium album: Their Antimicrobial and Anticancer Activities. *Nanomaterials* **2020**, *10*, 10. [CrossRef] [PubMed]

13.   Genaro, L.E.; Anovazzi, G.; Hebling, J.; Zuanon, A.C.C. Glass Ionomer Cement Modified by Resin with Incorporation of Nanohydroxyapatite: In Vitro Evaluation of Physical-Biological Properties. *Nanomaterials* **2020**, *10*, 1412. [CrossRef] [PubMed]

14.   Shin, T.H.; Lee, D.Y.; Ketebo, A.A.; Lee, S.; Manavalan, B.; Basith, S.; Ahn, C.; Kang, S.H.; Park, S.; Lee, G. Silica-Coated Magnetic Nanoparticles Decrease Human Bone Marrow-Derived Mesenchymal Stem Cell Migratory Activity by Reducing Membrane Fluidity and Impairing Focal Adhesion. *Nanomaterials* **2019**, *9*, 1475. [CrossRef] [PubMed]

15.   Matsuo, S.; Higashi, K.; Moribe, K.; Kimura, S.-I.; Itai, S.; Kondo, H.; Iwao, Y. Combination of Roll Grinding and High-Pressure Homogenization Can Prepare Stable Bicelles for Drug Delivery. *Nanomaterials* **2018**, *8*, 998. [CrossRef] [PubMed]

16.   Sánchez-López, E.; Guerra, M.; Dias-Ferreira, J.; Lopez-Machado, A.L.; Ettcheto, M.; Cano, A.; Espina, M.; Camins, A.; Garcia, M.L.; Souto, E.B. Current Applications of Nanoemulsions in Cancer Therapeutics. *Nanomaterials* **2019**, *9*, 821. [CrossRef] [PubMed]

Publisher's Note: MDPI stays neutral with regard to jurisdictional claims in published maps and institutional affiliations.

Article

Nanostructured Cellulose-Based Sorbent Materials for Water Decontamination from Organic Dyes

Laura Riva [1], Nadia Pastori [1], Alice Panozzo [1], Manuela Antonelli [2,*] and Carlo Punta [1,3,*]

[1] Department of Chemistry, Materials, and Chemical Engineering "G. Natta" and INSTM Local Unit, Politecnico di Milano, Piazza Leonardo da Vinci 32, I-20133 Milano, Italy; laura2.riva@polimi.it (L.R.); nadia.pastori@polimi.it (N.P.); alice.panozzo@mail.polimi.it (A.P.)

[2] Department of Civil and Environmental Engineering, Politecnico di Milano, Piazza Leonardo da Vinci 32, I-20133 Milano, Italy

[3] Centro Nazionale Ricerche (C. N. R.) Istituto di Chimica del Riconoscimento Molecolare (ICRM), 20131 Milan, Italy

* Correspondence: manuela.antonelli@polimi.it (M.A.); carlo.punta@polimi.it (C.P.); Tel.: +39-0223-996-407 (M.A.); +39-0223-993-026 (C.P.)

Received: 18 July 2020; Accepted: 5 August 2020; Published: 10 August 2020

Abstract: Nanostructured materials have been recently proposed in the field of environmental remediation. The use of nanomaterials as building blocks for the design of nano-porous micro-dimensional systems is particularly promising since it can overcome the (eco-)toxicological risks associated with the use of nano-sized technologies. Following this approach, we report here the application of a nanostructured cellulose-based material as sorbent for effective removal of organic dyes from water. It consists of a micro- and nano-porous sponge-like system derived by thermal cross-linking among (2,2,6,6-Tetramethylpiperidin-1-yl)oxyl (TEMPO)-oxidized cellulose nanofibers (TOCNF), branched polyethylenimine 25 kDa (bPEI), and citric acid (CA). The sorbent efficiency was tested for four different organic dyes commonly used for fabric printing (Naphthol Blue Black, Orange II Sodium Salt, Brilliant Blue R, Cibacron Brilliant Yellow), by conducting both thermodynamic and kinetic studies. The material performance was compared with that of an activated carbon, commonly used for this application, in order to highlight the potentialities and limits of this biomass-based new material. The possibility of regeneration and reuse of the sorbent was also investigated.

Keywords: nanostructured materials; sorption; organic dyes; wastewater treatment; nanocellulose

---

1. Introduction

The use of engineered nanomaterials (ENMs) to clean-up polluted media, including groundwater and wastewater, has attracted more and more attention in the last decade [1]. This approach offers the possibility to take advantage of the high reactivity and high surface area of nanomaterials, opening the way towards more effective and economically sustainable remediation processes.

Nevertheless, the use of ENMs generates concerns associated with the potential risk for humans and environment, as the (eco-)toxicological impact of these solutions is often underestimated [2].

Recently, we proposed a systematic approach for possibly overcoming this issue, which consists of the use of sustainable and bio-based nanomaterials as building blocks for the design of nano-structured and nano-porous sorbents, capable of taking advantage of the intrinsic nano-dimension of the network, while overpassing the risk of ENM release and migration [3].

In accordance with this strategy, we first identified polysaccharides, and cellulose in particular, as ideal sources, often derived from discharged biomass, for the extraction of nano-sized particles to be further processed [4].

Cellulose represents the most abundant biodegradable and renewable polymer source in the biosphere, with an annual production estimated as over 7.5×10^{10} tons. This almost inexhaustible sustainable polymer possesses unique chemical, physical, and mechanical properties, which have suggested its use for the production of a wide range of materials [5].

Moreover, it is possible to cleave cellulose hierarchical structure in order to obtain nanocellulose (NC) in the form of nanocrystals (cellulose nanocrystals (CNC)) and nanofibers (cellulose nanofibers (CNF)). NC has been widely proposed as building block for the design of a wide range of sorbent materials to be used in wastewater treatment [6–8].

Among the several mechanical and chemical approaches for NC extraction, in recent years we focused on the one we considered the most effective and economically convenient, consisting of the selective oxidation of C6 alcoholic groups of the cellulose glucopyranose units to the corresponding carboxylic acids [9,10]. This transformation is mediated by 2,2,6,6-tetramethylpiperidine 1-oxyl free radical (TEMPO)/NaClO/NaBr system [9,10], and allows to promote nanodefibrillation at basic pH, thanks to the electrostatic repulsion among negatively charged TEMPO-oxidized CNF (TOCNF), due to the deprotonation of carboxylic moieties. The obtained nanofibers present a diameter in a range of 5–100 nm and a length of several microns.

Following the strategy previously described, in 2015 we reported a new class of nanostructured sorbent cellulose nanosponges (CNS), obtained by thermally promoting the cross-linking between TOCNF and branched polyethyleneimine 25 kDa (bPEI), a polymer bearing a high amount of primary, secondary, and tertiary amino groups, thanks to which it is able to interact with a wide range of ions and molecules [11].

In recent years, we modified the CNS formulation by adding an optimized amount of citric acid (CA) as an additional source of carboxylic groups, in order both to strength the mechanical properties of the final material by increasing cross-linking nodes [12], and to better fix bPEI into the network, so that it was also possible to reduce the amount of the same polymer and to support an eco-safe and sustainable design [13,14]. Moreover, we also demonstrated the versatility of the system, with the possibility of grafting bPEI with suitable moieties before undergoing cross-linking, in order to give the material additional functional properties, such as the sensing of fluoride ions [15,16].

Due to the chelating action of amino-groups, CNS have shown superb performances as heavy metals sorbents from both fresh- [11] and seawater [13,17]. On the contrary, the investigation into the behavior of this sorbent material with organic pollutants was limited to few examples, namely phenols, and simply based on an acid–base interaction.

However, by considering the morphology of the sponge, characterized by a high micro- and nano-porosity, as evidenced in previous works [12,18] and better discussed later on, we envisioned the possibility of exploiting the potential of CNS, and to better clarify the possible mechanisms of interaction between the sorbent and the selected organic pollutants.

Anionic organic dyes were chosen as model molecules, due to the continuous request for new and more sustainable technologies for treating a multiplicity of colored industrial wastewater, whose discharge in superficial water bodies can significantly impair freshwater use.

Nowadays, over 100,000 dyes are commercially available, with 7×10^5 tons/year of dyestuff being produced [19]. Synthetic dyes are widely used in several industrial fields, including food, rubber, paper, cosmetic, pharmaceutical, automotive, and textile production chains. Inevitably, a certain fraction of dyes ends up in the washing water during the process and it has to be removed before release in the environment, due to the potential toxicity and negative effects on the aquatic ecosystem.

Due to their complex and variable chemical structure, many dyes are difficult to be removed by following a standard photo- and/or oxidative degradation approach, and sorption is often the process selected for wastewater treatment, with activated carbon mainly used for this purpose. In fact, sorption onto activated carbon is a well-established process, effective at removing a wide range of molecules, including color-responsible molecules from textile wastewater. Moreover, recently it has been reported that the efficiency of this process can be improved by driving the growth of nanoparticles into activated carbon pores, so that it is possible to combine sorption and degradation of the organic dye pollutant [20]. More generally, sorption represents a simply managed process, with no important drawbacks and a relative low cost, compared to other processes, such as ozonation.

In 2016, Zhu and coworkers first reported the grafting of cellulose with hyperbranched polyethyleneimine for the selective sorption of a wide range of anionic and cationic organic dyes [21]. The material was obtained by $NaIO_4$-mediated oxidation to afford dialdehyde cellulose, which in turn was reacted with bPEI in ethanol. While the approach is quite interesting, the synthetic procedure required controlled and anhydrous conditions and the use of flammable solvents, which could limit the scalability of the synthesis. One year later, Wang and coworkers reported the use of TEMPO oxidized cellulose membranes modified with linear PEI to remove both anionic (xylenol orange (XO)) and cationic (methylene blue (MB)) dyes from wastewater [22]. However, in this procedure the cross-linking is obtained by means of glutaraldehyde, which is known to be toxic. In both cases a good sorption performance was observed, and it was ascribed to an electrostatic interaction between the cationic polymer linked to cellulose and the dyes, with a sorption efficiency depending on the charge present on the surface. This aspect somehow limits the operating conditions of the process, as it would require wastewater pH adjustment before (and then, after) sorption treatment, with a consequent increase in the economic impact.

Herein, we report an investigation on the sorption efficiency of CNS towards four commercially available and highly used organic dyes (Naphthol Blue Black (NBB), Orange II Sodium Salt (OSS), Brilliant Blue R (BB), Cibacron Brilliant Yellow (CBY)). All the selected dyes presented a similar structure but differed in terms of molecular weight and the number of sulphate functional groups. The results in terms of sorption efficacy and regeneration efficiency were correlated with both the porosity of the sorbent sponge and the chemical and dimensional differences of the dyes, much more than on the charge present on the material. In fact, and quite surprisingly, CNS performed even better at a slightly basic pH, that is, one above the value of point of zero charge. Moreover, a comparison with an activated carbon could highlight the potentialities and limits of the proposed solution.

2. Materials and Methods

All the reagents were purchased from Sigma-Aldrich (Milano, Italy). Cotton linters was provided by Bartoli Spa paper mill (Capannori, Lucca, Italy). Deionized water was produced with a Millipore Elix® Deionizer with Progard® S2 ion exchange resins. UV spectra and data were recorded on a V-600 Series UV-vis spectrophotometer from JASCO (Cremella (LC), Italy). Other equipment used in the procedures include a Branson Sonifier 250 equipped with a 6.5 mm probe tip, a Heldolph multi-reax shaker (Schwabach, Germany) and a SP Scientific BenchTop Pro Lyophilizer (Perugia, Italy).

2.1. TOCNF Synthesis and Titration

Cellulose was oxidized to a degree of 1.546 $mmol_{COOH}/g_{TOCNF}$ according to a procedure previously reported in the literature [9,10]. Briefly, cotton linters (190 g) were dispersed in deionized water produced with a Millipore Elix® Deionizer with Progard® S2 ion exchange resins (Milan, Italy) (total volume 5.7 L), in the presence of TEMPO (2.15 g, 13.8 mmol) and KBr (15.42 g, 129 mmol) and a solution of NaClO (12.5% *w/w* aqueous solution, 437 mL) was slowly added under vigorous stirring. During oxidation, pH was maintained in the range of 10.5-11 by dripping 4 N NaOH water solution. The solution was maintained stirred for 16 h. Oxidized cellulose nanofibers were aggregated by using concentrated HCl and then washed with deionized water on a Büchner funnel to reach a neutral pH (see Supplementary Materials for further information).

To estimate the concentration of carboxyl groups on the cellulose structure after oxidation, a titration was performed with 0.1 N NaOH water solution, using phenolphthalein as a colorimetric indicator. The first step was the titration of NaOH 0.1 N. The detailed procedure and the equation used to calculate the concentration of the carboxyl groups are reported in Supplementary Materials.

2.2. Synthesis of CNS

CNS were synthesized according to the procedure previously reported [13]. First, 3.5 g of TOCNF were suspended in deionized water, adding a stoichiometric amount of granular NaOH. The suspension was ultrasonicated with a Branson Sonifier 250 equipped with a 6.5 mm probe tip to further promote the separation of the nanofibers, obtaining a homogeneous solution, which was then acidified with 12 M HCl, filtered on a Büchner funnel under vacuum, and washed with deionized water until neutrality. Then, two aqueous solutions of 25 kDa bPEI (3.5 g of bPEI in 10 mL) and anhydrous citric acid (CA) (0.896 g in 10 mL) were slowly added to the TOCNF solution, while continuously stirring until obtaining a white and homogeneous hydrogel, which was placed in well-plates, quickly frozen at −35 °C, freeze-dried for 48 h using a SP Scientific BenchTop Pro Lyophilizer (at −52 °C temperature and 140 µbar pressure) and then thermally treated in the laboratory oven at a maximum temperature of 102 °C for 16 h. At the end of the process, CNS was grinded with a mortar and then washed with water to remove the excess bPEI (for further information, see Supplementary Materials). The particle size distribution of the grinded CNS powder was measured in the Laboratory Chemical Analysis (LAC) of Politecnico di Milano by means of a Malvern Mastersizer 3000 Particle Size Analyzer (Malvern, UK) with Fraunhofer modeling, which considers opaque non-spherical particles. The CNS powder was suspended under stirring in 500 mL of water to reach an obscuration level in the range of 8–12%.

2.3. Point of Zero Charge (PZC) Calculation

The pH of the point of zero charge (pH_{PZC}), namely the pH above which the total surface of the sorbent material is negatively charged, was measured by the pH drift method [23]. For this purpose, 20 mL of a 0.01 M NaCl solution was placed in a Falcon vial and N_2 was bubbled through the solution to steady the pH by preventing the dissolution of CO_2 in the solution. The pH was then adjusted to selected initial values between 2 and 12, by dripping HCl 0.01 N or NaOH 0.01 N, and the sorbent (0.06 g) was added to the solution. The final pH, reached after 4 h, was measured and plotted against the initial pH. The pH at which the curve crosses the bisector pH(final) = pH(initial) is the pH_{PZC} of the given sorbent. The considered sorbents were CNS and the activated carbon SAE SUPER, provided by Norit (Italy) (see Supplementary Materials for general characteristics).

2.4. Preliminary Sorption Tests

The tests were carried out by dipping 12 mg of CNS powder in 15 mL of aqueous solutions of the selected dye for 24 h, under static conditions and at room temperature in a Falcon vial. The selected dyes are reported in Figure 1. The concentration of the buffer dye solution was selected considering the extinction coefficient of each dye (20 mg/L for OSS and NBB, 100 mg/L for BB and CBY) and the type of buffer was chosen according to the solubility of the dye in the buffer solution at room temperature (OSS: 116 g/L, NBB: 30 g/L, BB: 70 g/L, CBY: 50 g/L). Tests at pH 5.5 were carried out in piperazine and citrate buffer. Tests at pH 7.6 were conducted in deionized water using the normal buffering power of CNS. Each sample has been reproduced in triplicate. After 24 h, one collection was taken from each sample and analyzed by UV analysis. UV-Vis spectra (Figure S1), characteristic λ_{max} (Table S1), extinction coefficients (Table S2) and calibration lines (Figures S2–S5) for all the four dyes are reported in Supplementary Materials.

2.5. Isotherms and Kinetics

Isotherm and kinetic sorption tests were carried out under dynamic conditions (using the shaker at 450 rpm) at room temperature (25 °C). Isotherm tests were carried out by maintaining a constant sorbent quantity and solution volume throughout the data gathering (24 h), while changing the solution concentration. Eight different concentrations were tested for each dye and three trials were carried out for each concentration. The sponge-to-solution ratio used was 12 mg of CNS/15 mL of mono-contaminated dye solution (0.8 mg/mL). The selected concentrations for each dye are described in Table S3 in Supplementary Materials. The absorbance of the solutions was measured at time zero and after 24 h.

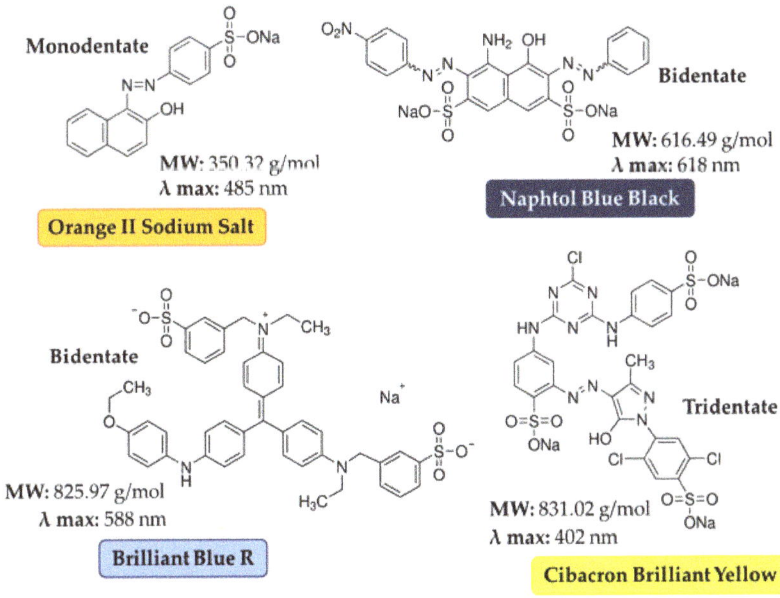

Figure 1. Chemical structure and properties of the dyes used for the sorption tests.

As for kinetic tests, the initial concentrations for each dye are reported in Table 1. The trials were performed under dynamic conditions and at room temperature, as described above. Samples were shaken and analyzed after 15, 30, 45, 60 and 90 min, and 2, 3, 4, 6, 8 and 24 h. A volume of 25 mL of solution was used for each test, to allow withdrawals from the solution 11 times, keeping the total diminution of the volume below 10%. The quantities of sorbent for each kinetic test were 20 mg for the lowest concentration and 40 mg for the highest concentration.

Table 1. Values of initial concentrations (C_0) and sponge quantities (m) for kinetic tests, where the solution volume was 25 mL. The same C_0 values are also used for comparison tests between cellulose nanosponges (CNS) and activated carbon (Section 2.7).

Dye	C_0 low		C_0 high	
	C_0 (mg/L)	m (mg)	C_0 (mg/L)	m (mg)
OSS	20 ± 0.5	20 ± 0.1	800 ± 5.0	40 ± 0.1
NBB	20 ± 0.5	20 ± 0.1	250 ± 2.0 [a]	40 ± 0.1
BB	100 ± 1.0	20 ± 0.1	320 ± 2.0	40 ± 0.1
CBY	100 ± 1.0	20 ± 0.1	320 ± 2.0	40 ± 0.1

[a] The value of C_0 high for comparison tests between CNS and activated carbon is 400 mg/L.

2.6. Desorption and Reusability Tests

Tests were conducted to evaluate the possibility of reusing the sponges. At first, colored CNS was produced by leaving white CNS powder in contact with a solution of the selected dye in static conditions (5 g/L concentration, 30 mL, 200 mg of CNS) for 24 h, then filtering it on a Büchner funnel and washing it with deionized water.

The desorption test was conducted with HCl 0.1 N and NaOH 0.1 N. A total of 20 mg of colored CNS was soaked in 20 mL of each solution under static conditions at room temperature for 24 h. After the first test, which decreed the efficiency for only the NaOH solution, three different molar concentrations of NaOH were compared—0.5, 0.1 and 0.05 N—following the same discoloration procedure. Other types of alkaline solutions were tested using the previously reported conditions: the test was repeated with triethylamine (TEA), NH_3 30% aqueous solution and KOH 0.1 N aqueous solution.

The reusability test consisted of a sorption test conducted on the decolored sponge. The test was carried out only on the OSS discolored CNS and following the same procedure as for previous sorption tests: 12 mg of sorbent in 15 mL of 20 mg/L solution of OSS dye for 24 h under static conditions at room temperature. Five sorption–desorption cycles were carried out. For this purpose, 100 mg of OSS colored CNS were soaked in 110 mL of NaOH 0.05 N. The resulting sponge was then vacuum-filtered with the aid of a Büchner funnel and washed with deionized water until neutrality. Once dried, the sponge was weighted for the next phase of sorption with a constant sponge-to-solution ratio of 0.8 mg/mL. Each sample was carried out in triplicate. After 24 h of static sorption at room temperature, absorbance was analyzed. The sponge was then gravity-filtered and air-dried before the new desorption phase, carried out with NaOH 0.05 M solution.

2.7. Comparison between CNS and Activated Carbons

A comparison test was carried out by evaluating the sorption capacity of the activated carbon SAE SUPER. The experimental setup was the same as the previous tests, using a sorbent-to-solution ratio of 0.8 mg/mL. Two initial concentrations were tested for each dye (C_0 low and C_0 high, reported in Table 1). This test was carried out in dynamic conditions and equilibrium was reached after 24 h in the multi reax shaker at 450 rpm. Filtration via syringe filter was required for this test due to the fine particulate dispersion of activated carbons in the solution.

Kinetic tests were carried out using SAE SUPER activated carbon in OSS solutions at high and low concentrations. For the high-concentration trial, 240 mg of SAE SUPER activated carbon were dispersed in 150 mL of 800 mg/L OSS solution. Three trials were prepared and agitated by magnetic stirring. A volume of 1 mL was withdrawn for each measurement, filtered through a cotton filter and opportunely diluted. Withdrawals were performed after 15, 30, 45, 60 and 90 min, 2, 3 and 24 h. The total amount of withdrawn solution after eight samplings was still lower than 10% of the total volume of the solution.

3. Results and Discussion

3.1. CNS Synthesis and Characterization

CNS were synthesized according to Scheme 1, following a two-step protocol. TOCNF and bPEI were first mixed in deionized water in a 1:1 weight ratio, and CA (18% with respect to primary amino groups of bPEI) was added, with the final aim to better trap bPEI in the final network by increasing the concentration of carboxylic groups. The same result cannot be achieved by raising the content of carboxylic units on the nanofiber, as more severe oxidation conditions lead to depolymerization rather than further selective conversion of C6 alcoholic groups. In a second thermal step, the resulting hydrogel was transferred in well-plates, used as molds, lyophilized and then heated in an oven at about 100 °C, in order to favor the cross-linking between the carboxylic groups of TOCNF and the primary amines of bPEI. Finally, CNS was grinded in a mortar before use, in order to increase sorption performance.

A complete chemical characterization of the resulting sponges was reported in a previous paper [12]. The formation of amide bonds after thermal treatment was evidenced by FT-IR analysis, with an increase in the peak at 1664 cm^{-1} (−C=O stretching of the amide bonds), directly related to an increase in CA content in the formulation, as also confirmed by ^{13}C CP-MAS solid-state NMR. The role of CA in better fixing bPEI in the network was also quantitatively confirmed by processing data derived from the elemental analysis of different nanosponges, prepared by varying the content of this tri-carboxylic molecule in starting solution. This optimization study, also supported by an eco-toxicology evaluation of the materials [13], led to the CNS formulation herein investigated.

CNS exhibit a high micro-porosity with pore sizes in the range of 10–100 μm, as observed by scanning electron microscopy (SEM) (Figure 2). Pores are characterized by a two-dimensional sheetlike morphology, often reported in the literature for cellulose-based aerogels prepared by freeze-drying aqueous suspensions. According to this approach, ice crystals act as templates for pores' generation, preventing the formation of occlusions and guaranteeing complete penetrability of the structure [24].

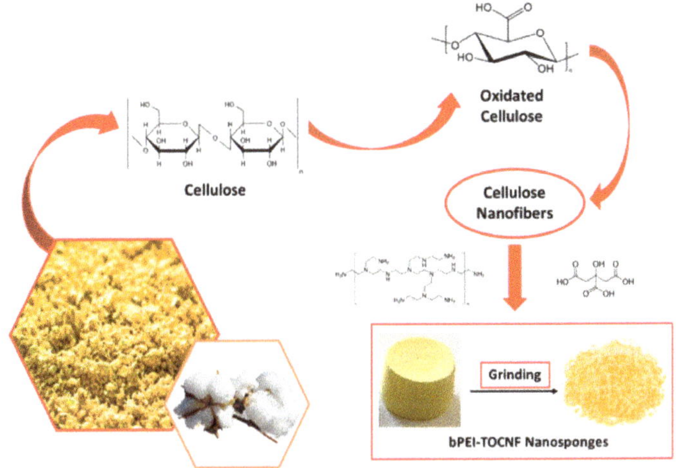

Scheme 1. Preparation of CNS starting from cellulose-based sources.

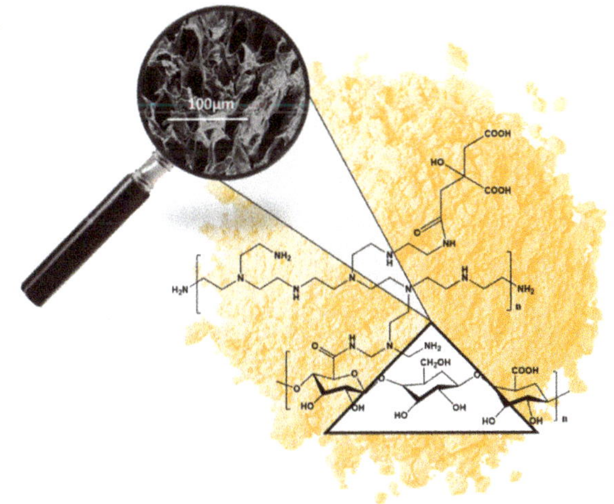

Figure 2. Inner structure of grinded CNS analyzed by Scanning Electron Microscopy.

Microcomputed tomography (μ-CT) analysis previously reported [12] also indicated that CNS has a porosity of 70–75%, and a trabecular inner structure with an average trabecular thickness of about 30–40 μm and a trabecular separation of about 70–75 μm.

Moreover, we recently provided experimental evidence of nano-porosity in the network, by means of small angle neutron scattering (SANS) analysis of water nanoconfinement geometries in the sorbent material [18]. The analysis of the experimental data allowed us to measure the short-range correlation length, which resulted in a range between 25 and 35 Å. In addition, a more recent combined investigation of the FTIR-ATR spectra of CNS hydrated with H_2O and D_2O allowed to detect a supercooled behavior of entrapped water molecules, supporting the idea of a nano-confinement for water in these systems.

The high porosity and the wide pores' dimensional dispersion would suggest a high diffusivity of solutes in the material, which could be however affected by their structure and dimension.

3.2. Sorption Tests

3.2.1. First Sorption Screening for All Dyes

Before starting sorption experiments, we determined the pH_{PZC} for CNS, which was pH 7.1 (Figure 3). This value indicates that at alkaline pH, the material is negatively charged.

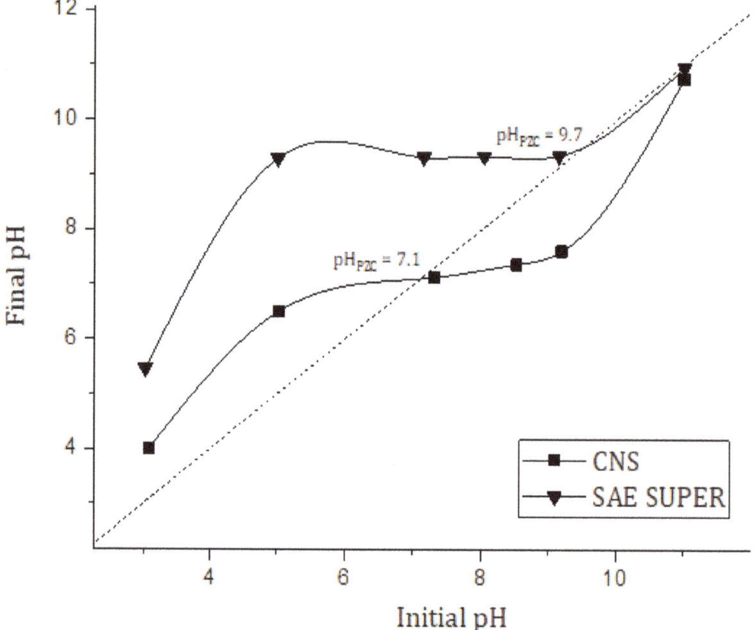

Figure 3. pH drift method plot to determine the pH_{PZC} for CNS and SAE SUPER-activated carbon.

According to this result, we selected two different pH ranges for performing preliminary sorption tests. At pH 7.5–7.8, CNS should be negatively charged. This condition limits the electrostatic interaction with the negatively charged dyes and could be considered as not ideal for our purpose. However, it falls into the typical pH ranges of textile wastewater. Moreover, this is the buffer directly generated by CNS powder once dispersed in deionized water. On the contrary, at pH 5.5, CNS should be positively charged. However, at this pH value, both OSS and NBB showed a poor solubility, which limited the interest in this experimental set.

Preliminary results at pH 7.6 confirmed a good sorption performance of CNS towards all the dyes under investigation (Table 2). We operated at two different dyes' concentrations (C_0) for NBB and OSS (20 mg/L) and BB and CBY (100 mg/L), due to the different extinction coefficients of each dye. Obviously, sorption capacity (q_e) value, which is the amount of contaminant taken up by the sorbent per unit mass of the sorbent, was dependent on C_0, so that a comparison among dyes is not correct at this stage. As expected, cellulose alone, regardless of its original form (cotton linters, TEMPO-oxidized cellulose (TOC), or TOCNF) did not perform any sorption (Figure 4), confirming the crucial role of bPEI in CNS network. Moreover, sorption tests at pH 5.5 conducted on BB and CBY provided a q_e value quite similar to that measured at pH 7.6 (96.31 and 110.53 mg/g, respectively), indicating that CNS charge is not crucial for the sorption performance.

Table 2. Preliminary sorption tests at pH 7.6 starting from solutions at different dyes' concentrations (C_0).

Dye	q_e (mg/g)	C_0 (mg/L)
OSS	22.58 ± 1.19	20 ± 0.5
NBB	23.85 ± 2.63	20 ± 0.5
BB	88.96 ± 4.23	100 ± 1.0
CBY	121.53 ± 4.59	100 ± 1.0

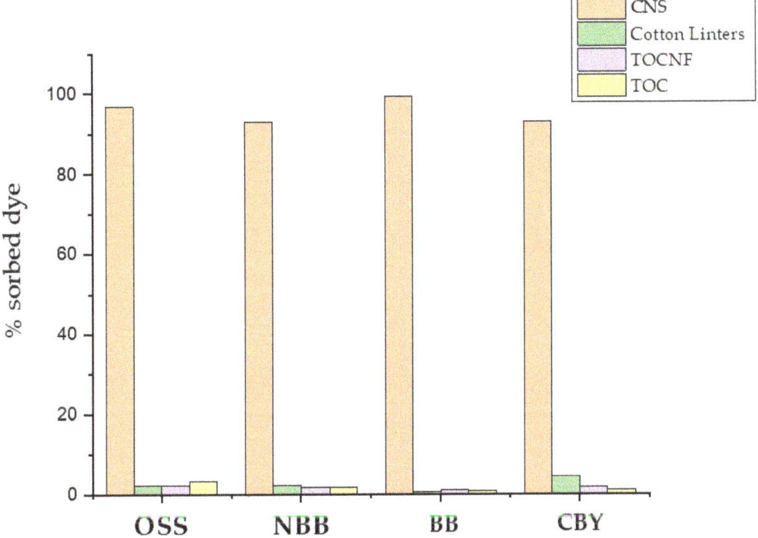

Figure 4. Comparison between sorption capacities of CNS and different cellulose forms (cotton linters, TEMPO-oxidised cellulose (TOC), TOC nanofibers (TOCNF)). Data are reported as the percentage of dye adsorbed at the end of the test with respect to the total amount of dye in solution at t_0.

3.2.2. Isotherm Models

Isotherm experiments were conducted maintaining constant the amount of CNS in solution, and progressively increasing dye concentration [25]. Data were modeled via non-linearized methods [26], originally considering three different models (Langmuir, Freundlich, and Dubinin–Radushkevic) [27], and then selecting the Langmuir one, which better fitted collected data. This model (Equation (1)), which assumes that the sorbent is coated by a monolayer of the adsorbate, correlates the sorption capacity of the sorbent material (q_e)—calculated as mg of pollutants sorbed per g of sorbent material—with the concentration of the pollutant in the solution at the equilibrium (C_e). All data collected and Langmuir fittings for OSS, NBB and BB are reported in Table 3 and Figure 5.

$$q_e = \frac{Q_{max}KC_e}{1 + KC_e} \tag{1}$$

Table 3. Estimation of Q_{max} and K parameters according to the Langmuir isotherm model for OSS, NBB and BB. The number of experimental data used for each dye is reported in column N.

Dye	Q_{max} (mg/g)	K (L/mol)	R^2	N
OSS	898.4 ± 15.6	0.06059 ± 0.00500	0.978	33
NBB	240.2 ± 10.0	0.35811 ± 0.07050	0.930	24
BB	228.7 ± 6.9	0.13381 ± 0.02544	0.875	39

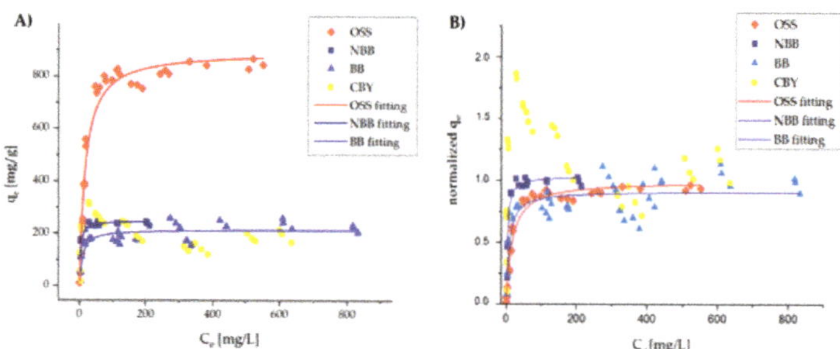

Figure 5. Graph showing the data collected for the calculation of isotherm curves. The x-axis shows the concentration at equilibrium (C_e) expressed in mg/L, while the y-axis shows the capacity at equilibrium (q_e) expressed in mg/g and not normalized (**A**) and normalized for the maximum capacity (Q_{max}) (**B**).

The graph's and the main fitting's statistics for Freundlich and Dubinin–Radushkevic models are reported in the Supplementary Materials (Tables S3–S5 and Figures S6–S8). Langmuir isotherm is a model for monolayer-localized sorption on a homogeneous surface containing a finite number of identical sites; probably for this reason, it is the one providing the better description of experimental data of dye removal by nanosponge, since neither Freundlich nor Dubinin–Radushkevic models assume a homogeneous surface or constant sorption potential. In fact, the Freundlich's empirical formula accounts for the sorption on heterogeneous surfaces as well as multilayer sorption on microporous structure, which is not the case of CNS, which is characterized by a homogenous nanoporous structure. Similarly, the Dubinin–Radushkevic model is more general than the Langmuir one, since it does not assume a homogeneous surface or constant adsorption potential, meaning that it assumes a Gaussian energy distribution onto a heterogeneous surface. Consequently, these assumptions do not fully describe CNS characteristics. As for the parameters of the Langmuir model, Q_{max} represents the maximum sorption capacity of the sorbent, while K is related to the free energy of sorption, directly related to the affinity of the compound for the solid phase.

The OSS dataset was fitted with a very high accuracy. The very steep slope of the first portion of the curve indicates that the sorption efficiency (q_e) would also be high at very low initial concentrations of pollutant, highlighting the high sorbent/solute affinity. Moreover, the maximum sorption capacity for this dye (Q_{max}) was estimated to be very high compared to other similar sorbents [28].

The NBB dataset was not highly coherent with the Langmuir model. When operating at a C_0 lower than 150 mg/L, the equilibrium concentration (C_e) was constant and very close to 0. This behavior is typical of the chemisorption processes, which can occur in the presence of sulfonated dyes. By excluding the data that deviate from the Langmuir model, a quite accurate modeling can be obtained, where the slope in the first section is still very steep (suggesting excellent sorption capacity also at low initial concentrations). Q_{max} was significantly lower than that measured for OSS, but still in line with the results obtained for other similar sorbents [28].

For the BB dataset, the modelling was not deemed reliable, because of a quite low accuracy. However, general considerations can still be made. It can be observed that, very similarly to NBB, the slope of the initial section of the curve is very steep and the maximum sorption capacity is in line with the results reported in the literature [28].

A particular case is represented by CBY, for which an anomalous isotherm data trend makes the Langmuir model's application impossible. As the equilibrium concentration grows, a very steep increase in the sorption capacity can be observed up to a maximum value, after which the sorption capacity decreases to a lower plateau value. This anomalous behavior can be attributed to the formation of solute–solute interactions stronger than the sorbent-solute ones at higher concentrations. The maximum sorption capacity value is of approximately 310 mg/g, but the sorption capacity at the plateau is of approximately 200 mg/g. Despite this irregular behavior, some similarities with the other dyes can be evidenced. In fact, in this case, the slope of the initial section of the curve is steep and the sorption capacity plateau value is comparable with those obtained for BB and NBB.

All these results suggest some preliminary considerations. A clear difference in Q_{max} values between OSS and the set of the other three dyes is observed. We identify two parameters which make these dyes different, namely the number of sulfonate units present into the molecular structure and the molecule dimension. The first parameter should affect the sorbent–solute interaction strength, as evidenced by the very steep slope at low C_0, especially for NBB, BB, and CBY, and the higher K values calculated for BB and NBB with respect to OSS (Table 4), with the latter bearing only one sulfonate group. Nevertheless, we assume also that the molecular size can play a key role in determining the diffusivity of the solute in the sponge. In fact, OSS (MW = 350.32 g/mol) is much smaller with respect to the other three dyes (NBB (MW = 616.49 g/mol), BB (MW = 825.97 g/mol) and CBY (MW = 831.02 g/mol)), and this probably favors its penetration in the sponge.

Table 4. Results of *pseudo second-order* fitting on all kinetic datasets. The number of experimental data used for each dye is reported in column N.

		Pseudo Second-Order Model			
Dye	C_0	k_2 $\mathrm{mg \cdot g^{-1} \cdot min^{-1}}$	q_{eq} $\mathrm{mg \cdot g^{-1}}$	R^2	**N**
OSS	20 mg/L	8.9×10^{-3}	25	0.9999	30
	800 mg/L	5.7×10^{-4}	500	0.9998	30
NBB	20 mg/L	1.3×10^{-2}	20	0.9997	30
	250 mg/L	2.6×10^{-5}	200	0.9887	30
BB	100 mg/L	4.0×10^{-4}	100	0.9989	24
	320 mg/L	1.9×10^{-4}	167	0.9989	24

To confirm our hypothesis, we carried out some sorption tests on solutions contaminated by Indigo Carmine (IC), [29] an organic dye with a structure quite similar to those previously analyzed (Figure 6), and which presents two sulfonate groups, but an MW of 466.35 g/mol, that is an intermediate value between OSS and the other dyes. A Q_{max} of 540 mg/g was obtained, which, as for the molecular weight, is an intermediate value between the Q_{max} of OSS and the Q_{max} of NBB, BB and CBY.

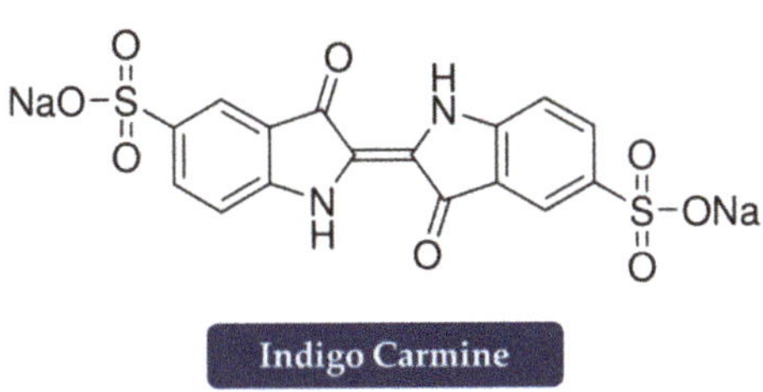

MW: 466.35 g/mol

λ max: 610 nm

Figure 6. Chemical and sorption properties of Indigo Carmine.

3.2.3. Sorption Kinetics

Kinetic study is of great importance for the use of granular sorbent in the water treatment field, since the solute removal rate affects the reactor residence time required for completing sorption reactions and therefore for achieving the selected quality standard for treated water [30,31].

Kinetic experiments were conducted at two different C_0 (see Table 1). The two initial concentrations were selected as the minimum UV detectable concentration (C_0 low) and the minimum C_0, which can reach the Q_{max} value (C_0 high) according to the Langmuir model for OSS, NBB and BB. As regards CBY, concentration and CNS amounts were selected in order to be in the same range as BB, due to its similar chemical structure.

All the graphs related to the sorption kinetic behavior for each dye are reported in Figures 7 and 8.

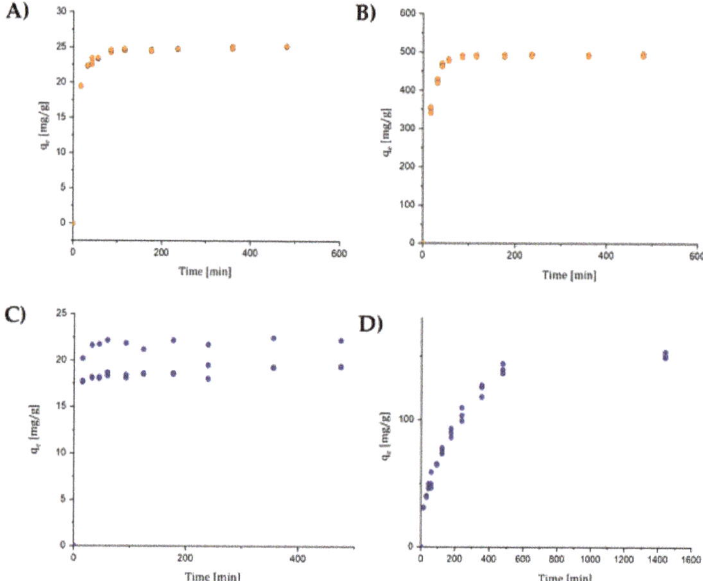

Figure 7. (**A**) OSS kinetic behavior. Initial concentration of 20 mg/L. (**B**) OSS kinetic behavior. Initial concentration of 800 mg/L. (**C**) NBB kinetic behavior. Initial concentration of 20 mg/L. (**D**) NBB kinetic behavior. Initial concentration of 250 mg/L.

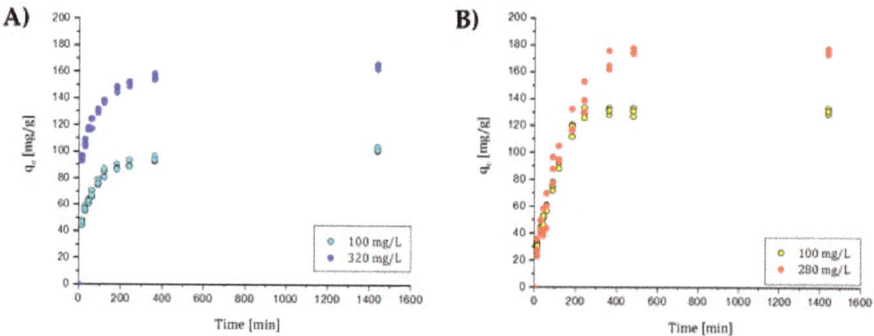

Figure 8. (**A**) BB kinetic behavior. Initial concentration of 100 and 320 mg/L. (**B**) CBY kinetic behavior. Initial concentration of 100 and 280 mg/L.

OSS was tested at C_0 low and C_0 high (20 and 800 mg/L). At both concentrations, more than 80% and 70%, respectively, of the total dye sorbed was reached in the first 15 min, while 90% was reached after 1 h. This dye was the one with the best performance for both initial conditions. NBB was also tested at C_0 low and C_0 high (20 and 250 mg/L). At C_0 low, more than 90% of the total dye sorbed was reached in the first 15 min. At C_0 high, 50% was adsorbed after 2 h, while it required 8 h to reach 90% sorption efficiency.

For BB C_0 low was 100 mg/L, while C_0 high was 320 mg/L. In the first case, 50% of the maximum efficiency was reached in the first 15 min, while in the second case, 70% was reached in the same time. To reach 90%, 4 h were required in the first case, while in the second case only 3 h were sufficient. C_0 low and C_0 high for CBY were 100 and 280 mg/L. At the lowest concentration, 50% was reached after less than 90 min, while to reach 90%, 3 h were required. In the other case, 90 min were required to reach 50%, while 6 h were necessary to reach 90%. This dye was the one with the worst kinetic behavior. However, in all cases the sorption kinetic was quite fast, and almost immediate for low concentrations, which are, however, much more similar to those expected in wastewater.

All kinetics were analyzed to determine whether the data reflected a *pseudo first-order* model or a *pseudo second-order* model. The model that best fits all the kinetics datasets was the *pseudo second-order* one (Equation (2)), as was expected considering the results from isotherm experiments (see Table S7 in Supplementary Materials for the comparison between the two models). In fact, the *pseudo second-order* kinetic usually applies to chemisorption processes, in which solutes can react with more than one active site [31,32]. This is associated with two main assumptions: (i) the kinetic rate limiting step is a chemical reaction involving valent forces through sharing or the exchange of electrons, (ii) the sorption follows the Langmuir equation. CBY, which does not follow the Langmuir fitting regarding the isotherm fit, cannot be modelled either with first-order kinetics or second-order kinetics. Figures S9 and S10 in Supplementary Materials report the *pseudo second-order fitting* for OSS sorption at 20 mg/L and 800 mg/L initial concentrations, respectively.

Table 4 shows schematically the values obtained using the *pseudo second-order* model on all the dyes considered. In Section S5 of Supplementary Materials, the procedure and *pseudo second-order* fitting graphs are also shown, taking as an example the kinetics of OSS at 20 and 800 mg/L.

$$\frac{dq_t}{dt} = k_2\left(q_{eq} - q_t\right)^2 \tag{2}$$

3.3. CNS Regeneration

To better exploit the potentialities of CNS as a sorbent towards anionic dyes, we wanted to evaluate the possibility to regenerate and re-use the material after first sorption experiments.

3.3.1. Desorption Tests

Alkaline washing of CNS sponges with NaOH water solution led to discoloration of the material, while desorption trials with acidic aqueous solutions provided poor results in terms of CNS regeneration. We hypothesized that the selected dyes, due to the presence of sulfonate groups, are much more soluble at alkaline pH and therefore, under these conditions, the solute–solvent interactions prevail on solute–sorbent ones. While dyes' sorption can occur also at a slight basic pH, which is higher than the pH_{PZC} of CNS, thanks to Van der Waals interactions between the amino-groups of the sponge and the sulfonate moieties of the dye; once they are sorbed, an acidic treatment seems to enforce the solute–sorbent interaction, rather than promoting sponge regeneration.

Interestingly, while OSS-CNS were completely discolored through alkaline washing, NBB-, BB- and CBY-CNS showed only partial discoloration under the same treatment conditions. In this case, the different behaviors should be ascribed to the different number of sulfonate groups present on the organic dye molecules, rather than to the molecular dimensions. Polydentate dyes NBB and BB (bidentate) and CBY (tridentate) seem to have a stronger interaction with the sorbent if compared with

OSS, which bears just one sulfonate group. Once again, this aspect can be explained by considering the K values reported in Table 4. As previously stated, this parameter is tightly related to the free energy of sorption, and corresponds to the affinity of the compound for the solid phase. By considering these values, we can notice that the K parameter for OSS is notably lower than those for NBB and BB, suggesting a stronger sorbent–molecule interaction for the latter, and so implying more difficulty in the desorption process, due to the higher energy required.

Once again, a similar desorption test conducted on CNS loaded with IC, a bidentate molecule with an MW lower than those of the other polydentate dyes, provided an incomplete regeneration, comparable with that of NBB, BB and CBY.

For OSS-CNS, the 0.05 N NaOH solution was sufficient to completely regenerate the material.

3.3.2. Reusability Tests

Tests were performed on the OSS-CNS, due to its complete regeneration by alkaline treatment. This test consisted of several sorption–desorption cycles to evaluate the impact of repeated decoloring on the sorption efficiency of the sponge. Reusability efficiency was evaluated through five cycles and the results reported in Table 5 clearly show how the sorption capacity of the sponge is maintained at a constant after several regeneration cycles.

Table 5. Sorption–desorption cycles: sorption capacity and equilibrium concentration after each regenerating cycle, considering OSS dye (C_0 = 20 mg/L).

Iterations	1	2	3	4	5
C_e (mg/L)	3.13 ± 0.22	2.59 ± 0.19	1.65 ± 0.16	2.06 ± 0.41	3.21 ± 0.35
q_e (mg/g)	21.78 ± 1.19	22.17 ± 1.06	24.41 ± 0.98	24.54 ± 1.61	21.84 ± 1.02

3.4. Comparison with Activated Carbons

A comparison test between sponges and an activated carbon was carried out. For the selected activated carbon, the PZC was calculated according to the procedure described in Section 2.3. Results are reported in Figure 3.

Contrary to the CNS, SAE SUPER is positively charged at the operating pH (7.5–7.8), thus favoring the interaction between the positive charge of the sorbent and the negative charge of the deprotonated SO_3^- groups of the dye. A detailed set up for the comparison test is reported in Section 2.7. The sorption comparative results for low and high concentrations of the dyes are reported in Table 6.

Table 6. Comparison of sorption capacities (q_e, (mg/g)) determined at C_0 low and C_0 high for SAE SUPER and CNS (for C_0 values see Table 1).

Sorbent	OSS		NBB		BB		CBY	
	C_0 Low	C_0 High	C_0 Low	C_0 High	C_0 Low	C_0 High	C_0 Low	C_0 High
SAE SUPER	25.6 ± 2.0	383.7 ± 3.6	23.8 ± 1.9	220.1 ± 2.8	108.7 ± 2.2	360.3 ± 3.6	118.0 ± 2.1	159.8 ± 3.6
CNS	22.6 ± 1.6	793.4 ± 5.1	23.9 ± 1.8	239.0 ± 3.2	89.0 ± 1.9	197.0 ± 3.8	121.5 ± 2.3	242.1 ± 2.9

It can be observed that the sorption capacity of the activated carbon at low concentrations is comparable to the one for CNS for all dyes. Regarding the high-concentration sorption tests, the sponge showed comparable performances to the activated carbon for NBB, while the sorption capacity towards OSS was double for the sponge compared to the activated carbon. CBY was sorbed better by the sponge than the activated carbon. A different behavior was displayed by BB, which showed clearly better performances for the activated carbon than for the nanosponge.

Kinetic studies were thus performed on SAE SUPER activated carbon for the sorption of 20 and 800 mg/L solutions of OSS dye. For the 800 mg/L trial with activated carbon, a plateau was reached in the first 30 min, while for the nanosponge one hour was required to reach the plateau. However, the plateau values obtained for the nanosponge were much higher than for the activated carbon.

As regards the 20 mg/L trial, the first measurement for the SAE SUPER after 7 min was registered to be already below the UV detection limit. On the contrary, at this concentration, the kinetic behavior of the nanosponge took approximately 90 min to reach the same result. Figure 9 shows graphically the kinetic behavior of SAE SUPER and CNS in contact with an 800 mg/L solution of OSS.

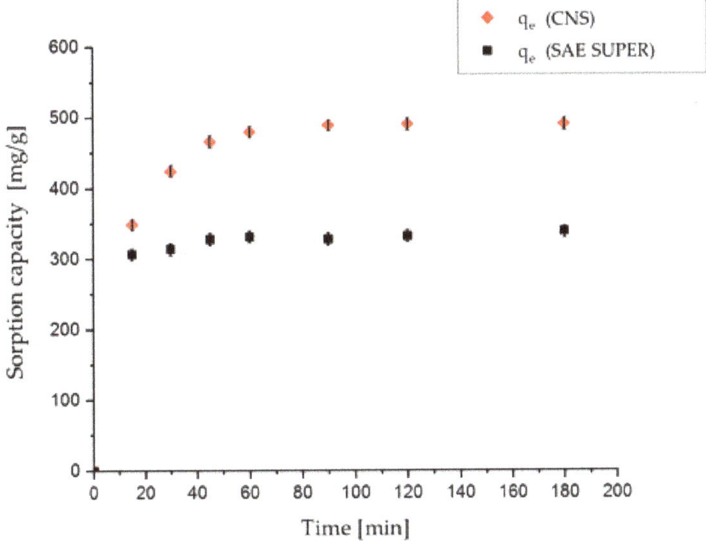

Figure 9. Comparison between SAE-activated carbon and sponge sorption kinetics for OSS dye at 800 mg/L concentration.

The faster kinetics observed for the activated carbon with respect to the nanosponge could be due to different intra-particle diffusion resistances, considering the different porous structures of these two sorbents. In fact, in batch experiments, usually performed in turbulent conditions, the limiting step is usually diffusion into pores [33].

4. Conclusions

In this work, we reported the use of a nanostructured-cellulose-based sorbent material for water decontamination from anionic organic dyes. The material was prepared starting from biomass sources, by combining TEMPO-oxidized cellulose nanofibers, branched polyethyleneimine, and citric acid, and following a simple thermal protocol. The sorption performance was tested on four commercial dyes (OSS, BB, NBB, and CBY), differing for both molecular dimension and the number of sulfonate groups present onto the molecular structure (one for OSS, two for BB and NBB, three for CBY). The sorbent was effective also at a slightly basic pH, even if under these conditions the nanostructured sponge is negatively charged. This result suggested that the sorbent–solute interaction should not be simply ascribed to electrostatic attraction between opposite charges, but other intermolecular interactions could occur between the sulfonate groups of the dyes and the amino groups present on the nano-sponge. The role of bPEI in the network was crucial, as cellulose alone was not able to reproduce significant sorption.

Isotherm and kinetic investigation revealed a molecular-size dependence of sorption performance, as the smallest OSS is much more trapped on the material, probably because of the possibility of it being more diffused in the nano-porous network. Nevertheless, these studies also showed that the strength of sorbent–solute binding was higher when two or more sulfonate groups were present on the dye. This evidence was also confirmed by conducting regeneration and reusability tests, as once again OSS was much more easily removed from the nano-sponge under alkaline conditions, so that the sorbent system could be reused several times, by maintaining its sorption efficiency.

The dye-removal efficiency of the material herein described was compared to that of the commercially available activated carbon SAE SUPER. While with the nano-sponge the sorption kinetic was slightly slower, probably due to the nano-porous structure with respect to a microporous structure of the activated carbon, the sorption capacity was higher for all dyes except BB. In any case, the advantages of the use of this cellulose-based material can be found in its easy handling, reusability, eco-safety, and sustainability, as it can be produced from wasted biomass, following the virtuous route of the circular economy.

Supplementary Materials: The following are available online at http://www.mdpi.com/2079-4991/10/8/1570/s1. Detailed TOCNF preparation; Detailed synthesis of CNS; Procedure for determination of calibration lines; Isotherm models (Langmuir model, Freundlich model, Dubinin-Radushkevich model); *Pseudo second-order* fitting; General characteristics of NORIT® SAE SUPER. Figure S1: UV-vis spectra for (A) OSS, (B) NBB, (C) BB and (D) CBY; Figure S2: Calibration line for Orange Sodium Salt; Figure S3: Calibration line for Naphtol Blue Black; Figure S4: Calibration line for Brilliant Blue R; Figure S5: Calibration line for Cibacron Brilliant Yellow; Figure S6: Isotherm fitting with Langmuir model; Figure S7: Isotherm fitting with Freundlich model; Figure S8: Isotherm fitting with Dubinin-Radushkevic model. Figure S9: OSS kinetic at 20 mg/L; Figure S10: OSS kinetic at 800 mg/L. Table S1: Characteristic UV-vis peaks of each dye; Table S2: Molar mass and extinction coefficient of each dye; Table S3: Values of initial concentrations (C_0) for isotherm curves expressed as mg/L; Table S4: Estimation of Qmax and K parameters according to the Langmuir isotherm model for OSS, NBB and BB; Table S5: Estimation of Qmax and K parameters according to the Freundlich isotherm model for OSS, NBB and BB; Table S6: Estimation of Qmax and K parameters according to the Dubinin-Radushkevic isotherm model for OSS, NBB and BB; Table S7: Summary of k and q_{eq} values obtained with *pseudo first-order* and *pseudo second-order* fitting of the kinetic data set.

Author Contributions: C.P.: Conceptualization, Supervision, Resources, Funding acquisition, Writing—Review and Editing; M.A.: Conceptualization, Supervision, Writing—Review and Editing; L.R.: Investigation, Writing—Original Draft, Writing—Review and Editing, Methodology; N.P.: Investigation, Writing—Original Draft, Writing—Review and Editing, Methodology; A.P.: Investigation, Writing—Review and Editing, Methodology. All authors have read and agreed to the published version of the manuscript.

Funding: This research was funded by Regione Toscana, NanoBonD (Nanomaterials for Remediation of Environmental Matrices associated to Dewatering) POR CReO FESR Toscana 2014–2020-30/07/2014- LA 1.1.5 CUP 3389.30072014.067000007. L.R. also thanks Innovhub—Stazioni Sperimentali per l'Industria –Area Carta, Cartoni e Paste per Carta for financial support.

Conflicts of Interest: The authors declare no conflict of interest.

References

1. Bonelli, B.; Freyria, F.S.; Rossetti, I.; Sethi, R. *Nanomaterials for the Detection and Removal of Wastewater Pollutants*, 1st ed.; Elsevier: Amsterdam, The Nederland, 2020.

2. Corsi, I.; Fiorati, A.; Grassi, G.; Rubin Pedrazzo, A.; Caldera, F.; Trotta, F.; Punta, C. *Ecosafe Nanomaterials for Environmental Remediation in Nanomaterials for the Detection and Removal of Wastewater Pollutants*, 1st ed.; Elsevier: Amsterdam, The Nederland, 2020.

3. Corsi, I.; Winther-Nielsenb, M.; Sethic, R.; Punta, C.; Della Torre, D.; Libralato, G.; Lofrano, G.; Sabatini, L.; Aiello, M.; Fiordi, L.; et al. Ecofriendly nanotechnologies and nanomaterials for environmental applications: Key issue and consensus recommendations for sustainable and ecosafe nanoremediation. *Ecotoxicol. Environ. Saf.* **2018**, *154*, 237–244. [CrossRef] [PubMed]

4. Corsi, I.; Fiorati, A.; Grassi, G.; Bartolozzi, I.; Daddi, T.; Melone, L.; Punta, C. Environmentally Sustainable and Ecosafe Polysaccharide-Based Materials for Water Nano-Treatment: An Eco-Design Study. *Materials* **2018**, *11*, 1228. [CrossRef] [PubMed]

5. Kargarzadeh, H.; Ahmad, I.; Thomas, S.; Dufresne, A. *Handbook of Nanocellulose and Cellulose Nanocomposites*; Wiley Online Library: Hoboken, NJ, USA, 2017.

6. Voisin, H.; Bergström, L.; Liu, P.; Mathew, A.P. Nanocellulose-based materials for water purification. *Nanomaterials* **2017**, *7*, 57. [CrossRef] [PubMed]

7. Mahfoudhi, N.; Boufi, S. Nanocellulose as a novel nanostructured adsorbent for environmental remediation: A review. *Cellulose* **2017**, *24*, 1171–1197. [CrossRef]

8. Carpenter, A.W.; de Lannot, C.-F.; Wiesner, M.R. Cellulose nanomaterials in water treatment technologies. *Environ. Sci. Technol.* **2015**, *49*, 5277–5287. [CrossRef]

9. Isogai, A.; Saitoa, T.; Fukuzumia, H. TEMPO-oxidized cellulose nanofibers. *Nanoscale* **2011**, *3*, 71–85. [CrossRef]

10. Pierre, G.; Punta, C.; Delattre, C.; Melone, L.; Dubessay, P.; Fiorati, A.; Pastori, N.; Galante, Y.M.; Michaud, P. TEMPO-mediated oxidation of polysaccharides: An ongoing story. *Carbohyd. Polym.* **2017**, *165*, 71–85. [CrossRef]

11. Melone, L.; Rossi, B.; Pastori, N.; Panzeri, W.; Mele, A.; Punta, C. TEMPO-oxidized cellulose cross-linked with branched polyethyleneimine: Nanostructured adsorbent sponges for water remediation. *ChemPlusChem* **2015**, *80*, 1408–1415. [CrossRef]

12. Fiorati, A.; Turco, G.; Travan, A.; Caneva, E.; Pastori, N.; Cametti, M.; Punta, C.; Melone, L. Mechanical and drug release properties of sponges from cross-linked cellulose nanofibers. *ChemPlushem* **2017**, *82*, 848–858. [CrossRef]

13. Fiorati, A.; Grassi, G.; Graziano, A.; Liberatori, G.; Pastori, N.; Melone, L.; Bonciani, L.; Pontorno, L.; Punta, C.; Corsi, I. Eco-design of nanostructured cellulose sponges for sea-water decontamination from heavy metal ions. *J. Clean. Prod.* **2019**, *246*, 1190092. [CrossRef]

14. Bartolozzi, I.; Daddi, T.; Punta, C.; Fiorati, A.; Iraldo, F. Life cycle assessment of emerging environmental technologies in the early stage of development: A case study on nanostructured materials. *J. Ind. Ecol.* **2020**, *24*, 101–115. [CrossRef]

15. Melone, L.; Bonafede, S.; Tushi, D.; Punta, C.; Cametti, M. Dip in colorimetric fluoride sensing by a chemically engineered polymeric cellulose/bPEI conjugate in the solid state. *RSC Adv.* **2015**, *5*, 83197–83205. [CrossRef]

16. Riva, L.; Fiorati, A.; Sganappa, A.; Melone, L.; Punta, C.; Cametti, M. Naked-eye heterogeneous sensing of fluoride ions by co-polymeric nanosponge systems comprising aromatic-imide-functionalized nanocellulose and branched polyethyleneimine. *ChemPlusChem* **2019**, *84*, 1512–1518. [CrossRef] [PubMed]

17. Liberatori, G.; Grassi, G.; Guidi, P.; Bernardeschi, M.; Fiorati, A.; Scarcelli, V.; Genovese, M.; Faleri, C.; Protano, G.; Frenzilli, G.; et al. Effect-based approach to assess nanostructured cellulose sponge removal efficacy of zinc ions from seawater to prevent ecological risks. *Nanomaterials* **2020**, *10*, 1283. [CrossRef]

18. Paladini, G.; Venuti, V.; Almásy, L.; Melone, L.; Crupi, V.; Majolino, D.; Pastori, N.; Fiorati, A.; Punta, C. Cross-linked cellulose nano-sponges: A small angle neutron scattering (SANS) study. *Cellulose* **2019**, *26*, 9005–9019. [CrossRef]

19. Hashem, A.; El-Shishtawy, R.M. Preparation and characterization of cationized cellulose for the removal of anionic dyes. *Adsorpt. Sci. Technol.* **2001**, *19*, 197–210. [CrossRef]

20. Pargoletti, E.; Pifferi, V.; Falciola, L.; Facchinetti, G.; Depaolini, A.R.; Davoli, E.; Marelli, M.; Cappelletti, G. A detailed investigation of MnO_2 nanorods to be grown onto activated carbon. High efficiency towards aqueous methyl orange adsorption/degradation. *Appl. Surf. Sci.* **2019**, *472*, 118–126. [CrossRef]

21. Zhu, W.; Liu, L.; Liao, Q.; Chen, X.; Qian, Z.; Shen, J.; Liang, J.; Yao, J. Functionalization of cellulose with hyperbranched polyethylenimine for selective dye adsorption and separation. *Cellulose* **2016**, *23*, 3785–3797. [CrossRef]

22. Wang, W.; Bai, Q.; Liang, T.; Bai, H.; Liu, X. Two-sided surface oxidized cellulose membranes modified with PEI: Preparation, characterization and application for dyes removal. *Polymers* **2017**, *9*, 455. [CrossRef]

23. Lopez-Ramon, M.V.; Stoeckli, F.; Moreno-Castilla, C.; Carrasco-Marin, F. On the characterization of acidic and basic surface sites on carbons by various techniques. *Carbon N. Y.* **1999**, *37*, 1215–1221. [CrossRef]

24. Panzella, L.; Melone, L.; Pezzella, A.; Rossi, B.; Pastori, N.; Perfetti, M.; D'Errico, G.; Punta, C.; d'Ischia, M. Surface-functionalization of nanostructured cellulose aerogels by solid state eumelanin coating. *Biomacromolecules* **2016**, *17*, 564–571. [CrossRef]

25. Bonilla-Petriciolet, A.; Mendoza-Castillo, D.I.; Reynel-Ávila, H.E. *Adsorption Processes for Water Treatment and Purification*; Springer International Publishing: Berlin, Germany, 2017.

26. Subramanyam, B.; Das, A. Linearised and non-linearised isotherm models optimization analysis by error functions and statistical means. *J. Environ. Heal. Sci. Eng.* **2014**, *12*, 92. [CrossRef]

27. Foo, K.Y.; Hameed, B.H. Insights into the modeling of adsorption isotherm systems. *Chem. Eng. J.* **2010**, *156*, 2–10. [CrossRef]

28. Abouzeid, R.E.; Khiari, R.; El-Wakil, N.; Dufresne, A. Current state and new trends in the use of cellulose nanomaterials for wastewater treatment. *Biomacromolecules* **2019**, *20*, 573–597. [CrossRef] [PubMed]

29. Lakshmi, U.R.; Srivastava, V.C.; Mall, I.D.; Lataye, D.H. Rice husk ash as an effective adsorbent: Evaluation of adsorptive characteristics for indigo carmine dye. *J. Environ. Manag.* **2009**, *90*, 710–720. [CrossRef] [PubMed]

30. Kumar, K.V. Linear and non-linear regression analysis for the sorption kinetics of methylene blue onto activated carbon. *J. Hazard. Mater.* **2006**, *137*, 1538–1544. [CrossRef]

31. Rudzinski, W.; Plazinski, W. Kinetics of solute adsorption at solid/solution interfaces: A theoretical development of the empirical pseudo-first and pseudo-second order kinetic rate equations, based on applying the statistical rate theory of interfacial transport. *J. Phys. Chem. B* **2006**, *110*, 16514–16525. [CrossRef]

32. El-Naggar, I.M.; Zakaria, E.S.; Ali, I.M.; Khalil, M.; El-Shahat, M.F. Kinetic modeling analysis for the removal of cesium ions from aqueous solutions using polyaniline titanotungstate. *Arab J. Chem.* **2012**, *5*, 109–119. [CrossRef]

33. Xie, Y.; Jing, K.J.; Lu, Y. Kinetics, equilibrium and thermodynamic studies of L-tryptophan adsorption using a cation exchange resin. *Chem. Eng. J.* **2011**, *171*, 1227–1233. [CrossRef]

 *nanomaterials*

Article

Suitability of a Cellulose-Based Nanomaterial for the Remediation of Heavy Metal Contaminated Freshwaters: A Case-Study Showing the Recovery of Cadmium Induced DNA Integrity Loss, Cell Proliferation Increase, Nuclear Morphology and Chromosomal Alterations on *Dreissena polymorpha*

Patrizia Guidi [1,*], Margherita Bernardeschi [1], Mara Palumbo [1], Massimo Genovese [2], Vittoria Scarcelli [1], Andrea Fiorati [3], Laura Riva [3], Carlo Punta [3], Ilaria Corsi [4] and Giada Frenzilli [1]

[1] Section of Applied Biology and Genetics and INSTM Local Unit, Department of Clinical and Experimental Medicine, University of Pisa, 56126 Pisa, Italy; margherita.bernardeschi@for.unipi.it (M.B.); m.palumbo@studenti.unipi.it (M.P.); vittoria.scarcelli@unipi.it (V.S.); giada@biomed.unipi.it (G.F.)

[2] Department of Experimental and Clinical Biomedical Sciences "Mario Serio", University of Florence, 50121 Florence, Italy; massimo.genovese@student.unisi.it

[3] Department of Chemistry, Materials, and Chemical Engineering "G. Natta" and INSTM Local Unit, Politecnico di Milano, 20131 Milano, Italy; andrea.fiorati@polimi.it (A.F.); laura2.riva@polimi.it (L.R.); carlo.punta@polimi.it (C.P.)

[4] Department of Physical, Earth and Environmental Sciences and INSTM Local Unit, University of Siena, 53100 Siena, Italy; ilaria.corsi@unisi.it

* Correspondence: patrizia.guidi@unipi.it; Tel.: +39-05-02219110

Received: 17 July 2020; Accepted: 10 September 2020; Published: 14 September 2020

Abstract: The contamination of freshwaters by heavy metals represents a great problem, posing a threat for human and environmental health. Cadmium is classified as carcinogen to humans and its mechanism of carcinogenicity includes genotoxic events. In this study a recently developed eco-friendly cellulose-based nanosponge (CNS) was investigated as a candidate in freshwater nano-remediation process. For this purpose, $CdCl_2$ (0.05 mg L^{-1}) contaminated artificial freshwater (AFW) was treated with CNS (1.25 g L^{-1} for 2 h), and cellular responses were analyzed before and after CNS treatment in *Dreissena polymorpha* hemocytes. A control group (AFW) and a negative control group (CNS in AFW) were also tested. DNA primary damage was evaluated by Comet assay while chromosomal damage and cell proliferation were assessed by Cytome assay. AFW exposed to CNS did not cause any genotoxic effect in zebra mussel hemocytes. Moreover, DNA damage and cell proliferation induced by Cd(II) turned down to control level after 2 days when CNS were used. A reduction of Cd(II)-induced micronuclei and nuclear abnormalities was also observed. CNS was thus found to be a safe and effective candidate in cadmium remediation process being efficient in metal sequestering, restoring cellular damage exerted by Cd(II) exposure, without altering cellular physiological activity.

Keywords: DNA damage; micronucleus; nuclear morphology alteration; cellular proliferation; cadmium; zebra mussel (*Dreissena polymorpha*); polysaccharide-based nanosponge; nanoremediation

---

1. Introduction

Metals represent one of the most widespread environmental contaminants. Among them, cadmium (Cd) is a persistent, non-essential metal coming from anthropogenic activities and natural processes and it is present in all the environmental matrices [1,2]. Cadmium enters the aquatic environment from diffuse and point sources, it is estimated that weathering and erosion contribute in 15,000 tons of cadmium every year while atmospheric fall-down (of anthropogenic and natural emissions) is estimated to contribute between 900–3600 tons to the global aquatic environment [3]. Cadmium concentration in industrial effluents may vary from 0 to 1000 mg L^{-1}, whereas that in municipal waste waters is commonly lower than 0.01 mg L^{-1} [4]. Although drinking water usually contains very low concentrations of cadmium (ranging from 0.00001 mg L^{-1} to 0.001 mg L^{-1}), levels up to 0.01 mg L^{-1} have been sometimes reported. In polluted areas, well-water cadmium concentration could exceed 0.025 mg L^{-1} [5,6]. On the other hand, speaking in terms of threshold values, while the European Union directive posed it 0.005 mg L^{-1} for drinking water [7], other countries like Brazil allowed the threshold to be set 0.01 mg L^{-1} for freshwater [8]. Moreover, besides smoking habit and alcohol consumption, diet plays an important role in humans being vegetables able to accumulate cadmium whose environmental levels need to be maintained under control [9].

Since 1993 the International Agency for Research on Cancer has classified cadmium and cadmium compounds as human carcinogens [10] and the contamination of surface and ground waters by cadmium represents a serious environmental problem which involves human health. An increased risk of cancer, including bladder cancer, was found to be statistically linked with cadmium contamination [11], thus indicating the development of remediation technology for the reduction of Cd content in freshwaters is highly recommended.

Cadmium carcinogenesis mechanisms seem to include oxidative DNA damage and subsequent apoptotic resistance, epigenetic DNA methylation status changes, aberrant gene expressions [12], and DNA repair systems interference [13]. Moreover, a possible dysfunction of mitotic apparatus resulting from incorrect segregation of the chromosomes at anaphase has been suggested [1].

In this context, the development of a sustainable methodology to safely remove cadmium from contaminated waters deserves much attention. Techniques of nano-remediation, which is the application of nanotechnology and the use of engineered nanomaterials (ENMs) to clean contaminated environmental matrices, including groundwater and wastewaters [14], can be considered the most appropriate tools.

For this purpose, the selected nanomaterial needs to be completely reliable, not exerting any degree of toxicity towards biota [15], thus overcoming the problems related for example with the use of zero valent iron (nZVI) [16,17] after injection into aquifers for in-situ remediation. The use of nZVI has shown great potential in the remediation of contaminated water [18], but it has also underlined its reactivity in the remediation of highly polluted areas, including toxic effects during environmental applications. Carbon nanotubes, nano-zinc and nano-titanium oxides have been tested in remediation processes [19–21], but conflicting results have been reported concerning their potential ecotoxicity [22]. For these reasons, sustainable nanomaterials, developed from renewable sources, arouse growing interest for their application in nano-remediation [22,23] being effective, biodegradable, and biocompatible. Among all the renewable sources, polysaccharides (and cellulose in particular) are emerging as starting materials for a wide plethora of applications, including the synthesis of engineered nanomaterials (ENMs) for environmental remediation [23–25].

In this context, a two-step protocol for the synthesis of a novel family of nanocellulose-based ENMs was set up. Among their features, these ENMs possessed superb performance in the adsorption of organic and inorganic pollutants. They were obtained by the cross-linking of 2,2,6,6-tetramethyl-piperidine-1-oxyl (TEMPO)-oxidized cellulose nanofibers (TOCNF) [26] with branched polyethyleneimine (bPEI), affording nanostructured cellulose nanosponges (CNS) [27].

These materials possess a 2D sheet-like morphology and are characterized by a high micro-porosity, as revealed by scanning electron microscopy (SEM) [28] and microcomputed tomography (μCT) analysis [27]. More interestingly, CNS also shows a nano-porosity, as revealed by in-depth small angle neutron scattering (SANS) and Fourier Transform Infrared Spectroscopy (FT-IR) studies [29], and can be also easily functionalized to confer the material additional properties, including ion sensing [30].

The high porosity allows to favor the contacting between metal ion analytes and active sites of the sponge, consisting in the cooperative chelating action of vicinal amino groups. To encourage the applicability of CNS as adsorbent materials for decontamination of heavy metal polluted waters, their environmental safety (ecosafety) was improved by exploiting an eco-design approach consisting into the combination of life cycle assessment (LCA) and environmental risks assessment [31–33]. In these works, the synthetic protocol was revised as function of the results achieved by a laboratory scale LCA study [31] and by their impact on marine biota (*Dunaliella tertiolecta* and *Mytilus galloprovincialis*), while employed for sea decontamination [31,32].

For the first time, in the present work we aimed to evaluate the safeness and efficacy of CNS when employed as adsorbent material in remediation of heavy metal contaminated freshwater by studying cellular responses in a freshwater organism.

The sedentary bivalve *Dreissena polymorpha* (zebra mussel) has been widely used to study the effects of classical and emerging contaminants due to its high filtration rate, suitable size for in vivo laboratory exposure, and maintenance [34].

The sensitivity of zebra mussel hemocytes to genotoxic compounds has been demonstrated through the induction of DNA strand breaks and DNA adducts, the increase of micronuclei and nuclear abnormalities [35,36] related with the level of contaminants in water [1,37].

In the present study, zebra mussel was selected as a biological model to evaluate the safety of CNS for freshwater organisms and to test CNS efficacy in the removal of Cd(II) from a laboratory freshwater environment, with consequent restoration of physiological cellular functions in exposed organisms.

DNA integrity loss, cellular proliferation, and chromosomal damage associated to $CdCl_2$ exposure were investigated in *D. polymorpha* hemocytes to evaluate the remediation capacity of CNS and to ascertain their safeness at a cellular level.

2. Materials and Methods

2.1. Chemical Reagents

$CdCl_2$ (CAS 7790-78-5) and Giemsa were purchased from Carlo Erba (Milano, Italy). Cotton linter cellulose was kindly provided by Bartoli Spa (Capannori, Lucca, Italy) paper mill. Cadmium analytical standard, low melting agarose, normal melting agarose, Neutral Red, PBS, and all the other reagents were purchased from Sigma Aldrich (Milano, Italy).

2.2. Preparation and Characterization of the Cellulose Nanosponges (CNS)

CNS were synthesized by reproducing a standardized protocol previously reported [28,31]. This procedure includes (a) the production of cellulose nanofibers, by means of TEMPO/NaClO/KBr mediated oxidation of cotton linters, (b) their cross-linking with 25 KDa branched polyethyleneimine (bPEI), (c) a freeze-drying process and a thermal treatment, in order to obtain a nanostructured xerogel. The obtained CNS undergo a washing step prior the use.

In detail, cotton linters cellulose (100 g) were minced and dispersed in deionized water (2 L) and then added to a solution (3.7 L) of TEMPO (2.15 g, 13.8 mmol) and KBr (15.42 g, 129 mmol) in demineralized water and kept under vigorous stirring. By means of two dropping funnels, NaClO (12.5% *w/w* aqueous solution, 437 mL) was added dropwise to the mixture, while sodium hydroxide solution (4 M) was added in order to maintain the pH value in the range of 10.5–11. The reaction was maintained under stirring for 16 h, and then acidified to pH 2 with concentrated HCl (37% *w/w* aqueous solution). TEMPO oxidized cellulose (84 g, 84% yield) was recovered as a white precipitate which was collected by filtration and washed extensively with deionized water (5 × 2 L) and acetone (2 × 0.5 L). The oxidation degree was evaluated by titration of the carboxylic groups introduced on the cellulose fibers using a NaOH solution and phenolphthalein as colorimetric indicator (1.5 mmol g^{-1} of -COOH units). After drying, TEMPO-oxidized cellulose was suspended in water, NaOH was added (1 equivalent respect to carboxylic acids), and the dispersion was ultrasonicated at 0 °C (Branson

Sonifier 250, Branson Ultrasonic SA, Carouge, Switzerland), 6.5 mm probe tip, 20 kHz in continuous mode, output power 50%) until a clear solution was obtained, indicating the complete defibrillation of TEMPO-oxidized cellulose into nano-dimensioned fibers (TOCNF). An excess of HCl (aq, 1 M) was then added, the precipitate was collected by filtration and further washed on the filter with deionized water until the filtrate reached a pH 6–7. TOCNF were dispersed (2 % *w/v*) in an aqueous solution containing bPEI and citric acid (mass ratio 1:1:3.53 respectively), then the mixture was sonicated for 10 min before being transferred into a 24-well plate as mold, frozen at −80 °C and freeze-dried for 48 h, affording the corresponding xerogels which were removed from the mold and thermally treated in an oven (103 °C, 12–16 h). The resulting sponge-like materials (Figure 1) were ground in a mortar (particles sizes range 50–400 μm, maximum distribution 130 μm) and washed with water (6 times, 1 h contact time for each cycle).

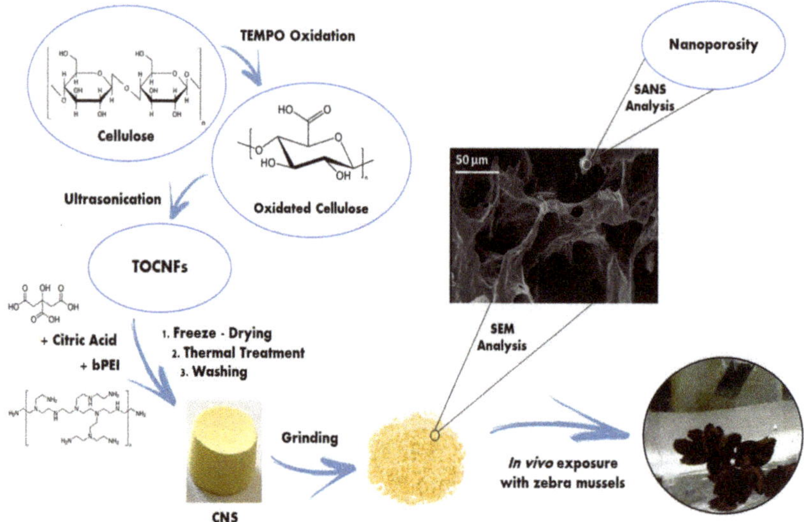

Figure 1. Schematic representation of synthetic and morphological aspects of cellulose-based nanosponge (CNS).

2.3. Sampling and Maintenance Condition

Independent experiments were conducted in freshwater aquaria for the preliminary $CdCl_2$ dose-exposure test and for experiments in which $CdCl_2$ contaminated artificial freshwater was treated with CNS. Adult specimens of *D. polymorpha* (medium valves length 2 ± 0.5 cm) were collected from Bilancino Lake, a pristine area in Tuscany (Florence, Italy) and carried to the laboratory in lake original water. For the acclimatization period mussels were placed in a 10-L aerated tank with artificial freshwater (AFW) obtained by mixing distilled (50%) and dechlorinated tap water (50%) 5 days before preliminary dose-exposure experiment and 2 days before the second set of experiments. Water temperature was 18 ± 1 °C and a natural photoperiod was maintained, pH values were 8.04 ± 0.10 for preliminary dose-exposure experiments and 8.05 ± 0.15 for experiments with CNS.

2.4. In Vivo Exposure

After an acclimatization period, mussels were placed on glass sheets suspended in small glass aerated aquaria. For each treatment tank, at least 30 zebra mussels were exposed to the experimental time of 48 h. Zebra mussels were not fed during the experiments, and only specimens that were able to re-attach themselves by their byssus filaments on glass sheets immersed in water were used for the research as suggested by Binelli and collaborators [36]. Mussels were preliminarily exposed to 0.005;

0.01; 0.05; 0.1 mg L^{-1} CdCl$_2$ for 48 h and to 7 days in order to set up sub-lethal concentration to be used for the CNS efficacy exposure study. The lowest dose was selected because established by United States Environmental Protection Agency (US EPA) as a maximum contaminant level for cadmium in drinking water [38] while the highest one is considered representative of highly impacted sites. CdCl$_2$ was dissolved in distilled water and a stock solution was prepared. These preliminary studies indicated that a loss of DNA integrity in *D. polymorpha* hemolymph was observable after 48 h exposure, starting from 0.005 mg L^{-1} and from 0.01 mg L^{-1} after 7 days displaying cytotoxicity only at the highest dose of 0.1 mg L^{-1} (data not shown). For this reason, in order to work with environmentally realistic exposure levels, the CdCl$_2$ dose selected for the experiments resulted to be 0.05 mg L^{-1}, being genotoxic but not cytotoxic (cytotoxicity appeared at 0.1 mg L^{-1}). Moreover, the dose of 0.05 mg L^{-1} (53.7 ± 0.5 μg L^{-1} nominal dose) allowed to assess CNS activity at a cadmium concentration which can be considered environmentally realistic in polluted sites [1], even avoiding mussel valves' closure with consequent cessation of filtering activity, a general defense mechanism occurring in mussels after exposure to very high concentrations pollutants, higher than 0.1 mg L^{-1} for Cd(II) concentrations [39,40]. To assess CNS ecosafety, i.e., absence of toxicity in freshwaters, and CdCl$_2$ adsorption efficacy, zebra mussels were in vivo exposed for 48 h to the following groups in AFW: CdCl$_2$ 0.05 mg L^{-1} (Cd(II)), CdCl$_2$ 0.05 mg L^{-1} after treatment with CNS (Cd-t CNS), AFW after treatment with CNS (CNS). Mussels in AFW were used as controls (AFW). The ratio of CNS able to adsorb Cd(II) from freshwaters was set up at 1.25 g CNS per L of AFW, based on our previous studies [31,33]. In particular, CNS samples were incubated with AFW for 2 h at room temperature, under vigorous magnetic stirring, reproducing the treatment process normally applied to contaminated water matrices. Mixtures were then filtered at 0.45 μm to remove CNS powder and the conditioned AFW was used for zebra mussels during the in vivo exposure study. The effects of CdCl$_2$ and CNS-treated AFW, alone and in combination, were evaluated in zebra mussel hemocytes gently aspirated from the posterior adductor muscle sinus, processed for Comet and Cytome assays [41].

2.5. Cadmium Concentration in Water

Water media samples were taken from each tank after 48 h and stored at 4 °C, for chemical analyses. Cadmium concentrations in AFW were determined by means of Inductively Coupled Plasma-Optical Emission Spectrometry (ICP-OES) using a Perkin Elmer Optica 8300 (Perkin Elmer, Waltham, MA, USA), equipped with a CrossFlow nebulizer and a Scott Spray Chamber, followed by a standard quartz torch. The instrument calibration was performed by dilution of a cadmium analytical standard (Honeywell FLUKA™, Fisher Scientific Italia, Rodano, MI, Italy), with MilliQ® water obtaining 5, 10, and 50 μg L^{-1} solutions, to each analyzed samples Y (2 mg L^{-1}) was added as internal standard.

2.6. Viability Assessment

The Neutral Red Retention Time (NRRT), used extensively as convenient and rapid measurement of cell viability [42], was performed to discriminate the healthy cells, whose lysosomes take up and retain longer the dye (neutral red), from damaged cells unable to keep the stain inside the organelles.

NRRT was set up according to Guidi and coworkers [43]. Briefly, 10 μL of hemolymph for each sample were incubated on a polylysine pre-coated microscope glass slide with neutral red working solution (0.1 mg mL^{-1}). Granular hemocytes were examined to light microscope at 15 min intervals, for up to 180 min to evaluate the time at which 50% of cells had leaked to the cytoplasm the dye previously trapped by lysosomes. The Comet assay was performed only with cell population that showed a viability >90% to avoid false positive results.

2.7. Comet Assay

The Comet Assay was performed on hemocytes from 10 specimens per tank at the end of the exposure, according to Singh and co-authors [44] with slight modifications [43]. Briefly, cells were embedded in 0.5% low-melting agarose (LMA), spread onto microscope slides pre-coated with 1%

normal-melting agarose (NMA) and covered with a further layer of LMA 0.5%. Slides were dipped into a lysing solution (NaCl 2.5 M, Na₂EDTA 100 mM, Trizma Base 10 mM, 10% DMSO, 1% Triton X-100, pH 10) and kept overnight at 4 °C in the dark, in order to solubilize cell membranes and cytoplasm. Successively, slides were treated with alkali (NaOH 300 mM, Na₂EDTA 1 mM, pH > 13) for 10 min and placed in a horizontal electrophoresis apparatus. Electrophoresis was performed for 5 min at 25 V and 300 mA. After the run, slides were neutralized with Tris-HCl (0.4 M, pH 7.5), stained with 100 µL ethidium bromide and observed under a fluorescence microscope (400×). The amount of DNA damage was quantified as the percentage of DNA migrated into the comet tail (tail DNA) [45], using an image analyzer (Kinetic imaging, Ltd., Komet, Version 5, Software for Live Cell Imaging, Tissue and Structure Quantification, Cytogenetics and Toxicology, Stereology. Bromborough, Wirral, Merseyside CH62 3NY, UK, 2005). At least 50 nuclei per slide and 2 slides per sample were scored, for a total of 100 nuclei per organisms and the mean calculated. At least 5 organisms per treatment were analyzed.

2.8. Cytome Assay

The genotoxic effects were evaluated at a chromosomal level by the micronuclei and nuclear abnormalities frequency assessment. The Cytome assay was carried out on hemocytes according to Guidi and co-authors [43]; 900 µL of PBS were added to100 µL of hemolymph, then centrifuged for 10 min at 1000 rpm. Cell pellet was prefixed for 20 min in a 5% acetic acid, 3% ethanol, 92% PBS, then centrifuged for 10 min at 1000 rpm. The supernatant was removed and 1 mL fixative (7:1, ethanol:acetic acid) was added to the suspended pellet; this process was repeated twice. After the last fixation cells were centrifuged, spread onto microscope slides (two slides per mussel), air dried and stained with 3% Giemsa solution for 10 min. Cells with well-preserved cytoplasm per specimen were scored (500 per slide) at light microscope to determine the frequency of micronuclei (MN) and nuclear abnormalities [46] according to the following criteria proposed by Fenech [47]. The presence of cells with morphologically altered nuclei (nuclear blebs (BL), nuclear buds (NBUD), notched nucleus (NT), lobed nucleus (LB)), and apoptotic cells (APO) were scored on the same slides in parallel and collectively reported. The frequency of total nucleus abnormalities (NA), was also evaluated in mussel hemocytes and binucleated cells (BN), with or without nuclear bridges (NPB), were scored for proliferation activity (Figure 2). Ten specimens, two slides per specimens, 500 cells per slide were scored for each experimental group.

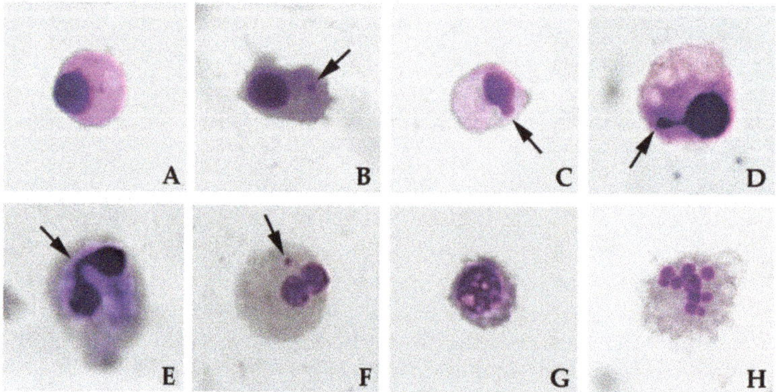

Figure 2. Hemocyte nuclear abnormalities in *D. polymorpha*. (**A**) Mononucleated healthy cell; (**B**) hemocyte displaying a micronuclei (MN); (**C**) hemocyte with a nuclear bleb (BL); (**D**) hemocyte with

a nuclear bud (NBUD); (**E**) Binucleated (BN) hemocyte with nuclear bridges (NPB); (**F**) BN with a MN; (**G**) vacuolated cell; (**H**) apoptotic cell.

2.9. Statistical Analysis

Results obtained are represented as mean ± SE from at least 5 specimens. Data were analyzed by the multifactor analysis of variance (MANOVA) or multiple regression analysis (MRA). The multiple range test (MRT) was performed in order to detect differences among experimental groups. For all data analysis, statistical significance was set at $p < 0.05$.

3. Results and Discussion

3.1. CNS Synthesis and Characterization

The aim of the present work was to assess the efficacy and the safety of specifically synthesized polysaccharide based nanosponges to prevent cadmium-induced genotoxicity in freshwaters in the framework of a bigger project on nanoremediation ("Nanomaterials for Remediation of Environmental Matrices associated to Dewatering", POR CReO FESR 2014–2020). Concerning results about CNS synthesis and characterization, CNS were produced according to the protocol reported in Figure 1. In a first step, the oxidation of native cellulose by means of TEMPO/NaClO/KBr system leads to the partial conversion of C6 alcoholic groups present on the glucopyranose units to the corresponding carboxylic acids [26]. The final content of acidic groups resulted to be 1.5 mmol g^{-1}, which was considered a limit value, as further oxidation would lead to partial depolymerization of cellulose chains. Carboxylic moieties, at basic pH, underwent deprotonation, promoting the electrostatic repulsion among single nanofibers (TOCNF), which could be filtered and recovered. The second step in the synthetic procedure consisted into mixing TOCNF and bPEI in deionized water in a 1:1 weight ratio. In order to increase the content of carboxylic groups and to further increase cross-linking, 18% of citric acid (CA) respect to primary amino groups of bPEI was added to the mixture. This formulation is the result of an eco-design performed since the early stage of production was at a laboratory scale [32]. The obtained hydrogel underwent lyophilization and then heating at about 100 °C, affording CNS, whose structural and chemical stability was derived by the formation of amide bonds between the carboxylic groups of TOCNF and the primary amines of bPEI, as evidenced in a previous work by FT-IR analysis (-C=O stretching of the amide bonds at 1664 cm^{-1}) [28]. ^{13}C CP-MAS solid-state NMR and elemental analysis also confirmed the role of CA in better fixing bPEI molecules [28], the latter playing a crucial role in the adsorption process by chelating the metal ions thanks to the multiple amino groups. CNS were grinded in a mortar before use, in order to increase sorption performance, by facilitating the diffusion of ions in the network [32]. Microcomputed tomography (μ-CT) analysis also revealed that CNS are characterized by a high porosity (70–75%), with pore sizes in the range of 10–100 μm, as evidenced by scanning electron microscopy (SEM) [28]. Pores are formed during the freeze-drying process, and are characterized by a 2D sheet-like morphology, often reported in literature for cellulose-based aerogels obtained following this procedure. This approach also prevents the formation of occlusions, guaranteeing a complete diffusivity of analytes in the network [48]. Besides the evidence of micro-porosity, a small angle neutron scattering (SANS) analysis of water nano-confinement geometries in CNS allowed to evidence the presence of nano-pores, measuring a short-range correlation length in a range between 25 and 35 Å [29]. More recently, the FTIR-ATR investigation of H_2O and D_2O in CNS revealed a supercooled behavior of these molecules, further supporting the concept a nano-confinement for water, and consequently the presence of nanopores in the network. Due to all these characteristics CNS represents a sustainable cellulose nanomaterial, developed from renewable sources, which appears to be eligible for nano-remediation strategies in heavy metal polluted environments.

As cadmium is a human carcinogen, its contamination of surface and ground waters represents a serious environmental problem, even if the mechanism of Cd-induced carcinogenesis is still under discussion. It was proposed that it is multi-factorial [49], and an important role may be played by the indirect increase of reactive oxygen species and the disruption of the cellular antioxidant system [12,50,51], leading to oxidative stress induction [52], which, in turn, causes lesions to the DNA, including potentially lethal DSBs [13]. It also competes with Zn (II) in cellular components and binds to the DNA bases, giving rise to single-stranded DNA breaks [53,54]. While the European Union directive posed 0.005 mg L^{-1} of cadmium as a threshold for drinking water, other countries allowed higher thresholds to be set for freshwater. For this reason, the ability of harmless polysaccharide based nanosponges to reduce both Cd(II) concentration in water and Cd(II)-induced genotoxicity in freshwater organisms was investigated.

3.2. Cadmium Concentration in Water

In order to assess the efficacy of cellulose-based nanosponges in removing Cd(II) from freshwater together with their safety evaluation for organisms, exposure experiments to cadmium contaminated AFW treated with CNS (Cd-t CNS) were carried out. ICP-OES analysis on exposure waters showed the strong Cd(II) adsorption capability of CNS in freshwater. Table 1 reports the results of ICP-OES analysis on all the waters to which zebra mussels have been exposed. In AFW and AFW treated with CNS (CNS) the Cd(II) concentration resulted negligible. On the contrary, artificially Cd(II) polluted AFW treated with 1.25 g L^{-1} CNS (Cd-t CNS) showed a remarkable decrease of Cd(II) concentration (up to 88.8%) when compared to artificially Cd polluted AFW (Cd(II)).

Table 1. Cd(II) concentration (mg L^{-1}) measured by plasma spectroscopy in treatment waters after 48 h of exposure to the following experimental groups: artificial freshwater (AFW) (control); Cd(II) (0.05 mg L^{-1} of CdCl$_2$ in AFW); Cd-t CNS (CdCl$_2$ 0.05 mg L^{-1} contaminated AFW treated with CNS); CNS (AFW treated with only CNS).

Experimental Group	Cd(II) mg L^{-1}
AFW	<0.001
CNS	<0.001
Cd(II)	0.0537 ± 0.005
Cd-t CNS	0.0060 ± 0.0001

This result was in line with what was recently found in pure water [27] and in artificial sea water [32,33]. It is interesting to note how starting from 0.05 mg L^{-1}, a dose associable with high contaminated rivers [55], the residual final cadmium concentration found in waters treated with CNS was 0.0060 ± 0.0001 mg L^{-1}, the same order of magnitude posed by the European Union directive for drinking water (0.005 mg L^{-1}).

3.3. Cellular Responses

Concerning biological responses water samples pretreated with CNS alone (CNS) or in combination with 0.05 mg L^{-1} CdCl$_2$ (Cd-t CNS), they did not result as cytotoxic in terms of lysosome membrane stability, always displaying values of Neutral Red Retention Time above 180 min. After both the exposures to CNS alone and cadmium contaminated AFW treated with CNS, data were statistically comparable to control levels also in terms of DNA integrity.

Cadmium per se is weakly genotoxic [13], and one of the advantages of the Comet assay is its demonstrated sensitivity for detecting low levels of DNA damage and in its alkaline version it is capable of detecting DNA single-strand breaks (SSB), alkali-labile sites (ALS), DNA-DNA/DNA-protein cross-linking, and SSB associated with incomplete excision repair sites [56]. Moreover, a negative association has been often showed between oxidative stress and DNA primary damage detected by Comet assay [57].

Zebra mussel hemocytes collected from the tank containing AFW exposed to polysaccharide-based nanosponges did not show any loss of DNA integrity. On the contrary, CdCl$_2$ 0.05 mg L^{-1} induced a statistically significant increase ($p < 0.05$) of DNA primary damage in zebra mussel hemocytes compared to controls. In specimens exposed to Cd-t CNS the DNA damage induced by CdCl$_2$

0.05 mg L^{-1} turned down to control level (Figure 3). Interestingly, DNA integrity loss was found to be related to cadmium nominal concentration (c.c. 0.46; R^2 = 21.7; $p < 0.001$).

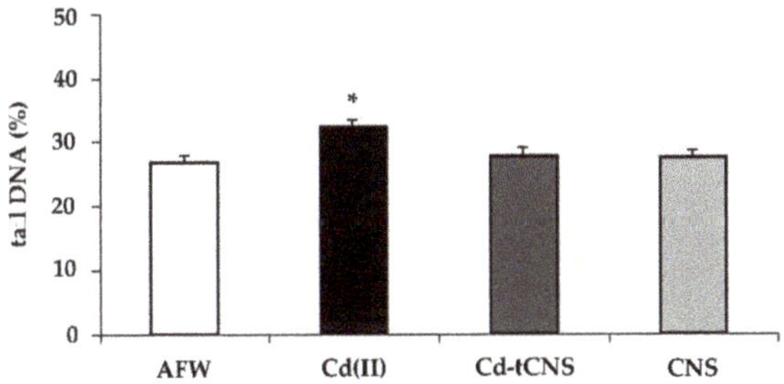

Figure 3. DNA primary damage (% tail DNA) in zebra mussel hemocytes after 48 h of exposure to the following experimental groups: AFW (control); Cd(II) (0.05 mg L^{-1} of CdCl$_2$ in AFW); Cd-t CNS (CdCl$_2$ 0.05 mg L^{-1} contaminated AFW treated with CNS); CNS (AFW treated with only CNS). Results are shown as mean ± SE. (*) indicates significant differences respect to the control group (AFW) ($p < 0.05$).

Although our background level of DNA damage in control mussels did not range among the recommended values suggested by Tice and co-authors [56], the results obtained from the control tank were able to discriminate the exposure effects. In literature, high background levels obtained by the Comet assay were reported in *D. polymorpha* as well as in other aquatic organisms [58] and baseline levels of DNA damage in mussel hemocytes seem to be correlated with animal maintenance temperature; resulting 4 °C to be the optimal experimental condition for lower Comet assay basal results [59]. However, 4 °C was not selected in the present work because it was not representative of natural water lake temperature required by our experimental design which aimed, instead, to recreate environmental realistic conditions to test CNS performance and potential cellular toxicity. On the other hand, we cannot exclude that results obtained from AFW experimental group might have been modulated by other factors such as water quality of sample site and/or origin population genetic background. CNS thus resulted to be not genotoxic in terms of DNA primary damage, speaking in favor of it being a harmless material for biological systems. The significant loss of DNA integrity of zebra mussel hemocytes induced by Cd was completely restored in organisms exposed to Cd-t CNS paralleling the observed reduction of cadmium bioavailability in the water, thus indicating CNS efficacy in Cd removal.

Besides the disruption in cellular signal transduction [60], cell proliferation activity appears to be affected by cadmium exposure [61] and our data seem to confirm in zebra mussel hemocytes what was reported for human and mammalian cell lines [62–64]. The dose selected for CdCl$_2$ treatment (0.05 ppm) was able to induce proliferation activity evaluated in terms of binucleated cell frequency increase ($p < 0.05$) after 48 h exposure. Interestingly, after 2 day-exposure the frequency of binucleated hemocytes turned down to control level in specimens exposed to cadmium in AFW treated with CNS (Cd-t CNS) (Figure 4).

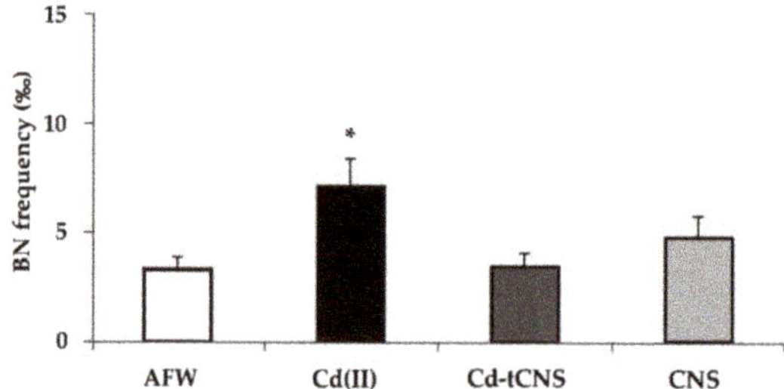

Figure 4. Frequency (‰) of binucleated cells (BN) in zebra mussel hemocytes after 48 h of exposure to the following experimental groups: AFW (control); Cd(II) (0.05 mg L^{-1} of CdCl$_2$ in AFW); Cd-t CNS (CdCl$_2$ 0.05 mg L^{-1} contaminated AFW treated with CNS); CNS (AFW treated with only CNS). Results are shown as mean ± SE. (*) indicates significant differences respect to the control group (AFW) ($p < 0.05$).

Cadmium was actually found to stimulate DNA synthesis and cell proliferation at low concentrations in mammalian cell lines [54], leading to relevant implications for carcinogenesis processes. An increased level of binucleated gill cells has been also observed in zebra mussel after 5 days of exposure and proposed to be related to an inhibition of cytokinesis activity more than a Cd proliferative effect [1]. Alteration of cytoskeleton function, resulting in reduced phagocytic processes, has been observed in *D. polymorpha* hemocytes [2] after 5×10^{-4} M Cd exposure. Moreover, an alteration in actyne fibers, resulting in altered hemocyte morphology, with reduced pseudopods formation, was reported after cadmium exposure in marine mussels [2,65,66]. In our study, the significant increase of binucleated cell frequency induced by Cd treatment was reduced to control levels in organisms exposed to cadmium contaminated AFW treated with CNS, showing that the Cd removal by CNS restored a physiological proliferation and/or cytokinesis activity. Among several mechanisms proposed to explain actin cytoskeleton alterations, an increase of ROS production plays an important role, which, in turn, might associate the two phenomena observed, i.e., loss of DNA integrity and cell proliferation increase.

Cd-induced biochemical changes may play roles in all the stages of carcinogenicity and the induction of oxidative stress in combination with decreased DNA repair can lead to DNA damage [54]. It has been proposed that Cd interferes with major DNA repair systems [13], leading to their inhibition [67,68], and unrepaired DNA lesions may give rise to chromosomal damage. The spontaneous MN frequency (mean ± SD) found in our preliminary studies (1.2 ± 1.1 for 2 days and 2.1 ± 1.05 for 7 days exposure, data not shown) is similar to the basal levels reported in zebra mussel hemocytes by Binelli and coauthors [36] (from 0 to 4). In the present study, Cd(II) exposure induced an increase of hemocyte MN basal level of 83%. AFW exposed to CNS did not cause any chromosomal damage in zebra mussel hemocytes, while CdCl$_2$ 0.05 mg L^{-1} confirmed the statistically significant induction of cells displaying both micronuclei and total nuclear morphology abnormalities (Figure 5A,B). No induction of apoptotic cells was observed at all the experimental points.

Mussels exposed to CNS alone and Cd-t CNS showed that Cd-induced MN and nuclear abnormality frequency increases, biomarkers associated with numerical and structural chromosomal mutations, were reduced by Cd removal activity of CNS, speaking in favor of the effectiveness of this newly synthetized nanosponge in restoring not only a Cd(II) induced repairable DNA damage, assessed by Comet assay, but also in reducing a consolidated chromosomal damage in freshwater organisms. Taking into consideration all the chromosomal damage, it is represented by both micronuclei, potentially coming from a possible dysfunction of mitotic apparatus, and nuclear abnormalities. NA include structural aberrations mainly representing the results of clastogenic events often induced by oxidative processes, like dicentric chromosomes (NPB), DNA repair system failure and gene amplification. In the present study, Multiple Regression Analysis also showed a moderate but significant positive correlation between the frequencies of micronuclei and total nuclear abnormalities ($r = 0.52$; R^2 = 27.17;

$p < 0.001$) (Figure 6), indicating the involvement of the whole chromosomal damage in response to cadmium-induced genotoxicity.

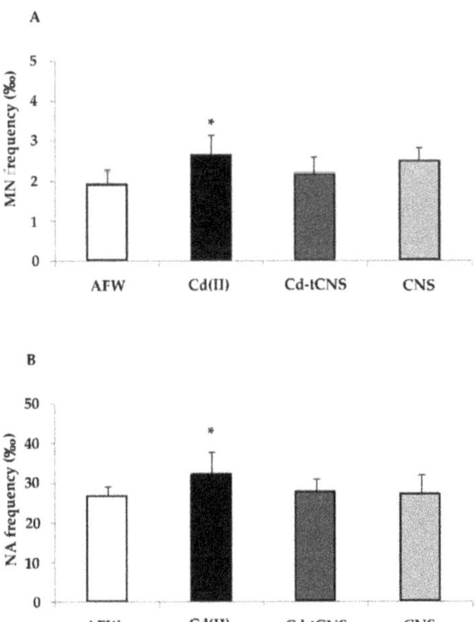

Figure 5. (**A**) Frequencies (‰) of micronuclei (MN) and (**B**) total nuclear abnormalities (NA) in zebra mussel hemocytes after 48 h of exposure to the following experimental groups: AFW (control); Cd(II) (0.05 mg L^{-1} of CdCl$_2$ in AFW); Cd-t CNS (CdCl$_2$ 0.05 mg L^{-1} contaminated AFW treated with CNS); CNS (AFW treated with only CNS). Results are shown as mean ± SE. (*) indicates significant differences respect to the control group (AFW) ($p < 0.05$).

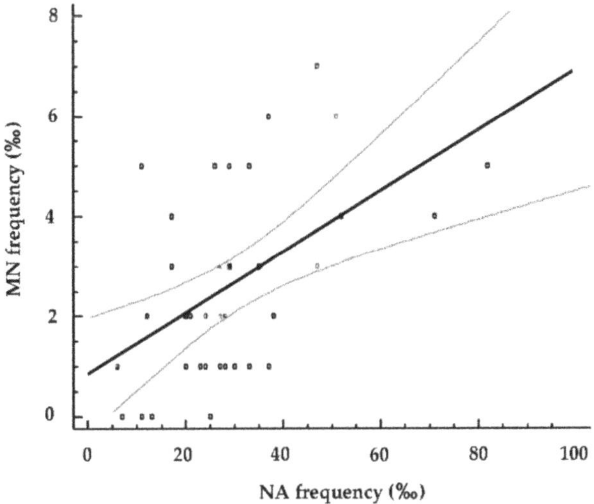

Figure 6. Relationship between MN and NA frequencies in zebra mussels from the different experimental conditions (control, Cd(II), Cd-t CNS, CNS). $r = 0.52$; R^2 = 27.17; $p < 0.001$. Black line is the regression line, which represents the regression equation. Gray lines represent 95% of the confidence interval.

Our analyses also supported the use of hemolymph as a useful tissue to perform Cytome assay, allowing the analysis of samples from single individuals and not from pooled cells taken from different specimens as frequently applied [69].

The principal novelty of the present work relies on the efficacy of specifically designed eco-safe CNS for nanoremediation, which were found to efficiently remove cadmium from contaminated waters without inducing cellular alterations in a model freshwater organism. Data from literature concerning nano-remediation procedures mainly report cadmium removal from solid environmental matrices, like river sediments, by the use of nZVI nanomaterials [11,70], making less fitting any comparison with the present study. In such cases, the remaining concentration of Cd(II) in the aqueous solution was enough to induce strong deleterious effects on the mechanisms of cellular respiration [70]. Nano TiO_2 sprays have been even used to reduce cadmium content in cowpea plant leaves [21] in order to reduce cadmium dietary intake, but the need of cadmium removal from freshwaters is still high and future research directions call for the assessment of CNS efficacy by using lower concentrations of cadmium, aiming to break down Cd(II) concentration close to zero level and to promote their potential large scale applications in different freshwaters and other environmental matrices.

4. Conclusions

We have proposed a cellulose-based nanostructured material derived from renewable sources as a valuable sorbent material for cadmium ions removal from freshwater. The high affinity for metal ions is ascribed to the chelating action of the amino-groups present in the inner structure, and derived from the bPEI polymer used for TOCNF cross-linking. The eco-designed material is characterized by a high micro- and nano-porosity, facilitating the contacting between the metal ions and the active sites of the sorbent. In our experimental model, this cellulose based nanosponge resulted to be a suitable safe candidate in cadmium remediation process being efficient in metal sequestering, without altering cellular physiological activity in freshwater organisms. Our data demonstrate, for the first time, the efficacy of CNS to carry out their remediation action in freshwater environment in terms of reduction of DNA primary and chromosomal damage induced by Cd to freshwater zebra mussel hemocytes. The efficacy of CNS was confirmed at a chromosomal level by the Cytome assay, which included the evaluation of different nuclear abnormalities, supporting the application of piscine Cytome assay [71] as a valuable tool for nano-remediation validating studies applied to freshwater models.

Author Contributions: G.F.: Conceptualization, Supervision, Resources, Funding acquisition; C.P.: Conceptualization, Supervision, Resources; Funding acquisition; I.C.: Conceptualization, Supervision, Resources; Funding acquisition; P.G.: Investigation, Writing—Original Draft, Writing—Review & Editing, Methodology; M.B.: Investigation, Writing—Review & Editing, Methodology; A.F.: Investigation, Writing—Review & Editing, Methodology; L.R. Investigation, Writing—Review & Editing, Methodology; M.P.: Investigation, Writing—Review & Editing; V.S.: Investigation; M.G.: Investigation. This work is part of the degree thesis of M.G. All authors have read and agreed to the published version of the manuscript.

Funding: This research was funded by Regione Toscana, NanoBonD (Nanomaterials for Remediation of Environmental Matrices associated to Dewatering) POR CReO FESR Toscana 2014–2020-30/07/2014- LA 1.1.5 CUP 3389.30072014.067000007.

Acknowledgments: Authors thank Giacomo Grassi for his useful suggestions and Claudio Ghezzani for graphic processing.

Conflicts of Interest: The authors declare no conflict of interest. The funders had no role in the design of the study; in the collection, analyses, or interpretation of data; in the writing of the manuscript, or in the decision to publish the results.

References

1. Vincent-Hubert, F.; Arini, A.; Gourlay-Francé, C. Early genotoxic effects in gill cells and haemocytes of *Dreissena polymorpha* exposed to cadmium, B(a)P and a combination of B(a)P and Cd. *Mutat. Res.* **2011**, *723*, 26–35. [CrossRef] [PubMed]
2. Evariste, L.; Rioult, D.; Brousseau, P.; Geffard, A.; David, E.; Auffret, M.; Fournier, M.; Betoulle, S. Differential sensitivity to cadmium of immunomarkers measured in hemocyte subpopulations of zebra mussel *Dreissena polymorpha*. *Ecotox. Envir. Saf.* **2017**, *137*, 78–85. [CrossRef] [PubMed]
3. United Nations Environment Programme Unep. Available online: http://wedocs.unep.org/bitstream/handle/20.500.11822/27636/Cadmium_Review.pdf (accessed on 17 July 2020).
4. Ravera, O.; Mislin, H. Cadmium in freshwater ecosystems. In *Cadmium in the Environment*; Birkhäuser: Basel, Switzerland, 1986; pp. 75–79.
5. World Health Organization. Chapter 6.3. In *Cadmium Air Quality Guidelines*, 2nd ed.; Regional Office for Europe: Copenhagen, Denmark, 2000; pp. 136–139.
6. Morin, S.; Duong, T.T.; Dabrin, A.; Coynel, A.; Herlory, O.; Baudrimont, M.; Delmas, F.; Durrieu, G.; Schäfer, J.; Winterton, F.; et al. Long-term survey of heavy-metal pollution, biofilm contamination and diatom community structure in the Riou Mort watershed, South-West France. *Environ. Pollut.* **2008**, *15*, 532–542. [CrossRef] [PubMed]
7. CE. CEE/CEEA/CE n.83. 1998. Available online: https://eur-lex.europa.eu/eli/dir/1998/83/oj (accessed on 17 July 2020).
8. Pereira, L.S.; Ribas, J.L.C.; Vicari, T.; Silva, S.B.; Stival, J.; Baldan, A.P.; Valdez Domingos, F.X.; Grassi, M.T.; Cestari, M.M.; Silvade Assis, H.C. Effects Eofecologically relevant concentrations of cadmium I na freshwater fish. *Ecotoxicol. Environ. Saf.* **2016**, *13*, 29–36. [CrossRef] [PubMed]
9. Martins, A.C.; Urbano, M.R.; Almeida Lopes, A.C.B.; Carvalho, M.F.H.; Buzzo, M.L.; Docea, A.O.; Mesas, A.E.; Aschner, M.; Silva, A.M.R.; Silbergeld, E.K.; et al. Blood cadmium levels and sources of exposure in an adult urban populationin southern Brazil. *Environ. Res.* **2020**, *187*, 109618. [CrossRef]
10. Aitio, A.; Alessio, L.; Axelson, O.; Coenen, J.; de Flora, S.; Grandjean, P.; Heinrich, U.; Huff, J.E.; Ikeda, M.; Kavlock, R.; et al. IARC monographs on the evaluation of carcinogenic risks to humans: Beryllium, cadmium, mercury, and exposures in the glass manufacturing industry. *IARC Monogr. Eval. Carcinog. Risks Hum.* **1993**, *58*, 1–415.
11. Xue, W.; Peng, Z.; Huang, D.; Zeng, G.; Wan, J.; Wan, J.; Xui, R.; Cheng, M.; Zhang, C.; Jiang, D.; et al. Nanoremediation of Cadmium Contaminated River Sediments: Microbial Response and Organic Carbon Changes. *J. Hazard. Mater.* **2019**, *359*, 290–299. [CrossRef]
12. Liu, J.; Qu, W.; Kadiiska, M.B. Role of oxidative stress in cadmium toxicity and carcinogenesis. *Toxicol. Appl. Pharmacol.* **2009**, *238*, 209–214. [CrossRef]
13. Lee, W.K.; Thévenod, F. Cell organelles as targets of mammalian cadmium toxicity. *Arch. Toxicol.* **2020**, *94*, 1017–1049. [CrossRef]
14. Karn, B.; Kuiken, T.; Otto, M. Nanotechnology and in situ remediation: A review of the benefits and potential risks. *Environ. Health Perspect.* **2009**, *117*, 1813–1831. [CrossRef]
15. Oughton, D.; Koin, P.; Bleyl, S.; Filip, J.; Skácelová, P.; Klaas, N.; von der Kammer, F.; Gondikas, A. *Development and Application of Analytical Methods for Monitoring Nanoparticles in Remediation: NanoRem 5*; CL:AIRE: London, UK, 2017; p. 6.
16. Fu, F.L.; Dionysiou, D.D.; Liu, H. The use of zero-valent iron for groundwater remediation and wastewater treatment: A review. *J. Hazard. Mater.* **2014**, *267*, 194–220. [CrossRef] [PubMed]
17. Kim, H.J.; Leitch, M.; Naknakom, B.; Tilton, R.D.; Lowry, G.V. Effect of emplaced nZVI mass and groundwater velocity on PCE dechlorination and hydrogen evolution in water-saturated sand. *J. Hazard. Mater.* **2017**, *322*, 136–144. [CrossRef] [PubMed]
18. Lei, C.; Sun, Y.; Tsang, D.C.W.; Lin, D. Environmental transformations and ecological effects of iron-based nanoparticles. *Environ. Pollut.* **2018**, *232*, 10–30. [CrossRef]
19. Hanna, S.K.; Miller, R.J.; Lenihan, H.S. Deposition of carbon nanotubes by a marine suspension feeder revealed by chemical and isotopic tracers. *J. Hazard. Mater.* **2014**, *279*, 32–37. [CrossRef] [PubMed]
20. Callaghan, N.I.; MacCormack, T.J. Ecophysiological perspectives on engineered nanomaterial toxicity in fish and crustaceans. *Comp. Biochem. Physiol. Part C. Toxicol. Pharmacol.* **2017**, *193*, 30–41. [CrossRef]

21. Ogunkunle, O.C.; Odulaja, D.A.; Akande, F.O.; Varun, M.; Vishwakarma, V.; Fatoba, O.P. Cadmium toxicity in cowpea plant: Effect of foliar intervention of nano-TiO$_2$ on tissue Cd bioaccumulation, stress enzymes and potential dietary health risk. *J. Biotechnol.* **2020**, *310*, 54–61. [CrossRef]

22. Corsi, I.; Fiorati, A.; Grassi, G.; Bartolozzi, I.; Daddi, T.; Melone, L.; Punta, C. Environmentally Sustainable and Ecosafe Polysaccharide-Based Materials for Water Nano-Treatment: An Eco-Design Study. *Materials* **2018**, *11*, 1228. [CrossRef]

23. Carpenter, A.W.; de Lannoy, C.F.; Wiesner, M.R. Cellulose nanomaterials in water treatment technologies. *Environ. Sci. Technol.* **2015**, *49*, 5277–5287. [CrossRef]

24. Voisin, H.; Bergström, L.; Liu, P.; Mathew, A.P. Nanocellulose-Based Materials for Water Purification. *Nanomaterials* **2017**, *7*, 57. [CrossRef]

25. Mahfoudhi, N.; Boufi, S. Nanocellulose as a novel nanostructured adsorbent for environmental remediation: A review. *Cellulose* **2017**, *24*, 1171–1197. [CrossRef]

26. Pierre, G.; Punta, C.; Delattre, C.; Melone, L.; Dubessay, P.; Fiorati, A.; Pastori, N.; Galante, Y.M.; Michaud, P. TEMPO-mediated oxidation of polysaccharides: An ongoing story. *Carbohyd. Polym.* **2017**, *165*, 71–85. [CrossRef]

27. Melone, L.; Rossi, B.; Pastori, N.; Panzeri, W.; Mele, A.; Punta, C. TEMPO-Oxidized Cellulose Cross-Linked with Branched Polyethyleneimine: Nanostructured Adsorbent Sponges for Water Remediation. *Chem.Plus.Chem.* **2015**, *80*, 1408–1415. [CrossRef] [PubMed]

28. Fiorati, A.; Turco, G.; Travan, A.; Caneva, E.; Pastori, N.; Cametti, M.; Punta, C.; Melone, L. Mechanical and Drug Release Properties of Sponges from Cross-linked Cellulose Nanofibers. *ChemPlusChem* **2017**, *82*, 848–858. [CrossRef] [PubMed]

29. Paladini, G.; Venuti, V.; Almásy, L.; Melone, L.; Crupi, V.; Majolino, D.; Pastori, N.; Fiorati, A.; Punta, C. Cross-linked cellulose nano-sponges: A small angle neutron scattering (SANS) study. *Cellulose* **2019**, *26*, 9005–9019. [CrossRef]

30. Riva, L.; Fiorati, A.; Sganappa, A.; Melone, L.; Punta, C.; Cametti, M. Naked-eye heterogeneous sensing of fluoride ions by co-polymeric nanosponge systems comprising aromatic-imide-functionalized nanocellulose and branched polyethyleneimine. *ChemPlusChem* **2019**, *84*, 1512–1518. [CrossRef]

31. Bartolozzi, I.; Daddi, T.; Punta, C.; Fiorati, A.; Iraldo, F. Life cycle assessment of emerging environmental technologies in the early stage of development: A case study on nanostructured materials. *J. Ind. Ecol.* **2020**, *24*, 101–115. [CrossRef]

32. Fiorati, A.; Grassi, G.; Pastori, N.; Graziano, A.; Liberatori, G.; Melone, L.; Punta, C.; Corsi, I. Eco-design of nanostructured cellulose sponges for sea-water decontamination from heavy metal ions. *J. Clean. Prod.* **2020**, *246*, 119009. [CrossRef]

33. Liberatori, G.; Grassi, G.; Guidi, P.; Bernardeschi, M.; Fiorati, A.; Scarcelli, V.; Genovese, M.; Faleri, C.; Protano, G.; Frenzilli, G.; et al. Effect-Based Approach to Assess Nanostructured Cellulose Sponge Removal Efficacy of Zinc Ions from Seawater to Prevent Ecological Risks. *Nanomaterials* **2020**, *10*, 1283. [CrossRef]

34. Binelli, A.; Della Torre, C.; Magni, S.; Parolini, M. Does zebra mussel (*Dreissena polymorpha*) represent the freshwater counterpart of *Mytilus* in ecotoxicological studies? A critical review. *Environ. Pollut.* **2015**, *196*, 386–403. [CrossRef] [PubMed]

35. Bolognesi, C.; Fenech, M. Mussel micronucleus cytome assay. *Nat. Protoc.* **2012**, *7*, 1125–1137. [CrossRef]

36. Binelli, A.; Cogni, D.; Parolini, M.; Riva, C.; Provini, A. In vivo experiments for the evaluation of genotoxic and cytotoxic effects of Triclosan in Zebra mussel hemocytes. *Aquat. Toxicol.* **2009**, *91*, 238–244. [CrossRef] [PubMed]

37. De Lafontaine, Y.; Gagné, F.; Blaise, C.; Costan, G.; Gagnon, P.; Chan, H. Biomarkers in zebra mussels (*Dreissena polymorpha*) for the assessment and monitoring of water quality of the St Lawrence River (Canada). *Aquat. Toxicol.* **2000**, *50*, 51–71. [CrossRef]

38. Environmental Protection Agency. Available online: https://www.epa.gov (accessed on 17 July 2020).

39. Naimo, T.J.; Atchison, G.J.; Holland-Bartels, L.E. Sublethal effects of cadmium on physiological responses in the pocketbook mussel, *Lampsilis ventricosa*. *Environ. Toxicol. Chem.* **1992**, *11*, 1013–1021. [CrossRef]

40. Ivankovic, D.; Pavicic, J.; Beatovic, V.; Sauerborn Klobucar, R.; Klobucar, G.I.V. Inducibility of Metallothionein Biosynthesis in the Whole Soft Tissue of Zebra Mussels *Dreissena polymorpha* Exposed to Cadmium, Copper, and Pentachlorophenol. *Environ. Toxicol.* **2010**, *25*, 198–211. [CrossRef] [PubMed]

41. Guidi, P.; Bernardeschi, M.; Scarcelli, V.; Cantafora, E.; Benedetti, M.; Falleni, A.; Frenzilli, G. Lysosomal, genetic and chromosomal damage in haemocytes of the freshwater bivalve (*Unio pictorum*) exposed to polluted sediments from the River Cecina (Italy). *Chem. Ecol.* **2017**, *33*, 516–527. [CrossRef]

42. Chiba, K.; Kawakami, K.; Tohyama, K. Simultaneous evaluation of cell viability by neutral red, MTT and crystal violet staining assays of the same cells. *Toxicol. Vitro* **1998**, *12*, 251–258. [CrossRef]

43. Guidi, P.; Frenzilli, G.; Benedetti, M.; Bernardeschi, M.; Falleni, A.; Fattorini, D.; Regoli, F.; Scarcelli, V.; Nigro, M. Antioxidant, genotoxic and lysosomal biomarkers in the freshwater bivalve (*Unio pictorum*) transplanted in a metal polluted river basin. *Aquat. Toxicol.* **2010**, *100*, 75–83. [CrossRef]

44. Singh, N.P.; McCoy, M.T.; Tice, R.R.; Schneider, E.L. A simple technique for quantitation of low levels of DNA damage in individual cells. *Exp. Cell Res.* **1988**, *175*, 184–191. [CrossRef]

45. Kumaravel, T.S.; Jha, A.N. Reliable Comet assay measurements for detecting DNA damage induced by ionising radiation and chemicals. *Mutat. Res. Genet. Toxicol. Environ. Mutagen.* **2006**, *605*, 7–16. [CrossRef]

46. Asker, N.; Carney Almroth, B.; Albertsson, E.; Coltellaro, M.; Bignell, J.P.; Hanson, N.; Scarcelli, V.; Fagerholm, B.; Parkkonen, J.; Wijkmark, E.; et al. A gene to organism approach—assessing the impact of environmental pollution in eelpout (*Zoarces viviparus*) females and larvae. *Environ. Toxicol. Chem.* **2015**, *34*, 1511–1523. [CrossRef]

47. Fenech, M. Cytokinesis-block micronucleus cytome assay. *Nat. Protoc.* **2007**, *2*, 1084–1104. [CrossRef] [PubMed]

48. Panzella, L.; Melone, L.; Pezzella, A.; Rossi, B.; Pastori, N.; Perfetti, M.; D'Errico, G.; Punta, C.; d'Ischia, M. Surface-functionalization of nanostructured cellulose aerogels by solid state eumelanin coating. *Biomacromolecules* **2016**, *17*, 564–571. [CrossRef]

49. Huff, J.; Lunn, R.M.; Waalkes, M.P.; Tomatis, L.; Infante, P.F. Cadmium-induced cancers in animals and in humans. *Int. J. Occup. Environ. Health.* **2007**, *13*, 202–212. [CrossRef] [PubMed]

50. Cuypers, A.; Plusquin, M.; Remans, T.; Jozefczak, M.; Keunen, E.; Gielen, H.; Opdenakker, K.; Nair, A.R.; Munters, E.; Artois, T.J. Cadmium stress: An oxidative challenge. *Biometals* **2010**, *23*, 927–940. [CrossRef]

51. Guo, S.N.; Zheng, J.L.; Yuan, S.S.; Zhu, Q.L.; Wu, C.W. Immunosuppressive effects and associated compensatory responses in zebrafish after full life-cycle exposure to environmentally relevant concentrations of cadmium. *Aquat. Toxicol.* **2017**, *88*, 64–71. [CrossRef]

52. Thévenod, F.; Lee, W.K. Cadmium and cellular signaling cascades: Interactions between cell death and survival pathways. *Arch. Toxicol.* **2013**, *87*, 1743–1786. [CrossRef]

53. McMurray, C.T.; Tainer, J.A. Cancer, cadmium and genome integrity. *Nat. Genet.* **2003**, *34*, 239–241. [CrossRef]

54. Apykhtina, O.L.; Dybkova, S.M.; Sokurenko, L.M.; Yu, B.; Chaikovsky, C. Cytotoxic and genotoxic effects of cadmium sulfide nanoparticles. *Exp. Oncol.* **2018**, *40*, 194–199. [CrossRef]

55. Oporto, C.; Vandecasteele, C.; Smolders, E. Elevated Cadmium Concentrations in Potato Tubers Due to Irrigation with River Water Contaminated by Mining in Potosí, Bolivia. *J. Environ. Qual.* **2007**, *36*, 1181–1186. [CrossRef]

56. Tice, R.R.; Agurell, E.; Anderson, D.; Burlinson, B.; Hartmann, A.; Kobayashi, H.; Miyamae, Y.; Rojas, E.; Ryu, J.C.; Sasaki, Y.F. Single cell gel/comet assay: Guidelines for in vitro and in vivo genetic toxicology testing. *Environ. Mol. Mutagen.* **2000**, *35*, 206–221. [CrossRef]

57. Regoli, F.; Winston, G.W.; Gorbi, S.; Frenzilli, G.; Nigro, M.; Corsi, I.; Focardi, S. Integrating enzymatic responses to organic chemical exposure with total oxyradical absorbing capacity and DNA damage in the European eel *Anguilla anguilla*. *Environ. Toxicol. Chem.* **2003**, *22*, 2120–2129. [CrossRef] [PubMed]

58. Nacci, D.E.; Cayula, S.; Jackim, E. Detection of DNA damage in individual cells from marine organisms using the single cell gel assay. *Aquat. Toxicol.* **1996**, *35*, 197–210. [CrossRef]

59. Buschini, A.; Carboni, P.; Martino, A.; Poli, P.; Ross, C. Effects of temperature on baseline and genotoxicant-induced DNA damage in haemocytes of *Dreissena polymorpha*. *Mutat. Res.* **2003**, *537*, 81–92. [CrossRef]

60. Thevenod, F. Cadmium and cellular signaling cascades: To be or not to be? *Toxicol. Appl. Pharmacol.* **2009**, *238*, 221–239. [CrossRef]

61. Templeton, D.M.; Liu, Y. Multiple roles of cadmium in cell death and survival. *Chem. Biol. Interact.* **2010**, *188*, 267–275. [CrossRef]

62. Ho, J.L.; Ju, H.L.; Seon, M.L.; Na, H.K.; Yeon, G.M.; Tae, K.T. Cadmium induces cytotoxicity in normal mouse renal MM55.K cells. *Int. J. Environ. Health Res.* **2020**, 1–11. [CrossRef]

63. Kant, U.; Govind, M.; Gomal, G.; Salvatore, A.; Pizzo, V. Cadmium-induced DNA synthesis and cell proliferation in macrophages: The role of intracellular calcium and signal transduction mechanisms. *Cell. Signal.* **2002**, *14*, 327–340. [CrossRef]
64. Yua, X.; Filardo, E.J.; Shaikha, Z.A. The membrane estrogen receptor GPR30 mediates cadmium-induced proliferation of breast cancer cells. *Toxicol. Appl. Pharmacol.* **2010**, *245*, 83–90. [CrossRef]
65. Olabarrieta, I.; L'Azou, B.; Yuric, S.; Cambar, J.; Cajaraville, M.P. In vitro effects of cadmium on two different animal cell models. *Toxicol. Vitro* **2011**, *15*, 511–517. [CrossRef]
66. Gómez-Mendikute, A.; Cajaraville, M.P. Comparative effects of cadmium, copper, paraquat and benzo[a]pyrene on the actin cytoskeleton and production of reactive oxygen species (ROS) in mussel haemocytes. *Toxicol. In Vitro* **2003**, *17*, 539–546. [CrossRef]
67. Hsu, T.; Huang, K.M.; Tsai, H.T.; Sung, S.T.; Ho, T.N. Cadmium (Cd) induced oxidative stress down-regulates the gene expression of DNA mismatch recognition proteins MutS homolog 2 (MSH2) 580 and MSH6 in zebrafish (*Danio rerio*) embryos. *Aquat. Toxicol.* **2013**, *126*, 9–16. [CrossRef] [PubMed]
68. Pierron, F.; Baillon, L.; Sow, M.; Gotreau, S.; Gonzalez, P. Effect of Low-Dose Cadmium Exposure on DNA Methylation in the Endangered European Eel Environ. *Sci. Technol.* **2014**, *48*, 797–803. [CrossRef] [PubMed]
69. Bourgeault, A.; Gourlay-Francé, C.; Vincent-Hubert, F.; Palais, F.; Geffard, A.; Biagiant-Risbourg, S.; Pain-Devin, S.; Tusseau-Vuillemin, M.H. Lessons from a transplantation of zebra mussels into a small urban river: An integrated ecotoxicological assessment. *Environ. Toxicol.* **2010**, *25*, 10. [CrossRef] [PubMed]
70. Fajardo, C.; Costa, G.; Nande, M.; Martín, C.; Martín, M.; Sánchez-Fortún, S. Heavy metals immobilization capability of two iron-based nanoparticles (nZVI and Fe_3O_4): Soil and freshwater bioassays to assess ecotoxicological impact. *Sci. Total Environ.* **2019**, *656*, 421–432. [CrossRef]
71. Vicari, T.; Ferraro, M.V.M.; Ramsdorf, W.A.; Mela, M.; Ribeiro, C.A.O.; Cestari, M.M. Genotoxic evaluation of different dosages of methylmercury (CH_3Hg^+) in *Hoplias malabaricus*. *Ecotoxicol. Environ. Saf.* **2012**, *82*, 47–55. [CrossRef] [PubMed]

 *nanomaterials*

Article

Effect-Based Approach to Assess Nanostructured Cellulose Sponge Removal Efficacy of Zinc Ions from Seawater to Prevent Ecological Risks

Giulia Liberatori [1], Giacomo Grassi [1], Patrizia Guidi [2], Margherita Bernardeschi [2], Andrea Fiorati [3], Vittoria Scarcelli [2], Massimo Genovese [2], Claudia Faleri [4], Giuseppe Protano [1], Giada Frenzilli [2,*], Carlo Punta [3] and Ilaria Corsi [1,*]

[1] Department of Physical, Earth and Environmental Sciences and INSTM Local Unit, University of Siena, 53100 Siena, Italy; giulia.liberatori@student.unisi.it (G.L.); grassi23@student.unisi.it (G.G.); giuseppe.protano@unisi.it (G.P.)

[2] Department of Clinical and Experimental Medicine-Section of Applied Biology and Genetics, University of Pisa, 56126 Pisa, Italy; patrizia.guidi@unipi.it (P.G.); margherita.bernardeschi@unipi.it (M.B.); vittoria.scarcelli@unipi.it (V.S.); massimo.genovese@unipi.it (M.G.)

[3] Department of Chemistry, Materials, and Chemical Engineering "G. Natta" and INSTM Local Unit, Politecnico di Milano, 20131 Milano, Italy; andrea.fiorati@polimi.it (A.F.); carlo.punta@polimi.it (C.P.)

[4] Department of Life Sciences, University of Siena, 53100 Siena, Italy; faleric@unisi.it

* Correspondence: giada@biomed.unipi.it (G.F.); ilaria.corsi@unisi.it (I.C.); Tel.: +30-050-2219111 (G.F.); +39-0577-232168 (I.C.)

Received: 17 June 2020; Accepted: 28 June 2020; Published: 30 June 2020

Abstract: To encourage the applicability of nano-adsorbent materials for heavy metal ion removal from seawater and limit any potential side effects for marine organisms, an ecotoxicological evaluation based on a biological effect-based approach is presented. $ZnCl_2$ (10 mg L^{-1}) contaminated artificial seawater (ASW) was treated with newly developed eco-friendly cellulose-based nanosponges (CNS) (1.25 g L^{-1} for 2 h), and the cellular and tissue responses of marine mussel *Mytilus galloprovincialis* were measured before and after CNS treatment. A control group (ASW only) and a negative control group (CNS in ASW) were also tested. Methods: A significant recovery of Zn-induced damages in circulating immune and gill cells and mantle edges was observed in mussels exposed after CNS treatment. Genetic and chromosomal damages reversed to control levels in mussels' gill cells (DNA integrity level, nuclear abnormalities and apoptotic cells) and hemocytes (micronuclei), in which a recovery of lysosomal membrane stability (LMS) was also observed. Damage to syphons, loss of cilia by mantle edge epithelial cells and an increase in mucous cells in $ZnCl_2$-exposed mussels were absent in specimens after CNS treatment, in which the mantle histology resembled that of the controls. No effects were observed in mussels exposed to CNS alone. As further proof of CNS' ability to remove Zn(II) from ASW, a significant reduction of >90% of Zn levels in ASW after CNS treatment was observed (from 6.006 to 0.510 mg L^{-1}). Ecotoxicological evaluation confirmed the ability of CNS to remove Zn from ASW by showing a full recovery of Zn-induced toxicological responses to the levels of mussels exposed to ASW only (controls). An effect-based approach was thus proven to be useful in order to further support the environmentally safe (ecosafety) application of CNS for heavy metal removal from seawater.

Keywords: cellulose-based nanosponges; zinc; seawater; ecotoxicity; effect-based approach; remediation

---

1. Introduction

The rapid industrialization and economic development of marine coastal areas have significantly increased the release of toxic pollutants, leading the marine environment to be considered a final sink [1]. Heavy metal pollution due to different sources and anthropogenic activities (sewages, ship and gas/oil platforms discharges, antifouling paints, landfill sites) still represents a critical issue in marine coastal waters, being responsible for ecological deterioration and for giving rise to health risks for humans and marine wildlife [2,3]. In order to achieve a Good Environmental Status following the Marine Strategy Framework Directive (MSFD, Directive 2008/56/EC) [4], heavy metals in European coastal waters should be kept below those concentrations that are able to cause toxicity to marine species (from mg L^{-1} down to µg L^{-1}) (Environmental Quality Standard Directive 2013/39/EU) [5]. Zinc levels very often exceed regulatory limits in marine coastal areas, with concentrations of up to 10 ppm in both sediments and the water column. Although Zn(II) is known to pose a serious threat to humans and the environment, its worldwide production has increased in the last ten years, so that its presence in the aquatic environment is expected to grow [6]. Further sources of Zn(II) release into the sea include Zn-based antifouling paints and sunscreens (e.g., zinc oxide), as well as discharges from desalination plants and bilge water disposal [7,8]. In the ocean, dissolved Zn(II) undergoes a nutrient-like vertical profile and sediments represent a source to the water column. This will in turn significantly affect Zn(II)'s bioavailability, bioaccumulation and toxicity for marine species belonging to different trophic levels [9–16]. Zn(II) is also an essential element for selected marine species, therefore its levels in surface waters should not decrease below a certain concentration (µg L^{-1}), which could limit phytoplankton growth [17].

The use of engineered nanomaterials (ENMs) was recently intended as a more attractive and convenient remediation solution compared to conventional ones, being less expensive and more effective, as well as environmentally and economically sustainable [18,19]. Their usage for cleaning up polluted soils and waters, termed "nanoremediation", has made enormous progress over the last few decades [20]. However, the risks associated with their mobility and transformations once released into the environment are almost unknown [18]. To fill these scientific gaps whilst going back to a perspective of environmental and economic sustainability, research is moving towards the design of environmentally friendly ENMs which do not pose any risk for living beings and overall ecosystems [21–23]. In this direction, several contributions have been made to develop eco-friendly bio-based ENMs which have a high capacity to reduce environmental pollution and are suitable to be used onsite [24–27]. Among renewable sources, cellulose is receiving great attention, being an abundant biopolymer with good mechanical properties and numerous possibilities for chemical functionalization. It also represents a promising sustainable material, since its synthesis does not require the use of high temperatures and CO_2 emissions are lower than those connected to the production of classical ENMs [22,23,28–33]. Moreover, due to the degree of branching that can be obtained, the area-to-volume ratio is very high, promoting a greater efficiency of adsorption, which can be increased by tailoring cellulose with suitable functional groups [34]. In 2015, for the first time, we reported a two-step thermal protocol for the synthesis of a novel family of cellulose-based ENMs which possessed superb performance in the adsorption of metals and organic contaminants [35]. It was obtained by the cross-linking of 2,2,6,6-tetramethyl-piperidine-1-oxyl (TEMPO)-oxidized and ultra-sonicated cellulose nanofibers (TOUS-CNFs) [36] in the presence of branched polyethyleneimine (bPEI), thus achieving a nanostructured cellulose sponge (CNS) suitable for heavy metal ion removal from water. CNS shows a 2D sheet-like morphology, characterized by a micro-porosity, which can be observed by the scanning electron microscopy (SEM) technique, and a nano-porosity, as revealed by an in-depth small angle neutron scattering (SANS) analysis [37]. More recently, the adsorption efficiency of CNS from seawater was revealed to be better than from freshwater, and their environmental safety (ecosafety) is achieved by exploiting an eco-design approach capable of combining life cycle assessment (LCA) and environmental risks [38–40]. To encourage the applicability of CNS as adsorbent materials for toxic heavy metal removal from seawater, here, we propose an ecotoxicological evaluation designed by an effect-based approach. Considering their proven ability to remove heavy metals from seawater [40], we aim to prove their efficacy in reducing Zn toxicity by measuring known Zn-induced toxicological responses, both at cellular and tissue levels, in a model marine species, namely the bivalve mollusk *Mytilus galloprovincialis*. Being a widely adopted marine model in ecotoxicology, *Mytilus sp.* has been historically used as a bioindicator in marine pollution monitoring (mussels watch programs) [41,42] due to its known ability to accumulate heavy metals in soft tissues, including Zn (ranging between μg-mg/kg wet weight), in proportion to their concentration in seawater [43–46]. The main pathway of Zn(II) uptake in *Mytilus* is through the gills, from which it is transported to the digestive system via the hemolymph [47–50]. Zn-induced biological responses/effects have been already well identified in marine mussels, both at cellular and tissue levels, and described in detail [51–57]. Being the first organs involved in Zn(II) uptake in bivalves, the gills and mantle are the main targets of Zn toxicity in mussels [58]. Zn damage to gills' cilia impairs their movement but also affects their sensory and secretory activities, as well as shell growth and repair mechanisms [59]. Zn waterborne exposure (125 μg L^{-1}) is able to induce histopathological alterations of mantle external epithelial cells and to increase mucous cells [60]. At the cellular level, lysosomal membrane destabilization has been documented in mussels' hemocytes upon waterborne exposure to $ZnCl_2$ (0.5 and 1 mg L^{-1}, 7 days) [61]. DNA damage was reported to be either induced by reactive oxygen species (ROS) or by direct binding to DNA in mussels' embryos which were exposed to Zn-contaminated sediments [62] and micronuclei and nuclear abnormalities in mussels' gill cells as cyto-genotoxic biomarkers in pollution monitoring [63].

With the aim of using an effect-based approach to assess CNS' ability to remove Zn(II) from seawater, lysosomal membrane stability (LMS) and DNA damage were investigated in mussels' hemocytes and gill cells, as well as histopathological alterations on the mantle edge and external epithelial cells. Specimens of *M. galloprovincialis* were exposed to $ZnCl_2$ contaminated artificial seawater (10 mg L^{-1} ASW) before and after CNS treatment, and two groups exposed to ASW (control) and CNS alone were also tested. Levels of Zn(II) were also measured in ASW in all experimental groups ($ZnCl_2$, Zn-t CNS, CNS, ASW) at time zero and after 24 h of exposure.

2. Materials and Methods

2.1. CNS Synthesis, Characterization and Zn(II) Adsorption

TOUS-CNFs were prepared according to a standardized protocol reported by [38]. Briefly, 190 g of cotton linters cellulose, provided by Bartoli Spa paper mill (Capannori, Lucca), were dispersed in deionized water (2 L) and added to an aqueous solution (3.7 L) of 2,2,6,6-tetramethyl-piperidine-1-oxyl (TEMPO 2.15 g, 13.8 mmol) and KBr (15.42 g, 129 mmol) under vigorous stirring. While maintaining the stirring, NaClO (12.5% w/w aqueous solution, 437 mL) was slowly added to the mixtures and the pH value maintained in the range of 10.5–11 by using sodium hydroxide solution (4 M). The reaction was left stirring overnight and then acidified to pH 1–2 with concentrated HCl (37% w/w aqueous solution). The oxidized cellulose was collected by filtration on a sintered glass funnel and washed extensively with deionized water (5 × 2 L) and acetone (2 × 0.5 L, 99.9% purity). TOUS-CNFs were recovered by 84%. The carboxylic content of the oxidized cellulose was then colorimetric titrated (1.5 mmol/g of –COOH units). The obtained oxidized cellulose was dispersed in deionized water (3% w/v), a stoichiometric amount of NaOH was added and the mixture was sonicated at 0 °C for 30 min, using a Branson Sonifier 250 equipped with a 6.5 mm probe tip, working at 20 kHz in continuous mode, with an output power 50% the nominal value (200 W). An excess of HCl (aq, 1 M) was added in order to precipitate TOUS-CNFs that were collected by filtration on Büchner funnel with a sintered glass disc and further washed on the filter with deionized water (450 mL × 3 times).

CNS were prepared according to the optimized procedure which was recently reported [40]. TOUS-CNFs were dispersed in an aqueous solution of 25 kDa bPEI (84 g of bPEI in 300 mL of water) and citric acid (23.8 g in 200 mL of water). The mixture was sonicated again for 10 min. The viscous gel obtained was transferred into a mold, frozen at −80 °C for 12 h and freeze-dried at −60 °C for 48 h. The cylindrical disks were thermally treated in an oven (103 °C, 12–16 h). The resulting xerogels were ground in a mortar and washed with water (6 × 150 mL, 1 h contact time for each cycle). The CNS synthetic procedure is summarized in Figure 1.

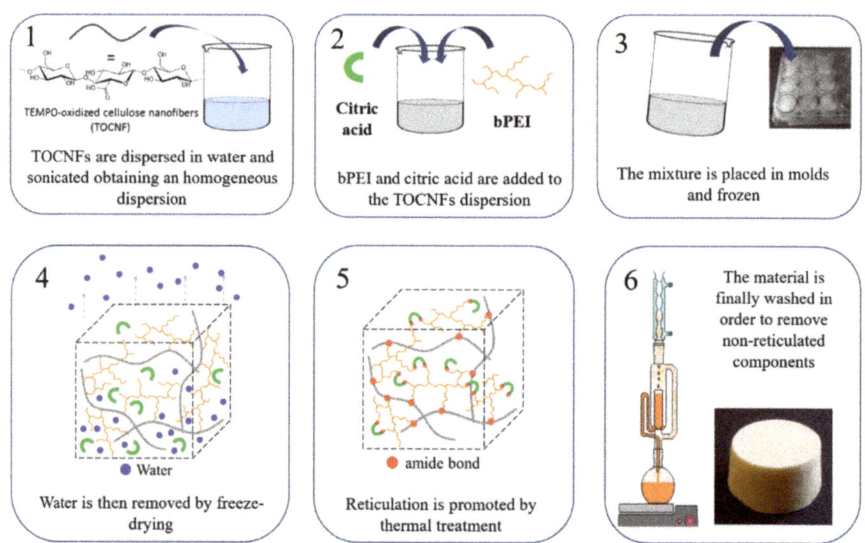

Figure 1. Synthetic procedure for the production of cellulose-based nanosponges CNS.

The particle sizes of the ground CNS powder were measured in the Laboratory Chemical Analysis (LAC) of Politecnico di Milano by means of a Malvern Mastersizer 3000 Particle Size Analyzer (Malvern, UK) with Fraunhofer modeling, which considers opaque non-spherical particles. The CNS powder was suspended under stirring in 500 mL of water to reach an obscuration level in the range of 8%–12%.

The adsorption ability of CNS towards Zn(II) from ASW was determined as follows. In a Falcon test tube, 12 mg (±0.2 mg) of CNS were dispersed in 15 mL of $ZnCl_2$-contaminated solution at different metal ion concentrations (1.5, 15, 30, 45 and 90 ppm). In all cases, the mass of sorbent material per volume of solution was 0.8 mg mL^{-1}. The test tubes were sealed and left at 25 °C and shacked for 24 h. Upon filtration (0.45 μm), the Zn(II) solutions were analyzed using Agilent Technologies Inductively Coupled Plasma Mass Spectrometry (ICP-MS, Agilent Technologies, Santa Clara, CA, USA) 7900, equipped with a MicroMist nebulizer and a spray chamber in a Peltier-cooled sample introduction system, in order to increase stability and consistency, followed by a shield torch system and temperature-controlled collision/reaction octopole ion guide cell, in order to provide interference removal in the helium collision mode. All analytical operations were carried out in compliance with method EPA 6020B 2014. Instrument calibration was carried out by standard solutions used as reference material, which were purchased from qualified suppliers (Exaxol Italia, Genova, Italy), and containing the analytes in a suitable concentration range.

The scanning electron microscopy (SEM) of the CNS was performed according to the method which has been described in detail [40]. Briefly, after Zn(II) adsorption, horizontal sections of CNS (half height) were fixed in aluminum stubs using graphite powder (Agar scientific, G3300 Leit C—conducting carbon cement) and placed in the oven at 50 °C (30 min) and coated with graphite (Emitech K450 apparatus, Quorum Technologies Ltd., Laughton, UK). SEM images were obtained using a Zeiss instrument Gemini Supra 40 model (accelerating voltage 20 kV, spot size 60 μm, Oberkochen, Germany) with an energy-dispersive electron probe X-ray (EDX) (Oxfod x-act).

2.2. Experimental Design of Effect-Based Study

2.2.1. In Vivo Waterborne Exposure

Adult specimens of *M. galloprovincialis* (medium length of the valves 6.5 ± 0.5 cm) were purchased from an aquaculture farm (Sardinia, Italy) and shipped to the marine aquarium facility of the Department of Physical, Earth and Environmental Sciences of the University of Siena (Siena, Italy). Based on our previous experience, an acclimatization period of 48 h was sufficient for reaching a stable physiological condition for mussels [64]. Natural seawater (NSW) collected from a pristine area in Tuscany (Italy) was used for acclimatization and the following parameters were kept constant during the entire period: salinity 40% ± 1%, pH 8 ± 0.1, density 1.025 ± 0.001 g cm^{-3}, conductivity 49 ± 1 mS cm^{-1} and temperature 18 ± 1 °C.

A set of two experiments was run: (i) a preliminary experiment was designed to select $ZnCl_2$ sub-lethal effects concentration for mussels; (ii) a second experiment aimed to assess the CNS' ability to reduce $ZnCl_2$-induced toxicological responses/effects in exposed mussels. Both experiments were run in artificial seawater (ASW), prepared following [65], in order to test CNS' performance in high ionic strength media and avoid inorganic and organic colloidal particles as well as other compounds naturally occurring in NSW.

In the preliminary experiment (i), mussels were exposed for 48 h to $ZnCl_2$ at concentrations of 1, 10 and 100 mg L^{-1} in ASW, which were chosen based on the LC$_{50}$ values (range 2–20 mg L^{-1}) available in the literature for bivalve species [51,52,66–70]. A parallel group of mussels exposed to ASW only was used as a control. Mussels were placed in 6 L glass tanks in groups of 6 individuals per experimental condition, using a ratio of 1 mussel: 1 L ASW, and they were not fed during the experiment. The exposure medium was renewed every 24 h to maintain constant Zn(II) nominal exposure concentrations. At the end of the 48 h period, the mussels were removed from the tanks, the hemolymph was collected for lysosomal membrane stability (LMS) in hemocytes by a neutral red retention time assay (NRRT) (method described in detail in Section 2.2.2), and chromosomal damage was measured through the micronucleus test (MN) (method described in detail in Section 2.2.2).

In the second experiment (ii), the mussels were exposed for 48 h to $ZnCl_2$ (10 mg L^{-1}) in ASW (Zn), before and after CNS treatment (Zn-t CNS), and two groups, one exposed to ASW only (ASW) and one with only CNS (CNS), were prepared by following the same experimental conditions used for the Zn-contaminated ASW treatment. We applied a ratio of 1.25 g of CNS: 1 L of ASW for the CNS treatment, based on our previous findings on heavy metal adsorption from ASW [40]. Briefly, 1.25 g L^{-1} of CNS powder was incubated with $ZnCl_2$-contaminated ASW (10 mg L^{-1}) for 2 h at room temperature, under vigorous magnetic stirring in order to increase the contact between CNS and Zn (II) in the water medium. CNS alone were also incubated in ASW at the same ratio (1.25 g powder L^{-1} ASW) and further tested for reference. Both CNS treated mediums were then filtered at 0.45 μm using a cellulose filter to remove excess CNS powder, and the obtained solutions (Zn-t CNS and CNS alone) were tested with mussels. The experimental groups of $ZnCl_2$ in ASW (10 mg L^{-1}) and ASW only were also set up and tested. In summary, mussels (1 specimen: 1 L of ASW) were placed in 6L tanks (total of 6 individuals) and exposed to the following experimental solutions: $ZnCl_2$, Zn-t CNS, CNS only and ASW for 48 h. Mussels were not fed during the experiment and all tested solutions were renewed every 24 h to maintain constant exposure conditions.

At the end of the 48 h period, the hemolymph was collected for the LMS NRRT assay (described in detail in Section 2.2.2) and chromosomal damage was measured through micronucleus (MN) frequency (described in detail in Section 2.2.2).

Gill arches and the full mantle were also removed with scissors, and single gill arches were processed for comet and cytome assays (described in detail in Section 2.2.2), while the mantle was processed for histological examination [71]. In addition, three independent samples of ASW (10 mL) from each tested solution ($ZnCl_2$, Zn-t CNS, CNS only and ASW) were collected at time zero and after 24 h, and they were stored at 4 °C for analysis of the total Zn(II) levels, according to the method reported below (Section 2.3).

2.2.2. Cellular Bioassays

For the NRRT assay [72], the hemolymph was withdrawn from the mussel's adductor muscle using a sterile syringe (1 mL) pre-loaded with a buffer solution (20 mM 4-(2-hydroxyethyl)-1-piperazineethanesulfonic acid (HEPES), 436 mM NaCl, 53 mM $MgSO_4$, 12 mM KCl, 10 mM $CaCl_2$) to prevent hemocyte clotting. A 200 µL volume of hemocyte suspension was then placed in a coverslip (22 × 22 mm), incubated in the dark at 18 °C by adding 200 µL of neutral red dye solution (0.1 mg mL^{-1} NR in dimethyl sulfoxide (DMSO)). Excess NR was removed before observing slides under optical light microscopy (Olympus BX51, 80× magnification), and the percentage of cells showing loss of the NR dye from lysosomes was scored every 15 min. Three replicates of three different slides were analyzed, each one made by pooling the hemolymph from at least five individuals for each experimental group. Final scores were set at the time when 50% of hemocytes showed NR leaking from lysosomes and became round-shaped [58,73].

The MN frequency assay [74] was run on withdrawn hemolymph diluted in 10 mM ethylenediaminetetraacetic acid (EDTA) buffer saline solution. A 200 µL volume of hemolymph was placed in cold slides and incubated at −20 °C in methanol for 20 min first and in 6% Giemsa dye in MilliQ for another 20 min. MN frequency was scored every 1000 hemocytes under an optical light microscope (Olympus BX51, 80× magnification). Three replicates of three different slides were analyzed, each one made by pooling hemolymph from at least five individuals for each experimental group.

DNA primary damage, including DNA single and double strand breaks (SB) and alkali labile sites, was evaluated by the comet assay in mussels' gill cells, according to the method previously described in [75]. Briefly, after dissection, gills were put in a tissue digestion solution (dispase/Hank's Balance Salt Solution (HBSS)) at 37 °C for 20 min; then, the enzyme was inactivated. The resulting digestion product was filtered through a 100 µm mesh nylon filter and the cell suspension obtained was centrifuged at 2000 rpm for 5 min. Cell viability was evaluated by trypan blue exclusion and only cells displaying a cell viability >95% were included. The slide preparation and electrophoresis procedures were performed according to the protocol reported in [76]. The amount of DNA damage, expressed as the percentage of DNA migrated into the comet tail (% tail DNA), was assessed from at least 50 nuclei per slide ($n = 2$) per specimen, using an image analyzer (Kinetic Imaging Ltd., Komet, Version 5).

The pellets obtained from mussels' gill cells were processed for the cytome assay according to [77]. One thousand cells with preserved cytoplasm per specimen were scored (500 per slide) in order to determine the frequency of MN, according to the criteria listed by Fenech [78]. Morphologically altered nuclei (i.e., incomplete MN, lobed or multiple nuclei, nuclei connected by chromatin bridges in the same cell) were scored on the same slides in parallel and were collectively reported. The frequency of total nuclear abnormalities (NA) (nuclear blebs (BL), included nuclear buds (NBUD), binucleated cells with nuclear bridges (NPB), notched nucleus (NT), circular nucleus (CIR), lobed nucleus (LB), and anisochromatic nuclei (AN) was observed (Figure S1). Apoptotic cells (APO) were also evaluated. For each experimental group, four specimens, with 2 slides per specimen and 500 cells per slide for a total of four thousand cells per experimental point, were scored.

2.2.3. Mantle Histology and Histochemistry

Mantles were gently removed from the internal shells of mussels using scissors, and small sections of mantle edge were obtained with a scalpel (average 1 cm) and freshly observed under a Zeiss Stemi SV6 (8×–50× magnification) binocular stereo microscope at magnification (8×, 10× and 20×) by placing them in glass Petri dish.

The mantle edge sections for histological analysis were obtained by cutting the most external part of the mantle (average 1 mm) and fixing it in a modified Karnovsky solution (3% glutaraldehyde in

0.1 M sodium cacodylate buffer 0.9% sucrose, pH 7.2) for 2 h at room temperature. They were fully dehydrated in ethanol and embedded in resin (Kulzer Technovit 7100) and polymerized at 27 °C for 3 h. Serial sections (2 μm) were obtained with the LKB Ultratome Nova microtome and stained with haematoxylin and eosin (H&E) (pH 2.4, 10 min at 25 °C). As a contrast dye able to stain eosinophilic structures in various shades of red, pink and orange, 0.5% eosin Y in aqueous solution, acidified with acetic acid and filtered at 0.45 μm, was used for 1 min. Sections were then rinsed in distilled water for 10 min, air dried, sealed and analyzed under a Zeiss Axiophot epifluorescent microscope (80X) with AxioCamMRc5.

In order to detect mucosubstances, the specific dye, periodic acid Schiff Alcian blue (PAS AB), was used. The reaction between the periodic acid (PA) and the Schiff reagent stains the neutral mucopolysaccharides in purple, while the AB identifies the acid mucopolysaccharides in blue. Before being colored, the sections were placed for 20 min in 100% acetone in order to remove the resin. The sections were rinsed in tap water for 2 min and then placed in 1% PA in distilled water for 5 min. Sections were rinsed in tap water for 5 min and then the Schiff reagent was added for 1 h. After the Schiff reagent was removed, 0.5% $Na_2S_2O_5$ in 1% HCl was placed for 10 min on the slides. Slides were rinsed in tap water for 5 min and 1% AB in 3% acetic acid were left for 30 min. Finally, the slides were rinsed in distilled water for 2 min, dried, sealed and analyzed under a Zeiss Axiophot epifluorescent microscope (80X) with AxioCamMRc5.

2.3. Quantification of Total Zn Levels in Tested Solutions

The total amount of Zn was measured in each tested solution ($ZnCl_2$, Zn-t CNS, CNS only and ASW), at time zero (T_0) (soon after preparation) and after 24 h (T_{24h}), as follows: 10 L of mussels' exposure water was removed from each tank, placed in polyethylene plastic tubes and stored at 4 °C. Analysis was performed by ICP-MS using the Perkin Elmer NexION 350 spectrometer (Waltham, MA, USA). The analytical accuracy was checked through the determination of Zn(II) concentration in SLRS-6 (river water certified reference materials for trace metals and other constituents) and CASS-4 (nearshore seawater certified reference materials for trace metals and other constituents) of the National Research Council of Canada. Analytical precision was evaluated by means of the percentage relative standard deviation (% RSD) of five replicate analyses of each exposure water sample. Zn levels were expressed as mg L^{-1}. Variations in pH in exposure waters were also recorded at time zero and after 24 h.

2.4. Statistical Analysis

The data that were obtained and represented as mean ± SD of at least 5 samples in triplicate were analyzed by GraphPad Prism software package 6. Statistical analysis for hemocyte tests was performed using one-way analysis of variance (ANOVA) plus the Bonferroni post-test. For the tests on gill cells, data from 4 specimens were analyzed by the multifactor analysis of variance (MANOVA) and the multiple range test (MRT) was performed in order to detect differences among experimental groups. For all data analyses, the statistical significance level was set at $p < 0.05$.

3. Results and Discussion

3.1. CNS Adsorption Performance of Zn(II) from ASW

The adsorption efficiency of Zn(II) from ASW was evaluated by contacting the mono-contaminated solution at different ion concentrations with ground CNS. The results reported in Figure 2 clearly show that, by operating at a CNS/ASW ratio of 0.8 g L^{-1}, the abatement of Zn(II) ions in a range between 1.5 and 45 ppm was higher than 90%, with a maximum adsorption capability of about 100 mg per g of adsorbent material [40]. As described in our previous works [35,40], the adsorption efficiency can be ascribed to the chelating action of free amino groups present in the CNS network. As a consequence, the slight basic pH (8) of seawater allows the prevention of the protonation of amino groups, permitting them to completely exploit their chelating action.

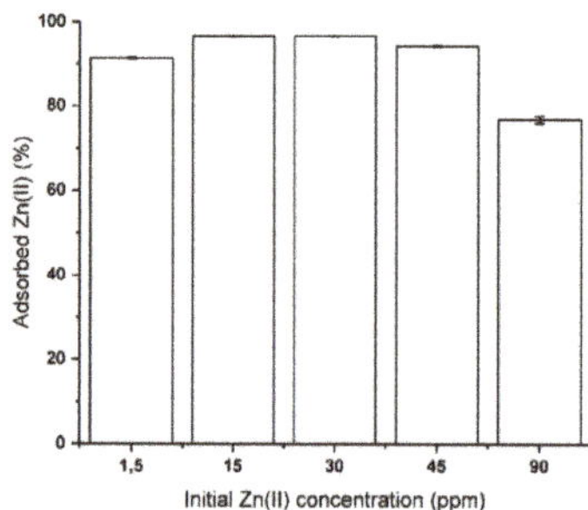

Figure 2. Cellulose-based nanosponges' (CNS) adsorption efficiency of Zn(II) from mono-contaminated artificial seawater (ASW) at different ion concentrations. CNS amount: 0.8 g L^{-1}.

No significant differences in pH were observed in ASW before and after CNS treatment during 24 h of exposure, and this was similar to the control and CNS only groups ($p < 0.05$) (Table S1). The SEM–EDX analysis evidences the limited diffusion kinetics of the ions in the intact sponge when operating at low metal salt concentrations (Figure 3a), while at higher concentrations, it is possible to verify the complete penetrability of the material, with active sites homogeneously distributed throughout the whole sponge (Figure 3b). CNS grinding allowed us to overcome this operative limit. The resulting powder consisted of particles with sizes in a range between 50 and 400 μm, with a maximum distribution at 130 μm.

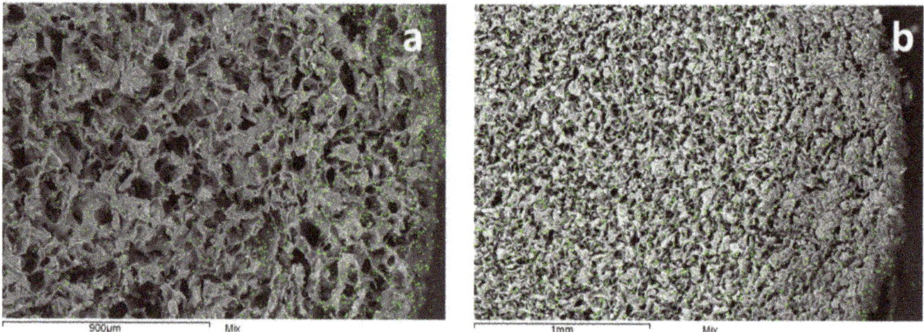

Figure 3. SEM–EDX (scanning electron microscopy–electron probe X-ray) image of a CNS cross-section after Zn(II) adsorption from 50 mg L^{-1} ASW (left, **a**) and 450 mg L^{-1} (right, **b**). Green dots represent metal ions.

The porosity of the CNS was determined in a previous work [38] by microcomputed tomography quantitative analysis and found to be about 70% of the bulk material. Moreover, in a more recent study [37], it was possible also to reveal the presence of nano-pores, thanks to a small angle neutron scattering (SANS) analysis, by investigating the water nanoconfinement geometries in the adsorbent material. The analysis of the experimental data allowed us to measure the short-range correlation length, which was measured to be in a range between 25 and 35 Å. We interpreted this data as the very first indirect evidence of the effective nano-dimension of the cavities, produced by the cross-linking of the reticulated cellulose nanofibers.

3.2. Effect-Based Study

3.2.1. Cellular Bioassays

In the preliminary study, a dose-dependent increase in lysosomal membrane destabilization was observed in the hemocytes of mussels (16%, 41%, 100%) upon $ZnCl_2$ exposure (1, 10 and 100 mg L^{-1}). This allowed us to set the sub-lethal Zn(II) concentration at 10 mg L^{-1} as changes in NRRT resulted in levels significantly below the 50% of exposed cells without generating cell death [58,73] (Figure S2).

In the CNS remediation study, a significant decrease in lysosomal membrane stability (LMS) (Figure 4a) was observed in the hemocytes of mussels exposed to $ZnCl_2$ (10 mg L^{-1}) compared to controls ($p < 0.0001$), while specimens exposed after CNS treatment showed a LMS similar to the controls (ASW) and to those exposed to CNS alone ($p < 0.05$).

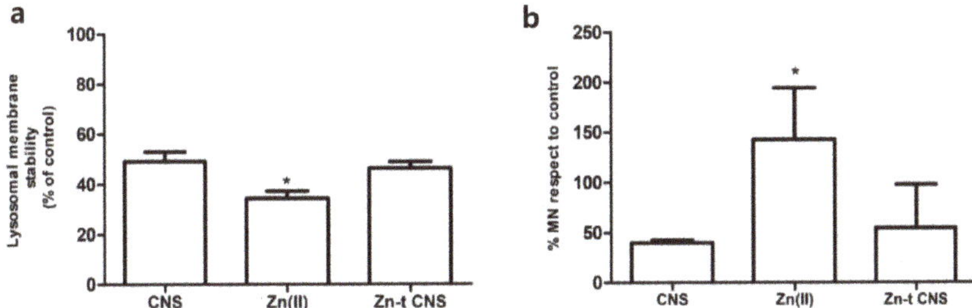

Figure 4. (**a**) Lysosomal membrane stability (LMS) shown as % vs. control (ASW) in mussel hemocytes at 30 min and (**b**) micronucleus frequency (MN) shown as % vs. control (ASW), after 48 h of exposure in the experimental groups: Zn(II) (10 mg L^{-1} of $ZnCl_2$ in ASW); Zn-t CNS ($ZnCl_2$ 10 mg L^{-1} contaminated ASW treated with CNS); CNS (ASW treated with only CNS). Results are shown as % towards controls (mussels exposed to ASW only). (*) indicates significant difference with respect to the Zn(II) group ($p < 0.01$).

Zn(II) at concentrations in the range of mg L^{-1} are known to cause severe damage in lysosomal membranes and therefore their stability has been widely used as a biomarker of exposure in field monitoring studies [61,79]. Dissolved Zn(II) is easily taken up through mussels' gills and diffused across membranes and mantle syphons, while particulate forms are generally mistaken as food and accumulate in the digestive gland [48,55]. Consequently, Zn ions' sequestration and toxic actions are closely connected to lysosomal function in bivalves, both in Zn accumulating organs as the digestive system and in single circulating immune cells (hemocytes) [47]. The recovery of lysosomal membranes' stability to the levels of the controls in the mussels' hemocytes after CNS treatment does confirm that the levels of Zn(II) in ASW were lowered until they were not able to destabilize the lysosomal membranes. In fact, the Zn level measured in exposure waters after CNS treatment after 24 h was 0.510 ± 0.0258 mg L^{-1}, for which no significant effect on LMS was previously observed in mussels in our preliminary study (Figure S1). Furthermore, although CNS treatment did not reduce Zn levels to those of ASW (0.005 ± 0.001 mg L^{-1}), such a concentration (0.510 ± 0.0258 mg L^{-1}) was proven not to be effective in causing lysosomal membrane destabilization in the hemocytes of exposed mussels. Background levels of Zn in marine coastal waters are commonly below 1 mg L^{-1}, for which no toxicity for marine bivalve species has been reported [15,80].

As observed for lysosomal membranes, a significant decrease in MN frequency ($p < 0.001$) was found in the hemocytes of mussels exposed after CNS treatment compared to those exposed to $ZnCl_2$ ($p < 0.0001$); MN frequencies were also similar to those of the controls (ASW) and of specimens exposed to CNS only (Figure 4b). The same was observed in hemocytes; DNA strand breaks were significantly lower in the gills of mussels exposed after CNS treatment when compared to those exposed to $ZnCl_2$ ($p < 0.05$), and DNA strand breaks were comparable in mussels exposed after CNS treatment, both in controls and in those exposed to CNS only (Figure 5). Approximately 3.5-fold more nuclear abnormalities (NA) were recorded in the gill cells of mussels exposed to $ZnCl_2$ ($p < 0.001$), compared to specimens exposed after CNS treatment, controls and CNS-exposed only (Figure 6a). In addition, a ten-fold higher frequency of apoptotic cells (APO) was found in $ZnCl_2$-exposed mussels' gill cells compared to the controls, and no further differences were observed among mussels exposed after CNS treatment and those exposed to CNS only (Figure 6b).

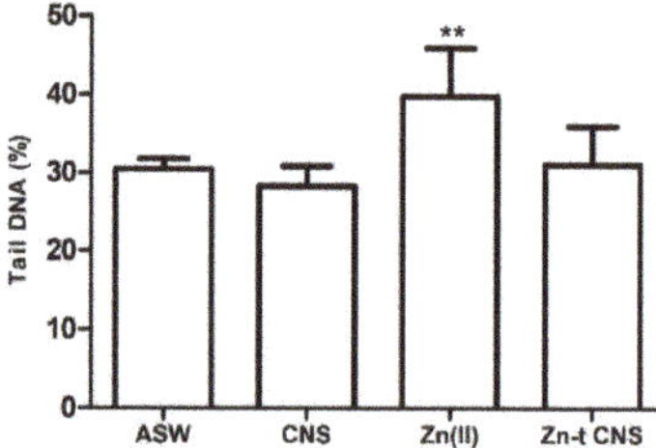

Figure 5. DNA primary damage (% tail DNA) in mussels' gill cells after 48 h of exposure to the following experimental groups: Zn(II) (10 mg L^{-1} of $ZnCl_2$ in ASW); Zn-t CNS ($ZnCl_2$ 10 mg L^{-1} contaminated ASW treated with CNS); CNS (ASW treated with only CNS). Results are shown as mean ± SD. (**) indicates significant differences with respect to the control group (ASW) ($p < 0.001$).

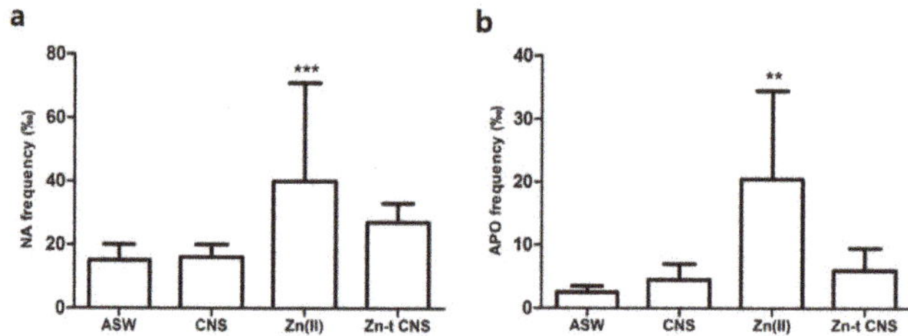

Figure 6. Frequencies (‰) of nuclear abnormalities (NA) (**a**) and apoptotic cells (APO) (**b**) in mussels' gills after 48 h of exposure to the following experimental groups: ASW (control); Zn(II) (10 mg L^{-1} of $ZnCl_2$ in ASW); Zn-t CNS ($ZnCl_2$ 10 mg L^{-1} contaminated ASW treated with CNS); CNS (ASW treated with only CNS). Results are shown as mean ± SD. (**) indicates significant differences with respect to the control group (ASW) ($p < 0.001$). (***) indicates significant differences with respect to the control group (ASW) ($p < 0.0001$).

$ZnCl_2$'s ability to induce DNA strand breaks has been documented in the circulating and tissue cells of *M. galloprovincialis* [56,57]. However, upon CNS treatment, the DNA integrity of mussels' gill cells was comparable to that of the controls and the mussels exposed to CNS alone.

As further confirmation, at the chromosomal level, the full recovery of DNA damage was observed in mussels exposed after CNS treatment. Heavy metals are known genotoxicants for marine bivalves, and NA in gill cells have been widely recognized as a suitable marker of DNA damage in *M. galloprovincialis* [63,81].

Genotoxicity can inhibit the cell cycle which influences the formation of MN [82]. The clastogenic activity of $ZnCl_2$, by increasing the MN frequencies in the gill cells of mussels after acute short-term exposure conditions (48 h) and lower concentrations (0.17 mg L^{-1}), has been reported [83]. The observed significant decrease in hemocyte MN and NA frequencies in mussels' gill cells upon CNS treatment confirms the efficient removal of Zn(II) from ASW to levels which are not able to cause genotoxicity to mussels.

3.2.2. Mantle Histology and Histochemistry

After 24 h, $ZnCl_2$-exposure waters were slightly pink in color and the presence of suspended small tissue debris was confirmed by filtrates recovered with cellulose filters (0.45 μm), as shown in Figure S3. Exposure waters after CNS treatment, as well as controls (ASW) and those treated with only CNS, were transparent and no tissue debris was found upon filtration with cellulose filters (Figure S3).

Mussels exposed to $ZnCl_2$ (10 mg L^{-1}) showed severe damage on mantle edges, with syphons reduced in length and/or wilted and fused together (Figure 7d–f). On the contrary, mussels exposed after CNS treatment (Figure 7g–i) showed an intact mantle margin and syphons were distended and vigorous. A similar mantle morphology was present in the control mussels (Figure 7l–n) and in those exposed to CNS alone (Figure 7a–c), suggesting the absence of any of those damages caused by $ZnCl_2$ exposure.

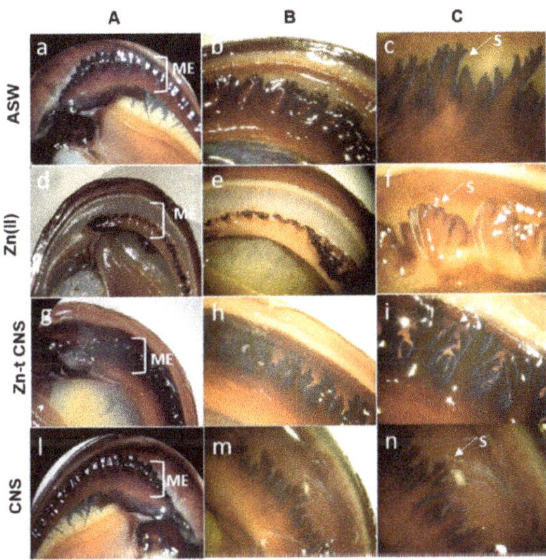

Figure 7. Mussels' mantle, mantle edges (ME) and syphons (S) of specimens exposed to the following experimental groups: ASW (**a–c**, controls); Zn(II) (**d–f**, 10 mg L^{-1} of $ZnCl_2$ in ASW); Zn-t CNS (**g–i**, $ZnCl_2$ 10 mg L^{-1} contaminated ASW treated with CNS); CNS (**l–n**, ASW treated with only CNS). Images obtained with a Zeiss Stemi SV6 (8–50× magnification) binocular stereo microscope, magnification A (8×) B (10×) C (20×).

External physical-chemical barriers, like shell, mantle and mucus, provide an important first line of defense in mollusks upon toxic chemical exposure. Zn(II) levels have been reported to exceed in the mantle edge compared to the rest of the mantle tissue, probably due to it being more directly in contact with seawater [43]. The lysosomes of mantle epithelial cells have been recorded to accumulate heavy metals [49,84], which can be stored in interstitial storage tissues and hemocytes [44,49,85]. Consequently, the mantle margin is more sensitive to Zn(II) exposure and, although acting as a physical barrier, it could be more damaged upon waterborne exposure than other internal organs such as the digestive gland [60]. By damaging syphons, which represent the mantle's functional structures, as observed in $ZnCl_2$-exposed mussels (ruptured, wilted or fused together), their protective role towards contaminant uptake can be seriously compromised [86]. The mantle of *Mytilus* also plays a major role in shell formation and the margin is considered the most active zone for shell deposition [87]. Cilia destruction in the outer epithelial cells of the mantle margin, as observed in mussels upon Zn-exposure, could affect the dynamic activity and movement of the organ itself [60].

The H&E staining method further confirmed the severe damage on the mantle edges caused by $ZnCl_2$ exposure. The columnar epithelium was fragmented along the entire margin and several holes were present (Figure 8d,e). In addition, epithelial cells were shortened, and the microvilli were fewer in number and were sagging and sparse (Figure 8f). Conversely, an intact columnar epithelium with elongated cells and with plicae along the edge was present in mussels exposed after CNS treatment (Figure 8g–i). A similar intact morphology was observed in the controls (Figure 8a–c) and in mussels exposed to CNS only, thus confirming a significant reduction in Zn damage on mantle edges in mussels exposed after CNS treatment (Figure 8l–n). In addition, columnar epithelial cells were well characterized by a monochromatic nucleus located at the basal area and several brown intracellular granules. Microvilli were homogeneously distributed, extended and elongated (Figure 8c,i,n).

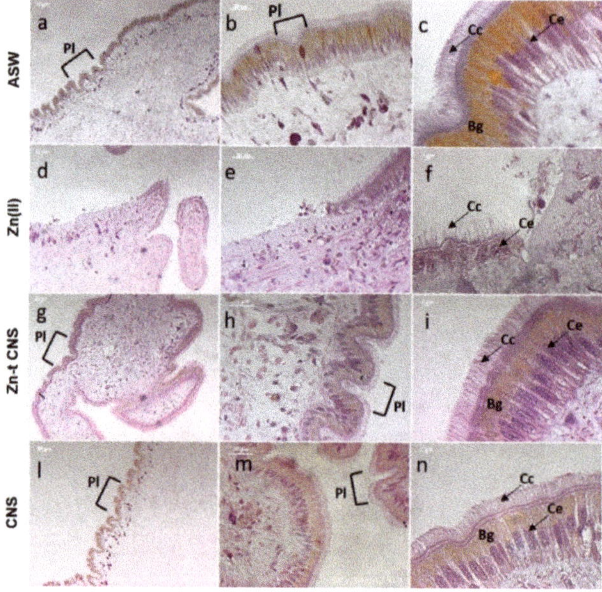

Figure 8. Light microscopy of mussels' mantle edges stained with haematoxylin and eosin (H&E) of specimens exposed to the following experimental groups: ASW (**a–c**, control); Zn(II) (**d–f**, 10 mg L^{-1} of $ZnCl_2$ in ASW); Zn-t CNS (**g–i**, $ZnCl_2$ 10 mg L^{-1} contaminated ASW treated with CNS); CNS (**l–n**, ASW treated with only CNS). **Cc**: Ciliated columnar cells; **Ce**: columnar epithelium; **Bg**: brown intracellular granules; **Pl**: plicae. Images obtained with a Zeiss Axiophot epifluorescent microscope with AxioCamMRc5.

All mussels, except those exposed to ASW only (controls), secreted a copious amount of mucus that settled to the bottom of the exposure tank. After 48 h of exposure, mussels' valves were slimy and sticky, particularly in those exposed to $ZnCl_2$ (data not shown). The characterization of mucus by PAS–AB highlighted the presence of two types of mucopolysaccharides in mussels' mantle edges (Figure 9). Neutral mucopolysaccharides (purple color) are PAS positive and AB negative, and they present low viscosity, while acid mucopolysaccharides (blue color) are PAS negative and AB positive, and they present high viscosity [88]. In $ZnCl_2$-exposed mussels, acidic mucosubstances were more present along the entire mantle margin, showing strong blue staining intensity, while the neutral ones were almost absent (purple staining) (Figure 9d–f). On the contrary, mussels exposed after CNS treatment showed an equal ratio (1:1) of acidic mucopolysaccharides and neutral ones (Figure 9g–i) and, furthermore, their distribution within the connective tissue was similar to that of the controls (Figure 9a–c) and of those exposed to CNS alone (Figure 9l–n).

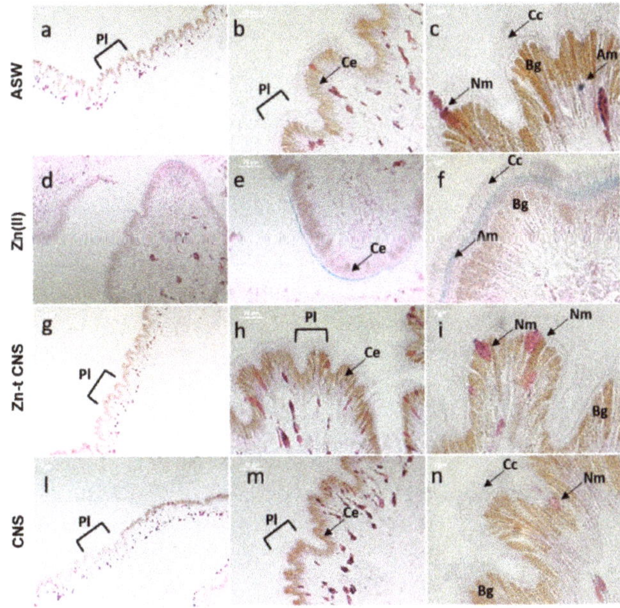

Figure 9. PAS–AB (periodic acid Schiff–Alcian blue) staining of mussels' mantle edge cells of specimens exposed to the following experimental groups: ASW (**a–c**, control); Zn(II) (**d–f**, 10 mg L^{-1} of ZnCl$_2$ in ASW); Zn-t CNS (**g–i**, ZnCl$_2$ 10 mg L^{-1} contaminated ASW treated with CNS); CNS (**l–n**, ASW treated with only CNS). **Cc**: Ciliated columnar cells; **Ce**: columnar epithelium; **Bg**: brown intracellular granules; **Nm**: neutral mucosubstances; **Am**: acidic mucosubstances; **Pl**: plicae. Images obtained with a Zeiss Axiophot epifluorescent microscope with AxioCamMRc5.

The continuous and unrestricted release of mucus secretion is considered a typical generalized stress response in bivalves that helps them to regulate metal levels in the tissues by acting as a purifier [11,89]. In bivalves, Zn(II) exposure has been described as inducing an overproduction of mucus, which is considered an inflammatory reaction that limits metal absorption or increases its excretion [43,52,90]. Mucus is a slimy and viscous secretory product that is involved in several physiological and behavioral functions, like the maintenance of internal water homeostasis, nutrition, immune defense and lubrification-related activities [86]. A role in isolating mussels from their environment and better counteracting external stressors has been also hypothesized [91]. Excessive mucus excretion may often lead to tissue desiccation and immune dysfunction due to the reduction of the internal mucosal sheath, which acts as a barrier to pathogens and toxins [86,92,93]. The higher secretion of acidic mucosubstances than of neutral ones in ZnCl$_2$-exposed mussels has already been described as a mussel's response to metal toxicity [94,95]. In addition, an increase in mucus-secreting cells and non-ciliated epithelial cells in the sub-epithelial region of the mantle margin, such as that observed in Zn-exposed mussels, have been linked to exposure to chemical stressors [48,60,89]. Therefore, the excess mucus production in mussels exposed after CNS treatment revealed their responsiveness to low levels of Zn, such as those still present in ASW after CNS treatment (0.510 ± 0.025 mg L^{-1}). Such evidence, together with other toxicological data obtained on mussels exposed after CNS treatment, supports the suitability of the proposed biological effect-based approach to assess the efficacy of CNS in reducing Zn toxicity in mussels.

Only very recently, aquatic ecotoxicology has begun to move toward the environmental safety of (nano)materials, including those for pollution remediation (nanoremediation), and related challenges in order to provide suitable testing strategies and methods to prevent side effects [21,96–104].

Such scientific gaps call for a thorough evaluation of their environmentally safe application and, therefore, a case-by-case analysis must be undertaken to assess their actual applicability [105,106]. While in our previous study [40], we aimed to develop an eco-design strategy to improve CNS' ecosafety by modifying both the formulation and purification protocols, here, we used a more complex effect-based approach to assess their efficacy in Zn(II) removal from seawater to levels which do not cause any harm to marine species [107,108]. The revealed sensitivity of the effect-based approach thus represents a promising solution to overcome the current limitations of nanotechnologies for environmental remediation and the reasonable policy concerns due to its in-situ application [19,21–23]. However, this clearly represents a first attempt to promote the applicability of an effect-based approach as ecotoxicological proof of the material's efficacy, which is commonly based on analytical chemistry validation. To this aim, in our study, the almost complete removal of Zn(II) upon CNS treatment and, respectively, of 91.7% and 91.5% at time zero and after 24 h, was confirmed by analyzing Zn levels in exposure waters (Table 1). A significant reduction in the Zn(II) concentration was generally observed in all experimental groups after 24 h, including ASW, except for waters treated with CNS, which showed a slight increase (Table 1).

Table 1. Zn(II) concentration (mg L^{-1}) in mussels' exposure waters at time zero (T_0) (soon after preparation) and after 24 h (T_{24h}) of the following experimental groups: Zn(II) (ZnCl$_2$ 10 mg L^{-1} in ASW); Zn-t CNS (ZnCl$_2$ 10 mg L^{-1} contaminated ASW treated with CNS), CNS (ASW treated with CNS only); ASW (control). Results are reported as mean ± SD. (*) indicates significant differences with respect to Zn(II) and (**) with respect to controls (ASW) ($p < 0.001$).

Exposure Groups	Zn(II) T_0	Zn(II) T_{24h}
Zn(II)	8.567 ± 0.185 **	6.006 ± 0.604 **
Zn-t CNS	0.710 ± 0.0568 *	0.510 ± 0.0258 *
CNS	0.00251 ± 0.0001 *	0.0245 ± 0.0008 *
ASW	0.0047 ± 0.0006	0.0005 ± 0.0001

The uptake of Zn(II) by mussels during the 24 h of exposure due to seawater filtration might have also reduced their levels in ASW. Adult mussels can easily filter up to 3 L in at least 1 h and reach a quick state-state in heavy metal accumulation in their soft tissues in proportion to their concentration in seawater [46,109]. This peculiar behavior of filter-feeders towards heavy metals in the water column demonstrates the need to further implement the effect-based approach by testing other species from different trophic levels (e.g., bottom-dwellers, benthic grazers), as well as the need to include more sensitive and/or target ones of specific pollutants to be remediated. Long-term exposure scenarios should also be included, which will allow us to reach a more comprehensive view of the potential ecological disturbances and assessment of the adverse effects on marine ecosystems. Field-scale application, for instance, by using natural seawater media and mesocosm settings, will also elucidate potential confounding factors which could affect the material's behavior and efficacy (i.e., dissolved organic carbon and colloids on adsorption ability), thus improving the environmental relevance of the assessment and linking ecotoxicological and chemical information.

4. Conclusions

Our findings demonstrate that after CNS treatment, Zn(II) toxicity was significantly reduced in marine mussels, both at cellular and tissue levels, which was in agreement with the significantly reduced Zn(II) nominal levels (>90%) in exposure waters. Upon CNS treatment, Zn-induced genotoxicity in circulating immune and gill cells was significantly reduced to that which was observed in specimens exposed to ASW (controls), as well as histological and cytological damage in mantle edge and epithelial cells. The proposed effect-based approach was thus proven to be useful in further supporting the environmental relevance of CNS' efficacy in heavy metal removal from seawater and their ecosafe application by linking ecotoxicological and chemical information.

Supplementary Materials: The following are available online at http://www.mdpi.com/2079-4991/10/7/1283/s1. Table S1: pH levels in mussel's exposure media. Figure S1: Lysosomal membrane destabilization in mussels exposed to ZnCl$_2$; Figure S2: Nuclear abnormalities detected in mussel's gill cells; Figure S3: water exposure media filtered using 0.45 µm cellulose filter paper.

Author Contributions: I.C.: Conceptualization, Supervision, Resources, Funding acquisition, G.F.: Conceptualization, Supervision, Resources; C.P.: Conceptualization, Supervision, Resources; G.L.: Investigation, Writing—Original Draft, Writing—Review & Editing, Methodology; G.G.: Investigation, Writing—Review & Editing, Methodology; A.F.: Investigation, Writing—Review & Editing, Methodology; P.G.: Investigation, Writing—Review & Editing; M.B.: Investigation, Writing—Review & Editing; V.S.: Investigation; M.G.: Investigation; G.P.: Investigation, C.F.: Investigation. All authors have read and agreed to the published version of the manuscript.

Funding: This research was funded by Regione Toscana, NanoBonD (Nanomaterials for Remediation of Environmental Matrices associated to Dewatering) POR CReO FESR Toscana 2014–2020-30/07/2014- LA 1.1.5 CUP 3389.30072014.067000007.

Conflicts of Interest: The authors declare no conflict of interest.

References

1. Furness, R.W.; Rainbow, P.S. Heavy Metals in Marine Environment. In *Heavy Metals in the Marine Environment*; CRC Press Taylor & Francis Group: Boca Raton, FL, USA, 1990.
2. Bebianno, M.J.; Pereira, C.G.; Rey, F.; Cravo, A.; Duarte, D.; D'Errico, G.; Regoli, F. Integrated approach to assess ecosystem health in harbor areas. *Sci. Tot. Environ.* **2015**, *514*, 92–107. [CrossRef] [PubMed]
3. Sharifuzzaman, S.M.; Rahman, H.; Ashekuzzaman, S.M.; Islam, M.M.; Chowdhury, S.R.; Hossain, M.S. Heavy Metals Accumulation in Coastal Sediments. In *Environmental Remediation Technologies for Metal-Contaminated Soils*; Hasegawa, H., Rahman, M.M., Rahman, I., Azizur, M., Eds.; Springer: Tokyo, Japan, 2016; pp. 21–42.
4. Marine Strategy Framework Directive. European Commission Directive 2008/56/EC. Available online: https://ec.europa.eu/environment/marine/eu-coast-and-marine-policy/marine-strategy-framework-directive/index_en.htm (accessed on 1 June 2020).
5. Environmental Quality Standard Directive 2013/39/EU. Directive of the European Parliament and of the Council on Environmental Quality Standards in the Field of Water Policy Amending and Subsequently Repealing Council Directives 82/176/EEC, 83/513/EEC, 84/156/EEC, 84/491/EEC, 86/280/EEC and Amending Directive 2000/60/EC of the European Parliament and of the Council. Available online: https://eur-lex.europa.eu/legal-content/EN/TXT/?uri=LEGISSUM:l28002b (accessed on 1 June 2020).
6. USGS. Mineral Commodity Summaries. 2015. Available online: https://minerals.usgs.gov/minerals/pubs/mcs/2015/mcs2015.pdf (accessed on 14 January 2017).
7. EPA (Environmental Protection Agency). Uniform National Discharge Standards for Vessels of the Armed Forces. Technical Development Document 821-R-99-001; 1999. Available online: https://nepis.epa.gov/Exe/ZyPURL.cgi?Dockey=20002KB0.TXT (accessed on 1 June 2020).
8. Lu, P.J.; Huang, S.C.; Chen, Y.P.; Chiueh, L.C.; Shih, D.Y.C. Analysis of titanium dioxide and zinc oxide nanoparticles in cosmetics. *J. Food Drug Anal.* **2015**, *23*, 587–594. [CrossRef] [PubMed]
9. Bruland, K.W. Oceanographic distributions of cadmium, zinc, nickel, and copper in the North Pacific. *Earth Planet. Sci. Lett.* **1980**, *47*, 176–198. [CrossRef]
10. Luoma, S.N. Can we determine the biological availability of sediment-bound trace elements? *Hydrobiologia* **1989**, *176/177*, 379–396. [CrossRef]
11. Neff, J.M. Zinc in the Ocean. In *Bioaccumulation in Marine Organisms*; Elsevier: Amsterdam, The Netherlands, 2002; pp. 175–189.
12. Millero, F.; Pierrot, D. Speciation of Metals in Natural Waters. In *Chemistry of Marine Waters and Sediments*; Gianguzza, A., Pelizzetti, E., Sammartano, S., Eds.; Springer: Berlin, Germany, 2002; pp. 193–220.
13. Kalnejais, L.H.; Martin, W.R.; Bothner, M.H. The release of dissolved nutrients and metals from coastal sediments due to resuspension. *Mar. Chem.* **2010**, *121*, 224–235. [CrossRef]
14. Sinoir, M.; Butler, E.C.V.; Bowie, A.R.; Mongin, M.; Nesterenko, P.N.; Hassler, C.S. Zinc marine biogeochemistry in seawater: A review. *Mar. Fresh. Res.* **2012**, *63*, 644–657. [CrossRef]
15. Burton, J.D.; Statham, P.J. Trace Metals in Seawater. In *Heavy Metals in the Marine Environment*; CRC Press Taylor & Francis Group: Boca Raton, FL, USA, 1990.

16. Li, L.; Zhen, X.; Wang, X.; Ren, Y.; Hu, L.; Bai, Y.; Liu, J.; Shi, X. Benthic trace metal fluxes in a heavily contaminated bay in China: Does the sediment become a source of metals to the water column? *Environ. Pollut.* **2020**, *257*, 113494. [CrossRef]

17. Sunda, W.G.; Huntsman, S.A. Feedback interactions between zinc and phytoplankton in seawater. *Limnol. Oceanogr.* **1992**, *37*, 25–40. [CrossRef]

18. Patil, S.S.; Shedbalkar, U.U.; Truskewycz, A.; Chopade, B.A.; Bal, A.S. Nanoparticles for environmental clean-up: A review of Potential risks and emerging solutions. *Environ. Technol. Innov.* **2016**, *5*, 10–21. [CrossRef]

19. Corsi, I.; Winther-Nielsen, M.; Sethi, R.; Punta, C.; Della Torre, C.; Libralato, G.; Lofrano, G.; Sabatini, L.; Aiello, M.; Fiordi, L.; et al. Ecofriendly nanotechnologies and nanomaterials for environmental applications: Key issue and consensus recommendations for sustainable and ecosafe nanoremediation. *Ecotox. Environ. Safe* **2018**, *154*, 237–244. [CrossRef]

20. Karn, B.; Kuiken, T.; Otto, M. Nanotechnology and in Situ remediation: A Review of the Benefits and Potential Risks. *Environ. Health Perspect.* **2009**, *117*, 1823–1831. [CrossRef] [PubMed]

21. Corsi, I.; Cherr, G.N.; Lenihan, H.S.; Labille, J.; Hassellov, M.; Canesi, L.; Dondero, F.; Frenzilli, G.; Hristozov, D.; Puntes, V.; et al. Common Strategies and Technologies for the Ecosafety Assessment and Design of Nanomaterials Entering the Marine Environment. *ACS Nano* **2014**, *8*, 9694–9709. [CrossRef]

22. Corsi, I.; Fiorati, A.; Grassi, G.; Bartolozzi, I.; Daddi, T.; Melone, L.; Punta, C. Environmentally Sustainable and Ecosafe Polysaccharide-Based Materials for Water Nano-Treatment: An Eco-Design Study. *Materials* **2018**, *11*, 1228. [CrossRef] [PubMed]

23. Corsi, I.; Grassi, G. The role of ecotoxicology in the eco-design of nanomaterials for water remediation. In *Ecotoxicology of Nanoparticles in Aquatic Systems*; Blasco, J., Corsi, I., Eds.; CRC Taylor & Francis Group: Boca Raton, FL, USA, 2019; pp. 219–229.

24. Gao, Y.; Guo, S.; Wang, J.; Li, D.; Wang, H.; Zeng, D. Effects of different remediation treatments on crude oil contaminated saline soil. *Chemosphere* **2014**, *117*, 486–493. [CrossRef] [PubMed]

25. Zhou, J.; Yong, Y.; Zhang, C. Toward Biocompatible Semiconductor Quantum Dots: From Biosynthesis and Bioconjugation to Biomedical Application. *Chem. Rev.* **2014**, *115*, 11669–11717. [CrossRef]

26. Zhang, L.; Chen, B.; Ghaffar, A.; Zhu, X. Nanocomposite Membrane with Polyethylenimine-Grafted Graphene Oxide as a Novel Additive to Enhance Pollutant Filtration Performance. *Environ. Sci. Technol.* **2018**, *52*, 5920–5930. [CrossRef]

27. Pedrazzo, A.R.; Smarra, A.; Caldera, F.; Musso, G.; Kumar Dhakar, N.; Cecone, C.; Corsi, I.; Trotta, F. Eco-friendly β-cyclodextrin and linecaps polymers for the removal of heavy metals. *Polymers* **2019**, *11*, 1658. [CrossRef]

28. Hokkanen, S.; Repo, E.; Suopajärvi, T.; Liimatainen, H.; Niinimaa, J.; Sillanpää, M. Adsorption of Ni(II), Cu(II) and Cd(II) from aqueous solutions by amino modified nanostructured microfibrillated cellulose. *Cellulose* **2014**, *21*, 1471–1487. [CrossRef]

29. Carpenter, A.W.; De Lannoy, C.; Wiesner, M.R. Cellulose Nanomaterials in Water Treatment Technologies. *Environ. Sci. Technol.* **2015**, *49*, 5277–5287. [CrossRef]

30. Suhas Gupta, V.K.; Carrott, P.J.M.; Singh, R.; Chaudhary, M.; Kushwaha, S. Cellulose: A review as natural, modified and activated carbon adsorbent. *Bioresour. Technol.* **2016**, *216*, 1066–1076. [CrossRef]

31. Bossa, N.; Carpenter, A.W.; Kumar, N.; De Lannoye, C.; Wiesner, M. Cellulose nanocrystal zero-valent iron nanocomposites for groundwater remediation. *Environ. Sci. Nano.* **2017**, *4*, 1294–1303. [CrossRef] [PubMed]

32. Mahfoudhi, N.; Boufi, S. Nanocellulose as a novel nanostructured adsorbent for environmental remediation: A review. *Cellulose* **2017**, *24*, 1171–1197. [CrossRef]

33. Chin, K.; Ting, S.; Ong, H.; Omar, M. Surface functionalized nanocellulose as a veritable inclusionary material in contemporary bioinspired applications: A review. *J. Appl. Polym.* **2018**, *135*, 46065. [CrossRef]

34. Mohammed, N.; Grishkewich, N.; Tam, K.C. Cellulose nanomaterials: Promising sustainable nanomaterials for application in water/wastewater treatment processes. *Environ. Sci. Nano* **2018**, *5*, 623–658. [CrossRef]

35. Melone, L.; Rossi, B.; Pastori, N.; Panzeri, W.; Mele, A.; Punta, C. TEMPO-Oxidized Cellulose Cross-Linked with Branched Polyethyleneimine: Nanostructured Adsorbent Sponges for Water Remediation. *ChemPlusChem* **2015**, *80*, 1408–1415. [CrossRef] [PubMed]

36. Pierre, G.; Punta, C.; Delattre, C.; Melone, L.; Dubessay, P.; Fiorati, A.; Pastori, N.; Galante, Y.M.; Michaud, P. TEMPO-mediated oxidation of polysaccharides: An ongoing story. *Carbohyd. Polym.* **2017**, *165*, 71–85. [CrossRef]

37. Paladini, G.; Venuti, V.; Almásy, L.; Melone, L.; Crupi, V.; Majolino, D.; Pastori, N.; Fiorati, A.; Punta, C. Cross-linked cellulose nano-sponges: A small angle neutron scattering (SANS) study. *Cellulose* **2019**, *26*, 9005–9019. [CrossRef]

38. Fiorati, A.; Turco, G.; Travan, A.; Caneva, E.; Pastori, N.; Cametti, M.; Punta, C.; Melone, L. Mechanical and Drug Release Properties of Sponges from Cross-linked Cellulose Nanofibers. *ChemPlusChem* **2017**, *82*, 848–858. [CrossRef] [PubMed]

39. Bartolozzi, I.; Daddi, T.; Punta, C.; Fiorati, A.; Iraldo, F. Life cycle assessment of emerging environmental technologies in the early stage of development. A case study on nanostructured materials. *J. Ind. Ecol.* **2020**, *24*, 101–115. [CrossRef]

40. Fiorati, A.; Grassi, G.; Pastori, N.; Graziano, A.; Liberatori, G.; Melone, L.; Punta, C.; Corsi, I. Eco-design of nanostructured cellulose sponges for sea-water decontamination from heavy metal ions. *J. Clean. Prod.* **2020**, *246*, 119009. [CrossRef]

41. OSPARCOM. JAMP Guidelines for Monitoring Contaminants in Biota. OSPAR Commission, London, 2012, 122. Available online: https://mcc.jrc.ec.europa.eu/documents/OSPAR/Guidelines_forMonitoring_of_ContaminantsSeawater.pdf (accessed on 1 June 2020).

42. Farrington, J.W.; Tripp, B.W.; Tanabe, S.; Subramanian, A.; Sericano, J.L.; Wade, T.L.; Knap, A.H.; Edward, D. Goldberg's proposal of "the mussel watch": Reflections after 40 years. *Mar. Pollut. Bull.* **2016**, *110*, 501–510. [CrossRef] [PubMed]

43. Lobel, R.B. Short-Term and Long-Term Uptake of Zinc by the Mussel, *Mytilus edulis*: A Study in Individual Variability. *Arch. Environ. Con. Toxicol.* **1987**, *16*, 723–732. [CrossRef] [PubMed]

44. Regoli, F.; Orlando, E. Accumulation and subcellular distribution of metals (Cu, Fe, Mn, Pb and Zn) in the Mediterranean Mussel *Mytilus galloprovincialis* during a field transplant experiment. *Mar. Pollut. Bull.* **1994**, *28*, 592–600. [CrossRef]

45. Vercauteren, K.; Blust, R. Bioavailability of dissolved zinc to the common mussel *Mytilus edulis* in complexing environments. *Mar. Ecol. Prog. Ser.* **1996**, *137*, 123–132. [CrossRef]

46. Beyer, J.; Green, N.W.; Brooks, S.; Allan, I.J.; Ruus, A.; Gomes, T.; Bråte Schøyen, M. Blue mussels (*Mytilus edulis spp.*) as sentinel organisms in coastal pollution monitoring: A review. *Mar. Environ. Res.* **2017**, *130*, 338–365. [CrossRef]

47. Lowe, D.M.; Moore, M.N. The cytochemical distributions of Zinc (Zn II) and Iron (Fe III) in the common mussel, *Mytilus edulis*, and their relationship with lysosomes. *J. Mar. Biol. Assoc. UK* **1979**, *59*, 851–858. [CrossRef]

48. George, S.G.; Pirie, B.J.S. Metabolism of Zinc in the mussel, *Mytilus edulis* (L.): A combined ultrastructural and biochemical study. *J. Mar. Biol. Assoc. UK* **1980**, *60*, 575–590. [CrossRef]

49. Soto, M.; Cajaraville, M.P.; Marigómez, I. Tissue and cell distribution of copper, zinc and cadmium in the mussel, *Mytilus galloprovincialis*, determined by autometallography. *Tissue Cell* **1996**, *28*, 557–568. [CrossRef]

50. Marigómez, I.; Soto, M.; Cajaraville, M.P.; Angulo, E.; Giamberini, L. Cellular and Subcellular Distribution of Metals in Molluscs. *Microsc. Res. Techniq.* **2002**, *56*, 358–392. [CrossRef]

51. Amiard-Triquet, C.; Berthet, B.; Metayer, C.; Amiard, J.C. Contribution to the ecotoxicological study of cadmium, copper and zinc in the mussel *Mytilus edulis* II. Experimental study. *Mar. Biol.* **1986**, *92*, 7. [CrossRef]

52. Hietanen, B.; Sunila, I.; Kristoffersson, R. Toxic effects of zinc on the common mussel *Mytilus edulis* L. (Bivalvia) in brackish water. I. Physiological and histopathological studies. *Ann. Zool. Fenn.* **1988**, *25*, 341–347.

53. Viarengo, A. Heavy metals in marine invertebrates: Mechanisms of regulation and toxicity at the cellular level. *Aquat. Sci. Rev.* **1989**, *1*, 295–317.

54. Livingstone, D.R.; Pipe, R.K. Mussels and environmental contaminants: Molecular and cellular aspects. In *The Mussel Mytilus: Ecology, Physiology, Genetics and Culture*; Gosling, E.M., Ed.; Elsevier Science Publishers B.V.: Amsterdam, The Netherlands, 1992; pp. 425–464.

55. Burbidge, F.J.; Macey, D.J.; Webb, J.; Talbot, V. A Comparison Between Particulate (Elemental) Zinc and Soluble Zinc (ZnCl$_2$) Uptake and Effects in the Mussel, *Mytilus edulis*. *Arch. Environ. Con. Toxicol.* **1994**, *26*, 466–472. [CrossRef]

56. Zhang, Y.F.; Chen, S.Y.; Qu, A.O.; Adeleye, A.O.; Di, Y.N. Utilization of isolated marine mussel cells as an in vitro model to assess xenobiotics induced genotoxicity. *Toxicol. Vitro* **2017**, *44*, 219–229. [CrossRef] [PubMed]

57. Pearson, H.B.; Dallas, L.J.; Comber, S.D.W.; Braungardt, C.B.; Worsfold, P.J.; Jha, A.N. Mixtures of tritiated water, zinc and dissolved organic carbon: Assessing interactive bioaccumulation and genotoxic effects in marine mussels, *Mytilus galloprovincialis*. *J. Environ. Radioact.* **2018**, *187*, 133–143. [CrossRef]

58. ICES. Report of the ICES\OSPAR Workshop on Lysosomal Stability Data Quality and Interpretation (WKLYS), 13–17 September 2010, Alessandria, Italy. 2010. Available online: http://ices.dk/sites/pub/CM%20Doccuments/CM-2010/ACOM/ACOM6110.pdf (accessed on 1 June 2020).

59. Bjärnmark, N.A.; Yarra, T.; Churcher, A.M.; Felix, R.C.; Clark, M.S.; Power, D.M. Transcriptomics provides insight into *Mytilus galloprovincialis* (Mollusca: Bivalvia) mantle function and its role in biomineralisation. *Mar. Genomics.* **2016**, *27*, 37–45. [CrossRef]

60. Moëzzi, F.; Javanshir, A.; Eagderi, S.; Poorbagher, H. Histopathological effects of zinc (Zn) on mantle, digestive gland and foot in freshwater mussel, *Anodonta cygnea* (Linea, 1876). *Int. J. Aquat. Biol.* **2013**, *1*, 61–67.

61. Pagano, M.; Porcino, C.; Briglia, M.; Fiorino, E.; Vazzana, M.; Silvestro, S.; Faggio, C. The Influence of Exposure of Cadmium Chloride and Zinc Chloride on Haemolymph and Digestive Gland Cells from *Mytilus galloprovincialis*. *Int. J. Environ. Res.* **2017**, *11*, 207–216. [CrossRef]

62. Jha, A.N.; Cheung, V.V.; Foulkes, M.E.; Hill, S.J.; Depledge, M.H. Detection of genotoxins in the marine environment: Adoption and evaluation of an integrated approach using the embryo-larval stages of the marine mussel, *Mytilus edulis*. *Mutat. Res.* **2000**, *464*, 213–228. [CrossRef]

63. Fernández, B.; Campillo, J.A.; Martínez-Gómez, C.; Benedicto, J. Micronuclei and other nuclear abnormalities in mussels (*Mytilus galloprovincialis*) as biomarkers of cyto-genotoxic pollution in mediterranean waters. *Environ. Mol. Mutagen.* **2011**, *52*, 479–491. [CrossRef]

64. Ale, A.; Liberatori, G.; Vannuccini, M.L.; Bergami, E.; Ancora, S.; Mariotti, G.; Bianchi, N.; Galdoporpora, J.M.; Desimone, M.; Cazenave, J.; et al. Exposure to a nanosilver-enabled consumer product results in similar accumulation and toxicity of silver nanoparticles in the marine mussel *Mytilus galloprovincialis*. *Aquat. Toxicol.* **2019**, *211*, 46–56. [CrossRef]

65. ASTM. International Standard Guide for Conducting Static Acute Toxicity Tests Starting with Embryos of Four Species of Salt Water Bivalve Molluscs. Available online: https://www.astm.org/DATABASE.CART/HISTORICAL/E724-98.htm (accessed on 1 June 2020).

66. D'Silva, C.; Kureishy, T.W. Experimental studies on the accumulation of copper and zinc in the green mussel. *Mar. Pollut. Bull.* **1978**, *9*, 187–190. [CrossRef]

67. Akberali, H.B.; Wong, T.; Trueman, E. Behavioural and siphonal tissue responses of *Scrobicularia plana* (bivalvia) to zinc. *Mar. Environ. Res.* **1981**, *5*, 251–264. [CrossRef]

68. Neuberger-Cywiak, L.; Achituv, Y.; Garcia, E.M. Effects of Zinc and Cadmium on the Burrowing Behavior, LC 50, and LT 50 on *Donax trunculus* Linnaeus (Bivalvia-Donacidae). *Environ. Contam. Toxicol.* **2003**, *70*, 713–722. [CrossRef]

69. Vlahogianni, T.H.; Valavanidis, A. Heavy-metal effects on lipid peroxidation and antioxidant defence enzymes in mussels *Mytilus galloprovincialis*. *J. Chem. Ecol.* **2007**, *23*, 361–371. [CrossRef]

70. Ramakritinan, C.M.; Chandurvelan, R.; Kumaraguru, A.K. Acute Toxicity of Metals: Cu, Pb, Cd, Hg and Zn on Marine Molluscs, *Cerithedia cingulata* G., and *Modiolus philippinarum* H. *Indian J. Geo-Mar. Sci.* **2012**, *41*, 141–145.

71. Howard, D.W.; Lewis, E.J.; Keller, B.J.; Smith, C.S. *Histological Techniques for Marine Bivalve Mollusks and Crustaceans*, 2nd ed.; NOAA/National Centers for Coastal Ocean Science: Oxford, UK, 2004.

72. Lowe, D.M.; Soverchia, C.; Moore, M.N. Lysosomal membrane responses in the blood and digestive cells of mussels experimentally exposed to fluoranthene. *Aquat. Toxicol.* **1995**, *33*, 105–112. [CrossRef]

73. Davies, I.M.; Gubbins, M.; Hylland, K.; Maes, T.; Martínez-Gómez, C.; Giltrap, M.; Burgeot, T.; Wosniok, W.; Lang, T.; Vethaak, A.D. Technical annex: Assessment criteria for biological effects measurements. In *Integrated*

Monitoring of Chemicals and Their Effects; Davies, I.M., Vethaak, A.D., Eds.; ICES Cooperative Research Report; ICES: Copenhagen, Denmark, 2012; p. 315.

74. Venier, P.; Maron, S.; Canova, S. Detection of micronuclei in gill cells and haemocytes of mussels exposed to benzo(a)pyrene. *Mutat. Res.* **1997**, *390*, 33–34. [CrossRef]

75. Guidi, P.; Lyons, B.P.; Frenzilli, G. The Comet Assay in Marine Animals. In *Genotoxicity Assessment. Methods in Molecular Biology*; Dhawan, A., Bajpayee, M., Eds.; Humana: New York, NY, USA, 2019; Volume 2031.

76. Frenzilli, G.; Nigro, M.; Scarcelli, V.; Gorbi, S.; Regoli, F. DNA integrity and total oxyradical scavenging capacity in the Mediterranean mussel, *Mytilus galloprovincialis*: A field study in a highly eutrophicated coastal lagoon. *Aquat. Toxicol.* **2001**, *53*, 19–32. [CrossRef]

77. Rocco, L.; Santonastaso, M.; Nigro, M.; Mottola, F.; Costagliola, D.; Bernardeschi, M.; Guidi, P.; Lucchesi, P.; Scarcelli, V.; Corsi, I.; et al. Genomic and chromosomal damage in the marine mussel *Mytilus galloprovincialis*: Effects of the combined exposure to titanium dioxide nanoparticles and cadmium chloride. *Mar. Environ. Res.* **2015**, *111*, 144–148. [CrossRef]

78. Fenech, M. Cytokinesis-block micronucleus cytome assay. *Nat. Protoc.* **2007**, *2*, 1084–1104. [CrossRef]

79. Lowe, D.M.; Fossato, V.U. The influence of environmental contaminants on lysosomal activity in the digestive cells of mussels (*Mytilus galloprovincialis*) from the Venice Lagoon. *Aquat. Toxicol.* **2000**, *48*, 75–85. [CrossRef]

80. EEA (European Environment Agency). *Hazardous Substances in Europe's Fresh and Marine Waters. An Overview*; EEA Technical Report No 8; European Environment Agency: Copenhagen, Denmark, July 2011.

81. Varotto, L.; Domeneghetti, S.; Rosani, U.; Manfrin, C.; Cajaraville, M.P.; Raccanelli, S.; Pallavicini, A.; Venier, P. DNA Damage and Transcriptional Changes in the Gills of *Mytilus galloprovincialis* Exposed to Nanomolar Doses of Combined Metal Salts (Cd, Cu, Hg). *PLoS ONE* **2013**, *8*, e54602. [CrossRef]

82. Gherras Touahri, H.; Boutiba, Z.; Benguedda, W.; Shaposhnikov, S. Active biomonitoring of mussels *Mytilus galloprovincialis* with integrated use of micronucleus assay and physiological indices to assess harbour pollution. *Mar. Pollut. Bull.* **2016**, *110*, 52–64. [CrossRef]

83. Majone, F.; Beltrame, C.; Brunetti, R. Frequencies of micronuclei detected on *Mytilus galloprovincialis* by different staining techniques after treatment with zinc chloride. *Mutat. Res.* **1988**, *209*, 131–134. [CrossRef]

84. Fowler, B.A.; Wolte, D.A.; Hettler, W.F. Mercury and iron uptake by cytosomes in mantle epithelial cells of quahog clams (*Mercenaria mercenaria*) exposed to mercury. *J. Fish. Res. Board. Can.* **1975**, *32*, 1767–1775. [CrossRef]

85. Al-Subiai, S.N.; Moody, A.J.; Mustafa, S.A.; Jha, A.N. A multiple biomarker approach to investigate the effects of copper on the marine bivalve mollusc, *Mytilus edulis*. *Ecotoxicol. Environ. Saf.* **2011**, *74*, 1913–1920. [CrossRef]

86. Larramendy, M.L. *Ecotoxicology and Genotoxicology: Non-traditional Aquatic Models (Issues in Toxicology)*; The RSC: Cambridge, UK, 2017.

87. Calvo-Iglesias, J.; Pérez-Estévez, D.; Lorenzo-Abalde, S.; Sánchez-Correa, B.; Quiroga, M.I.; Fuentes, J.M.; González-Fernández, A. Characterization of a Monoclonal Antibody Directed against Mytilus spp Larvae Reveals an Antigen Involved in Shell Biomineralization. *PLoS ONE* **2015**, *11*, e0152210. [CrossRef]

88. Beninger, P.G.; St-Jean, S.D. The role of mucus in particle processing by suspension-feeding marine bivalves: Unifying principles. *Mar. Biol.* **1997**, *129*, 389–397. [CrossRef]

89. Sze, P.W.C.; Lee, S.Y. The potential role of mucus in the depuration of copper from the mussels *Perna viridis* (L.) and *Septifer virgatus* (Wiegmann). *Mar. Poll. Bull.* **1995**, *31*, 390–393. [CrossRef]

90. Moraes, R.B.; Silva, M.L.P. Toxicity of zinc in the mussel *Perna perna* (Linne, 1758). *Arq. Biol. Tecnol.* **1995**, *38*, 541–547.

91. Davies, M.S.; Hawkins, S.J. Mucus from Marine Molluscs. *Adv. Mar. Biol.* **1998**, *34*, 1–71.

92. Gerdol, M.; Gomez-Chiarri, M.; Castillo, M.G.; Figueras, A.; Fiorito, G.; Moreira, R.; Novoa, B.; Pallavicini, A.; Ponte, G.; Roumbedakis, K.; et al. Immunity in Molluscs: Recognition and Effector Mechanisms, with a Focus on Bivalvia. In *Advances in Comparative Immunology*; Cooper, E.L., Ed.; Springer International Publishing AG, part of Springer Nature: Cham, Switzerland, 2018; pp. 225–341.

93. Loker, E.S.; Bayne, C.J. Molluscan immunobiology: Challenges in the Anthropocene epoch. In *Advances in Comparative Immunology*; Cooper, E.L., Ed.; Springer International Publishing AG, Springer Nature: Cham, Switzerland, 2018; pp. 343–407.

94. Fasulo, S.; Mauceri, A.; Giannetto, A.; Maisano, M.; Bianchi, N.; Parrino, V. Expression of metallothionein mRNAs by in situ hybridization in the gills of *Mytilus galloprovincialis*, from natural polluted environments. *Aquat. Toxicol.* **2008**, *88*, 62–68. [CrossRef] [PubMed]

95. D'Agata, A.; Fasulo, S.; Dallas, L.J.; Fisher, A.S.; Maisano, M.; Readman, J.W.; Jha, A.N. Enhanced toxicity of 'bulk' titanium dioxide compared to 'fresh' and 'aged' nano-TiO$_2$ in marine mussels (*Mytilus galloprovincialis*). *Nanotoxicology* **2013**, *1–10*, 1743–5404.

96. Rickerby, D.; Morrison, M. Report from the Workshop on Nanotechnologies for Environmental Remediation. Presented at the Joint Research Centre, Ispra, Italy, 16–17 April 2007. Available online: http://www.nanowerk.com/nanotechnology/reports/reportpdf/report101.pdf (accessed on 1 June 2020).

97. Grieger, K.D.; Wickson, F.; Andersen, H.B.; Renn, O. Improving risk governance of emerging technologies through public engagement: The neglected case of nano-remediation? *Int. J. Emerging Technol. Soc.* **2012**, *10*, 61–78.

98. Trujillo-Reyes, J.; Peralta-Videa, J.R.; Gardea-Torresdey, J.L. Supported and unsupported nanomaterials for water and soil remediation: Are they a useful solution for worldwide pollution? *J. Hazard. Mater.* **2014**, *280*, 487–503. [CrossRef]

99. Pedersen, K.B.; Lejon, T.; Ottosen, L.M.; Jensen, P.E. Screening of variable importance for optimizing electrodialytic remediation of heavy metals from polluted harbour sediments. *Environ. Toxicol.* **2015**, *36*, 2364–2373. [CrossRef]

100. Elliott, J.; Rösslein, M.; Song, N.; Toman, B.; Kinsner-Ovaskainen, A.; Maniratanachote, R.; Salit, M.; Petersen, E.; Sequeira, F.; Romsos, E.; et al. Toward achieving harmonization in a nanocytotoxicity assay measurement through an interlaboratory comparison study. *ALTEX—Altern. Anim. Exp.* **2017**, *34*, 201–218. [CrossRef]

101. Holden, P.A.; Gardea-Torresdey, J.L.; Klaessig, F.; Turco, R.F.; Mortimer, M.; Hund-Rinke, K.; Cohen Hubal, E.A.; Avery, D.; Barceló, D.; Behra, R.; et al. Considerations of environmentally relevant test conditions for improved evaluation of ecological hazards of engineered nanomaterials. *Environ. Sci. Technol.* **2016**, *50*, 6124–6145. [CrossRef]

102. Selck, H.; Handy, R.D.; Fernandes, T.F.; Klaine, S.J.; Petersen, E.J. Nanomaterials in the aquatic environment: A European Union–United States perspective on the status of ecotoxicity testing, research priorities, and challenges ahead. *Environ. Toxicol. Chem.* **2016**, *35*, 1055–1067. [CrossRef]

103. Oomen, A.G.; Steinhäuser, K.G.; Bleeker, E.A.J.; van Broekhuizen, F.; Sips, A.; Dekkers, S.; Wijnhoven, W.P.S.; Sayre, P.G. Risk assessment frameworks for nanomaterials: Scope, link to regulations, applicability, and outline for future directions in view of needed increase in efficiency. *Nanoimpact* **2018**, *9*, 1–13. [CrossRef]

104. Zhu, Y.; Liu, X.; Hu, Y.; Wang, R.; Chen, M.; Wu, J.; Wang, Y.; Kang, S.; Sun, Y.; Zhu, M. Behavior, remediation effect and toxicity of nanomaterials in water environments. *Environ. Res.* **2019**, *174*, 54–60. [CrossRef]

105. Sánchez, A.; Recillas, S.; Font, X.; Casals, E.; González, E.; Puntes, V. Ecotoxicity of, and remediation with, engineered inorganic nanoparticles in the environment. *TrAC-Trend. Anal. Chem.* **2011**, *30*, 507–516. [CrossRef]

106. Garner, K.L.; Keller, A.A. Emerging patterns for engineered nanomaterials in the environment: A review of fate and toxicity studies. *J. Nanopart. Res.* **2014**, *16*, 2503. [CrossRef]

107. Hjorth, R.; van Hove, L.; Wickson, F. What can nanosafety learn from drug development? The feasibility of "safety by design". *Nanotoxicology* **2017**, *11*, 305–312. [CrossRef]

108. Singh, A.V.; Laux, P.; Luch, A.; Sudrik, C.; Wiehr, S.; Wild, A.M.; Santomauro, G.; Bill, J.; Sitti, M. Review of emerging concepts in nanotoxicology: Opportunities and challenges for safer nanomaterial design. *Toxicol. Mech. Methods* **2019**, *29*, 378–387. [CrossRef]

109. Seed, R.; Suchanek, T.H. The mussel *Mytilus*: Ecology, Physiology, Genetics and Culture. In *Development in Aquaculture and Fisheries Science*; Gosling, E., Ed.; Elsevier: Amsterdam, The Netherlands, 1992; p. 25.

Article

Toxicity, Bioaccumulation and Biotransformation of Glucose-Capped Silver Nanoparticles in Green Microalgae *Chlorella vulgaris*

Stefania Mariano [1], Elisa Panzarini [1], Maria D. Inverno [2], Nick Voulvoulis [2] and Luciana Dini [3,4,*]

[1] Department of Biological and Environmental Science and Technology, University of Salento, 73100 Lecce, Italy; stefania.mariano@unisalento.it (S.M.); elisa.panzarini@unisalento.it (E.P.)
[2] Centre for Environmental Policy, Imperial College London, London SW7 2AZ, UK; mjoaoinverno@gmail.com (M.D.I.); n.voulvoulis@imperial.ac.uk (N.V.)
[3] Department of Biology and Biotechnology "Charles Darwin", Sapienza University of Rome, 00185 Rome, Italy
[4] CNR Nanotec, 73100 Lecce, Italy
* Correspondence: luciana.dini@uniroma1.it; Tel.: +39-064-991-2306; Fax: +39-064991

Received: 19 June 2020; Accepted: 13 July 2020; Published: 15 July 2020

Abstract: Silver nanoparticles (AgNPs) are one of the most widely used nanomaterials in consumer products. When discharged into the aquatic environment AgNPs can cause toxicity to aquatic biota, through mechanisms that are still under debate, thus rendering the nanoparticles (NPs) effects evaluation a necessary step. Different aquatic organism models, i.e., microalgae, mussels, *Daphnia magna*, sea urchins and *Danio rerio*, etc. have been largely exploited for NPs toxicity assessment. On the other hand, alternative biological microorganisms abundantly present in nature, i.e., microalgae, are nowadays exploited as a potential sink for removal of toxic substances from the environment. Indeed, the green microalgae *Chlorella vulgaris* is one of the most used microorganisms for waste treatment. With the aim to verify the possible involvement of *C. vulgaris* not only as a model microorganism of NPs toxicity but also for the protection toward NPs pollution, we used these microalgae to measure the AgNPs biotoxicity and bioaccumulation. In particular, to exclude any toxicity derived by Ag^+ ions release, green chemistry-synthesised and glucose-coated AgNPs (AgNPs-G) were used. *C. vulgaris* actively internalised AgNPs-G whose amount increases in a time- and dose-dependent manner. The internalised NPs, found inside large vacuoles, were not released back into the medium, even after 1 week, and did not undergo biotransformation since AgNPs-G maintained their crystalline nature. Biotoxicity of AgNPs-G causes an exposure time and AgNPs-G dose-dependent growth reduction and a decrease in chlorophyll-a amount. These results confirm *C. vulgaris* as a bioaccumulating microalgae for possible use in environmental protection.

Keywords: *Chlorella vulgaris*; silver nanoparticles; ecotoxicity; growth inhibition; chlorophyll-a content; morphological changes; bioaccumulation; crystalline structure

1. Introduction

In the last few decades, nanoparticles (NPs) have attracted great attention due to their chemical, physical, optic and biological properties. Accordingly, safety assessment becomes an important issue for the beneficial usage of these new materials [1]. NPs chemical and physical properties (chemical composition, size, shape) and the complex interactions occurring at various biological levels (organelle, cell, tissue, organ, organ system, organism) can potentially impact on human health [2]. Some studies have shown that NPs cause toxic effects which are instead not induced by similar but larger particles. These effects are probably due to the intrinsic characteristics of the NPs which permit targets that cannot be reached by their larger and chemically identical counterparts to be achieved [3]. The human body can interact with nanomaterials (NMs) mainly through different ways: inhalation through the respiratory system, ingestion through the gastrointestinal tract, and absorption through the skin. Once inside, they manage to overcome further barriers, such as the brain–blood barrier, causing toxic effects on human health [2]. Several studies demonstrated toxic effects of NPs on human health, including those associated with cardiovascular disease derived from titanium dioxide, metal oxide and metal nanoparticles exposure [4], and pulmonary inflammation induced by carbon containing NMs [5,6]. These effects include among others inflammation, granuloma formation, and fibrosis of the lungs [5,7].

Together with the increasing nanobiotechnological application, exposure to NPs of all living organisms and the environment enhances and justifies the need to identify, measure and manage the risks.

The environmental fate and transport models demonstrated that NPs/NMs can enter, as nanowaste, directly or indirectly, soil and waterways [8]. Thus, through washing, rain and other routes, these NPs/NMs can be released especially into the aquatic environment, where they can be potentially toxic to biota causing an ecological impact as well as to humans with socioeconomic consequences [9–11].

Silver nanoparticles (AgNPs) are frequently used in consumer products or medical devices for their antibacterial and antifungal activities [12,13], being the most studied in the field of nanoecotoxicology. Their widespread use in commercial products, mainly because of their bacterial power, has led to a steadily increasing amount of AgNPs in environment [14]. Most of the environmental concerns are raising the fate of AgNPs in washing machines, textile industry and similar applications. The release of AgNPs and Ag^+ into the water by simply immersing commercial socks containing AgNPs into shaken water was revealed by Benn et al. [15]. Consumables containing AgNPs have been subjected to different treatments, such as interaction with surfactants, oxidising agents, different pH, to test the release of AgNPs during the washing processes and the passage of these NPs in the sweat of human skin. These treatments greatly increased the release of Ag^+ and AgNPs [16,17] and became an important route for increasing the NPs in the environment.

In addition, industrial treatment of NPs can lead to the release of AgNPs into the sewage system or wastewater [18]. Shafer et al. [19] measured total silver concentrations (non-nanospecific) of up to 105 µg/L in the liquid inflow of a wastewater treatment plant. It has been proven that many of the AgNPs is retained in the sewage sludge during the wastewater treatment process. However, a smaller part of the AgNPs can still reach the environment via the effluent [20].

Against this background, a wide variety of organisms, i.e., bacteria, plants, fungi, algae, invertebrates and fish have been considered to evaluate the behavior of AgNPs with aquatic organisms.

Knowledge gaps on how AgNPs interact with a living-organisms remain an issue at all levels of organisation, in particular at a cellular and molecular level (genes, transcripts, metabolites, proteins, enzymes and soluble factors). However, the effects of AgNPs on aquatic algal microorganisms was reported to induce time and concentration changes in speciation of microalgae *Raphidocelis subcapita* [21]; inhibit growth and cellular viability of the diatom *Thalassiosira pseudonana*, cyanobacterium *Synechococcus* sp. [22] but also of the aquatic plant *Lemna gibba* [23]; and favour superoxide production in the marine raphidophyte *Chattonella marina* [24]. Again, AgNPs affect the photosynthesis process, leading to a change in the chlorophyll content of algae *Chlamydomonas reinhardtii* [25] or cyanobacterium *Microcystis aeruginosa* [26], algae *Pithophora oedogonia* and algae *Chara vulgaris* [27], microalgae *Acutodesmus dimorphus* [28] and green algae *Chlamydomonas reinhardtii* [29].

Among the different strategies for reducing the NPs' environmental impact and preventing the potentially toxic effects of AgNPs due to the release of Ag^+ or to the agglomeration of particles in aqueous systems, green chemistry has been introduced in the synthesis of AgNPs and surface coating for stabilisation [30–32]. Taken together these two technological approaches that use innovative principles in the design of industrial chemical processes, could be fundamental for achieving sustainable industrial development, preventing and reducing industrial pollution and environmental impact. Indeed, green chemistry promotes the design, manufacture and use of chemicals and processes that abolish or reduce the use or generation of substances injurious to environment and health [33]. To this purpose, the use of natural sources, non-hazardous solvents, biodegradable and biocompatible materials, such as cellulose, chitosan, dextran or tree gums, and energy-efficient processes are the main NPs' preparation innovation [34–36].

Considering that the zero release of AgNPs into water is not realistic, synergistic approaches to the technology of NPs synthesis should be considered, such as the use of microorganisms as bioaccumulators and/or biotransformators. In fact, during the last two decades, several methods have been developed for environmental removal of hazardous substances like precipitation, evaporation, ion-exchange etc., even if these methods have several disadvantages [37,38]. One alternative strategy is the use of microorganisms abundantly present in nature, i.e., microalgae, that are already used to remove heavy metals and in wastewater treatment facilities; in fact, the microalgae reduce the amount of toxic chemicals needed to clean and purify water [39], being able either to accumulate, adsorb or metabolise these noxious elements into a substantial level.

However, studies on the ability of microalgae to remedy NPs aquatic pollution are still very limited [40]. In this study, we used green chemistry-synthesised AgNPs, that were capped with glucose-G (AgNPs-G) to ensure AgNPs stability [41–43] and *Chlorella vulgaris*, one of the most widely used microorganism in testing NPs/NMs effects on aquatic biota but also known to reduce heavy metals from waters [37,38]. Among aquatic organisms, algae are an important model as they are primary producers, i.e., they fix CO_2 to produce oxygen in the presence of light. Also, they are at the base of the food chain, serving as a food to, e.g., the water flea but also fish. The microalgae *C. vulgaris* was chosen in the present study because of its easy growth in commercial culture. This unicellular green species with a cell diameter of 5 μm, has been utilised for varied purposes, ranging from nutrient removal from wastewater to their use as a food source. Its ability to survive in adverse conditions and its ubiquitous nature also make it a potentially useful algae for industrial wastewater treatment. Specifically, the microalgae *Chlorella vulgaris* is known for its robustness, in conjunction with being one of the fastest growing species and is easily cultivated [44].

To exploit the ability of *C. vulgaris* in the removal of AgNPs-G from water, we investigated the efficiency of the microalgae to uptake and retain NPs. Studies of AgNPs-G characterisation and nanotoxicology were also performed.

2. Materials and Methods

2.1. Chemicals

All chemicals were of analytical grade and were purchased from Sigma-Aldrich (St. Louis, MO, USA) unless otherwise indicated.

2.2. Synthesis of Glucose-Capped Silver Nanoparticles (AgNPs-G)

AgNPs-G were obtained by adding 2 mL of a 10^{-2} M aqueous solution of $AgNO_3$ to 100 mL of 0.3 M β-D-glucose water solution. The mixture was boiled for 30 min under vigorous stirring. The deep yellow colour of the solutions indicated the formation of AgNPs-G. Deionised ultra-filtered 18.2 MΩ water prepared with a Milli-Q Integral Water Purification System (Merck Millipore, Billerica, MA, USA) was used for all preparations. All glassware was washed in an ultrasonic bath of deionised water and not ionic detergent, followed by thorough rinsing with Milli-Q water and ethanol (Carlo Erba, Milan, Italy) to completely remove not ionic detergent contaminants. Finally, glassware was dried prior to use.

2.3. AgNPs-G Characterisation

Transmission electron microscopy (TEM) and ultraviolet–visible (UV–Vis) analysis were used to evaluate the average and distribution size and morphology of the NPs.

TEM analysis was performed by a Hitachi 7700, at 100 kV (Hitachi, Dallas, TX, USA). A drop of AgNPs-G solution diluted in complete Bold's basal medium (BBM) [45] was placed onto standard carbon-supported 600-mesh copper grid. Particle size distribution has been obtained using the ImageJ program (National Institutes of Health (NIH), Bethesda, MD, USA). A histogram was created by counting 500 particles. Optical spectra were obtained by measuring the absorption of the solution in the range between 300 and 800 nm by using a T80 spectrophotometer (PG Instruments Ltd., Leicester, UK) in a quartz cuvette with a 1 cm optical path.

The stability of AgNPs-G was assayed in BBM. In particular, the dissolution of AgNPs-G, in terms of release of Ag^+, up to 10 days at r.t. (room temperature) in BBM culture medium was determined by atomic absorption spectroscopy (AAS; Thermo Electron Corporation, M-Series, Waltham, MA, USA) after precipitation of AgNPs-G by ultracentrifugation (24,900× g; 30 min at 4 °C). The detection limit was 1 µg/L. Triplicate readings were analysed and control samples of known Ag concentration were analysed in parallel generating data with the standard deviation of three independent samples. Silver ions dissolution degree was expressed as percentage (%) of total Ag^+, as $AgNO_3$, used to reach the highest concentration of NPs solution during treatment.

Bold's basal medium composition: $NaNO_3$ 250 mg/L, K_2HPO_4 75 mg/L, $MgSO_4.7H_2O$ 75 mg/L, $CaCl_2.2H_2O$ 25 mg/L, KH_2PO_4 175 mg/L, NaCl 25 mg/L, Alkaline Ethylenediaminetetraacetic Acid (EDTA) solution 1 mg/L (alkaline EDTA solution: 5 g Na_2-EDTA and 3.1 g KOH in 100 mL distilled water), acidified iron solution 1 mg/L (acidified iron solution $FeSO_4.7H_2O$ 498 g and 0.1 mL H_2SO_4 in 100 mL distilled water) trace metal solution 1 mL/L (trace metal solution: $MnCl_2.4H_2O$ 1.44 g/L, $ZnSO_4.7H_2O$ 8.82 g/L, $(NH_4)_6 Mo_7O_{24}.2H_2O$ 0.88 g/L, $Co(NO_3)_2.6H_2O$ 0.49 g/L, $CuSO_4.5 H_2O$ 1.57 g/L).

2.4. Chlorella Vulgaris Culture

The freshwater microalga *C. vulgaris* was obtained from the Culture Collection of Algae and Protozoa (Argyll, UK). The algae were cultured in 250 mL flasks containing 100 mL of BBM and covered with loose cotton. The flasks were placed on a shaker to keep the turbulence of culture medium simulating the natural stream of water. The cultures were kept at 23 ± 1 °C under illumination of approximately 73.6 µmol m^{-2} s^{-1} with daily cycles of 12 h light and 12 h dark. The culture cell density was monitored with a spectrophotometer (Pharmacia Biotech, Stockholm, Sweden) at 684 nm every 24 h. Cells in the exponential phase were used for all experiments.

2.5. Growth-Inhibition Test

The evaluation was performed following the Organisation for Economic Co-operation and Development (OECD) 201 algal growth inhibition test guidelines [46]. Algae were incubated for 24 h and a week with Ag ions (0.1 µg/L and 1 mg/L of silver nitrate) and with different concentrations of AgNPs-G: 0.1, 1, 10, 100 µg/L and 1 mg/L with three replicates for each concentration. The inhibitory rate of growth was obtained by using the Equation (1):

$$\text{Inhibitory Rate } (IR)\% = (1 - N/N_0) \times 100 \tag{1}$$

where N is the density of cells/mL in the samples treated with AgNPs-G, N_0 is the density of cells/mL in the control samples. The test was performed with three independent experiments (with three technical replicates for each repeated experiment) by using the same batch of algae and AgNPs-G.

2.6. Chlorophyll Content

Treated samples were centrifuged to remove culture media. Then, 90% acetone was added to tubes. Sealed tubes were shaken to ensure that microalgae cells are in the whole solvent volume and centrifuged at 5000 rpm (5236× g) for 5 min. Chlorophyll-a concentration was determined by measuring the optical density (OD) of supernatant by spectrophotometer (Pharmacia Biotech,

Stockholm, Sweden). Absorbance values of extracts were measured at 645 and 663 nm in 1 cm pathlength cuvettes. Quantitative determination was undertaken according to Arnon (1949) [47].

2.7. Biodistribution and Subcellular Localisation of AgNPs: Transmission Electron Microscope (TEM) Analysis

The ultrastructural analysis of *C. vulgaris* treated with different concentrations of Ag ions and AgNPs-G for one day and one week was performed by TEM (Hitachi HT 7700 transmission electron microscopy) analysis.

Algae were centrifuged to remove culture media and then fixed with glutaraldehyde (2.5% in sodium cacodilate buffer 0.1 M, pH 7.2) for 2 h at 4 °C. Then, samples were washed twice for 15 min in sodium cacodilate buffer, postfixed in osmium tetraoxide (1% in sodium cacodilate buffer 0.1 M, pH 7.2) and washed twice for 30 min in deionised H_2O. Samples were stained with 0.5% uranyl acetate o.n. (over night) at 4 °C. Samples were dehydrated in a graded series of ethanol, from 30% to 100%. After dehydration, samples were embedded in Spurr resin (TAAB, Berks, UK).

Ultrathin sections of 50 nm in thickness were then cut using an ultramicrotome PowerTome PT-PC (RMC, Tucson, AZ, USA). Sections were picked up in 200 mesh copper grids and examined under a Hitachi HT7700 transmission electron microscope (Tokyo, Japan) at 75 kV.

Samples were analyzed by energy-dispersive X-ray spectroscopy (EDX) microanalysis with the TEM module of the Auriga 405 microscope (Carl Zeiss AG, Oberkochen, Germany) for the elemental analysis of the electron-dense particles inside the cells.

2.8. X-ray Diffraction (XRD) Analysis

To determine the amount of Ag^+ inside algal cells, X-ray diffraction (XRD) analysis was performed with samples of algae treated for a week with AgNPs-G. Only AgNPs-G were used as positive control and a culture of only *C. vulgaris* as negative control. Samples were collected, dried at 60 °C and then sintered at 650 °C for 4h under nitrogen protection. The analysis was performed with X-ray diffractometers (Malvern Pananalytical, Malvern, UK).

2.9. Inductively Coupled Plasma–Optical Emission Spectrometry (ICP–OES) Analysis

A series of AgNPs-G stocks (0.1, 1, 10, 100 µg/L and 1 mg/L) were prepared in BBM. Algal samples with different AgNPs-G exposure concentrations and times were vacuum filtered with a 0.45 µm Millipore filter to separate algae from the culture medium. Samples were acidified with HNO_3 and analysed by inductively coupled plasma–optical emission spectroscopy (ICP–OES, Perkin Elmer Optima 7300 V HF version, Waltham, MA, USA) to determine Ag content. ICP–OES is a technique commonly used for the analysis of metals in various fields based on Atomic Emission Spectroscopy, where the sample at high temperature plasma up to 8000 K is converted to free, excited or ionised ions. The ions emit a radiation when go back to ground state, whose intensities are optically measured and indicate the amount of ions. The absorbed Ag by algal cells was calculated by the total Ag (T_{Ag}, also determined by ICP–OES by measuring stock solutions) minus the Ag in filtrates (F_{Ag}). Therefore, the percentage of absorbed Ag was calculated as $(T_{Ag} - F_{Ag})/T_{Ag} \times 100$.

2.10. Statistical Analysis

Data were analysed by performing one-way analysis of variance (ANOVA) at the 95% confidence level. *p* values less than 0.05 were considered significant. The results are reported as mean ± standard deviation (SD) of 3 technical replicates in each of the 3 independent experiments.

3. Results and Discussion

3.1. Characterisation of AgNPs-G: Shape, Size and Stability

Uptake and/or toxic effects rely on the shape, size and dispersion of the NPs. AgNPs-G shape, average size and size distribution, evaluated by TEM and UV–visible spectra, are reported in

Figure 1. The AgNPs-G UV–visible absorbance spectrum (Figure 1A) shows a characteristic absorption wavelength of spheroidal AgNPs, as suggested by a strong extinction band with a maximum at 420 nm. TEM showed spherical shape and good monodispersity of AgNPs-G (Figure 1B). The size distribution ranges from 14 to 28 nm and the average size is d = 20 nm with a standard deviation of 5 nm (Figure 1C).

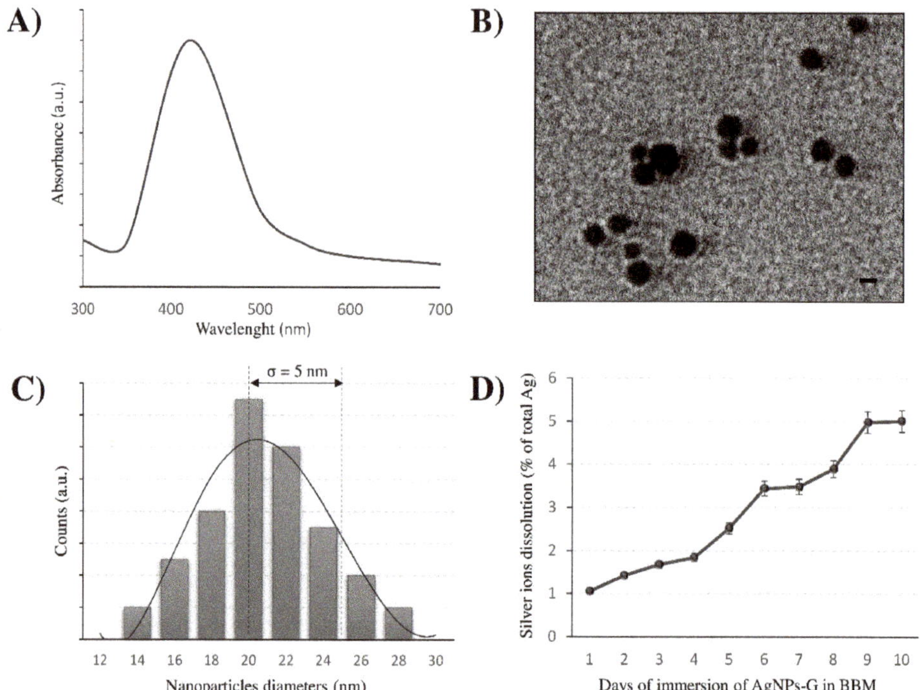

Figure 1. (**A**) Ultraviolet (UV)–visible spectra of glucose-capped silver nanoparticles (AgNPs-G)/mL in Bold's basal medium (BBM) culture medium reported as absorbance in arbitrary unit (a.u., y axis) vs. wavelength (nm, x axis). (**B,C**) Size distribution and transmission electron microscopy (TEM) micrograph of AgNPs-G. Size distribution is reported as arbitrary unit (a.u., y axis) vs. longitudinal diameter (nm, x axis). Bars = 20 nm. (**D**) Kinetic of Ag^+ dissolution. The dissolution of AgNPs-G in complete BBM culture medium was evaluated by atomic absorption spectroscopy. Data were analysed by performing one-way analysis of variance (ANOVA) at the 95% confidence level. Each value represents the mean ± standard deviation (SD) of 3 technical replicates in each of the 3 independent experiments. Ag^+ dissolution degree is expressed as percentage (%) of total $AgNO_3$ used to obtain the highest concentration of NPs solution during treatment.

It is known that particle toxicity could depend on Ag^+ released from NPs, thus the stability of AgNPs in the culture medium was estimated up to 10 days, in terms of Ag^+ release, by atomic absorption spectroscopy (Figure 1D). AgNPs-G were very stable in culture medium over time, since the dissolution degree, expressed as a percentage of total Ag^+ ranges between 1% and 5% at 1 and 10 days respectively. Since β-D-glucose capping ensures very low dissolution of Ag^+ from AgNPs and no loss of glucose was observed, the toxicity is due only to NPs.

In aquatic environments, dissolved oxygen in water oxidises the AgNPs surface causing Ag^+ ions release [48], identified as one of the most phytotoxic metal ions [49] for their cationic property and for the ability to associate with a variety of ligands present in natural waters. The toxicity of NP-released Ag^+ ions was reported for the alga *Chlamydomonas reinhardtii* [50,51], while Turner et al. [52] reported that AgNPs are only indirectly toxic to marine algae *Ulva lactuca* through the dissolution of Ag^+ ions into bulk seawater. However, whether AgNPs toxicity is due to the nanosized structure or to the released silver ions is still a matter of debate, and the results seem to be contingent mainly on the features of the AgNPs considered.

In our experiments, biotoxicity is not due to the Ag^+ ions release or the nanoparticle aggregates. To reduce as much as possible the Ag^+ release we used β-D-glucose for the green chemistry synthesis of AgNPs on the base of our previous data indicating that AgNPs-G are stable, well-dispersed with a minimum Ag^+ release in culture medium [42]. Indeed, in our experiments we measured the release of only 5% of the amount of the NPs after 10 days in seawater, thus confirming the effectiveness of the synthesis based on β D glucose as a reducing agent. The reduction of NPs toxicity by surface functionalisation with different coatings was also observed in several other studies [53,54]. Possible explanations could be attributed to the reduced nanoparticle dissolution as well as to the limited interactions between nanoparticles and organisms. For example, dexpanthenol, polyethylene glycol and polyvinyl polypyrrolidone coatings caused a similar toxic effect as $AgNO_3$ on *C. reinhardtii*, while carbonate, chitosan and citrate decreased the Ag effect on photosynthesis [29]. *C. vulgaris* were exposed to Ag^+ ions to understand if AgNP-G toxicity is driven by dissolved silver. The highest concentration of Ag^+ given as $AgNO_3$, was 100 times more the estimated release of AgNPs to the aquatic environment, that is about 0.01 mg/L^{-1} [55] and undoubtedly underestimated since this amount will increase in the near future for the forecast usage of these nanoparticles [56]. Our data showed that Ag^+ ions have only minimal effects on cell growth, morphological alteration, chlorophyll-a content and that the high doses of AgNPs-G only significantly reduced these parameters.

Toxicity of AgNPs has been a controversial topic for a long time. The open question is still the understanding of the toxicity mechanism of AgNPs. It seems not to be limited to the Ag^+ ions release but to different factors including the nanostructure [1]. According to Domingo et al. [57], AgNPs toxicity is not fully attributable to released ions since in photosynthetic organisms Ag^+ ions and AgNPs caused similar effects, although Ag^+ ions were often active at lower concentrations. Possible transformations of AgNPs-G mainly due to the aquatic chemistry cannot be excluded. Data in literature show that the aggregation of NPs in water depends on different parameters such as the pH or the surface charge of the NPs involved and by the specific type of organic matter or other natural particles present in freshwater [58]. In addition, AgNPs toxicity may depend on the species and on the type of growth medium in which the organisms are cultivated [59]. However, some adverse effects can also be attributed to specific properties of NPs, such as the size and the degree of aggregation, that in seawater is increased when compared to freshwater [60,61], and that in turn affect the capacity of NPs to cross biological membranes or bind the cell surface [62,63]. Cell wall, in fact, constitutes a primary site for interaction and serves as a barrier for the entrance of AgNPs into algal cells. In our hands, even after 1 week from the synthesis, AgNPs-G diluted in BBM were stable and well dispersed.

3.2. AgNPs-G are Bio-Absorbed by C. vulgaris *Maintaining Their Crystalline Structure*

C. vulgaris, at the exponential growth phase, was exposed to Ag^+ (0.1 µg/L and 1 mg/L of silver nitrate) or to different AgNPs-G concentrations (0.1, 1, 10, 100 µg/L and 1 mg/L) for 1 day or 1 week. In order to ensure that the cytotoxic effect (in terms of cell viability, chlorophyll content and ultrastructural changes) of silver nanoparticles is not due to the presence of Ag^+ ions in the suspension, $AgNO_3$, the salt of which the nanoparticles are made, has been used to prepare two solutions with a range that covers the amounts of Ag released in 10 days (from 1% to 5%) in the nanoparticles stability analysis. Moreover, in order to test our glucose-capped AgNPs on *C. vulgaris*, we chose scalar dilutions of NPs following data in literature reporting the same range of commercial AgNPs to investigate their effects on microalgae. The Ag content of algae filtrates measured by ICP–OES is reported in Figure 2B as percentage of internalised Ag. *C. vulgaris* is able to efficiently take up the AgNPs-G. The Ag content correlates with AgNPs-G amounts used for treatments and with exposure time. Moreover, internalisation of AgNPs increased of about 10% after a week for every treatment, raising the 75% and 86% of internalised NPs at the higher AgNPs-G concentrations (100µg/L and 1 mg/L, respectively).

This ability to internalise the AgNPs-G was confirmed by TEM observations (Figure 3(Cd)). AgNPs-G were observed inside large vacuoles or crossing the cell wall (Figure 3(Cd–g)).

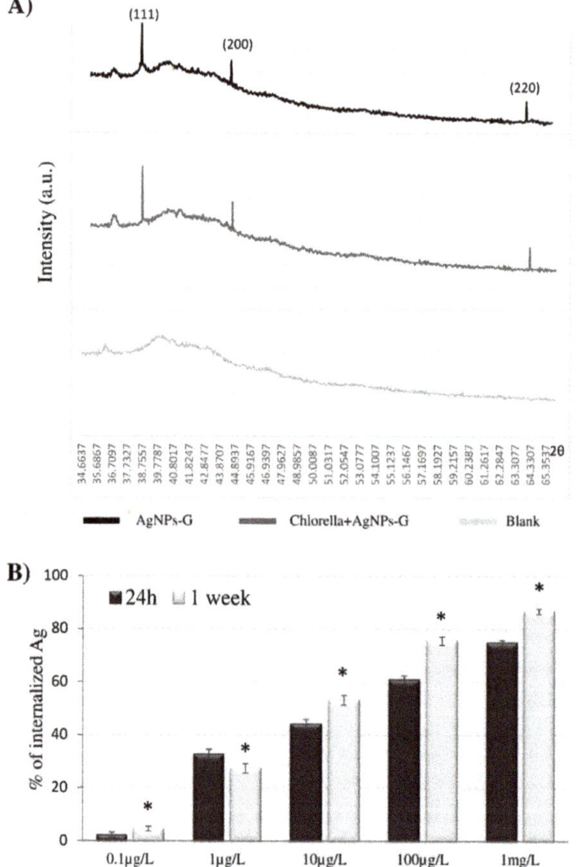

Figure 2. (**A**) X-ray diffraction (XRD) spectrum of AgNPs-G before and after the interaction with algae. A culture of *C. vulgaris* is used as negative control. Numbers refer to diffraction peaks of Ag in its crystalline form. (**B**) Inductively coupled plasma–optical emission spectrometry (ICP–OES) to determine Ag internalisation by algal cells treated with five concentrations of AgNPs-G. The absorbed Ag was calculated by the total Ag (T_{Ag}, also determined by ICP–OES by using stocks at five concentrations) minus the Ag in filtrates (F_{Ag}). Therefore, the percentage of absorbed Ag = $(T_{Ag} - F_{Ag})/T_{Ag} \times 100$. Data were analysed by performing one-way ANOVA at the 95% confidence level. Each value represents the mean ± SD of 3 technical replicates in each of the 3 independent experiments. Asterisks indicate significant differences from respective values at 24 h at the same concentration ($p < 0.05$).

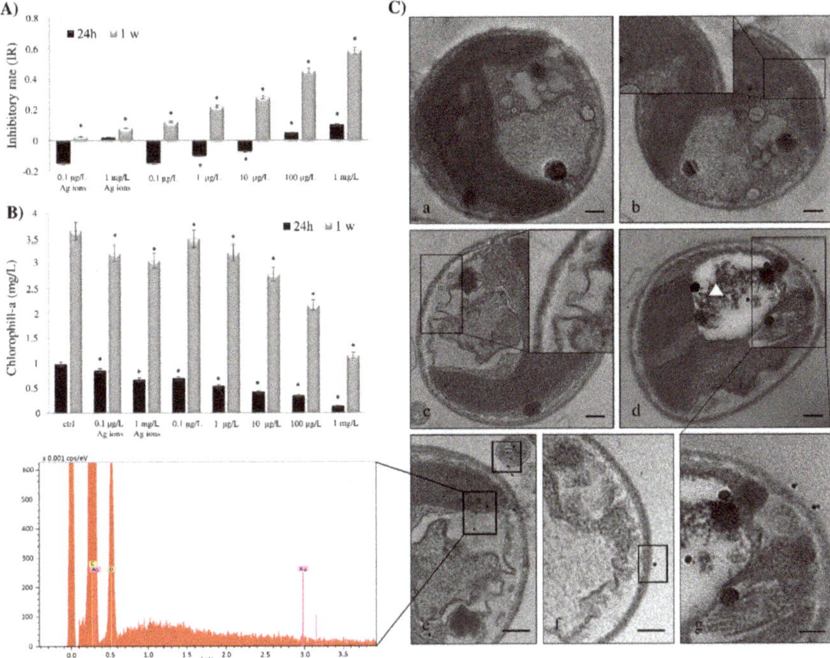

Figure 3. (**A**) Analysis of inhibitory rate. Algae were incubated for 24 h and a week with Ag ions and with five concentrations of AgNPs-G. Data were analysed by performing one-way ANOVA at the 95% confidence level. Each value represents the mean ± SD of 3 technical replicates in each of the 3 independent experiments. Asterisks indicate significant differences from the control values ($p < 0.05$); (**B**) Analysis of chlorophyll-a content by spectrophotometric analysis of centrifuged samples. Quantitative determination was done according to Arnon et al. (1949). The experiments were conducted in triplicate and results are the mean with standard deviation. Asterisks indicate significant differences from the respective untreated samples ($p < 0.05$) (**C**) TEM micrographs of algal cells and elemental X-ray spectrum (lower panel) of the square area of micrograph (**e**) containing black spots. (**a**) control cell; (**b**) algal cells treated with Ag ions. (**c**) algal cells treated with AgNPs-G for 24 h. Plasma membrane detaches from the cell wall, as indicated in the magnification; (**d–g**) Algal cells treated with AgNPs-G for a week. AgNPs-G were observed inside large vacuoles (**d**, white triangle), inside algae (**d–e**) or crossing the cell wall (**f–g**). Bar = 500 nm.

EDX microanalysis (Figure 3C) confirms that the electron-dense particles observed inside the microalgae correspond to AgNPs. Interestingly, once inside the microalgae, the NPs are not released back into the medium, either as an active secretion or cell ruptures.

The Ag content of algae filtrates analysed by ICP–OES correlated to the AgNPs-G amounts used for treatments and the time of exposure. The continuous internalisation of AgNPs-G particles observed in our experiments could be dependent on the size and dispersion of our NPs preparation. Data in literature report that bio-adsorption of heavy metal particles to algae is dependent on different properties of NPs, such as surface charge or size, chemical composition, and by the cell walls pore sizes, spanning through the thickness of the walls, ranging from 5 to 20 nm [62,64]. Thus, small nanostructures are highly diffusible, and only NPs up to 20 nm can reach the cell membrane. Other physicochemical properties of AgNPs can influence the internalisation, the rate of entrance and the biological response. Once the cell wall is penetrated, endocytic passage through plasma membrane may be possible and internalised NPs enhance biological effects. Sendra et al. [61] found that the attachment of AgNPs on the surfaces of freshwater and marine microalgae *Chlamydomonas reinhardtii* and *Phaeodactylum tricornutum,* and the presence of AgNPs inside cells directly drives the toxic effects. NPs can also enter into the cells via ion channels, transport proteins and endocytosis mechanism [65].

Also, AgNPs-G enter the algal cells maintaining their crystalline structure once inside even after 1 week. The lack of changes in the crystalline structure of AgNPs was investigated with XRD analysis. Figure 2A shows the XRD pattern of the three Bragg reflections with 2θ values of 38.1°, 44.3° and 64.4° which correspond to the (111), (200), and (220) sets of Bragg's reflections planes of the metallic AgNPs in a sample containing only AgNPs-G and in a sample of *C. vulgaris* treated with AgNPs-G for a week. A sample of only *C. vulgaris* was used as negative control. The spectrum confirmed the face-centred cubic crystalline structure of AgNPs-G with a spherical morphology as characterised by TEM. When AgNPs-G were added to *C. vulgaris* culture, no new diffraction peaks appeared, suggesting that AgNPs-G maintain their crystalline nature. Data in literature report that the microalgae can change the NPs crystalline structure. Studies demonstrated that living *C. vulgaris* showed a capacity to reduce nickel oxide nanoparticles (NiONPs) for zero valence nickel, changing their crystalline structure. The reduction from nanosized NiO to nanosized Ni led to weakened toxicity [40,66]. The maintenance of the crystalline structure of the NPs once inside the microalgae should be analysed as a positive or adverse outcome. In our case the presence of silver in the NPs shape could be a positive aspect as the microalgae can hold silver inside by removing it from the outside environment.

3.3. Cell Viability, Chlorophyll Content and Ultrastructure of AgNPs-G Treated C. vulgaris

C. vulgaris, at the exponential growth phase, was exposed to Ag ions or AgNPs-G at different concentrations (0.1, 1, 10, 100 µg/L and 1 mg/L) for 1 day and 1 week. The concentration-inhibition graph is reported in Figure 3A. Exposure of algae to AgNPs-G causes a reduction of cell metabolism. The inhibitory rate of growth (IR) increased in a significant way with the increasing time of exposure and doses. In fact, the IR increases up to 6 folds after 1 week of culture in the presence of 1 mg/L of AgNPs-G. Significant growth inhibition was observed in the presence of 100 µg/L and 1 mg/L of AgNPs-G for 24 h. Ag ions exposure induces no effect on cell growth. The negative values of IR at 24 h exposure indicate the so-called hormesis effects of poisoning, both in Ag ions and AgNPs-G treatments.

In line with the growth reduction, the chlorophyll-a concentration reduction (Figure 3B) was dependent on the NPs doses and time of treatment. Statistical analysis revealed a significant difference ($p < 0.05$) between control and treated samples at 24 h in the presence of 1, 10, 100 µg/L and 1 mg/L with a reduction of chlorophyll amount of about 80% at the higher AgNPs-G concentration. Conversely, the treatment of *C. vulgaris* for 1 week induces the decrement in chlorophyll amount in the presence of 10, 100 µg/L and 1 mg/L. The reduction is of about 50% than control cells after culture in the presence of 1 mg/L AgNPs-G. Ag ions exposure induces only a moderate effect on chlorophyll-a content.

TEM ultrastructure of *C. vulgaris* is reported in Figure 3C. In control cells, the plasma membrane was close to the cell wall. Chloroplasts contain well-compartmentalised thylakoids, which are fundamental structures involved in photosynthesis (Figure 3(Ca)). A morphology not different from control cells was observed upon Ag ions treatment with (Figure 3(Cb)). Cells cultured in the presence of the highest AgNPs-G concentration, showed the plasma membrane detaching from the cell wall, (Figure 3(Cc)). The morphological alterations also correlate with AgNPs-G incubation time (Figure 3(Cd)). Large vacuoles with degraded materials were observed (Figure 3(Cd–g), white triangle). A partial structural disorder of thylakoids suggesting a reduced photosynthesis activity is present.

The work reported here confirms *C. vulgaris* an useful microalgae for the detection of the biotoxicity of AgNPs-G, as demonstrated by the fact that culture time and amount of AgNPs influenced growth inhibition, morphological alteration, reduction in chlorophyll-a concentration and photosynthesis perturbation due to structural disorders of thylakoids. The reduced chlorophyll content is in accordance with the data of Hazeem et al., [63] who demonstrated the AgNPs have a negative effect on viability, chlorophyll-a concentration, and increased reactive oxygen species (ROS) formation in *C. vulgaris*. Comparable effects have been demonstrated for other algae species, e.g., *Pithophora oedogonia* and *Chara vulgaris* [27]. Several studies have shown that AgNPs caused inhibition of growth in freshwater green microalgae [67], a reduction in chlorophyll content and morphological changes in the freshwater green alga *Pithophora oedogonia* [27], a decrease of photosynthesis activity in the unicellular green algae *Chlamydomonas reinhardtii* [29], and an increase of antioxidant activities in the marine flagellate *Chattonella marina* [24]. Decrease in chlorophyll content, viable algal cells, increased ROS formation, and lipid peroxidation in the freshwater microalga *Chlorella vulgaris* and marine microalga *Dunaliella tertiolecta* were observed after exposure to AgNPs for 24 h [68].

4. Conclusions

In conclusion, it should be kept in mind that the continued increase in the use of AgNPs is a consistent hazard in aquatic ecosystems, where microalgae are key actors, and actions to prevent/reduce this hazard cannot be postponed. Many critical points have to be overcome as the identification of the best biological model for risk assessment, because of species response, exposure conditions and environment-particle chemical interactions. It is thus important to choose the best NPs to be commercialised according to their safety by design synthesis (green chemistry, coatings). In our case, silver nanoparticles are coated with glucose, that ensures stability, in terms of morphology, dispersion and dissolution (release of silver ions). They are stable in several culture media, both in experiments with human cell lines and those with aquatic organisms, as reported in our previous works [42,43]. In BBM culture medium they kept their shape, size and stability, although the medium contained EDTA, which is known to promote the dissolution and dispersion of silver nanoparticles through Ag chelation [69,70]. However, EDTA concentration reported in literature is higher (or not indicated) than concentration used in our experiment, so EDTA effects could be irrelevant on AgNPs-G. Also, it is possible that the presence of the glucose coating interferes in the interaction between EDTA and silver, avoiding the impact of EDTA on the dispersion/dissolution of silver nanoparticles. However, this is a pilot in vitro experiment. Further investigations are needed to understand what can happen on large scale.

Knowledge gaps remain because of the enormous number of nanomaterials (in terms of shape, size, materials coatings, etc.) and the scarce possibility of drawing generalised conclusions.

Microalgal biomass has been applied as a simple and effective alternative to remove heavy metals from aquatic environments. The capacity to adsorb/absorb and accumulate heavy metals in microalgal cells depends on many biotic factors, in particular, the cell density and how algal cells are pretreated before use. The application of suspended algal systems can be limited by the difficulty of removing algae from wastewater after the treatment. The effectiveness of microalgal cells to remove heavy metals can be further enhanced by immobilisation which eliminates the necessity for separating the cells from treated wastewater [71].

Furthermore, the various knowledge gaps are also related to the assessment of functionalised coating toxicity and NPs are still lacking the introduction of a safe approach concept for the use of nanomaterials. Further evaluation is needed to provide suitable methods and procedures to overcome the existing gaps that need to be addressed for the design and production of eco-safe NMs to ensure at the same time marine ecosystem sustainability and remediation.

Our results indicated that exposure to AgNPs-G of *C. vulgaris* caused significant bioaccumulation of nanoparticles and a consequent reduction of microalgae growth and chlorophyll-a content. The internalised NPs were not released back into the medium, even after 1 week, and did not undergo biotransformation since AgNPs-G maintained their crystalline nature. It should be considered that we used an NPs amount that is several times more than the AgNPs released into water. *C. vulgaris* was able to efficiently internalise the AgNPs inside vacuoles and to avoid any volunteer leakages of particles or massive discharge back to the medium for cell disruptions. This bioaccumulation ability of *C. vulgaris* for AgNPs should be taken into consideration for environmental safety and further investigated.

Author Contributions: S.M. contributed to perform experiments, to the analysis of the results and to the writing of the manuscript; E.P. contributed to the supervision and the writing of the manuscript; M.D.I. contributed to the writing of the manuscript; N.V. contributed to the critical revision of the results and L.D. contributed to the design of the research, to the critical revision of the results and to the writing of the manuscript. All authors have read and agreed to the published version of the manuscript.

Funding: Part of this work was supported by the COST Action ES1205 "ENTER—The transfer of engineered nanomaterials from wastewater treatment and stormwater to rivers".

Conflicts of Interest: The authors claim that they have no affiliations with or involvement in any organisation or entity with any financial interest (such as honoraria; educational grants; participation in speakers' bureaus; membership, employment, consultancies, stock ownership, or other equity interest; and expert testimony or patent-licensing arrangements), or non-financial interest (such as personal or professional relationships, affiliations, knowledge or beliefs) in the subject matter or materials discussed in this manuscript.

References

1. Khan, I.; Khalid, S.; Idrees, K. Nanoparticles: Properties, applications and toxicities. *Arab. J. Chem.* **2019**, *12*, 908–931. [CrossRef]
2. Pietroiusti, A.; Stockmann-Juvala, H.; Lucaroni, F.; Savolainen, K. Nanomaterial exposure, toxicity, and impact on human health. *Wiley Interdiscip. Rev. Nanomed. Nanobiotechnol.* **2018**. [CrossRef]
3. Kreyling, W.G.; Semmler-Behnke, M.; Seitz, J.; Szymczak, W.; Wenk, A.; Mayer, P.; Oberdörster, G. Size dependence of the translocation of inhaled irid-ium and carbon nanoparticle aggregates from the lung of rats to the blood and secondary target organs. *Inhal. Toxicol.* **2009**, *21*, 55–60. [CrossRef]
4. Saber, A.T.; Lamson, J.S.; Jacobsen, N.R.; Ravn-Haren, G.; Hougaard, K.S.; Nyendi, A.N.; Wahlberg, P.; Madsen, A.M.; Jackson, P.; Wallin, H.; et al. Particle-induced pulmonary acute phase response correlates with neutrophil influx linking inhaled particles and cardiovascular risk. *PLoS ONE* **2013**, *8*, e69020. [CrossRef]
5. Kinaret, P.; Ilves, M.; Fortino, V.; Rydman, E.; Karisola, P.; Lähde, A.; Koivisto, J.; Jokiniemi, J.; Wolff, H.; Savolainen, K.; et al. Inhalation and oropharyngeal aspiration exposure to rod-like carbon nanotubes induce similar airway inflammation and biological responses in mouse lungs. *ACS Nano* **2017**, *11*, 291–303. [CrossRef] [PubMed]
6. Mercer, R.R.; Scabilloni, J.F.; Hubbs, A.F.; Battelli, L.A.; McKinney, W.; Friend, S.; Wolfarth, M.G.; Andrew, M.; Castranova, V.; Porter, D.W. Distribution and fibrotic response following inhalation exposure to multi-walled carbon nanotubes. *Part. Fibre Toxicol.* **2013**, *10*, 33. [CrossRef]
7. Rossi, E.M.; Pylkkänen, L.; Koivisto, A.J.; Nykäsenoja, H.; Wolff, H.; Savolainen, K.; Alenius, H. Inhalation exposure to nanosized and fine TiO_2 particles inhibits features of allergic asthma in a murine model. *Part. Fibre Toxicol.* **2010**, *7*, 35. [CrossRef]
8. Meesters, J.; Peijnenburg, W.; Hendriks, J.; van de Meent, D.; Quik, J. A Model Sensitivity Analysis to Determine the Most Important Physicochemical Properties Driving Environmental Fate and Exposure of Engineered Nanoparticles. *Environ. Sci. Nano.* **2019**, *6*, 2049–2060. [CrossRef]
9. Xiaojia, H.; Peter, F.; Aker, W.G.; Hwang, H.M. Toxicity of engineered nanomaterials mediated by nano–bio–eco interactions. *J. Environ. Sci. Health Part. C* **2018**, *36*, 1–22.
10. Panzarini, E.; Tenuzzo, B.; Vergallo, C.; Dini, L. Biological systems interact with Engineered NanoMaterials (ENMs): Possible environmental risks. *Il Nuovo Cimento. C* **2013**, *36*, 111–116.
11. Quigg, A.; Chin, W.C.; Chen, C.S.; Zhang, S.; Jiang, Y.; Miao, A.J.; Schwehr, K.A.; Xu, C.; Santschi, P.H. Direct and indirect toxic effects of engineered nanoparticles on algae: Role of natural organic matter. *ACS Sustain. Chem. Eng.* **2013**, *1*, 686–702. [CrossRef]
12. Burduşel, A.C.; Gherasim, O.; Grumezescu, A.; Mogoantă, L.; Ficai, A.; Andronescu, E. Biomedical applications of silver nanoparticles: An up-to-date overview. *Nanomaterials* **2018**, *8*, 681. [CrossRef]
13. Siddiqi, K.S.; Husen, A.; Rao, R.A. A review on biosynthesis of silver nanoparticles and their biocidal properties. *J. Nanobiotechnol.* **2018**, *16*, 14. [CrossRef] [PubMed]
14. Piccinno, F.; Gottschalk, F.; Seeger, S.; Nowack, B. Industrial production quantities and uses of ten engineered nanomaterials in Europe and the world. *J. Nanoparticle Res.* **2012**, *14*, 1109. [CrossRef]
15. Benn, T.M.; Westerhoff, P. Nanoparticle silver released into water from commercially available sock fabrics. *Environ. Sci. Technol.* **2008**, *42*, 4133–4139. [CrossRef] [PubMed]
16. Egorova, E.M.; Kaba, S.I. The effect of surfactant micellization on the cytotoxicity of silver nanoparticles stabilized with aerosol-OT. *Toxicol. Vitr.* **2019**, *57*, 244–254. [CrossRef]

17. Spielman-Sun, E.; Zaikova, T.; Dankovich, T.; Yun, J.; Ryan, M.; Hutchison, J.E.; Lowry, G.V. Effect of silver concentration and chemical transformations on release and antibacterial efficacy in silver-containing textiles. *Nanoimpacticle* **2018**, *11*, 51–57. [CrossRef]

18. Brar, S.K.; Verma, M.; Tyagi, R.D.; Surampalli, R.Y. Engineered nanoparticles in wastewater and wastewater sludge-evidence and impacts. *Waste Manag.* **2010**, *30*, 504–520. [CrossRef]

19. Shafer, M.M.; Hoffmann, S.R.; Overdier, J.T.; Armstrong, D.E. Physical and kinetic speciation of copper and zinc in three geochemically contrasting marine estuaries. *Environ. Sci. Technol.* **2004**, *38*, 3810–3819. [CrossRef]

20. Kaegi, R.; Voegelin, A.; Sinnet, B.; Zuleeg, S.; Hagendorfer, H.; Burkhardt, M.; Siegrist, H. Behavior of metallic silver nanoparticles in a pilot wastewater treatment plant. *Environ. Sci. Technol.* **2011**, *45*, 3902–3908. [CrossRef]

21. Kleiven, M.; Macken, A.; Oughton, D.H. Growth inhibition in Raphidocelis subcapita-Evidence of nanospecific toxicity of silver nanoparticles. *Chemosphere* **2019**, *221*, 785–792. [CrossRef]

22. Burchardt, A.D.; Carvalho, R.N.; Valente, A.; Nativo, P.; Gilliland, D.; Garcìa, C.P.; Passarella, R.; Pedroni, V.; Rossi, F.; Lettieri, T. Effects of silver nanoparticles in diatom Thalassiosira pseudonana and cyanobacterium Synechococcus sp. *Environ. Sci. Technol.* **2012**, *46*, 11336–11344. [CrossRef]

23. Oukarroum, A.; Barhoumi, L.; Pirastru, L.; Dewez, D. Silver nanoparticle toxicity effect on growth and cellular viability of the aquatic plant Lemna gibba. *Environ. Toxicol. Chem.* **2013**, *32*, 902–907. [CrossRef]

24. He, D.; Dorantes-Aranda, J.J.; Waite, T.D. Silver nanoparticle-algae interactions: Oxidative dissolution, reactive oxygen species generation and synergistic toxic effects. *Environ. Sci Technol.* **2012**, *46*, 8731–8738. [CrossRef]

25. von Moos, N.; Maillard, L.; Slaveykova, V.I. Dynamics of sub-lethal effects of nano-CuO on the microalga Chlamydomonas reinhardtii during short-term exposure. *Aquat. Toxicol.* **2015**, *161*, 267–275. [CrossRef]

26. Xiang, L.; Fang, J.; Cheng, H. Toxicity of silver nanoparticles to green algae M. aeruginosa and alleviation by organic matter. *Environ. Monit. Assess.* **2018**, *190*, 667. [CrossRef] [PubMed]

27. Dash, A.; Singh, A.P.; Chaudhary, B.R.; Singh, S.K.; Dash, D. Effect of silver nanoparticles on growth of eukaryotic green algae. *Nano-Micro Lett.* **2012**, *4*, 158–165. [CrossRef]

28. Chokshi, K.; Pancha, I.; Ghosh, T.; Paliwal, C.; Maurya, R.; Ghosh, A.; Mishra, S. Green synthesis, characterization and antioxidant potential of silver nanoparticles biosynthesized from de-oiled biomass of thermotolerant oleaginous microalgae Acutodesmus dimorphus. *RSC Adv.* **2016**, *6*, 72269–72274. [CrossRef]

29. Navarro, E.; Wagner, B.; Odzak, N.; Sigg, L.; Behra, R. Effects of differently coated silver nanoparticles on the photosynthesis of Chlamydomonas reinhardtii. *Environ. Sci. Technol.* **2015**, *49*, 8041–8047. [CrossRef]

30. Lekamge, S.; Miranda, A.F.; Ball, A.S.; Shukla, R.; Nugegoda, D. The toxicity of coated silver nanoparticles to *Daphnia carinata* and trophic transfer from alga *Raphidocelis subcapitata*. *PLoS ONE* **2019**, *14*, e0214398. [CrossRef]

31. Yang, X.Y.; Gondikas, A.P.; Marinakos, S.M.; Auffan, M.; Liu, J.; Hsu-Kim, H.; Meyer, J.N. Mechanism of Silver Nanoparticle Toxicity Is Dependent on Dissolved Silver and Surface Coating in Caenorhabditis elegans. *Environ. Sci. Technol.* **2012**, *46*, 1119–1127. [CrossRef] [PubMed]

32. Panacek, A.; Kvitek, L.; Prucek, R.; Kolar, M.; Vecerova, R.; Pizurova, N.; Sharma, V.K.; Nevecna, T.; Zboril, R. Silver colloid nanoparticles: Synthesis, characterization, and their antibacterial activity. *J. Phys. Chem. B* **2006**, *110*, 16248–16253. [CrossRef] [PubMed]

33. de Marco, B.A.; Rechelo, B.S.; Tótoli, E.G.; Kogawa, A.C.; Salgado, H.R.N. Evolution of green chemistry and its multidimensional impacts: A review. *Saudi Pharm. J.* **2019**, *27*, 1–8. [CrossRef] [PubMed]

34. Padil, V.V.T.; Wacławek, S.; Černík, M. Green synthesis: Nanoparticles and nanofibres based on tree gums for environmental applications. *Ecol. Chem. Eng. S* **2016**, *23*, 533–557. [CrossRef]

35. Cinelli, M.; Coles, S.R.; Nadagouda, M.N.; Błaszczyński, J.; Słowiński, R.; Rajender, S.V.; Kirwan, K. A green chemistry-based classification model for the synthesis of silver nanoparticles. *Green Chem.* **2015**, *17*, 2825–2839. [CrossRef]

36. Vinod, V.T.; Saravanan, P.; Sreedhar, B.; Devi, D.K.; Sashidhar, R.B. A facile synthesis and characterization of Ag, Au and Pt nanoparticles using a natural hydrocolloid gum kondagogu (Cochlospermum gossypium). *Colloids Surf. B Biointerfaces* **2011**, *83*, 291–298. [CrossRef]

37. Carolin, C.F.; Kumar, P.S.; Saravanan, A.; Joshiba, G.J.; Naushad, M. Efficient techniques for the removal of toxic heavy metals from aquatic environment: A review. *J. Environ. Chem. Eng.* **2017**, *5*, 2782–2799. [CrossRef]

38. Mishra, R.K.; Sharma, V. Biotic Strategies for Toxic Heavy Metal Decontamination. *Recent Pat Biotechnol.* **2017**, *11*, 218–228. [CrossRef]

39. Rath, B. Microalgal bioremediation: Current practices and perspectives. *J. Biochem. Technol.* **2012**, *3*, 299–304.

40. Gong, N.; Shao, K.; Feng, W.; Lin, Z.; Liang, C.; Sun, Y. Biotoxicity of nickel oxide nanoparticles and bio-remediation by microalgae *Chlorella vulgaris*. *Chemosphere* **2011**, *83*, 510–516. [CrossRef]

41. Panzarini, E.; Mariano, S.; Dini, L. Glycans coated silver nanoparticles induces autophagy and necrosis in HeLa cells. In *AIP Conference Proceedings*; Rossi, M., Dini, L., Passeri, D., Terranova, M.L., Eds.; AIP Publishing LLC: Melville, NY, USA, 2015; Volume 1667, p. 020017. [CrossRef]

42. Panzarini, E.; Mariano, S.; Vergallo, C.; Carata, E.; Fimia, G.M.; Mura, F.; Rossi, M.; Vergaro, V.; Ciccarella, G.; Corazzari, M.; et al. Glucose capped silver nanoparticles induce cell cycle arrest in HeLa cells. *Toxicol. Vitr.* **2017**, *41*, 64–74. [CrossRef]

43. Manno, D.; Serra, A.; Buccolieri, A.; Panzarini, E.; Carata, E.; Tenuzzo, B.; Izzo, D.; Vergallo, C.; Rossi, M.; Dini, L. Silver and carbon nanoparticles toxicity in sea urchin Paracentrotus lividus embryos. *Bio. Nano. Mater.* **2013**, *14*, 229–238.

44. Amenorfenyo, D.K.; Huang, X.; Zhang, Y.; Zeng, Q.; Zhang, N.; Ren, J.; Huang, Q. Microalgae brewery wastewater treatment: Potentials, benefits and the challenges. *Int. J. Environ. Res. Pub. He.* **2019**, *16*, 1910. [CrossRef]

45. Nichols, H.W. Growth media-freshwater. In *Handbook of Phycological Methods. Culture Methods and Growth Measurements*; Stein, J.R., Ed.; Cambridge University Press: New York, NY, USA, 1973; pp. 7–24.

46. OECD. No. 201, freshwater algae and cyanobacteria, growth inhibition test. In *OECD Guideline for Testing of Chemicals*; OECD: Paris, France, 2011. [CrossRef]

47. Arnon, D.I. Copper enzymes in isolated chloroplasts polyphenol oxidase in Beta vulgaris. *Plant. Physiol.* **1949**, *24*, 1–15. [CrossRef] [PubMed]

48. Zhang, W.; Li, Y.; Niu, J.; Chen, Y. Photogeneration of reactive oxygen species on uncoated silver, gold, nickel, and silicon nanoparticles and their antibacterial effects. *Langmuir* **2013**, *29*, 4647–4651. [CrossRef]

49. Tkalec, M.; Štefanić, P.P.; Balen, B. Phytotoxicity of silver nanoparticles and defence mechanisms. Analysis, Fate, and Toxicity of Engineered Nanomaterials. In *Comprehensive Analytical Chemistry*; Sandeep, K.V., Ashok, K.D., Eds.; Elsevier: Amsterdam, The Netherlands, 2019; Volume 84, pp. 145–148.

50. Navarro, E.; Piccapietra, F.; Wagner, B.; Marconi, F.; Kaegi, R.; Odzak, N.; Sigg, L.; Behra, R. Toxicity of silver nanoparticles to Chlamydomonas reinhardtii. *Environ. Sci. Technol.* **2008**, *42*, 8959–8964. [CrossRef] [PubMed]

51. Leclerc, S.; Wilkinson, K.J. Bioaccumulation of Nanosilver by Chlamydomonas reinhardtii-Nanoparticle or the Free Ion? *Environ. Sci. Technol.* **2014**, *48*, 358–364. [CrossRef]

52. Turner, A.; Brice, D.; Brown, M.T. Interactions of silver nanoparticles with the marine macroalga, Ulva lactuca. *Ecotoxicology* **2012**, *21*, 148–154. [CrossRef]

53. Tejamaya, M.; Römer, I.; Merrifield, R.C.; Lead, J.R. Stability of citrate, PVP, and PEG coated silver nanoparticles in ecotoxicology media. *Environ. Sci. Technol.* **2012**, *46*, 7011–7017. [CrossRef]

54. Rybak-Smith, M.J. Effect of surface modification on toxicity of nanoparticles. *Encycl. Nanotechnol.* **2012**, 645–652.

55. Tiede, K.; Hassellöv, M.; Breitbarth, E.; Chaudhry, Q.; Boxall, A.B. Considerations for environmental fate and ecotoxicity testing to support environmental risk assessments for engineered nanoparticles. *J. Chromatogr. A* **2009**, *1216*, 503–509. [CrossRef] [PubMed]

56. Bilberg, K.; Malte, H.; Wang, T.; Baatrup, E. Silver nanoparticles and silver nitrate cause respiratory stress in Eurasian perch (Perca fluviatilis). *Aquat. Toxicol.* **2010**, *96*, 159–165. [CrossRef] [PubMed]

57. Domingo, G.; Bracale, M.; Vannini, C. Phytotoxicity of silver nanoparticles to aquatic plants, algae, and microorganisms. In *Nanomaterials in Plants, Algae and Microorganisms*; Durgesh, K.T., Parvaiz, A., Shivesh, S., Devendra, K.C., Nawal, K.D., Eds.; Academic Press: Cambridge, MA, USA, 2019; pp. 143–168.

58. Wagner, S.; Gondikas, A.; Neubauer, E.; Hofmann, T.; von der Kammer, F. Spot the Difference: Engineered and Natural Nanoparticles in the Environment-Release, Behavior and Fate. *Angew. Chem. Int. Ed.* **2014**, *53*, 12398–12419. [CrossRef]

59. Lekamge, S.; Miranda, A.F.; Abraham, A.; Li, V.; Shukla, R.; Bansal, V.; Nugegoda, D. The toxicity of Silver Nanoparticles (AgNPs) to three freshwater invertebrates with different life strategies: *Hydra vulgaris, Daphnia carinata*, and *Paratya australiensis. Front. Environ. Sci.* **2018**, *6*, 152. [CrossRef]

60. Angel, B.M.; Batley, G.E.; Jarolimek, C.V.; Rogers, N.J. The impact of size on the fate and toxicity of nanoparticulate silver in aquatic systems. *Chemosphere* **2013**, *93*, 359–365. [CrossRef]

61. Sendra, M.; Yeste, M.P.; Gatica, J.M.; Moreno-Garrido, I.; Blasco, J. Direct and indirect effects of silver nanoparticles on freshwater and marine microalgae (Chlamydomonas reinhardtii and Phaeodactylum tricornutum). *Chemosphere* **2017**, *179*, 279–289. [CrossRef]

62. Wang, F.; Guan, W.; Xu, L.; Ding, Z.; Ma, H.; Ma, A.; Terry, N. Effects of Nanoparticles on Algae: Adsorption, Distribution, Ecotoxicity and Fate. *Appl. Sci.* **2019**, *9*, 1534. [CrossRef]

63. Hazeem, L.J.; Kuku, G.; Dewailly, E.; Slomianny, C.; Barras, A.; Hamdi, A.; Boukherroub, R.; Culha, M.; Bououdina, M. Toxicity Effect of Silver Nanoparticles on Photosynthetic Pigment Content, Growth, ROS Production and Ultrastructural Changes of Microalgae *Chlorella vulgaris. Nanomaterials* **2019**, *9*, 914. [CrossRef]

64. Zemke-White, W.L.; Clements, K.D.; Harris, P.J. Acid lysis of macroalgae by marine herbivorous fishes: Effects of acid pH on cell wall porosity. *J. Exp. Mar. Biol. Ecol.* **2000**, *245*, 57. [CrossRef]

65. Chang, Y.N.; Zhang, M.Y.; Xia, I.; Zhang, J.; Xing, G.M. The toxic effects and mechanisms of CuO and ZnO nanoparticles. *Materials* **2012**, *5*, 2850–2871. [CrossRef]

66. Li, Y.; Xiao, R.; Liu, Z.; Liang, X.; Feng, W. Cytotoxicity of NiO nanoparticles and its conversion inside *Chlorella vulgaris. Chem. Res. Chin. Univ.* **2017**, *33*, 107–111. [CrossRef]

67. Ribeiro, F.; Gallego-Urrea, J.A.; Jurkschat, K.; Crossley, A.; Hasselöv, M.; Taylor, C.; Soares, A.M.; Loureiro, S. Silver nanoparticles and silver nitrate induce high toxicity to *Pseudokirchneriella subcapitata, Daphnia magna* and *Danio rerio. Sci. Total Environ.* **2014**, *466–467*, 232–241. [CrossRef] [PubMed]

68. Oukarroum, A.; Bras, S.; Perreault, F.; Popovic, R. Inhibitory effects of silver nanoparticles in two green algae, *Chlorella vulgaris* and *Dunaliella tertiolecta. Ecotoxicol. Environ. Saf.* **2012**, *78*, 80–85. [CrossRef] [PubMed]

69. Martinez-Andrade, J.M.; Avalos-Borja, M.; Vilchis-Nestor, A.R.; Sanchez-Vargas, L.O.; Castro-Longoria, E. Dual function of EDTA with silver nanoparticles for root canal treatment-A novel modification. *PLoS ONE* **2018**, *13*, e0190866. [CrossRef]

70. Chappell, M.A.; Miller, L.F.; George, A.J.; Pettway, B.A.; Price, C.L.; Porter, B.E.; Bednar, A.J.; Seiter, J.M.; Kennedy, A.J.; Steevens, J.A. Simultaneous dispersion-dissolution behavior of concentrated silver nanoparticle suspensions in the presence of model organic solutes. *Chemosphere* **2011**, *84*, 1108–1116. [CrossRef]

71. Ahmad, A.; Bhat, A.H.; Buang, A. Enhanced biosorption of transition metals by living *Chlorella vulgaris* immobilized in Ca-alginate beads. *Environ. Technol.* **2019**, *40*, 1793–1809. [CrossRef]

Article

Spectroscopic Study of the Salicyladazine Derivative–UO$_2^{2+}$ Complex and Its Immobilization to Mesoporous Silica

Sujin Park [1], Jaehyeon Park [1], Ji Ha Lee [2,*], Myong Yong Choi [1,*] and Jong Hwa Jung [1,*]

[1] Department of Chemistry and Research Institute of Natural Sciences, Gyeongsang National University, Jinju 52828, Korea; 0165255562@naver.com (S.P.); parkjae@gnu.ac.kr (J.P.)

[2] Department of Chemistry and Biochemistry, The University of Kitakyushu, Hibikino, Kitakyushu 808-0135, Japan

* Correspondence: l-ji@kitakyu-u.ac.jp (J.H.L.); mychoi@gnu.ac.kr (M.Y.C.); jonghwa@gnu.ac.kr (J.H.J.); Tel.: +82-55-772-1488 (J.H.J.)

Received: 2 April 2019; Accepted: 26 April 2019; Published: 2 May 2019

Abstract: Uranyl ion, the most soluble toxic uranium species, is recognized as an important index for monitoring nuclear wastewater quality. The United States Environmental Protection Agency (US EPA) and the World Health Organization (WHO) prescribed 30 ppb as the allowable concentration of uranyl ion in drinking water. This paper reports on a nanohybrid material that can detect uranyl ions spectroscopically and act as a uranyl ion absorbent in an aqueous system. Compound **1**, possessing a salicyladazine core and four acetic acid groups, was synthesized and the spectroscopic properties of its UO$_2^{2+}$ complex were studied. Compound **1** had a strong blue emission when irradiated with UV light in the absence of UO$_2^{2+}$ that was quenched in the presence of UO$_2^{2+}$. According to the Job's plot, Compound **1** formed a 1:2 complex with UO$_2^{2+}$. When immobilized onto mesoporous silica, a small dose (0.3 wt %) of this hybrid material could remove 96% of UO$_2^{2+}$ from 1 mL of a 100-ppb UO$_2^{2+}$ aqueous solution.

Keywords: UO$_2^{2+}$; salicyladazine; fluorescence; mesoporous silica

---

1. Introduction

The development of nuclear technology leads to new environmental concerns, such as radiation exposure and accidents resulting therefrom. Of special concern is the uranyl cation (UO$_2^{2+}$), a highly toxic neurotoxin that is very mutable in biological systems and can cause radioactive poisoning if proper containment rules are violated [1–4]. Therefore, the development of technologies that can measure the exact amount of UO$_2^{2+}$ exposed to the environment is an important safety priority.

Several studies on UO$_2^{2+}$ sensors have been reported to date [5–10]. L. S. Natrajan et al. reported a method for detecting UO$_2^{2+}$ via a unique fluorescence energy transfer process to a water-soluble europium (III) lanthanide complex triggered by UO$_2^{2+}$ [11]. Yi Lu et al. developed colorimetric uranium sensors based on a UO$_2^{2+}$-specific DNAzyme and gold nanoparticles using both labeled and label-free methods [12]. Julius Rebek Jr. et al. investigated a tripodal receptor capable of extracting uranyl ion from aqueous solutions. In their system, at a uranyl concentration of 400 ppm, the developed ligand extracted approximately 59% of the UO$_2^{2+}$ into the organic phase [13].

While various detection methods for UO$_2^{2+}$ have been developed based on fluorogenic and colorimetric methods, studies on UO$_2^{2+}$ adsorbents have received much less attention. We aim to synthesize an adsorbent, or uranyl-capture agent based on an organic–inorganic hybrid material because such compounds tend to have higher stability and controllable homogenous pore sizes. Therefore, we designed and synthesized compound **1** (Figure 1). Compound **1** possesses four acetic acid groups as ligands for UO_2^{2+}. The spectroscopic properties of compound **1** were observed upon adding UO_2^{2+} via fluorometry, IR spectroscopy, and ^{1}H NMR spectroscopy. Furthermore, compound **1** was immobilized to **MPS** (mesoporous silica nanoparticles) to create an adsorbent for UO_2^{2+}. Herein, we report the spectroscopic properties of the compound **1**–UO_2^{2+} complex and the adsorption capacity of mesoporous silica nanoparticles loaded with compound **1** for UO_2^{2+} capture.

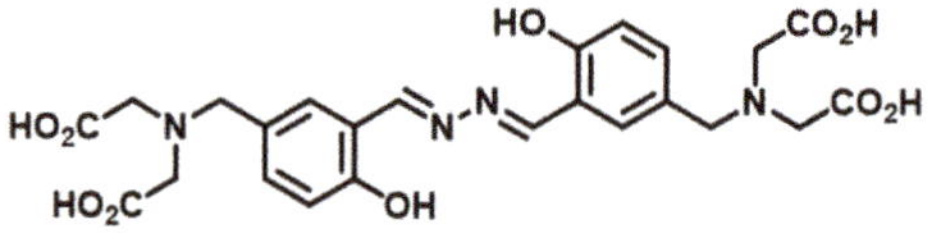

Figure 1. Chemical structure of compound **1**.

2. Materials and Methods

2.1. Reagents and Instruments

All reagents were purchased from Sigma-Aldrich (Buchs, Switzerland) and Tokyo chemical industry (Fukaya, Japan). The solvent was purchased from Samchun Pure Chemicals (Pyeongkaek, Korea) and used with further purification. ^{1}H and ^{13}C NMR spectra were obtained with a Bruker DRX 300 apparatus (Rheinstetten, Germany). The IR spectra were measured on a Shimadzu FT-IR 8400S instrument (Kyoto, Japan) by KBr pellet method in the range of $4000-1000$ cm^{-1}. A JEOL JMS-700 mass spectrometer (Kyoto, Japan) was used to obtain the mass spectra. The UV–vis absorption and fluorescence spectra were obtained at 298 K with a Thermo Evolution 600 spectrophotometer (Waltham, MA, USA) and a RF-5301PC spectrophotometer (Kyoto, Japan), respectively. A PerkinElmer 2400 series (Waltham, MA, USA) was employed for the elemental analyses. The quantitative analysis was performed using ICP-DRC-MS (ELAN DRC II, PerkinElmer, Waltham, MA, USA). The morphological images were observed using a TEM (TECNAI G2 F30, FEI, Hillsboro, OR, USA).

2.2. Synthesis of Compound 1

The compounds **1–4** were synthesized according to the reported method (Scheme S1) [14,15]. Compound **2** (2.56 mg, 0.4 mmol) was dissolved in THF (10 mL), followed by the addition of sodium hydroxide solution (10 mL, 0.8 M). The mixture was stirred at room temperature for 2 h. After completion of the reaction, the mixture was added to aq HCl solution (1 wt %) to give yellow precipitation, which was filtered off and dried under vacuum to yield compound **1** (121 mg, 57%). IR (KBr pellet) cm^{-1} 3424, 3014, 1728, 1495 1629, 1389, 1277, 1164; ^{1}H NMR (300 MHz, DMSO-d_6) δ 12.25 (s, 4H), 11.07 (s, 2H), 8.98 (s, 2H), 7.63 (d, J = 2.2 Hz, 2H), 7.40 (dd, J = 8.5, 2.2 Hz, 2H), 6.96 (d, J = 8.4 Hz, 2H), 3.78 (s, 4H), 3.42 (s, 8H); ^{13}C NMR (75 MHz, DMSO-d_6) δ 172.74, 162.82, 158.30, 134.37, 131.08, 130.12, 118.33, 116.99, 56.80, 53.96.; ESI-MS: calculated for $C_{24}H_{26}N_4O_{10}$, $[M - H]^-$ 529.16; found, 529.15; Anal. calcd for $C_{24}H_{26}N_4O_{10}$: C, 54.34; H, 4.94; N, 10.56; found: C 54.31, H 4.97, N 10.51.

2.3. Preparation of MPS

The **MPS** was synthesized according to the reported method [16]. 8.2 g of (1-Hexadecyl) trimethyl-ammonium bromide was dissolved in H_2O (600 mL). After stirring for 10 min, 1.6 mL of triethanolamine was added and the reaction mixture was heated. When the reaction temperature reached 80 °C, TEOS (tetra ethyl ortho silicate) (60 mL) was added and stirred for 1 h. The solvent of the reaction was removed by rotary evaporator and the resulting solid (including some water) was heated at 500 °C for 5 h by using a furnace.

*2.4. Preparation of **MPS-1***

0.1 g of compound **1** and 1 g of **MPS** in of acetonitrile (20 mL) were stirred for 10 min. The reaction mixture was refluxed at 80 °C for 24 h. After the reaction mixture was cooled to room temperature, the solid product was filtered and washed with 200 mL acetonitrile.

2.5. Photophysical Studies

The UV−vis absorption and fluorescence spectra were determined over the range 200–800 nm. The samples were prepared by dispersion in H_2O solution. The concentration of standard UO_2^{2+} solution was 100 ppb.

2.6. NMR Measurement

Compound **1** (2.65 mg, 0.005 mmol) and uranyl acetate (8.48 mg, 0.02 mmol) were dissolved in 0.5 mL and 0.1 mL of of DMSO-d_6 (0.5 mL) in DMSO-d_6 (0.1 mL), respectively. To NMR titration, the different amount of the uranyl acetate solution (12.5 μL, 25 μL, 37.5 μL, 50 μL) was added to compound **1** of DMSO-d_6 (0.5 mL).

2.7. ICP-MS

MPS-1 (1 mg, 3 mg, and 5 mg) were dispersed in aqueous solution containing UO_2^{2+}, Na^+, Mg^{2+}, Ca^{2+}, Cu^{2+}, Ag^+, Co^{2+}, Ni^{2+}, Mn^{2+} and Pb^{2+} (100 ppb) for 10 min. Mixture was added to H_2O (4 mL). The mixture solution was centrifuged and the supernatant solution was filtrated with syringe filter (PTFE, 0.45 μm). The collected solution was measured 3 times by ICP-MS.

3. Results and Discussion

*3.1. Spectroscopic Properties of Complex **1** with UO_2^{2+}*

Binding between UO_2^{2+} and specific ligands, such as cyclic peptides [17,18], porphyrins [19,20], and naphthobipyrrole [21,22], is well known. We prepared a salicyladazine derivative as a ligand for UO_2^{2+}. The salicyladazine derivative was synthesized starting from 2-hydroxybenzaldehyde. The diethyl 2,2′-azanediyldiacetate groups were designed on both sides of the compound to create a symmetric structure. As the final step, hydrochloric acid treatment of the precursor yielded the desired compound **1**. Compound **1** contains the acetic acid end group ($–CH_2COOH$) to form the binding site for UO_2^{2+} and was characterized via FT IR, ^1H and ^{13}C NMR, mass spectrometry, and elemental analysis (Figures S1, S2 and S3 in Supplementary Materials).

UV−vis spectroscopy was performed to confirm that a coordination bond between compound **1** and UO_2^{2+} resulted in a colorimetric change. Compound **1** was dissolved in an aqueous solution containing 1% of DMSO, and the UV−vis absorption spectrum was measured (Figure 2A). Before adding UO_2^{2+}, the π–π * absorption band of compound **1** appeared at around 300–350 nm. When UO_2^{2+} (from 0.5 to 3 equivalents in 0.5 equivalent steps) was added to compound **1**, the absorption peak intensities at 300 and 350 nm decreased until up to 2 equivalents of UO_2^{2+} were added. At 2.5 or more equivalents of UO_2^{2+}, the ligand-to-metal charge transfer absorption wavelength between compound **1** and UO_2^{2+} was observed at around 365–380 nm [23].

Figure 2B shows the photographs of the cuvettes used when UO_2^{2+} was added to compound **1** and irradiated under UV light. The change in fluorescence after more than 2 equivalents of UO_2^{2+} were added was visible to the naked eye. Compound **1** yielded an emission wavelength around 560 nm (excitation = 365 nm). When 0.5 equivalents of UO_2^{2+} were added to compound **1**, the fluorescence intensity decreased. The decrease in fluorescence was noticeable from 0.5 to 2 equivalents of UO_2^{2+}; however, the fluorescence intensity remained constant thereafter (Figure 2C). Figure 2D presents a plot of the fluorescence intensity vs. amount of UO_2^{2+} added. To determine the stoichiometric ratio between compound **1** and UO_2^{2+}, we constructed a Job's plot using the fluorescence data and found a 1:2 binding ratio for compound **1**:UO_2^{2+} (Figure S4).

To investigate the chemical interactions between compound **1** and UO_2^{2+}, we used nuclear magnetic resonance (NMR) spectroscopy. We measured the ^{1}H NMR signal of compound **1** in DMSO-d_6 while increasing the UO_2^{2+} content (Figure 3). When 0.5 equivalent of UO_2^{2+} was added to compound **1** in DMSO-d_6, the proton peak (aromatic OH: 11.08 ppm) of compound **1** decreased and multiple new peaks were observed. When we added 1 equivalent of UO_2^{2+}, the ratio of the proton peaks of compound **1** and the resulting complex was 1:1. Because compound **1** has C2 symmetry, coordination of UO_2^{2+} occurs on one side of compound **1** (bound to two acetic acid ligands) and no coordination occurs on the other side of compound **1**. Upon additional UO_2^{2+} input (over 2 equivalents), all the peaks representing free compound **1** disappeared entirely. Therefore, compound **1** has a binding capacity of two UO_2^{2+} molecules in DMSO-d_6 (forms a 1:2 complex). This result is consistent with the Job's plot.

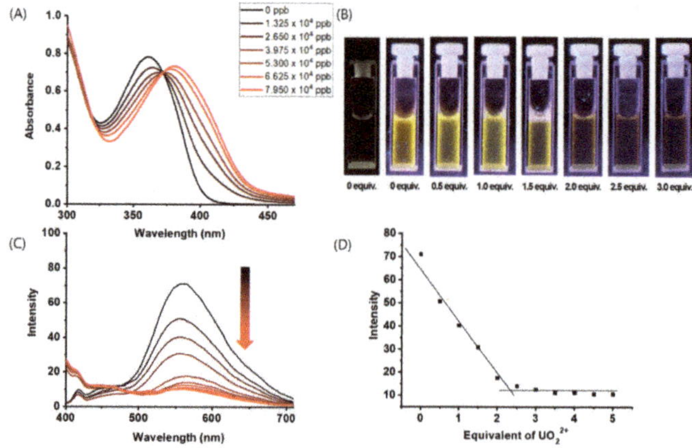

Figure 2. (**A**) UV–vis spectra of compound **1** (2.65×10^4 ppb) in DMSO/H_2O (1:99 *v/v*) containing various amounts of UO_2^{2+}. (**B**) Photographed cuvettes from (**A**) under UV light irradiation. (**C**) Fluorescence spectra of compound **1** (2.65×10^4 ppb) in DMSO/H_2O (1:99 *v/v*) containing various amounts of UO_2^{2+} (0–5 equivalents). (**D**) Plot of fluorescence intensity of (**C**) vs. amount of UO_2^{2+}.

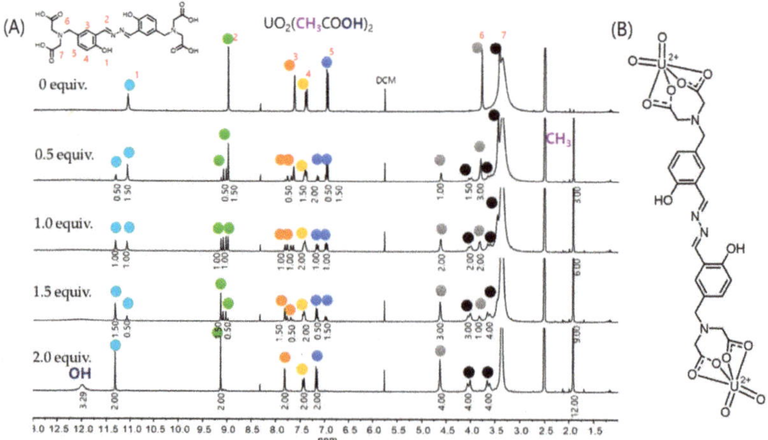

Figure 3. (**A**) ^{1}H nuclear magnetic resonance (NMR) spectra of compound **1** (10 mM) in DMSO-d_6 containing various equivalents of uranyl acetate. (**B**) Proposed structure of complex **1** with UO_2^{2+}.

The carbonyl oxygen (C=O) of the end group of acetic acids (–CH$_2$COOH) in compound **1** is well known to bind strongly to a radionuclide ion such as UO$_2$$^{2+}$ by supplying electrons [23,24]. IR spectroscopy was used to classify the complex formation of compound **1** with UO$_2$$^{2+}$. As shown in Figure S5, the oxygen of the carboxylic acid carbonyl (C=O) in compound **1** before the addition of UO$_2$$^{2+}$ produced a peak at 1728 cm^{-1}, whereas the C=O peak after the addition of UO$_2$$^{2+}$ was shifted to 1557 cm^{-1}. This shift to a lower wavenumber is indicative of the C=O in compound **1** providing electrons in a dative bond to UO$_2$$^{2+}$ and confirms that the carboxylic acid groups are employed in complex formation with UO$_2$$^{2+}$.

3.2. Immobilization of Compound **1** to Mesoporous Silica Nanoparticles

The morphology of mesoporous silica nanoparticle immobilized with **1** (**MPS-1**) was observed via transmission electron microscopy (TEM). The TEM image of **MPS-1** revealed a spherical structure with a narrow size distribution (circa 40 nm) (Figure 4A). Thermogravimetry analysis (TGA) was performed to determine the amount of compound **1** immobilized onto **MPS-1** (Figure 4B). At approximately 150 °C, the weight of **MPS-1** decreased by 4.2%. This mass reduction was attributed to moisture. At ~500 °C, compound **1** was pyrolyzed and the weight of **MPS-1** decreased to 86.8% (Figure S6). Thus, the amount of compound **1** introduced into **MPS-1** was 9% by weight (Figure 4B). Figure S7 presents the IR spectra of **MPS** and **MPS-1**; a C-H vibration peak of 2940 cm^{-1} was confirmed. This further supported the presence of compound **1** on the surface of the mesoporous silica nanoparticles. Under UV light irradiation, the filtered silica nanoparticles fluoresced blue, indicating that compound **1** was present on the mesoporous silica nanoparticle surface. Fluorescence spectra of **MPS-1** (2 mg) in water (2 mL) and **MPS-1** (2 mg) in 100 ppb UO$_2$$^{2+}$ solution (2 mL) were also measured. (Figure 4C). In 3.5% NaCl aqueous solution of 2mL, we measured the fluorescence spectra of **MPS-1** (2 mg) and **MPS-1** (2 mg) in 100 ppb UO$_2$$^{2+}$ solution, respectively (Figure S8). Fluorescence changes of **MPS-1** in the present of 100 ppb UO$_2$$^{2+}$ in water or 3.5% NaCl aqueous solution were shown similar results. This means that **MPS-1** can bind UO$_2$$^{2+}$ not only in water but also NaCl aqueous solution.

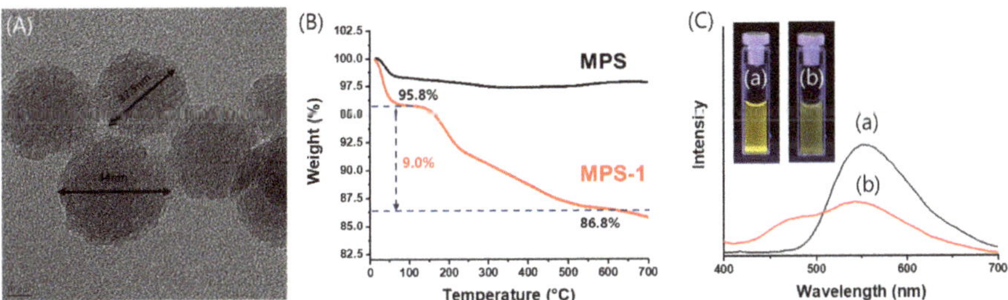

Figure 4. (**A**) Transmission electron microscopy (TEM) image of mesoporous silica (**MPS-1**) and (**B**) thermogravimetry analysis (TGA) thermogram of **MPS** (black) and **MPS-1** (red). (**C**) The photograph and the Fluorescence spectra of (**a**) **MPS-1** (2 mg) in water (2 mL) and (**b**) **MPS-1** (2 mg) in 100 ppb UO$_2$$^{2+}$ solution (2 mL).

3.3. The Adsorption Capacity of **MPS-1** for UO$_2$$^{2+}$

The United States Environmental Protection Agency (US EPA) and the World Health Organization (WHO) have prescribed safe limits of UO$_2$$^{2+}$ in drinking water at 30 ppb [25,26]. The adsorption capacity of **MPS-1** was tested by adding 1, 3, and 5 mg of **MPS-1** to 1 mL of 100 ppb UO$_2$$^{2+}$ solution. After standing for 10 min, the solution was filtered through a 0.45-μm syringe filler and the UO$_2$$^{2+}$ levels were determined by ICP-MS (experiment performed in triplicate). Calibration curves were obtained with dilute UO$_2$$^{2+}$ solutions (0.1, 1, 10, 50, and 100 ng/L), and the linearity of the calibration curve was confirmed (correlation coefficient was 0.9974) (Figure S9). Figure 5 and Table S1 present the results of the UO$_2$$^{2+}$ adsorption experiment using various amounts of **MPS-1**. The percentage of

UO_2^{2+} removed was 70%, 96%, and 95% for 1, 3, and 5 mg of **MPS-1**, respectively (RSD values all less than 15%). 1 mg of **MPS-1** was not sufficient for absorbing 100 ppb UO_2^{2+}. Using 3 or 5 mg of **MPS-1** removed 95% or more of the UO_2^{2+}; there was no statistically significant difference between the two absorbent dose amounts. In conclusion, **MPS-1** (even a small amount: 0.3 wt %) could reduce 100 ppb of UO_2^{2+} <5 ppb of UO_2^{2+}; this result would satisfy both EPA and WHO drinking water standards.

3.4. Adsorption of UO_2^{2+} and Other Cations onto **MPS-1**

The elemental analysis of **MPS**, **MPS-1**, and UO_2^{2+}-adsorbed **MPS-1** was performed via TEM dispersive X-ray spectroscopy (EDX) (Figure S10). Nitrogen was observed in the EDX spectrum of **MPS-1**, thus providing evidence for the presence of compound **1** in **MPS-1**. In the EDX spectrum of UO_2^{2+}-adsorbed **MPS-1**, uranium was detected. This confirms our rationale in designing this adsorbent: UO_2^{2+} was bound to the compound **1** attached onto the surface of **MPS**.

We also confirmed the adsorption capacity of **MPS-1** (5 mg) for other metal ions, such as Na^+, Mg^{2+}, Ca^{2+}, Cu^{2+}, Ag^+, Co^{2+}, Ni^{2+}, Mn^{2+}, and Pb^{2+} (100 ppb) under the same conditions. Among the metal ions tested, 42.3% of Ca^{2+} was adsorbed onto the surface of **MPS-1**; for the remaining metal ions, <30% were adsorbed (Table S2). These findings suggest that **MPS-1** would be useful as an adsorbent for UO_2^{2+}.

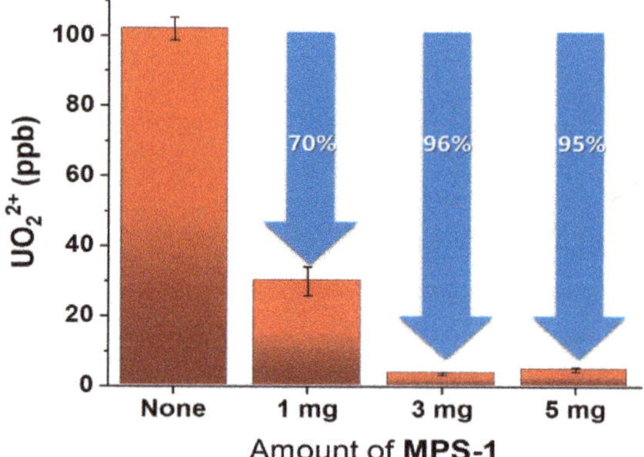

Figure 5. Result of UO_2^{2+} adsorption of **MPS-1** in 100 ppb UO_2^{2+} solution.

4. Conclusions

We synthesized the salicyladazine-based compound **1**, designed to be a uranyl ion capture ligand. Compound **1** formed a 1:2 complex with UO_2^{2+} as confirmed by the Job's plot. A fluorescence change was observed when UO_2^{2+} was bound to compound **1**. IR and NMR measurements were performed to identify compound **1** and the two UO_2^{2+} coordination sites. Compound **1** was immobilized into mesoporous silica (**MPS-1**); the resulting sorbent could remove 96% of the UO_2^{2+} from 1 mL of a 100-ppb UO_2^{2+} aqueous solution. A material was successfully developed that was capable of simultaneously absorbing uranyl ions and detecting their presence by fluorescence. We believe that this organic–inorganic hybrid material paradigm for detecting UO_2^{2+} will have a broad impact for the study on porous materials and their application.

Supplementary Materials: The following are available online at http://www.mdpi.com/2079-4991/9/5/688/s1. Scheme S1. Synthesis route of compound **1**; Figure S1. FT-IR spectrum of compound **1**; Figure S2. ^{1}H NMR spectrum of compound **1** in DMSO-d_6; Figure S3. ^{13}C NMR spectrum of compound **1** in DMSO-d_6; Figure S4. Job's plot for complex formed between compound **1** and UO_2^{2+}; Figure S5. FT-IR spectra of **1** and **1** with UO_2^{2+}; Figure S6. TGA thermogram of compound **1**; Figure S7. FT-IR spectra of (A) **MPS** and (B) **MPS-1**; Figure S8. Fluorescence spectra of (a) **MPS-1** (2 mg) in 3.5% NaCl solution (2mL) and (b) **MPS-1** (2 mg) with UO_2^{2+} solution (100 ppb) in 3.5% NaCl (2 mL); Figure S9. Linear equation of various concentrations of UO_2^{2+}; Figure S10. TEM EDX mapping of (A) **MPS** (B) **MPS-1** and (C) **MPS-1** with UO_2^{2+}; Table S1. Adsorption Capacities of **MPS-1** for UO_2^{2+} (100 ppb) solution; Table S2. Adsorption Capacities of **MPS-1** (5 mg) with various metal ions (100 ppb) solution.

Author Contributions: Conceptualization, S.P. and J.P.; Data curation, S.P.; Formal analysis, S.P. and J.P.; Investigation, S.P. and J.P.; Methodology, J.P.; Project administration, J.H.L. and J.H.J.; Supervision, J.H.L., M.Y.C. and J.H.J.; Writing—Original draft, J.H.L.; Writing—Review & editing, J.H.L., M.Y.C. and J.H.J.

Funding: This research was supported by the NRF (2018R1A2B2003637, 2017M2B2A9A02049940 and 2017R1A4A1014595) supported by the Ministry of Education, Science and Technology, Korea. In addition, this work was partially supported by a grant from the Next-Generation BioGreen 21 Program (SSAC, Grant no. PJ013186052019), Rural Development Administration, Korea.

Conflicts of Interest: The authors declare no conflict of interest.

References

1. O'Loughlin, E.J.; Kelly, S.D.; Cook, R.E.; Csencsits, R.; Kemner, K.M. Reduction of Uranium(VI) by Mixed Iron(II)/Iron(III) Hydroxide (Green Rust): Formation of UO_2 Nanoparticles. *Environ. Sci. Technol.* **2003**, *37*, 721–727. [CrossRef] [PubMed]

2. Feng, M.-L.; Sarma, D.; Qi, X.-H.; Du, K.-Z.; Huang, X.-Y.; Kanatzidis, M.G. Efficient Removal and Recovery of Uranium by a Layered Organic–Inorganic Hybrid Thiostannate. *J. Am. Chem. Soc.* **2016**, *138*, 12578–12585. [CrossRef] [PubMed]

3. Asiabi, H.; Yamini, Y.; Shamsayei, M. Highly efficient capture and recovery of uranium by reusable layered double hydroxide intercalated with 2-mercaptoethanesulfonate. *Chem. Eng. J.* **2018**, *337*, 609–615. [CrossRef]

4. Wu, P.; Hwang, K.; Lan, T.; Lu, Y. A DNAzyme-Gold Nanoparticle Probe for Uranyl Ion in Living Cells. *J. Am. Chem. Soc.* **2013**, *135*, 5254–5257. [CrossRef]

5. Cao, X.-H.; Zhang, H.-Y.; Ma, R.-C.; Yang, Q.; Zhang, Z.-B.; Liu, Y.-H. Visual colorimetric detection of UO_2^{2+} using o-phosphorylethanolamine-functionalized gold nanoparticles. *Sens. Actuators B* **2015**, *218*, 67–72. [CrossRef]

6. Elabd, A.A.; Attia, M.S. A new thin film optical sensor for assessment of UO_2^{2+} based on the fluorescence quenching of Trimetazidine doped in sol gel matrix. *J. Lumin.* **2015**, *165*, 179–184. [CrossRef]

7. Xiao, S.J.; Zuo, J.; Zhu, Z.Q.; Ouyang, Y.Z.; Zhang, X.L.; Chen, H.W.; Zhang, L. Highly sensitive DNAzyme sensor for selective detection of trace uranium in ore and natural water samples. *Sens. Actuators B* **2015**, *210*, 656–660. [CrossRef]

8. Elabd, A.A.; Attia, M.S. Spectrofluorimetric assessment of UO_2^{2+} by the quenching of the fluorescence intensity of Clopidogrel embedded in PMMA matrix. *J. Lumin.* **2016**, *169*, 313–318. [CrossRef]

9. Zheng, S.; Wang, H.; Hu, Q.; Wang, Y.; Hu, J.; Zhou, F.; Liu, P. "Turn-On" fluorescent chemosensor based on β-diketone for detecting Th^{4+} ions in Aqueous Solution and application in living cell imaging. *Sens. Actuators B* **2017**, *253*, 766–772. [CrossRef]

10. Wen, J.; Huang, Z.; Hu, S.; Li, S.; Li, W.; Wang, X. Aggregation-induced emission active tetraphenylethene-based sensor for uranyl ion detection. *J. Hazard. Mater.* **2016**, *318*, 363–370. [CrossRef]

11. Harvey, P.; Nonat, A.; Platas-Iglesias, C.; Natrajan, L.S.; Charbonniere, L.J. Sensing Uranyl(VI) Ions by Coordination and Energy Transfer to a Luminescent Europium(III) Complex. *Angew. Chem. Int. Ed.* **2018**, *57*, 9921–9924. [CrossRef]

12. Lee, J.H.; Wang, Z.; Liu, J.; Lu, Y. Highly sensitive and selective colorimetric sensors for uranyl (UO_2^{2+}): Development and comparison of labeled and label-free DNAzyme-gold nanoparticle systems. *J. Am. Chem. Soc.* **2008**, *130*, 14217–14226. [CrossRef]

13. Sather, A.C.; Berryman, O.B.; Rebek, J. Selective recognition and extraction of the uranyl ion from aqueous solutions with a recyclable chelating resin. *Chem. Sci.* **2013**, *4*, 3601–3605. [CrossRef]

14. Zhang, S.; Sun, M.; Yan, Y.; Yu, H.; Yu, T.; Jiang, H.; Zhang, K.; Wang, S. A turn-on fluorescence probe for the selective and sensitive detection of fluoride ions. *Anal. Bioanal. Chem.* **2017**, *409*, 2075–2081. [CrossRef] [PubMed]

15. Gao, M.; Li, Y.; Chen, X.; Li, S.; Ren, L.; Tang, B.Z. Aggregation-Induced Emission Probe for Light-Up and in Situ Detection of Calcium Ions at High Concentration. *ACS Appl. Mater. Interfaces* **2018**, *10*, 14410–14417. [CrossRef]

16. Zhang, K.; Xu, L.L.; Jiang, J.G.; Calin, N.; Lam, K.F.; Zhang, S.J.; Wu, H.H.; Wu, G.D.; Albela, B.; Bonneviot, L.; et al. Facile large-scale synthesis of monodisperse mesoporous silica nanospheres with tunable pore structure. *J. Am. Chem. Soc.* **2013**, *135*, 2427–2430. [CrossRef] [PubMed]

17. Yang, C.-T.; Han, J.; Gu, M.; Liu, J.; Li, Y.; Huang, Z.; Yu, H.-Z.; Hu, S.; Wang, X. Fluorescent recognition of uranyl ions by a phosphorylated cyclic peptide. *Chem. Commun.* **2015**, *51*, 11769–11772. [CrossRef] [PubMed]

18. Starck, M.; Sisommay, N.; Laporte, F.A.; Oros, S.; Lebrun, C.; Delangle, P. Preorganized Peptide Scaffolds as Mimics of Phosphorylated Proteins Binding Sites with a High Affinity for Uranyl. *Inorg. Chem.* **2015**, *54*, 11557–11562. [CrossRef] [PubMed]

19. Sessler, J.L.; Seidel, D.; Vivian, A.E.; Lynch, V.; Scott, B.L.; Keogh, D.W. Hexaphyrin(1.0.1.0.0.0): An Expanded Porphyrin Ligand for the Actinide Cations Uranyl (UO_2^{2+}) and Neptunyl (NpO^{2+}). *Angew. Chem. Int. Ed.* **2001**, *40*, 591–594. [CrossRef]

20. Sessler, J.L.; Gorden, A.E.V.; Seidel, D.; Hannah, S.; Lynch, V.; Gordon, P.L.; Donohoe, R.J.; Drew Tait, C.; Webster Keogh, D. Characterization of the interactions between neptunyl and plutonyl cations and expanded porphyrins. *Inorg. Chim. Acta* **2002**, *341*, 54–70. [CrossRef]

21. Anguera, G.; Brewster, J.T.; Moore, M.D.; Lee, J.; Vargas-Zúñiga, G.I.; Zafar, H.; Lynch, V.M.; Sessler, J.L. Naphthylbipyrrole-Containing Amethyrin Analogue: A New Ligand for the Uranyl (UO_2^{2+}) Cation. *Inorg. Chem.* **2017**, *56*, 9409–9412. [CrossRef]

22. Brewster, J.T.; He, Q.; Anguera, G.; Moore, M.D.; Ke, X.-S.; Lynch, V.M.; Sessler, J.L. Synthesis and characterization of a dipyriamethyrin–uranyl complex. *Chem. Commun.* **2017**, *53*, 4981–4984. [CrossRef] [PubMed]

23. Marten-Ramos, P.; Costa, A.L.; Silva, M.R.; Pereira, L.C.J.; Pereira da Silva, P.S.; Seixas de Melo, J.S.; Martin-Gil, J. Luminescent properties of [$UO_2(TFA)_2(DMSO)_3$], a promising material for sensing and monitoring the uranyl ion. *Mater. Res.* **2016**, *19*, 328–332. [CrossRef]

24. Du, N.; Song, J.; Li, S.; Chi, Y.-X.; Bai, F.-Y.; Xing, Y.-H. A Highly Stable 3D Luminescent Indium-Polycarboxylic Framework for the Turn-off Detection of $UO_2(^{2+})$, $Ru(^{3+})$, and Biomolecule Thiamines. *ACS Appl. Mater. Interfaces* **2016**, *8*, 28718–28726. [CrossRef] [PubMed]

25. 2018 Drinking Water Standards and Advisory Tables. Available online: https://www.epa.gov/dwstandardsregulations/2018-drinking-water-standards-and-advisory-tables (accessed on 29 April 2019).

26. Guidelines for Drinking-Water Quality, 4th ed. Available online: https://www.who.int/water_sanitation_health/publications/2011/dwq_guidelines/en/ (accessed on 29 April 2019).

 *nanomaterials*

Article

Pitch-Derived Activated Carbon Fibers for Emission Control of Low-Concentration Hydrocarbon

Hye-Min Lee [1,2], Byeong-Hoon Lee [1,3], Soo-Jin Park [2,*], Kay-Hyeok An [4,*] and Byung-Joo Kim [1,*]

[1] Research Center for Carbon Convergence Materials, Korea Institute of Carbon Convergence Technology, Jeonju 54852, Korea; hmlee2014@hanmail.net (H.-M.L.); byeonghoon1004@naver.com (B.-H.L.)
[2] Department of Chemistry, Inha University, Incheon 22212, Korea
[3] Department of Chemical Engineering, Chonbuk National University, Jeonju 54896, Korea
[4] Department of Nano & Advanced Materials Engineering, Jeonju University, Jeonju 55069, Korea
* Correspondence: sjpark@inha.ac.kr (S.-J.P.); khandragon@jj.ac.kr (K.-H.A.); kimbj2015@gmail.com (B.-J.K.); Tel.: +82-63-219-3710 (B.-J.K.)

Received: 19 August 2019; Accepted: 11 September 2019; Published: 14 September 2019

Abstract: The unburned hydrocarbon (HC) emissions of automobiles are subject to strong regulations because they are known to be converted into fine dust, ozone, and photochemical smog. Pitch-based activated carbon fibers (ACF) prepared by steam activation can be a good solution for HC removal. The structural characteristics of ACF were observed using X-ray diffraction. The pore characteristics were investigated using N_2/77K adsorption isotherms. The butane working capacity (BWC) was determined according to ASTM D5228. From the results, the specific surface area and total pore volume of the ACF were determined to be 840–2630 m^2/g and 0.33–1.34 cm^3/g, respectively. The butane activity and butane retentivity of the ACF increased with increasing activation time and were observed to range between 15.78–57.33% and 4.19–11.47%, respectively. This indicates that n-butane adsorption capacity could be a function not only of the specific surface area or total pore volume but also of the sub-mesopore volume fraction in the range of 2.0–2.5 nm of adsorbents. The ACF exhibit enhanced BWC, and especially adsorption velocity, compared to commercial products (granules and pellets), with lower concentrations of n-butane due to a uniformly well-developed pore structure open directly to the outer surface.

Keywords: pitch; activated carbon fiber; evaporated fuel; hydrocarbon emissions

---

1. Introduction

As the interest in environmental pollution increases, regulations of automotive emissions are being made stronger around the world [1,2]. Automotive emissions are divided into evaporation gas and exhaust gas. The exhaust gas is generated by the combustion of fuel and consists of harmful substances such as CO_x, NO_x, SO_x, and particulate matter (PM) [3–5]. The evaporation of unburned fuel from the fuel tank generally forms evaporation gas [6]. Its main component is hydrocarbon (HC) molecules, and there are many studies that indicate that HC is one of the critical materials causing smog, mist, and lung cancer [7,8].

Evaporation gas is normally generated even when the car is parked or being refueled; thus, it is difficult to prevent this using a separate actuator [9]. The evaporation gas is systematically collected and concentrated by means of fuel-vapor-emission control systems (a carbon canister and hydrocarbon trap sheet composed of activated carbon pellets and an activated carbon sheet, respectively; ACP and ACS) [10–12]. When the engine starts, the collected evaporation gas is transferred with air to the engine and used to enhance the fuel efficiency [13,14].

The diurnal regulation value for unburned HC emissions in cars was previously 500 mg/test (LEV-II), but recently it was further strengthened to 300 mg/test (LEV-III) [2]. Especially, canister bleed emission limits have been set at 20 mg/test for passenger car [2]. In the past, fuel-vapor-emission control systems were studied to enhance the pore characteristics of activated carbon or to increase the apparent density in order to adsorb larger amounts of evaporation gas [15–17]. However, for the fuel-vapor-emission control systems that meet the new stringent regulations (LEV-III), it is necessary to develop activated carbon capable of adsorbing a larger amount of evaporation gas and of completely adsorbing low concentrations of evaporation gas at the same time [2]. In order to solve this problem, a method for applying a material for the selective adsorption of low-concentration evaporation gas by an existing carbon canister, and a method for adding a separate device such as a ceramic honeycomb, have been discussed [10].

Activated carbon (AC) occurs in shapes classified as granules [18], pellets [19], and fibers [20–22]. Granular and pellet AC have a higher bulk density than fibrous activated carbon (activated carbon fiber, ACF), and so they can absorb a larger amount of harmful substances because of the larger input mass with the same volume [16,23]. Pellet AC usually exhibits lower pressure drop performance compared to granular AC [24,25], and so it is widely used in adsorption towers, canisters, and air cleaners. The ACF has a faster adsorption rate and better adsorption for low-concentration harmful substances than granular and pellet AC due to the excellent micropores on its surface [22,26].

In this study, the HC adsorption characteristics of AC (granular and pellet) and ACF were investigated to remove the evaporation gas of an automobile. The ACF was fabricated using various H_2O activation times for pitch fibers to observe the relationship between the adsorption capacity of low-concentration evaporation gas and the pore structure of the ACF. The pore development mechanism of ACF was confirmed through the pore characteristics and crystal structure. The HC adsorption characteristics of AC and ACF were analyzed using various concentrations of n-butane according to the ASTM D5228 standard.

2. Experimental

2.1. Preparation of Activated Carbon Fibers

ACF was fabricated from stabilized isotropic pitch fibers (GS Caltex, Jeonju, Korea). The precursor has a low ash content of less than 500 ppm and consists mostly of carbon. Three grams of pitch fiber was heated to 900 °C in nitrogen gas (500 mL/min of feeding rate) at a heating rate of 10 °C/min in a self-produced cylindrical steel tube furnace (diameter 130 mm × length 1000 mm) with SiC heaters, then activated with H_2O (0.5 mL/min of feeding rate by a micro-feeder) for 20–60 min (holding or reaction time). The H_2O was evaporated through a pre-heater heated at 200 °C and then introduced into a furnace. After activation, the flow of nitrogen was used to cool the furnace to room temperature to prepare the ACF. Each sample was named in the form ACF-(activation method)-(activation temperature)-(activation time).

The activation reaction of carbon crystallite with H_2O is endothermic and takes the following stoichiometric form [27]:

$$C + H_2O \rightarrow CO + H_2 \Delta H = +117 \text{ kJ/mol} \tag{1}$$

Commercial AC of the type used for canisters such as the BAX1500, BAXLBE (Pellet AC, Ingevity, North Charleston, SC, USA) and WVA1100 (Granular AC, Ingevity, North Charleston, SC, USA) were used for comparison with the ACF prepared in this work.

2.2. Characterizations

The pore characteristics of the ACF were analyzed using N_2/77K adsorption–desorption isotherm curves by Bel-MAX (BEL Japan, Inc., Tokyo, JAPAN). The specific surface area of the ACF was calculated by the Brunauer–Emmett–Teller (BET) equation [28], and the pore size distribution curves of the ACF were measured by both a non-local density functional theory (NLDFT) [29,30] and the Barrett–Joyner–Jalenda (BJH) equation [31] according to the pore range. The crystallite structure of the

ACF was analyzed using their X-ray diffraction (XRD, PANalytical, Almelo, The Netherlands) spectra. XRD patterns were measured using Cu Kα (1.542 A) in the range of 10–60° at a rate of 2°/min.

The butane working capacity (BWC) of the ACF was determined according to the ASTM D5228-16 standard [32]. The ACF was used to fill a U-shaped sample tube with a fixed volume of 16.7 mL. The sample tube was placed in a water bath maintained at 25 °C, and the butane was fed into the sample tube at 250 mL/min for 15 min. The mass change of the sample tube was measured using a balance, and the butane was adsorbed again for an additional 10 min. When a constant sample weight was determined, the sample tube was purged by helium gas at a rate of 300 mL/min for 40 min. BWC, butane activity, and butane retentivity were calculated using Equations (2)–(5). The change in the real-time butane concentration was measured using a quadrupole mass spectrometer (Q-mass) at the vent of the sample tube. A schematic diagram of the experimental setup is exhibited in Figure 1.

	Unit	Equation	
Butane working capacitor	g/100 mL	$\frac{(D-E)}{(C-B)} \times A \times 100$	(2)
Butane working capacitor	%	$\frac{(D-E)}{(C-B)} \times 100$	(3)
Butane activity	%	$\frac{(D-C)}{(C-B)} \times 100$	(4)
Butane retentivity	%	$\frac{(E-C)}{(C-B)} \times 100$	(5)

where A is the apparent density, B is the weight of the sample tube and stoppers, C is the weight of the sample, sample tube, and stoppers; D is the weight of the saturated sample, sample tube, and stoppers; and E is the weight of the purged sample, sample tube, and stoppers.

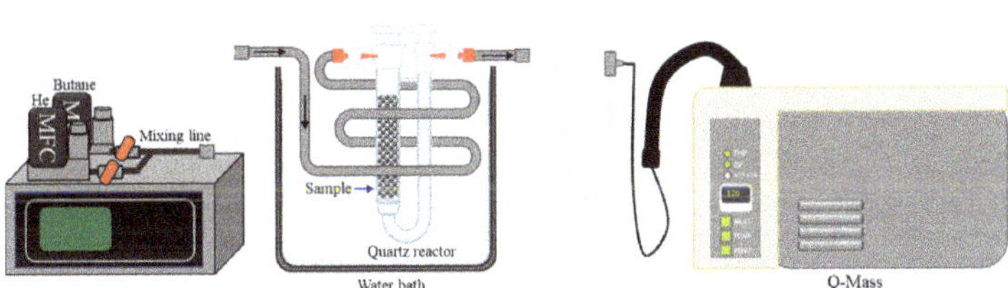

Figure 1. Schematic diagram of the butane working capacity test.

3. Results and Discussion

Isothermal adsorption–desorption curves are the most powerful method for analyzing the pore characteristics of the ACF. Figure 2 exhibits the N_2/77 K isothermal adsorption–desorption curves of the ACF. The curves of most ACF variants (ACF-H-9-2 to ACF-H-9-5) were classified as Type I by the IUPAC classification and were found to include mainly micropores [33]. Meanwhile, the curve of ACF-H-9-6 was classified as Type IV, and thus the volume ratio of mesopores to micropores was relatively high [33]. The hysteresis patterns were hardly observed in the curves of all the ACF variants, and only a few hysteresis patterns were observed in ACF-H-9-6. Therefore, it is considered that the pores of all the ACF are wedge-shaped, and that the pores are well-developed on the surface of the ACF. The adsorption–desorption isotherm curves of the AC (BAX1500, BAXLBE, and WVA1100) for canisters were determined to be Type IV, and very large hysteresis was observed. It is recognized that the AC has a high mesopore volume fraction within the total pore volume, and pot-shaped pores are present inside the pore structure.

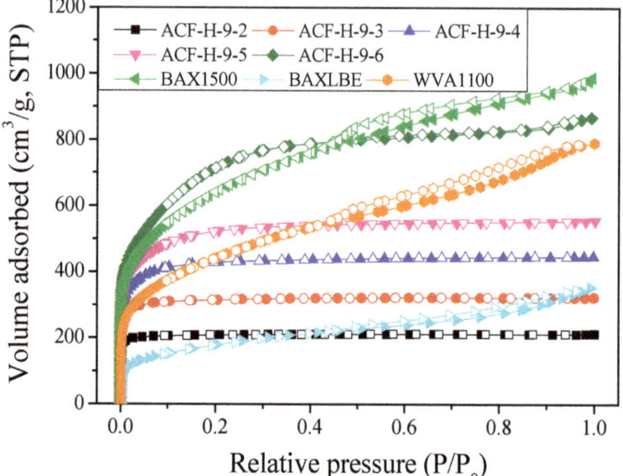

Figure 2. N_2/77 K adsorption–desorption isotherm curves of pitch-based activated carbon fibers as a function of various H_2O activation times and commercial activated carbons. ACF: activated carbon fibers.

Table 1 shows the textural properties of the ACF and AC. The specific surface area and total pore volume of the ACF increased from 840 to 2630 m^2/g and from 0.33 to 1.34 cm^3/g, respectively, with increased activation time. From ACF-H-9-2 to ACF-H-9-5, most of the porous structure was found to consist of micropores. As the activation time increased, the mesopore volume continuously increased. ACF-H-9-6 was observed to consist of about 37% of mesopores (of the total pore volume) and to have the highest specific surface area of all the ACF variants.

The AC for canisters has completely different pore structures compared to the ACF. The AC mesopores accounted for more than 57% of the total pore volume. The specific surface areas of the AC were in the order of BAX 1500 > WVA 1100 > BAX LBE. The pore ratio that ACF-H-9-6 exhibited was similar to that of the AC for canisters. ACF-H-9-6 had higher specific surface area and micropore volume than BAX 1500, but BAX 1500 had a higher mesopore volume than that of ACF-H-9-6.

Table 1. Textural properties of pitch-based activated carbon fibers as a function of various H_2O activation conditions and commercial activated carbon.

Sample	S_{BET} [1] (m^2/g)	V_{Total} [2] (cm^3/g)	V_{Micro} [3] (cm^3/g)	V_{Meso} [4] (cm^3/g)	Mesopore [5] Ratio (%)	Yield [6] (%)
ACF-H-9-2	840	0.33	0.32	0.01	3.0	55
ACF-H-9-3	1250	0.50	0.47	0.03	6.0	45
ACF-H-9-4	1670	0.69	0.63	0.06	8.7	38
ACF-H-9-5	1980	0.86	0.74	0.12	14.0	26
ACF-H-9-6	2630	1.34	0.85	0.49	36.6	15
BAX1500	2350	1.52	0.64	0.88	57.9	-
BAXLBE	650	0.55	0.17	0.38	69.1	-
WVA1100	1610	1.22	0.40	0.82	67.2	-

[1] S_{BET}: Specific surface area; Brunauer–Emmett–Teller (BET) method; $\frac{P}{v(P_0-P)} = \frac{1}{v_m c} + \frac{c-1}{v_m c}\cdot\frac{P}{P_0}$. [2] V_{Total}: Total pore volume; BET method. [3] V_{Meso}: Mesopore volume; Barrett–Joyner–Halender (BJH) method: $r_p = r_k + t$. (r_p = actual radius of the pore, r_k = Kelvin radius of the pore, t = thickness of the adsorbed film). [4] V_{Micro}: Micropore volume; V_{Total}-V_{Meso}. [5] Mesorpore ratio: $\frac{V_{Meso}}{V_{Total}} \times 100$. [6] Yield: $\frac{\text{Weight of activated sample}}{\text{Weight of carbonized sample input}} \times 100$.

Figure 3a exhibits the micropore size distribution curves of the ACF by an NLDFT method. The micropore size diameter was determined by gas sorption and estimated using nonlocalized density functional theory (NLDFT) and the grand canonical Monte Carlo method (GCMC) using the BELSORP evaluation software from the computer simulation. Figure 3a shows the typical pore size distribution of porous carbonaceous materials. The pore size distribution curves indicate that as the activation time increases to 40 min (ACF-H-9-4), the pore diameter increases and the width of the curve gradually increases. The pore size distribution curves of ACF-H-9-5 were observed to possess a narrow (~1 nm) and a broad (>2 nm) curve. ACF-H-9-6 had a broad curve, and the pore diameters increased from micropores to sub-mesopores. Therefore, it was confirmed that the mesopores were formed by the alteration of the micropores to mesopores by the widening of the pore diameters by further oxidation or by the collapse of the micropore walls. It was also confirmed that new micropore development continued in the variants up to ACF-H-9-6, as observed by the increase in micropore volume.

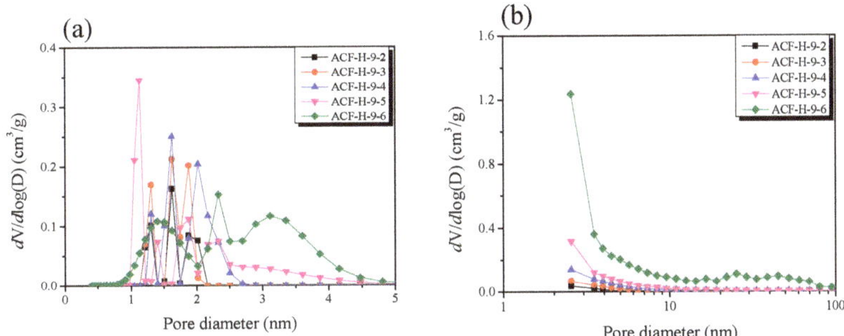

Figure 3. Pore size distribution of pitch-based activated carbon fiber as a function of various H_2O activation time: (**a**) micropore size distribution by the nonlocalized density functional theory (NLDFT) method; (**b**) nesopore size distribution by the BJH equation.

Figure 3b shows the mesopore size distribution of the ACF calculated using the BJH equation. The mesopore volume of the ACF was found to increase with increasing activation time. ACF-H-9-6 had the most massive mesopore volume among the ACF variants. From ACF-H-9-2 to ACF-H-9-5, mesopores of less than 10 nm in pore diameter were mainly developed, whereas ACF-H-9-6 had mesopores of pore diameter <50 nm. Because hysteresis of the ACF is hardly observed in Figure 2 (which means that the pore shape is probably a wedge-shape, as shown in Figure 2), it can be inferred that the location micropores of the ACF gradually shift to deep inside as the activation time increases, while maintaining a wedge-shaped pore structure (Figure 4a). On the other hand, it is assumed that the AC has a pore structure of a unique jar shape, resulting in a high mesopore volume ratio and a large hysteresis curve (Figure 4b).

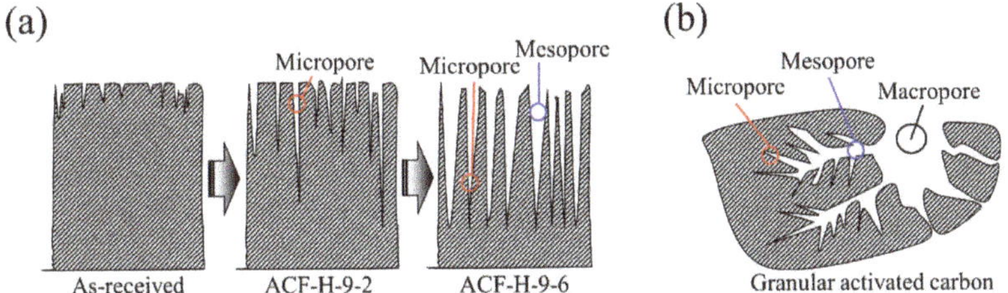

Figure 4. Schematic pore structure images of activated carbon fiber (**a**) and granular activated carbon (**b**).

Physical activation is the process of oxidizing crystallites to form various pores. Therefore, the changes of crystallite and pore structure with increasing activation time are very closely related. XRD is an advantageous method for observing the crystallite structure of ACFs.

In Figure 5, the XRD curve of the pitch fiber (as-received) exhibits a clear 002 peak and a well-developed domain characteristic of the inherent crystallite structure of the pitch. The XRD curves of the ACF exhibit the typical appearance of an isotropic carbon material. The 002 peak diffraction angle of the graphite crystallite is 26.56°, but the 002 peaks of the ACF samples are located at about 23°, indicating that the crystallite structures are considerably different from the structure of graphite. Besides this, the widely spread 10l peak indicates that each atomic layer is disordered and imperfectly laminated.

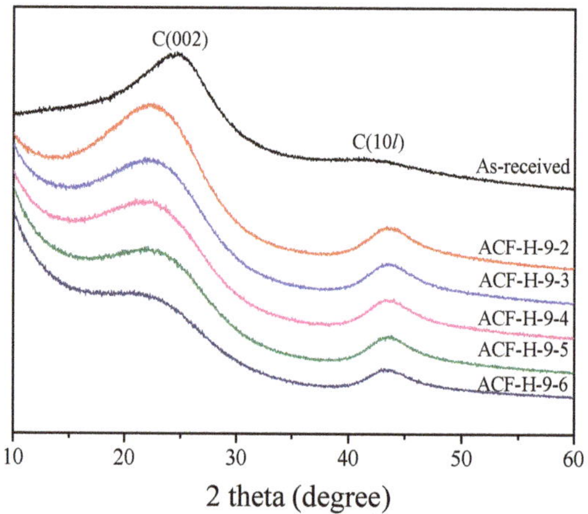

Figure 5. X-ray diffraction of pitch-based activated carbon fiber as a function of various H_2O activation times.

The XRD curves in Figure 3 were determined using the Bragg and Scherrer equations, and the interplanar spacings (d_{002} and d_{10l}) and the crystallite sizes (L_c and L_a) were measured. The structural parameters are shown in Table 2 and Figure 5.

Table 2. Structural parameters of pitch-based activated carbon fibers as function of various H_2O activation conditions. FWHM: full width half maximum.

Sample	002 Peak				10l Peak			
	2θ (°)	d_{002} (Å)	FWHM (2θ)	L_c (Å)	2θ (°)	d_{10l} (Å)	FWHM (2θ)	L_a (Å)
ACF-H-9-2	22.62	3.93	8.44	9.61	43.78	2.07	4.73	37.03
ACF-H-9-3	22.83	3.89	8.16	9.95	43.79	2.07	4.69	37.34
ACF-H-9-4	22.73	3.91	8.03	10.10	43.75	2.07	4.62	37.92
ACF-H-9-5	23.25	3.82	7.63	10.64	43.76	2.07	4.44	39.46
ACF-H-9-6	22.92	3.88	7.25	11.19	43.69	2.07	4.40	39.78

In Figure 6a, the L_c (crystallite height) and L_a (crystallite size) of the ACF increases with increasing activation time. Generally, the XRD data of carbonaceous materials provides statistical data about the number of crystallite aggregates. It is known that amorphous and small crystallites are preferentially oxidized, compared to larger crystallites, during the activation process; that is, as the activation of a pitch fiber proceeds, amorphous parts and small crystallite are easily oxidized, which may appear to increase the relative size of the entire crystallite. Therefore, it is considered that L_c and L_a are increased by the oxidation of amorphous areas or small crystallites as the activation time increases. Moreover, a steady increase of L_c and L_a during long activation times seems to result in the sustained oxidation of amorphous regions or small crystallite. The low increase of L_a in ACF-H-9-6 is considered to be highly correlated with the increase of the mesopore volume (Table 2). The increase in the micropore volume is the result of the oxidation of amorphous or small crystallites, leading to an increase in the L_c

and L_a. By contrast, the increase in the mesopore volume is formed by the oxidation of the micropore walls (the edge of the large crystallite), resulting in a decrease in L_c and L_a. Therefore, ACF-H-9-6 is considered to result in a tiny increase in L_a because this sample shows an increase in the micropore volume and a significant increase in the mesopore volume.

In Figure 6b, d_{002} decreases, and d_{10l} does not change significantly with increasing activation time. In the graphite crystal structure, the 002 planes are composed of strong hybridized sp^2 bonds, and the vertical π bond on 002 planes displays weak interlayer bonding. The decrease of d_{002} is considered to be the result of the continuous oxidation of amorphous parts and small crystallites.

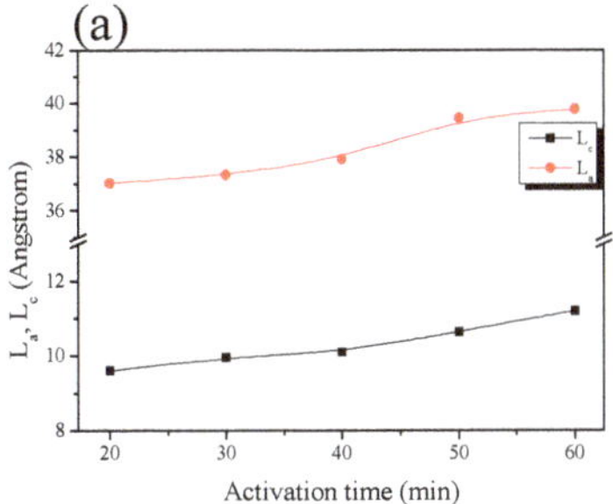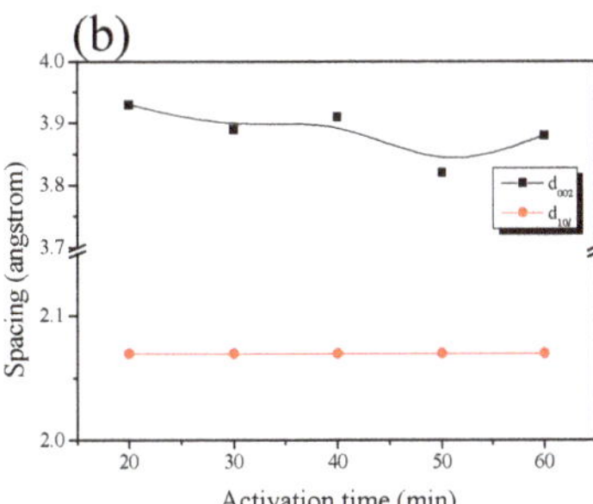

Figure 6. Structural characteristics of pitch-based activated carbon fibers as a function of various H_2O activation conditions: (**a**) structural parameters; (**b**) interplanar distance.

The butane working capacity (BWC) is a useful analytical method for evaluating the canister performance of ACF and AC. Table 3 lists the BWC, butane activity (BA), and butane retentivity (BR) of the ACF and AC measured according to ASTM D5228. The BA and BR indicate the butane adsorption capacity and the residual butane rate, respectively, after the desorption of the butane from the ACF. The BWC is defined as the difference between the butane adsorbed at saturation and that retained per unit volume of the ACF after a specified purge.

Table 3. Butane working capacity of pitch-based activated carbon fibers as function of various H_2O activation conditions.

Sample	Density (g/mL)	BWC [1] (g/100 mL)	BWC (%)	BA [2] (%)	BR [3] (%)
ACF-H-9-2	0.15	1.59	10.63	20.03	5.41
ACF-H-9-3	0.14	2.18	15.56	18.83	3.27
ACF-H-9-4	0.14	3.60	25.69	39.97	9.19
ACF-H-9-5	0.13	4.01	30.83	40.01	9.18
ACF-H-9-6	0.13	6.00	46.13	57.33	11.47
BAX 1500	0.31	15.03	50.11	50.11	7.43
BAX LBE	0.39	5.03	12.91	12.91	2.41
WVA 1100	0.28	8.73	31.17	31.17	5.87

[1] BWC: butane working capacity. [2] BA: butane activity. [3] BR: butane retentivity.

The BA and BR of the ACF increased with increasing activation time and were observed to range between 15.78–57.33% and 4.19–11.47%, respectively. The butane activity of the ACF was increased by the increase of the micropore and mesopore volume. It is generally known that AC with a high mesopore ratio can desorb the adsorbate more easily in the purging process. For example, as shown in Table 3, the BR of the AC for the canister was found to be lower with the increase in the mesopore volume ratio. However, the mesopore volume ratio of the ACF increased from ACF-H-9-3 to ACF-H-9-6, but the BR decreased. As shown in Figure 2, micropore development is more probable during pore development in the ACF. The micropores of the ACF were then transformed to mesopores, and new micropores developed deep inside the mesopores. Therefore, the micropores are uniformly formed on the surface of the ACF (up to ACF-H-9-5) and most of the pores are micropores. The ACF exhibited higher BA and BR values than those of AC with a similar specific surface area. The ACF is known to have faster adsorption characteristics than AC because of the formation of pores on its surface. This means that the ACF has a higher heat of adsorption than that of AC, resulting in desorption difficulty. Therefore, it is considered that ACF has better adsorption characteristics even if the specific surface area is similar to that of AC. On the other hand, because the AC has a higher mesopore volume ratio than the ACF, desorption can be easily performed during the purging process.

The BA of the AC was investigated by exploring its correlation with textural properties. In Figure 7, the BA of the ACF and AC appears to be linearly dependent on the specific surface area and microporosity. On the other hand, the BA was not found to be linearly related to the mesoporosity. These results are consistent with the results in Table 3 and suggest that the pore size may play an essential role in determining the BA of AC.

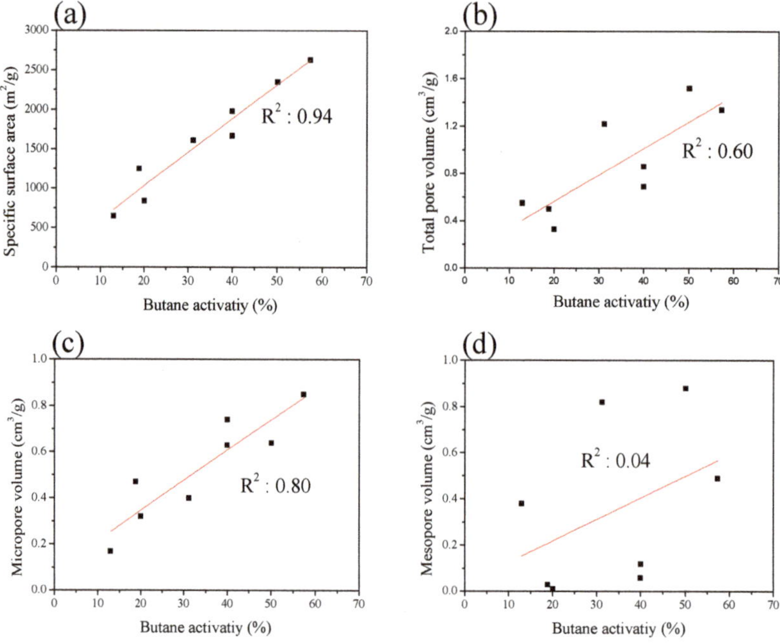

Figure 7. Correlations between the butane activity with specific surface area (**a**), total pore volume (**b**), micropore volume (**c**), and mesopore volume (**d**).

Figure 8 exhibits the result of plotting the pore volume according to pore diameter in 0.5 nm units using the NLDFT method and then plotting the coefficient of determination with BA. It is considered that the BA of the ACF is determined by pores with diameters from 2.0 to 3.5 nm. Primarily, the BA was strongly dependent on pores with diameters from 2.0 to 2.5 nm. This result is consistent with those in previous studies [15] showing that mesopores (2.0–2.5 nm) are essential for providing the butane adsorption capacity.

In order to meet the enhanced regulation of evaporation gas, it was necessary to develop an adsorbent with excellent adsorption capacity for HC at low concentrations. Because ASTM D5228 is used to evaluate the adsorption capacity of butane (100%) by mass change, it was impossible to conduct an experiment measuring butane at low concentration because the change in mass would be minimal. Figure 9 exhibits the change in concentration of the AC and ACF at various concentrations of butane using a Q-mass for the ASTM D5228 measurement.

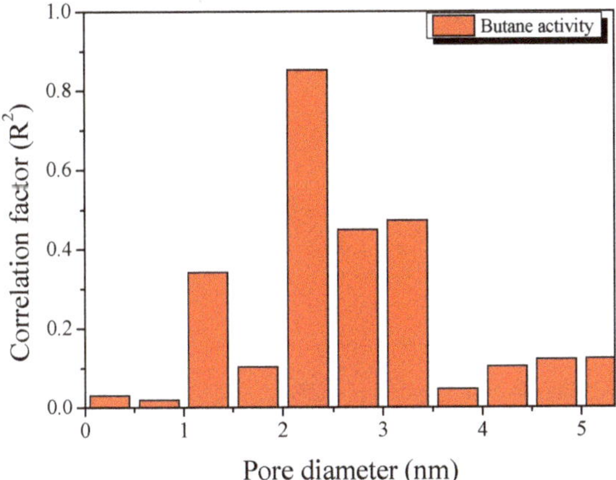

Figure 8. Correlations between the butane activity of activated carbon with pore volume. The X-axis exhibits the average pore size distribution after plotting the pore volume according to the pore diameter in 0.5 nm units using the average value of each pore size distribution.

As the butane concentration decreased, the slope of the curve near the breakthrough time of the AC decreased. On the other hand, the decrease in butane concentration did not significantly affect the curve slope of the ACF near the breakthrough time. In general, as the gas concentration decreases, the adsorption rate of the adsorbent decreases because the adsorption and the desorption reaction occur simultaneously.

The ACF has a faster adsorption rate than AC because micropores develop on the surface due to its inherent pore structure characteristics. Therefore, the adsorption rate of AC decreased as the concentration of butane decreased, suggesting that the slope of the curve decreased. On the other hand, because the adsorption rate of the ACF is faster than that of AC, it is considered that a change of the curved ACF slope does not appear, even if the concentration of butane decreases.

Figure 9a exhibits the change in concentration caused by the adsorption of 100% butane by the AC and ACF. In Table 3, the butane activity of the ACF is higher than that of AC. However, in Figure 9a, it can be seen that the breakthrough times of AC are longer than those of ACF. The apparent density of AC is about 1.9–2.3 times higher than that of ACF, so the mass of AC is higher, given the same volume. The butane adsorption capacity of the ACF was found to be proportional to the pore characteristics. In particular, ACF-H-9-6 exhibited the highest adsorption performance in the 100% butane-adsorption test because it had the highest specific surface area and highest total pore volume among the ACF variants.

Figure 9b shows the change in concentration caused by the adsorption of 10% butane by AC and ACF. In Figure 5b, the AC exhibits better butane adsorption characteristics than the ACF (Figure 5a), but the breakthrough times between the AC and ACF are very close. The breakthrough times of the ACF variants were almost the same, except for ACF-H-9-2. This result is attributed to the decrease in the adsorption rate of the ACF and AC due to the lower butane concentration.

Figure 9c shows the change in concentration caused by the adsorption of 1.0% butane by the AC and ACF. The slope of the curve near the breakthrough time of the AC was further reduced, while the curve of the ACF maintained a high slope. Moreover, ACF-H-9-4 and ACF-H-9-5 show breakthrough times similar to those of the AC. These results indicate that the adsorption rate of the AC is significantly lower than that of the ACF at very low butane concentrations. Among the ACF variants, although ACF-H-9-4 had a lower specific surface area and total pore volume than those of ACF-H-9-5 and ACF-H-9-6, it exhibited the best adsorption capacity for 1.0% butane. As shown in Figure 4, it can be recognized that ACF-H-9-4 may have micropores which formed mostly on the surface. In conclusion, ACF-H-9-4 exhibited excellent adsorption characteristics for 1.0% butane due to its well-developed micropore structure.

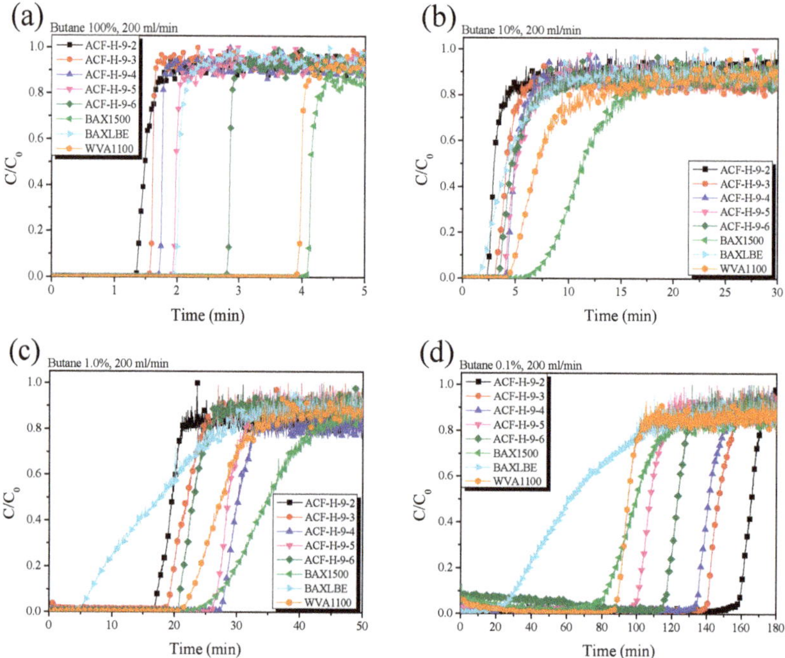

Figure 9. Breakthrough curves of pitch-based activated carbon fibers as a function of various H_2O activation time and butane concentration: (**a**) 100% butane, (**b**) 10% butane, (**c**) 1.0% butane, and (**d**) 0.1% butane.

Figure 9d exhibits the change in concentration caused by the adsorption of 0.1% butane by AC and ACF. At this butane concentration, the ACF exhibited better adsorption characteristics than the AC. Moreover, the shorter the activation time of the ACF, the better the adsorption capacity observed. In Figure 2, the isothermal curves of the ACF prepared in this work showed Type 1 curves, and the hysteresis of the samples was negligible. This could mean that the pore shapes of the samples might be wedge-shaped or cylindrical slit-shaped. Moreover, as the activation time increases, the diameter of the pore inlet located on the surface of the ACF increases gradually. Therefore, as the activation time increases, the positions of pores with diameters of 2.0–2.5 nm, which affect the butane adsorption, are expected to gradually shift to the inside of the pore structure, meaning that the adsorption rate decreases.

4. Conclusions

In this study, the HC adsorption characteristics of AC and ACF were studied according to their pore structures. The ACF was fabricated using various H_2O conditions to activate stabilized isotropic pitch fibers. From the results, the specific surface area and total pore volume of the ACF were determined to be 840–2630 m^2/g and 0.33–1.34 cm^3/g, respectively. A close relationship between BA and specific surface area was determined. It is also inferred that the sub-mesopore volume fraction in the range 2.0–2.5 nm primarily controls n-butane adsorption. The ACF exhibited higher butane activity than the AC, but the adsorption capacity was lower than AC for butane at a concentration of 10–100% due to the low apparent density. However, at a butane concentration of 0.1–1.0%, the intrinsic pore characteristics of the ACF were observed to have excellent adsorption capacity. With butane at 0.1% concentration, a higher butane adsorption capacity was observed with a lower specific surface area of the ACF.

Under the enhanced environmental regulations for automobile exhaust gas, an effective adsorbent is required for unburned-HC removal systems (canisters and ACS) with an excellent butane adsorption capability and a perfect adsorption property of butane at low concentrations. The results of this study show that AC exhibits a good butane adsorption capacity at high concentrations, and that ACF has an adsorption characteristic perfect for low concentrations of butane. Therefore, it is expected that an HC removal system incorporating AC with ACF would exhibit excellent performance without external accessories such as a honeycomb device.

Author Contributions: Conceptualization, S.-J.P. and B.-J.K.; methodology, S.-J.P. and B.-J.K.; software, H.-M.L.; validation, B.-H.L. and K.-H.A.; formal analysis, H.-M.L. and B.-H.L.; investigation, S.-J.P. and K.-H.A.; resources, S.-J.P. and K.-H.A.; data curation, H.-M.L. and B.-H.L.; writing—original draft preparation, H.-M.L.; writing—review and editing, B.-J.K.; visualization, B.-J.K.; supervision, B.-J.K.; project administration, B.-J.K.; funding acquisition, B.-J.K.

Funding: This research was supported by the "A Study on the Design of Protective Woven Fabrics Using ACF Materials" funded by the Agency for Defense Development, Republic of Korea (UD170102ID). This research was supported by Nano Material Technology Development Program through the National Research Foundation of Korea (NRF) funded by Ministry of Science and ICT (Project no. 2019M3A7B9071501).

Conflicts of Interest: The authors declare no conflict of interest.

References

1. Council of the European Union (CEU). *Annex to the Commission Regulation Amending Regulation (EC) No 692/2008 as Regards Emissions from Light Passenger and Commercial Vehicles (Euro 6)*; Council of the European Union: Brussels, Belgium, 2015.

2. California Air Resources Board. Amendments to the Low-Emission Vehicle Program—LEV III. Available online: http://www.arb.ca.gov/msprog/levprog/leviii/leviii.htm (accessed on 19 August 2019).

3. Manetas, C.; Sharifian, L.; Alexiadou, P.; Lafossas, F.A.; Mohammadi, A.; Koltsakis, G. Modeling of the Competitive Storage of NOx and Sulfur in Automotive Exhaust Catalysts. *Top. Catal.* **2019**, *62*, 10–17. [CrossRef]

4. Kim, H.G.; Yu, Y.; Yang, X.; Ryu, S.K. Carbon dioxide (CO_2) concentrations and activated carbon fiber filters in passenger vehicles in urban areas of Jeonju, Korea. *Carbon Lett.* **2019**, *26*, 74–80.

5. Saliba, G.; Saleh, R.; Zhao, Y.; Presto, A.A.; Lambe, A.T.; Frodin, B.; Sardar, S.; Maldonado, H.; Maddox, C.; May, A.A.; et al. Comparison of Gasoline Direct-Injection (GDI) and Port Fuel Injection (PFI) Vehicle Emissions: Emission Certification Standards, Cold-Start, Secondary Organic Aerosol Formation Potential, and Potential Climate Impacts. *Environ. Sci. Technol.* **2017**, *51*, 6542–6552. [CrossRef] [PubMed]

6. Yue, T.; Yue, X.; Chai, F.; Hu, J.; Lai, Y.; He, L.; Zhu, R. Characteristics of volatile organic compounds (VOCs) from the evaporative emissions of modern passenger cars. *Atmos. Environ.* **2017**, *151*, 62–69. [CrossRef]

7. Faith, W.L.; Goodwin, J.T.; Morriss, F.V.; Bolze, C.J. Automobile Exhaust and Somog Formation. *Air Pollut. Control Assoc.* **1957**, *7*, 9–12. [CrossRef]

8. Kotin, P.; Falk, H.L. Atmospheric Factors in Pathogenesis of Lung Cancer. *Adv. Cancer Res.* **1963**, *7*, 475–514. [PubMed]

9. United States Environmental Protection Agency (EPA). *Evaporative Emissions from On-Road Vehicles in MOVES2014*; EPA: Washington, DC, USA, 2014.

10. Hiltzik, L.H.; Jagiello, J.Z.; Tolles, E.D.; Williams, R.S. Method for Reducing Emissions from Evaporative Emissions Control Systems. U.S. Patent 6540815 B1, 1 April 2003.

11. Ishikawa, K.; Abe, S.; Ishimura, S. Activated Carbon Product in Sheet form and Element of Device for Preventing Transpiration of Fuel Vapor. U.S. Patent 7666507 B2, 23 February 2010.

12. Moyer, D.; Khami, R.; Bellis, A.; Luley, T. *Evolution of Engine Air Induction System Hydrocarbon Traps*; 2017-01-1014; SAE Technical Paper; SAE International: Warrendale, PA, USA, 2017.

13. Jamrog, J.R.; Pierce, M.A.; Johnson, P.J. Evaporative Emission Canister for an Automotive Vehicle. U.S. Patent 6237574 B1, 29 May 2001.

14. Vaishnav, D.; Ehteshami, M.; Collins, V.; Ali, S.; Gregory, A.; Werner, M. CFD Driven Parametric Design of Air-Air Jet Pump for Automotive Carbon Canister Purging. *SAE Int. J. Passeng. Cars-Mech. Syst.* **2017**, *10*, 474–486. [CrossRef]

15. Lee, H.M.; Baek, J.; An, K.H.; Park, S.J.; Park, Y.K.; Kim, B.J. Effects of Pore Structure on n-Butane Adsorption Characteristics of Polymer-Based Activated Carbon. *Ind. Eng. Chem. Res.* **2019**, *58*, 736–741. [CrossRef]

16. Tolles, E.D.; Dimitri, M.S.; Matthews, C.C. High Activity, High Density Activated Carbon. U.S. Patent 5204310, 20 April 1993.

17. Brown, T.C. Adsorption Properties from Pressure-Varying Langmuir Parameters: N-Butane and Isobutane on Activated Carbon. *Energy Fuels* **2017**, *31*, 2109–2117. [CrossRef]

18. Yuan, H.; Jin, B.; Meng, L.Y. Effect of the SBA-15 template and KOH activation method on CO_2 adsorption by N-doped polypyrrole-based porous carbons. *Carbon Lett.* **2018**, *28*, 116–120.

19. Jeguirim, M.; Belhachemi, M.; Limousy, L.; Bennicim, S. Adsorption/reduction of nitrogen dioxide on activated carbons: Textural properties versus surface chemistry—A review. *Chem. Eng. J.* **2018**, *347*, 493–504. [CrossRef]

20. Kim, M.J.; Lee, S.; Lee, K.M.; Jo, H.; Choi, S.S.; Lee, Y.S. Effect of CuO introduced on activated carbon fibers formed by electroless plating on the NO gas sensing. *J. Ind. Eng. Chem.* **2018**, *60*, 341–347. [CrossRef]

21. Ohc, J.Y.; You, Y.W.; Park, J.; Hong, J.S.; Heo, I.; Lee, C.H.; Suh, J.K. Adsorption characteristics of benzene on resin-based activated carbon under humid conditions. *J. Ind. Eng. Chem.* **2019**, *71*, 242–249.

22. Das, D.; Gaur, V.; Verma, N. Removal of volatile organic compound by activated carbon fiber. *Carbon* **2004**, *42*, 2949–2962. [CrossRef]

23. Miura, K.; Nakagawa, H.; Okamoto, H. Production of high density activated carbon fiber by a hot briquetting method. *Carbon* **2000**, *38*, 119–125. [CrossRef]

24. McCue, J.C.; Repik, A.J.; Miller, C.E. Shaped Wood-Based Active Carbon. U.S. Patent 4677086, 30 June 1987.

25. Happel, J. Pressure Drop Due to Vapor Flow through Moving Beds. *Ind. Eng. Chem.* **1949**, *41*, 1161–1174. [CrossRef]

26. Zhang, X.; Gao, B.; Creamer, A.E.; Cao, C.; Li, Y. Adsorption of VOCs onto engineered carbon materials: A review. *J. Hazard. Mater.* **2017**, *338*, 102–123. [CrossRef]

27. Wigmans, T. Industrial aspects of production and use of activated carbons. *Carbon* **1989**, *27*, 13–22. [CrossRef]

28. Brunauer, S.; Emmet, P.H.; Teller, W.E. Adsorption of gases in multimolecular layers. *J. Am. Chem. Soc.* **1938**, *60*, 309–319. [CrossRef]

29. Lastoskie, C.; Gubbins, K.E.; Quirke, N. Pore size distribution analysis of microporous carbons: A density functional theory approach. *J. Phys. Chem.* **1993**, *97*, 4786–4796. [CrossRef]

30. Olivier, J.P.; Conklin, W.B.; Szombathely, M.V. Determination of pore size distribution from density functional theory: A comparison of nitrogen and argon results. *Stud. Surf. Sci. Catal.* **1994**, *87*, 81–89.

31. Barrett, E.P.; Joyner, L.G.; Halenda, P.P. The determination of pore volume and area distributions in porous substances. I. Computations from nitrogen isotherms. *J. Am. Chem. Soc.* **1951**, *73*, 373–380. [CrossRef]

32. American Society for Testing and Materials. *Standard Test Method for Determination of Butane Working Capacity of Activated Carbon*; ASTM Standard D5228-16; ASTM: West Conshohocken, PA, USA, 1992.

33. Sing, K.S.W.; Everett, D.H.; Haul, R.A.W.; Moscou, L.; Pierotti, R.A.; Rouquerol, J.; Siemieniewska, T. Reporting physisorption data for gas/solid systems with special reference to the determination of surface area and porosity. *Pure Appl. Chem.* **1985**, *57*, 603–619. [CrossRef]

 *nanomaterials*

Article

Antagonistic Interactions between Benzo[a]pyrene and Fullerene (C$_{60}$) in Toxicological Response of Marine Mussels

Audrey Barranger [1,†,‡], Laura M. Langan [1,†,§], Vikram Sharma [2], Graham A. Rance [3,4], Yann Aminot [5], Nicola J. Weston [4], Farida Akcha [6], Michael N. Moore [1,7,8], Volker M. Arlt [9,10], Andrei N. Khlobystov [3,4], James W. Readman [5] and Awadhesh N. Jha [1,*]

[1] School of Biological and Marine Sciences, University of Plymouth, Plymouth PL4 8AA, UK
[2] School of Biomedical Sciences, University of Plymouth, Plymouth PL4 8AA, UK
[3] School of Chemistry, University of Nottingham, University Park, Nottingham NG7 2RD, UK
[4] Nanoscale and Microscale Research Centre, University of Nottingham, University Park, Nottingham NG7 2RD, UK
[5] Centre for Chemical Sciences, University of Plymouth, Plymouth PL4 8AA, UK
[6] Ifremer, Laboratory of Ecotoxicology, F-44311, CEDEX 03 Nantes, France
[7] Plymouth Marine Laboratory, Prospect Place, The Hoe, Plymouth PL1 3HD, UK
[8] European Centre for Environment & Human Health (ECEHH), University of Exeter Medical School, Knowledge Spa, Royal Cornwall Hospital, Cornwall TR1 3LJ, UK
[9] Department of Analytical, Environmental and Forensic Sciences, King's College London, MRC-PHE Centre for Environmental & Health, London SE1 9NH, UK
[10] NIHR Health Protection Research Unit in Health Impact of Environmental Hazards at King's College London in partnership with Public Health England and Imperial College London, London SE1 9NH, UK
*   Correspondence: a.jha@plymouth.ac.uk; Tel.: +44-1752-584-633
†   These authors contributed equally to this work.
‡   Present Address: Université de Rennes 1/Centre National de la Recherche Scientifique, UMR 6553 ECOBIO, F-35000 Rennes, France.
§   Present Address: Department of Environmental Science, Baylor University, Waco, TX 76706, USA.

Received: 15 May 2019; Accepted: 28 June 2019; Published: 8 July 2019

Abstract: This study aimed to assess the ecotoxicological effects of the interaction of fullerene (C$_{60}$) and benzo[a]pyrene (B[a]P) on the marine mussel, *Mytilus galloprovincialis*. The uptake of nC$_{60}$, B[a]P and mixtures of nC$_{60}$ and B[a]P into tissues was confirmed by Gas Chromatography–Mass Spectrometry (GC–MS), Liquid Chromatography–High Resolution Mass Spectrometry (LC–HRMS) and Inductively Coupled Plasma Mass Spectrometer (ICP–MS). Biomarkers of DNA damage as well as proteomics analysis were applied to unravel the interactive effect of B[a]P and C$_{60}$. Antagonistic responses were observed at the genotoxic and proteomic level. Differentially expressed proteins (DEPs) were only identified in the B[a]P single exposure and the B[a]P mixture exposure groups containing 1 mg/L of C$_{60}$, the majority of which were downregulated (~52%). No DEPs were identified at any of the concentrations of nC$_{60}$ ($p < 0.05$, 1% FDR). Using DEPs identified at a threshold of ($p < 0.05$; B[a]P and B[a]P mixture with nC$_{60}$), gene ontology (GO) and Kyoto encyclopedia of genes and genomes (KEGG) pathway analysis indicated that these proteins were enriched with a broad spectrum of biological processes and pathways, including those broadly associated with protein processing, cellular processes and environmental information processing. Among those significantly enriched pathways, the ribosome was consistently the top enriched term irrespective of treatment or concentration and plays an important role as the site of biological protein synthesis and translation. Our results demonstrate the complex multi-modal response to environmental stressors in *M. galloprovincialis*.

Keywords: Trojan horse effect; B[a]P; nC$_{60}$; co-exposure; mussels; DNA damage; proteomics

1. Introduction

There have been concerns regarding the potential for manufactured nanomaterials to cause unpredictable environmental health or hazard impacts, including deleterious effects across differing organismal levels, for over a decade. Despite numerous years of study, it is still unclear at what quantity manufactured nanomaterials can be found in the aquatic environment, along with their fate, potential bioavailability and subsequent hazardous effects to biological systems. This is surprising given the growing concern in the field of aquatic toxicology regarding their availability and potential toxicity [1]. Fullerenes are the smallest known stable carbon nanostructures and lie on the boundary between molecules and nanomaterials, with fullerenes generally exhibiting strong hydrophobicity in aqueous media [2]. Buckminsterfullerene (C_{60}) is the only readily soluble carbon nanostructure, although graphene is dispersible in specific organic solvents [3]. Non-functionalised C_{60} possesses a measurable, but extremely low solubility in water (1.3×10^{-11} mg/mL), but can exist in the aqueous phase as aggregates (nC_{60}) [4] and is quantifiable in aqueous environmental samples [5]. nC_{60} can be formed in water when fullerenes are released into the aquatic environment, increasing the transport and potential risk of this nanomaterials to the ecosystem ecology.

The toxicity associated with C_{60} is controversial and largely unclear [6]. The ability of C_{60} to both generate and quench reactive oxygen species (ROS) has recently been recognised as a particularly important property in the interaction of fullerenes with biological systems [7], with many aquatic studies demonstrating that fullerenes are capable of eliciting toxicity via oxidative stress [8–10]. Numerous studies have investigated the beneficial and toxicological effects of fullerenes [11–17]. However, the toxicity of nanomaterials has been shown to be dependent on numerous factors, including surface area, chemical composition and shape [18,19]. In specific cases, such as aqueous fullerenes (nC_{60}), the physiochemical structure is influenced by different preparation methods [15,20,21]. Altered physiochemical properties induced through the different methods of solubilisation have been shown to profoundly influence the observed toxicological effects of fullerene exposure, thus making a consensus assessment of environmental toxicity difficult [20]. While the environmental toxicity of fullerenes is still being investigated, an emerging concern is whether fullerene aggregates can act as contaminant carriers (Trojan horse effects) in aquatic systems, and whether this confirms the reduction or enhancement of toxicity with these compounds. Current evidence suggests a mixture of effects dependent on chemical properties. Under combined aquatic exposure conditions (viz. nC_{60} and contaminant), it has been demonstrated that 17α-ethinylestradiol (EE2) has a decreased bioavailability [14], altered toxicity [11,22] and localised increases in mercury bioavailability [23]. Finally, when compared to other anthropogenic contaminants, Velzeboer et al. established that the absorption of polychlorinated biphenyls (PCBs) to nC_{60} was 3–4 orders of magnitude stronger than to organic matter and polyethylene [24]. This enhanced absorption and modifications to toxicity responses may have significant impacts on the fate, transport and bioavailability of co-contaminants already in the aquatic environment. However, more research is necessary to establish which co-contaminants bioavailability is impacted when co-exposed with nC_{60}.

The aquatic environment is often the ultimate recipient of an increasing range of anthropogenic contaminants, and likely in all probable combinations. Organisms which are exposed to complex mixtures of differing compounds and substances can interact in many ways to induce biological responses be it additively, synergistically or antagonistically. These interactions can and do change the organismal response compared with single compound exposures [2,25,26]. Bivalves have highly developed processes for the cellular internalization of nano- and microscale particles (viz. endocytosis and phagocytosis) that are integral to key physiological functions such as cellular immunity [27]. These organisms are also useful bio-indicators because as suspension feeders they filter large volumes of water which facilitates uptake and bio-concentration of toxic chemicals [28], in addition to microalgae, bacteria, sediments, particulates and natural nanoparticles. This high filtration rate has been shown to be associated with the high potential accumulation of different chemicals in their tissues. A variety of mussel species have been used to elucidate both physiological and molecular mechanisms of action to nanoparticles [29,30] making them an ideal model to investigate how organisms respond to environmental stressors such as chemical mixtures [27]. This study aims to test the hypothesis that C_{60} fullerenes and B[a]P can interact with each other to differentially modify their potential toxicity. To confirm this hypothesis, a set of biomarkers or biological responses including proteomic analysis were employed. In this study, we hypothesized that C_{60} would act as a contaminant carrier for B[a]P and would modify the toxicity of B[a]P due to the high adsorption of B[a]P molecules onto C_{60} nano-aggregates. This hypothesis has been verified through the measurement of B[a]P and C_{60} in water and tissue. As B[a]P and C_{60} are known or potential genotoxic contaminants, a change in genotoxic effect was evaluated through the measurements of 8-oxodGuo, DNA strand breaks and DNA adducts in the digestive gland. Proteomics analysis was also performed to evaluate changes of mussels' proteome profile under co-exposure, and to try to unravel the molecular mechanisms of the potential interactive effects.

2. Materials and Methods

2.1. Animal Collection and Husbandry

Mussels (*Mytilus galloprovincialis*; 45–50 mm) were collected from the intertidal zone at Trebar with Strand, Cornwall, UK (50° 38′ 40″ N, 4° 45′ 44″ S) in October 2016. The site has previously been used as a reference location for ecotoxocological studies and is considered relatively clean with a minimum presence of disease [31,32]. Following collection, mussels were transported to the laboratory in cool boxes and placed in an aerated tank at a ratio of 1 mussel L^{-1} with natural seawater from Plymouth Sound (filtered at 10 μm). Mussels were maintained at 15 °C, fed with micro-algae (*Isochrysis galbana*, Interpret, UK) every 2 days with a 100% water change 2 h post feeding.

2.2. Preparation of Stock Solutions

2.2.1. Fullerenes (C_{60})

C_{60} and $Er_3N@C_{80}$ were obtained from Sigma Aldrich (Gillingham, UK) and Designer Carbon Materials Ltd. (Oxford, UK), respectively. In order to better replicate the conditions of the experiment during analysis, 2 mussels were maintained in 2 L glass beakers for 24 h with natural seawater from Plymouth Sound (filtered at 10 μm). Subsequently, fullerenes (1 mg) were added to the mussel-exposed seawater (10 mL) and the suspension homogenised by ultrasonication (Langford Sonomatic 375, Bromsgrove, UK, 40 kHz) for 1 h at room temperature. The suspension was allowed to settle for at least 4 h at room temperature prior to analysis of the aggregate size. Dynamic light scattering (DLS) was performed using a Malvern Zetasizer Nano-ZS (Malvern, UK) at room temperature. Quoted values are the average of 3 measurements. Bright field transmission electron microscopy (TEM) and dark-field scanning transmission electron microscopy (STEM) were performed using the JOEL 2100+ microscope (Welwyn Garden City, UK) operated at 200 keV. Energy dispersive X-ray (EDX) spectra were acquired using an Oxford Instruments INCA X-ray microanalysis system (Oxford, UK) and processed using Aztec software (version 3.1 SP1, Oxford, UK). Samples were prepared by casting several drops of the respective suspensions onto copper grid-mounted lacey carbon films.

2.2.2. Benzo[a]pyrene (B[a]P)

B[a]P (≥96%, B1760, Sigma Aldrich) is not water soluble and was previously dissolved in dimethyl sulfoxide (DMSO) after having determined its solubility limit. Chemical solutions were prepared so that the DMSO concentration in the sea water was 0.001%.

2.3. In Vivo Exposure of M. galloprovincialis to B[a]P and C$_{60}$: Experimental Design

Following depuration, mussels were separated (2 per beaker) into 2 L glass beakers containing 1.8 L of seawater and allowed to acclimatize for 48 h. A photoperiod of 12 h light: 12 h dark was maintained throughout the experiment. Oxygenation was provided by a bubbling system. Seawater was monitored in each of the beakers by measuring salinity (36.45 ± 0.19‰). Mussels were exposed for 3 days with no water changes to B[a]P (5, 50 and 100 µg/L), C$_{60}$ alone (0.01, 0.1 and 1 mg/L) and a combination of B[a]P (5, 50 and 100 µg/L) and C$_{60}$ (1 mg/L). Control groups received only DMSO at the same concentrations as used in the other exposure groups (0.001% DMSO). A total of 26 individuals were used per treatment. Following exposure, tissue samples were collected as follows: gill and digestive gland (DG) tissue was collected from 3 mussels for chemical analysis, digestive tissue was collected from 9 mussels and pooled (3 mussels per one biological replicate) for shotgun proteomics, DG tissue from 10 mussels was collected for comet assay and DNA adducts, with a further 5 DG collected for DNA oxidation. Water samples from 3 beakers were randomly collected during each treatment for B[a]P and C$_{60}$ analyses.

2.4. Gas Chromatography–Mass Spectrometry (GC–MS) Analyses of B[a]P in Water and Tissue

Water and tissue extracts were analysed using an Agilent Technologies (Stockport, UK) 7890A Gas Chromatography (GC) system interfaced with an Agilent 5975 series Mass Selective (MS) detector as described in [33].

2.5. Analyses of C$_{60}$ in Water and Tissue

The analyses of C$_{60}$ were performed on the toluene extracts common to the B[a]P analyses. The water extracts were analysed with an Agilent 1100 high-performance liquid chromatography-ultraviolet-visible instrument (HPLC-UV, Stockport, UK). The separation was performed on a Shimadzu XR-ODS column (particle size 2.2 µm, 3.0 × 50 mm, Milton Keynes, UK) using an acetonitrile-toluene gradient starting at 40% toluene, at a flow rate of 1 mL/min and a column temperature set at 40 °C. The detection wavelength was set at 330 nm and the fullerene absorption at maximum. Quantification was performed by external calibration using authentic fullerene standards. Because of their lower concentrations, the tissue extracts were analysed by ultrahigh performance liquid chromatography coupled with high resolution mass spectrometry following a protocol adapted from [34].

2.6. Proteomics

2.6.1. Sample Collection and Quality Check

Tissue was removed from the −80 °C, weighed (100 mg) and twice washed in phosphate buffered saline (PBS) prior to being homogenised on ice for 60 s in radioimmunoprecipitation assay (RIPA) buffer. The lysed homogenate was centrifuged at 14,000 RPM for 60 min at 4 °C, the supernatant collected and aliquoted. Protein concentration was determined using the Pierce BCA protein assay kit (Thermo Scientific, Loughborough, UK) according to manufacturers' instructions with bovine serum albumin as standard. Reproducibility of protein extraction was carried out using sodium dodecyl sulfate polyacrylamide gel electrophoresis (SDS-PAGE). Briefly, 30 µg of protein from each sample was loaded on a polyacrylamide gradient gel (4–12%) and stained with Coomassie protein stain (Expedeon, UK) and destained with ELGA water. Quality checked protein samples were then processed for downstream liquid chromatography-mass spectrometry (LC-MS, Stockport, UK) analysis.

2.6.2. Sample Preparation for LC-MS

Equal amounts of intestinal protein (100 µg) were processed using the Filter Aided Sample Preparation (FASP) method as described by [35]. The digested proteins were subsequently purified using the Stop-and-go-extraction (STAGE) tip procedure as previously described [36]. Tryptic peptides were analysed using LC-MS.

2.6.3. Mass Spectrometry

Peptides were separated on a Dionex Ultimate 3000 RSLC nano flow system (Dionex, Camberly, UK) and analysed as described in [37].

2.6.4. Analysis

Peptide identification and quantification. Data analysis and quantification was performed using R (Version 3.5.0, Vienna, Austria) [38]. Thermo .raw files were imported into ProteoWizard [39] and converted to .mzML format before identification using the MS-GF+ algorithm which is implemented in R via the MSGFplus package [40]. MS-GF+ was chosen due to its known sensitivity in identifying more peptides than most other database search tools and its ability to work well with diverse types of spectra, configurations of instruments and experimetnal protcols [41]. The protein database utilised in this study consisted of the UniProt KnowledgeBase (KB) sequences from all organisms from the taxa Mollusca, sub category Bivalvia (84,410 sequences released 1/10/2018). This was cocatenated with a common contaminants list downloaded from ftp://ftp.thegpm.org/fasta/cRAP (Version: January 30th, 2015) using the R package seqRFLP [42]. Searches were carried out using the following criteria: mass tolerance of 10 ppm, trypsin as the proteolytic enzyme, maximum number of clevage sites = 2 and cysteine carbamidomethylation and oxidation as a fixed modification. Target decoy approach (TDA) was applied as it is the dominant strategy for false discovery rate (FDR) estimation in mass-spectrometry-based proteomics [43]. A 0.1% peptide FDR threshold was applied in accordance with standard practice, with a 1% protein FDR applied after protein identification (via aggregation). The resulting .mzid files were converted to MSnSet and quantified using label free spectral counts. The mass spectrometry proteomics data have been deposited to the ProteomeXchange Consortium [44] via the PRIDE [45] partner repository with the dataset identifier PXD013805 and 10.6019/PXD013805.

Data processing and quantification. Data processing was undertaken as follows: each sample was run individually and then regionally combined before all samples were amalgamated into a large dataset. Quantification of proteins occurred via spectral index (SI) [46]. For identification of proteins, the common practice of requiring three peptides per protein was used in order to reduce the number of false positives [47]. Peptides were subsequently aggregated using sum and the protein intensities scaled based on the actual number of proteins summed. Mussel samples were grouped based on biological replicate, exposure and concentration and the resulting data filtered to keep proteins which were identified in more than two biological replicates. To quantitatively describe reliable and biologically relevant protein expression changes based on single exposure to B[a]P, C_{60} or to a combination of the two, the data analysis was split into three distinct sections. As per recent recommendations, normalisation was carried out first [48]. Based on systematic evaluations of normalisation methods in label free proteomics, normalisation between technical replicates was carried out using variance stabilization normalisation (Vsn) [49]. Based on a study by Lazar et al. [48], it was hypothesized the most likely cause of missing values will be due to a mixture of MAR (missing at random), MCAR (missing completely at random) and MNAR (missing not at random) data. As such, missing value imputation was carried out using a mixed methodology in the form of KNN (K nearest neighbours, biological replicates) and QRILC (left censor method for MNAR data; whole dataset) [50,51]. Following normalisation, differential expression was carried out using msmsTests [52] with p-value less than 0.05 considered significant and Q-values (FDR: <1%) calculated for p-value target matches with the Benjamini–Hochberg procedure. Enrichment of function among up- or downregulated proteins was calculated using GOfuncR using gene ontologies associated with differentially expressed proteins (P-adj = 0.01, calculated using Benjamini–Hochberg method and q-value = 0.05). Kyoto Encyclopaedia of Genes and Genomes (KEGG) analysis was carried out on the identified unique proteins per treatment ($p < 0.05$) using the clusterProfiler package [53].

KEGG annotation was performed using GhostKOALA [54] and pathways with significant enrichment identified using ClusterProfiler (hypergeometric test, $q < 0.05$ following Benjamini correction). Unique and common proteins based on toxicant were graphically represented through Venn diagrams with the software Venny (http://bioinfogp.cnb.csic.es/tools/venny/index.html) [55]. The R script outlining project analysis for this study can be found in supplementary materials (R script S1).

2.7. DNA Damage

2.7.1. Measurement of 8-oxodGuo Levels Using HPLC/UV-ECD

DNA extraction was performed using 20 mg of digestive gland tissue according to the chaotropic NaI method derived from Helbock et al. [56], slightly modified by Akcha et al. [57]. In addition, 8-oxodGuo levels were determined by HPLC (Agilent 1200 series, Les Ulis, France) coupled to electrochemical (Coulochem III, ESA, Illkirch, France) and UV (Agilent 1200 series) detection as described in [58].

2.7.2. Comet Assay

The comet assay on digestive gland tissue was performed as previously described in [33].

2.7.3. DNA Adducts

For each sample, DNA from gills and DG tissues was isolated using a standard phenol-chloroform extraction procedure. We used the nuclease P1 enrichment version of the thin-layer chromatography (TLC) ^{32}P-postlabelling assay [59] to detect B[a]P-derived DNA adducts (i.e., 10-(deoxyguanosin-N^2-yl)7,8,9-trihydroxy-7,8,9,10-tetrahydro-B[a]P [dG-N^2-BPDE]). The procedure was essentially preformed as described [59]. After chromatography, TLC sheets were scanned using a Packard Instant Imager (Dowers Grove, IL, USA) and DNA adduct levels (RAL, relative adduct labelling) were calculated as reported [60]. An external BPDE-modified DNA standard was used as a positive control [61].

2.8. Confirmation of Uptake of Fullerenes by Mussels

2.8.1. Experimental Design

Mussels were exposed to a single treatment, 1 mg/L Er$_3$N@C$_{80}$ for 3 days (static exposure). For each treatment (control and labelled fullerenes), 2 mussels were exposed into 2 L glass beakers containing 1.8 L of seawater.

2.8.2. Bulk Spectroscopic Analysis

For the determination of erbium concentration in the digestive gland, 2 mussels per treatment were analysed using an X Series II ICP-MS (Thermo Fisher Scientific Inc., Waltham, MA, USA) with PlasmaLab software (Thermo Fisher Scientific Inc., Waltham, MA, USA) as described in [32].

2.8.3. Mussel Sectioning and Electron Microscopy Analysis

Following the exposures detailed above, a small piece (~5 mm^2) was dissected out of the centre of the digestive gland and fixed in 2% paraformaldehyde, 2.5% glutaraldehyde, 2.5% NaCl, 2 mM CaCl$_2$ in 0.1 M PIPES, pH 7.2 for 3 h. The tissue was then stored in 2.3 M sucrose (in 0.1M PIPES) until analysis. Two mussels were analysed per treatment. Electron transparent sections for scanning transmission electron microscope (STEM) analysis were prepared by cutting ~1 mm^2 pieces from the washed whole tissues and sectioning to a thickness of ~180–200 nm at −80 °C using the RMC Products PowerTome with the CR-X cryochamber (Tucson, AZ, USA). The cross-sections were transferred onto copper-grid mounted graphene oxide films using the Tokuyasu technique and imaged in dark field STEM using the JOEL 2100+ microscope operating at 200 keV.

2.9. Statistical Analysis

Statistical tests were conducted using R software (Version 3.6.0, Vienna, Austria) [62]. Normality and variance homogeneity were evaluated using Lilliefor's test and Bartlett's test, respectively. When necessary, raw data were mathematically transformed (Ln) to achieve normality before proceeding with an ANOVA. When significant, a posteriori Tukey test was performed. When data could not be normalized, statistical differences between treatments were tested using the non-parametric Kruskal–Wallis test.

Analysis of Interactions

Further analysis of the combined effects of C_{60} and BaP on DNA Damage (based on Comet Assay) was performed by calculating the Interaction Factor (IF) in order to test for evidence of additivity, synergism and antagonism [63–65]:

$$IF = (G_{(C60 + BaP)} - C) - [(G_{(C60)} - C) + (G_{(BaP)} - C)] = G_{(C60 + BaP)} - G_{(C60)} - G_{(BaP)} + C, \quad (1)$$

$$SEM_{(IF)} = \sqrt{(SEM^2_{(C60 + BaP)} + SEM^2_{(C60)} + SEM^2_{(BaP)} + SEM^2_{(C)})}, \quad (2)$$

where IF is the interaction factor: negative IF denotes antagonism, positive IF denotes synergism, and zero IF denotes additivity. G is the mean cell pathological reaction to toxicants (BaP, C_{60} and BaP + C_{60}), and C is the mean cellular response under control conditions. $SEM_{(x)}$ is the standard error of the mean for group X. Results were expressed as IF, and the 95% confidence limits were derived from the SEM values.

In order to test the mixture IF values against predicted additive values (assumed to have an IF = 0), the predicted additive mean values (A) were calculated:

$$A = (G_{(C60)} - C) + (G_{(BaP)} - C). \quad (3)$$

The Pythagorean theorem method for combining standard errors was used to derive combined standard errors for the predicted mean additive values (A) of C_{60} and BaP (http://mathbench.org.au/statistical-tests/testing-differences-with-the-t-test/6-combining-sds-for-fun-and-profit/). The standard errors for the three C60 and BaP treatments (predicted additive) were derived using the following equation:

$$SEM_{(add)} = \sqrt{(SEM^2_{(C60)} + SEM^2_{(BaP)} + SEM^2_{(C)})}. \quad (4)$$

This enabled the 95% confidence limits to be derived for the predicted additive values. The confidence limits were used to test the predicted additive values having an IF = 0 against the IF values for the mixtures.

3. Results and Discussion

Bivalves are ideal organisms for evaluating the adverse effects caused by various environmental stressors including polycyclic aromatic hydrocarbons (PAHs) and nanomaterials. PAHs such as B[a]P have a ubiquitous aquatic distribution and are known to cause several adverse effects in a diverse range of aquatic organisms. Nanomaterials, both as solids and colloids, are ingested by many organisms and bio-accumulate in large quantities, especially in molluscs. The mussel digestive gland is one of the principal detoxification organs with an acknowledged concentration of phase I detoxification enzymes [66]. As such, it is unsurprising that a mussel digestive gland has been used as model tissue for eco-toxicological studies of various NPs [67–69], with Di et al. reporting that the digestive gland in *Mytilus edulis* accumulates more C_{60} than other tissues [67].

There is considerable debate in the literature regarding the actual toxicological impact of nanomaterials in the aquatic environment, with fullerene toxicity controversial. In the aquatic system, Kahru et al. compiled fullerene toxicological data for fourteen organisms and classified C_{60} as "very toxic" [70]. Using mouse and human cell lines, Isakovic et al. demonstrated that pristine C_{60} and aqueous suspensions of C_{60} are more toxic than its hydroxylated derivatives [71]. In marked contrast, other studies have demonstrated that pristine C_{60} has low or limited toxicity to cells and various organisms [10,72–74]. The lack of consensus regarding C_{60} toxicity may be partly due to limited studies which incorporate both a physiological and ecological approach. As a consequence, little is still known about NP bioavailability, mode of uptake, ingestion rates and actual internal concentrations related to Absorption, Distribution, Metabolism and Excretion (ADME) [27]. Despite the contradictory reports, there is consensus that some nanomaterials may potentially affect biological systems directly but also through interactions with other compounds that may be available in the environment (reviewed in [6]). Studies that investigate co-exposure with carbon-based nano-compounds, such as nanotubes or fullerenes are limited, especially in aquatic systems. Using *Danio rerio* (zebrafish) hepatocytes, Ferreira et al. investigated the co-exposure of C_{60} with B[a]P and provided evidence of toxicological interactions, whereby C_{60} increased the uptake of B[a]P into cells, decreased cell viability and impaired detoxification responses [75], while Baun et al. reported that co-exposure with fullerene C_{60} enhanced toxicity of phenanthrene to *Daphnia magna* and *Pseudokirchneriella subcapitata* [22]. With respect to B[a]P and C_{60}, Di et al. demonstrated organ specific response to both single and combined mixtures with no observation of cytoxicity and duration dependent and condition specific genotoxic response in *M. galloprovincialis* [67]. Importantly, the observed genotoxic response was reversible after a recovery period. While single exposure studies are more common, bivalve species have already been used as biological models in proteomics to assess the effects of complex mixtures [22,76,77]. However, proteomics analyses on combined exposure with carbon nanomaterials in aquatic organisms are still very scarce [9].

3.1. Characterization of C_{60} in Seawater

Dynamic light scattering and electron microscopy analysis (Figures S1–S3 and Table S1) of C_{60} dispersed in mussel-exposed seawater (~100 µg/mL) with brief ultrasonication followed by equilibration indicates the formation of stable aggregates measuring 653 ± 87 nm (nC_{60} where $n = ~2 \times 10^8$) in mean hydrodynamic diameter. No significant change in the size of nC_{60} aggregates was observed upon addition of B[a]P.

3.2. Assessement of the Interaction between C_{60} and B[a]P through Bioaccumulation in Gills and the Digestive Gland

Changes in the bioavailability of contaminants co-exposed with carbon nanomaterial have been reported, from a decrease in bioavailability [78,79] to its enhancement, also called the "carrier effect" [80,81]. It has been demonstrated that carbon nanopowder helps B[a]P uptake by zebrafish embryos and very interestingly also affected the distribution of the pollutant in the organism [82]. However, in the same species, in zebrafish larvae, it has been shown that bioavailability of 17α-ethynylestradiol (EE2) was reduced with increasing concentration of nC_{60} nanoparticles [14].

In our study, we analyzed for the first time in a marine bivalve, B[a]P uptake (at different exposure concentrations) in the digestive gland in the presence of C_{60} fullerene in order to highlight a possible role of contaminant carrier of C_{60}. Regarding analyses of B[a]P in seawater, nominal concentrations were matched to stock concentrations (Table S2). No difference was observed between the presence or absence of C_{60}. As already established [33], there was a rapid disappearance of B[a]P over time in seawater and B[a]P accumulated preferentially in the digestive gland tissue. Interestingly, comparable B[a]P tissue concentrations in the presence or absence of C_{60} were observed indicating that, despite the expected strong sorption of B[a]P on C_{60} [83], no Trojan horse effect was observed and C_{60}-sorbed B[a]P remains also bioavailable to *M. galloprovincialis* (Figure 1). In gills, a significantly higher uptake is observed in the presence of C_{60} at the highest concentration of B[a]P (Figure 1). In general, high variability would conceal subtle changes. It appears that the bioavailability of nanomaterials and their co-contaminants depend on many factors such as their size, shape, surface coating and aggregation state and on the metabolism of the species investigated [78,84].

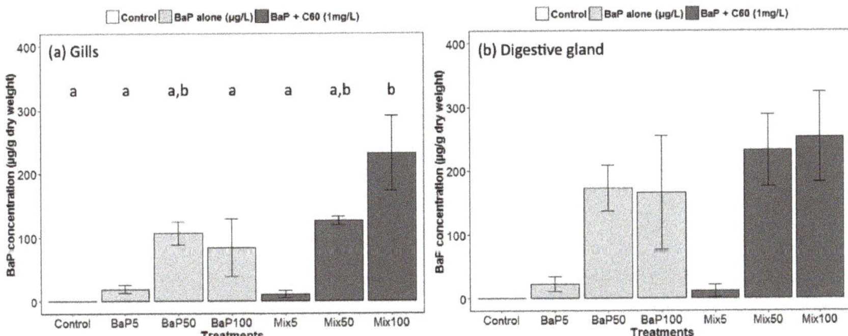

Figure 1. Gas Chromatography–Mass Spectrometry (GC–MS) analyses of B[a]P in (**a**) gills and (**b**) digestive gland of *M. galloprovincialis*. Data marked with different letters differed significantly (Tukey post-hoc test; $p < 0.05$).

A rapid decline in the concentration of C_{60} in seawater was observed with time, with no quantifiable amounts after day 1 (Table S3). At t_0, the measured water concentrations are in reasonable agreement with the nominal concentrations (427.6 ± 45.3, 63.8 ± 11.9 and 7.3 ± 1.8 µg/L for nominal concentrations of 1000, 100 and 10 µg/L, respectively). Low but quantifiable amounts of C_{60} in *M. galloprovincialis* tissues indicate active uptake, with adsorption on the outside of the tissue ruled out due to external washes with toluene prior to analysis (Figure 2). High variability in C_{60} concentrations in gills and DG makes it difficult to detect a difference in accumulation between treatments and to conclude regarding the uptake of C_{60} by mussels.

To provide further insight into the uptake of fullerenes by marine mussels, it was necessary to use a form labelled with a diagnostic marker. In our experiments, we explored the application of the endohedral fullerene $Er_3N@C_{80}$, fabricated using the trimetallic nitride template (TNT) process, as it represents a good structural analogue to C_{60}, possessing similar surface chemistry, and contains a rare earth element, shielded from the external environment within the fullerene cage, which is not found in nature. The presence of erbium in the mussel digestive gland, as a diagnostic of the uptake of labelled fullerenes, was thus quantified using Inductively Coupled Plasma Mass Spectrometer (ICP–MS) and found with a mean concentration of 151.5 µg/kg (236.5 and 66.4 µg/kg for each mussel). However, despite an exhaustive electron microscopy investigation of whole and cross-sectioned DG tissues (Figure S4, Supplementary materials), no direct visualisation of labelled fullerenes was observed. This result indicates that the fullerenes are likely distributed within the tissues at the near molecular level (i.e., highly dispersed) and therefore below the sensitivity of either microscopy or in situ spectroscopy approaches in complex materials such as these. In a previous study in *M. galloprovincialis* [85], it has been showed that mussels exposed to C_{60} alone exhibited higher accumulation of C_{60} in the digestive gland compared to the gill. Interestingly, co-exposure to fluoranthene modified accumulation of C_{60}, with higher accumulation of C_{60} when animals are exposed to C_{60} alone compared to combined exposure.

When comparing water and tissue concentrations for B[a]P and C_{60}, the bioconcentration observed in our conditions was much lower for C_{60} compared to B[a]P: the uptake in the DG of mussels exposed to a similar aqueous concentration of B[a]P and C_{60} was about 2000 times more important for B[a]P. However, non-constant concentrations in the aqueous phase, attributed to sorption and/or sedimentation, did not allow the calculation of bioaccumulation factors, which also requires reaching a steady-state in the tissues. The difference between B[a]P and C_{60} tissue concentration could also be attributed to different kinetics of uptake, which could only be explored through longer exposure periods and regular sampling. Recent work indicated a continuous increase of C_{60} concentrations in whole mussels over at least three weeks [86].

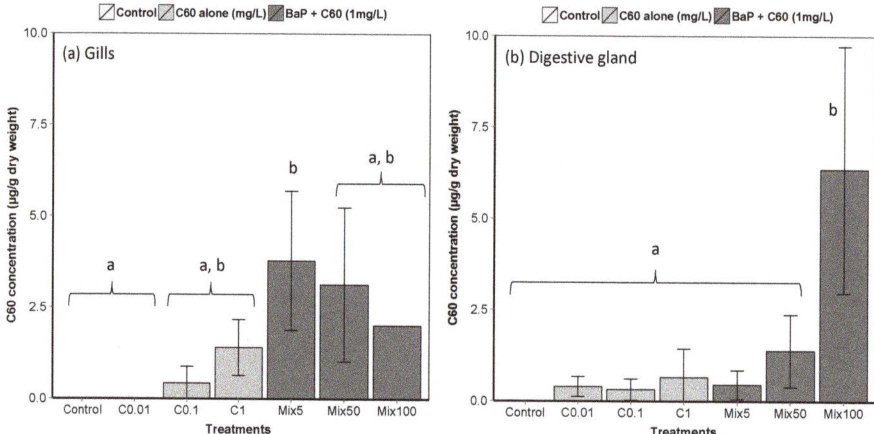

Figure 2. Liquid chromatography-mass spectrometry (LC–MS) analyses of C_{60} in *M. galloprovincialis* (**a**) gills and (**b**) digestive gland (means ± SE). Data marked with different letters differed significantly (Tukey post-hoc test; $p < 0.05$). An analytical problem led to the loss of two samples of the gills from mussels exposed to Mix100 explaining the absence of standard error.

3.3. Assessment of the Interactive Effect of C_{60} and B[a]P through Genotoxicity

B[a]P is a known genotoxic, mutagenic and carcinogenic [87]. According to a review by Johnston et al., fullerene toxicity has been suggested to involve oxidant-driven response and suggests evaluating toxicity by including oxidative stress and related consequences including inflammation or genotoxicity [88]. We assessed the interactive effect of C_{60} and B[a]P through three different genotoxicity assays. Regarding single exposures, B[a]P induced DNA strand breaks in the digestive gland at the intermediate and highest concentrations (50 and 100 µg/L) after three days of exposure (Figure 3). No effect on DNA strand breaks was observed at the lowest concentration. No effect was also observed on the level of 8-oxodGuo for B[a]P treatment. These results could be due to the short exposure time (three days). In [57], an increase in the level of 8-oxodGuo was observed after 10 days of B[a]P exposure in the digestive gland of *M. galloprovincialis*. Regarding exposure to C_{60} only, higher DNA strand breaks compared to the controls were observed only at the highest concentration (1 mg/L, $p < 0.001$). A significant increase ($p = 0.00108$) in 8-oxo-dGuo levels was also detected in the digestive gland of mussels exposed to C_{60} (15.3 ± 2.3) compared to control (5.9 ± 1.3) (Figure 4). Lower C_{60} concentration did not appear to have any genotoxic effects (DNA strand breaks) on mussel digestive gland at the concentrations tested. Whatever the exposure concentration of B[a]P and mixture of B[a]P and C_{60}, no DNA adducts were detectable in DNA samples from the digestive gland of *M. galloprovincialis*.

As observed for the bioaccumulation of contaminants co-exposed with carbon nanomaterial, controversial results are also obtained in the literature regarding genotoxicity. In aquatic organisms, co-exposure to C_{60} and organic contaminants induced a range of responses, to no effect until synergistic and antagonistic responses compared to single exposure [9,67,85]. In our study, no significant differences in DNA strand breaks were observed between exposure to B[a]P or C_{60} alone compared to co-exposure (Figure 3). Interestingly, the analysis of interactions performed on the comet assay and the oxidative DNA damage results revealed an antagonistic interaction only at the highest concentration between C_{60} and B[a]P (Table 1). This antagonistic effect may be caused by a reduction in ROS generation, or more effective scavenging of ROS by C_{60}, when C_{60} and B[a]P are present together in close association, as previously described by [9,67]. C_{60} fullerenes are both scavengers and generators of reactive oxygen species (ROS) [89]; and when C_{60} and B[a]P are closely associated or bound together within the lysosomal compartment of the mussel digestive cells, their ROS scavenging and generating properties may be altered.

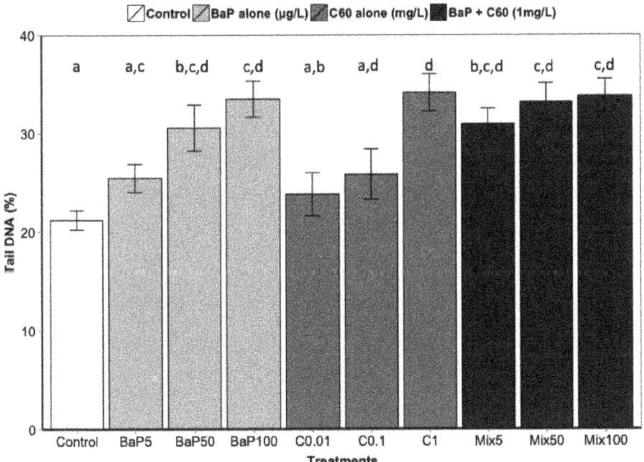

Figure 3. DNA strand break level following 3 days of exposure to C_{60}, B[a]P and mixture of both in the digestive gland. Data marked with different letters differed significantly (Tukey post-hoc test; $p < 0.05$).

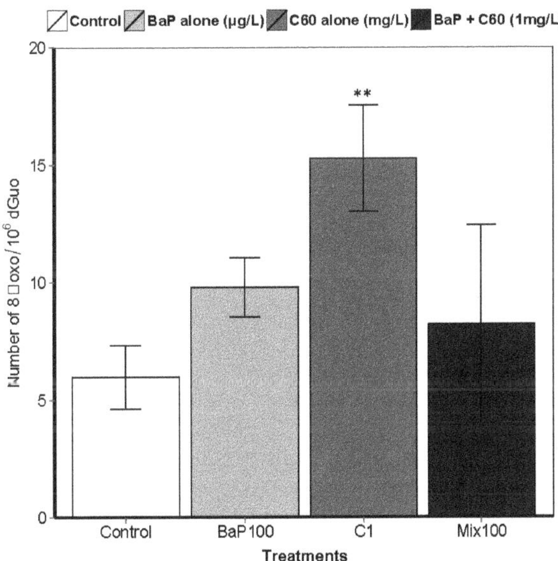

Figure 4. 8-oxodGuo levels in the digestive gland of mussels. Asterisks indicate the statistical differences observed between control and exposed groups. (**) $p < 0.01$.

Table 1. Analysis of combined effects of B[a]P and C_{60} on DNA damage based on Interaction Factors (IF).

Treatments	IF for DNA Damage (Comet Assay)
BaP 5 µg/L + C_{60} 1 mg/L	-7.48 ± 6.63
BaP 50 µg/L + C_{60} 1 mg/L	-10.39 ± 3.50
BaP 100 µg/L + C_{60} 1 mg/L	-12.69 ± 6.05 *

Interaction Factor $\pm$ 95% Confidence Limit $/\sqrt{2}$. * indicates significance at the 5% level. A negative IF indicates antagonism; an IF of 0 indicates additivity; and a positive IF indicates synergism. Statistical significance was determined by testing for overlap between the mixture IF $\pm$ 95% CL$/\sqrt{2}$ and the predicted additive value for C_{60} and B[a]P, assumed to have an IF = 0 $\pm$ 95% CL$/\sqrt{2}$, where the confidence limit is derived from the SEM$_{(add)}$ value for the additive C_{60} and B[a]P.

3.4. Assessment of the Interactive Effect of C_{60} and B[a]P on the Proteome Profile of the Digestive Gland

Investigations into proteome responses of marine organisms to various stressors is comparatively small when compared to other model laboratory organisms, both aquatic and terrestrial. Proteomic analysis represents a fundamental step in extending understanding of the physiological processes involved in organismal responses to environmental stressors. In addition, proteomics also provides better qualitative data on post-translational modifications without interference from mRNA instability [90]. A major limitation in the field has been the lack of available annotated genomes for a broad diversity of marine organisms. As a consequence, it has been considered a widely under utilised tool [91]. The lack of genome information has not stopped studies on proteome characterisation in bivalvia/mollusca species using broad protein databases limited to either the phylum, class or specific combination of species [92–95]. However, studies investigating proteome response to environmental stressors or injury are less abundant [30,77,96]. In the current study, a label free shot-gun proteomics approach was performed for the first time to our knowledge in aquatic organisms to investigate proteome alterations following treatment with B[a]P and C_{60} alone and a combination of B[a]P with 1 mg/L of C_{60}. This untargeted method was specifically chosen to identify molecular pathways involved in the interaction of C_{60} and B[a]P without a priori assumptions.

3.4.1. Identification of Differentially Expressed Proteins

In order to identify differentially expressed proteins in the digestive gland proteome of controls, B[a]P, nC_{60} and mixture (B[a]P and 1 mg/L nC_{60}), a label free LC-MS/MS approach was used with trypsinised tissue homogenates. Following removal of common contaminants in each dataset, peptide mapping quantified 3125, 3428 and 3475 unique proteins following identification from the Universal Protein Resource (UNIPROT) database distinct to B[a]P, C_{60} and mixture (B[a]P and 1 mg/L C_{60}) treatments, respectively. Irrespective of treatment, protein sequences from the Pacific oyster *Crassostrea gigas* (Organism ID = 94323) were highly represented in the samples at approximately 38%, followed by Japanese scallop *Mizuhopecten yessoensis* (Organism ID = 6573) at 34%. Surprisingly, sequences from the genus *Mytilus* were less represented in the search at approximately 3% with the Mediterranean mussel *Mytilus galloprovincialis* (Organism ID = 29158) representing approximately 1% of identified sequences. This may be due to a lack of genomic information available for this genus in the UNIPROT database, even though a genome sequence is available [97].

Differentially expressed proteins (DEPs) were determined using a quasi-likelihood GLM. Comparison of each dose per treatment (B[a]P: 5, 50 and 100 µg/L, nC_{60}: 0.01, 0.1 and 1 mg/L, and a mixture: 5, 50 and 100 µg/L B[a]P and 1 mg/L nC_{60}) with the control group was visualised using Venn diagrams (Figure 5). Minimal overlap between varying concentrations was observed for the mixture treatment (average of 2%) (Figure 5c) when compared to B[a]P (Figure 5a, 9%) or nC_{60} (Figure 5b, 8%). Volcano plots were used to visualise statistically significant changes in protein abundance for varying concentrations of the above treatments following comparison to controls (Figure S5). Applying a 1% FDR threshold, 401 differentially expressed proteins were identified following B[a]P treatment (all concentrations) and 297 differentially expressed proteins were identified following treatment with the mixture of B[a]P and nC_{60}. No differentially expressed proteins ($p < 0.05$) were identified in C_{60} treated samples. The identified DEPs can be further broken down based on treatment with 42, 50 and 164 DEPs identified at 5, 50 and 100 µg/L B[a]P. Following exposure to a mixture solution, 95, 108 and 94 DEPs were identified at each concentration respectively (1 mg/L of C_{60} and 5, 50 and 100 µg/L of B[a]P) with Figure 6 representing a visual comparison of commonalities between single exposure versus combined exposure. A subset of DEP based on the top three unique proteins per concentration is displayed in Table 2, with the full list of unique proteins and associated *p*-value and FDR correction (Spreadsheet S1). The majority of differentially expressed proteins detected in this study (B[a]P and mixture exposure) were downregulated (52%) between the treatment and control conditions irrespective of concentration.

The trend towards higher protein alterations in single exposures versus co-exposures suggests a non-additive combine effect and is in agreement with prior studies which reported generally higher protein alterations of B[a]P and Cu under single exposure then when co-exposed together [77]. The data in this study suggest that an interaction occurs between B[a]P and C_{60} whereby the effect of the mixture is different from the presumption of additivity (were by dose response relationships of mixtures are enhanced in comparison to the individual components) as outlined in Rosa et al. [98]. In this case, the data suggests an antagonistic relationship between B[a]P and C_{60} at the higher concentrations of 50 and 100 µg/L. This observation has previously been observed in *Mytilus edilus* digestive gland [67]. However, this trend is not replicated at the lowest concentration of 5 µg/L whereby mixture exposure resulted in higher DEPs than single exposure. This difference in DEPs may potentially be related to reduced accumulation of B[a]P at the higher concentrations due to saturation of mussel tissue and thereby limiting protein changes. In previous studies, increased impact and accumulation of B[a]P at lower concentrations in *M. galloprovincialis* have been attributed to tissue saturation [99]. The increase in differentially expressed proteins at the lower concentration may also reflect the inability of membrane transporters such as p-glycoprotein to efflux this particular nanoparticle [100] and as such acts to bypass typical protective mechanisms initiated to protect the organism from PAH stress.

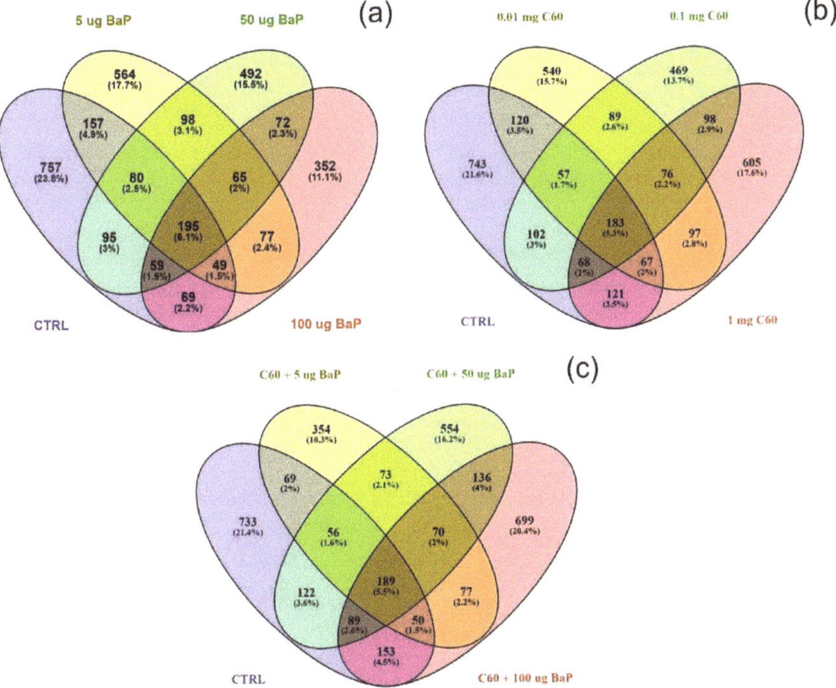

Figure 5. Venn diagram visualising the overlap between the control sample and varying concentrations of B[a]P (**a**), C_{60} (**b**) or a mixture of the two (5–50–100 µg/L B[a]P 1 mg/L C_{60}) (**c**) following exposure for three days. Note that overlap is based on a threshold of $p < 0.05$ and does not include FDR correction.

Table 2. Significantly expressed proteins of B[a]P, C_{60} and mixture (5–100 µg/L and 1 mg/L C_{60}). Species id's are as follows: 6573 = *Mizuhopecten yessoensis*, 6551 = *Mytilus trossulus*, 29159 = *Crassostrea gigas* and 94323 = *Crassostrea ariakensis*.

Treatment	Species	Protein Name	UNIPROTKB	GO Annotation	Regulation
B[a]P (5 µg/L)	6573	Arrestin domain-containing protein 3	A0A210PE39		Up
B[a]P (5 µg/L)	6573	Orexin receptor type 2	A0A210PSC6	GO:0004930, GO:0016021	Up
B[a]P (5 µg/L)	6573	Ran-specific GTPase-activating protein	A0A210Q6H5	GO:0005622, GO:0046907	Up
B[a]P (50 µg/L)	6573	5-hydroxytryptamine receptor 1A-alpha	A0A210R4M3	GO:0004993, GO:0005887, GO:0008283, GO:0042310, GO:0046883, GO:0050795	Down
B[a]P (50 µg/L)	6573	Adenylate kinase isoenzyme 5	A0A210QMB2	GO:0005524, GO:0006139, GO:0019205	Up
B[a]P (50 µg/L)	6573	Uncharacterised protein	A0A210Q912		Up
B[a]P (100 µg/L)	6573	Helicase with zinc finger domain 2	A0A210PQ46	GO:0004386, GO:0030374	Up
B[a]P (100 µg/L)	29159	Peroxiredoxin-4	K1QLH0	GO:0005623, GO:0045454, GO:0051920	Up
B[a]P (100 µg/L)	29159	Hypoxia up-regulated protein 1	K1QBF7	GO:0005524	Up
B[a]P (5 µg/L) + C60 (1 mg/L)	94323	Ras-like GTP-binding protein RHO	H9LJA2	GO:0003924, GO:0005525, GO:0005622, GO:0007264	Up
B[a]P (5 µg/L) + C60 (1 mg/L)	29159	Zinc finger CCCH domain-containing protein 13	K1PKC9	GO:0046872	Up
B[a]P (5 µg/L) + C60 (1 mg/L)	29159	Myosin heavy chain, non-muscle (Fragment)	K1QXX7	GO:0003774 GO:0003779, GO:0005524 GO:0016459	Up
B[a]P (50 µg/L) + C60 (1 mg/L)	6551	Ribosomal protein S20	A0A077H0N2	GO:0003723, GO:0003735, GO:0006412, GO:0015935	Down
B[a]P (50 µg/L) + C60 (1 mg/L)	6573	Nucleolar and coiled-body phosphoprotein 1	A0A210Q9W0	GO:0005730	Down
B[a]P (50 µg/L) + C60 (1 mg/L)	29159	Tripartite motif-containing protein 2	K1QBD4	GO:0005622, GO:0008270	Up
B[a]P (100 µg/L) + C60 (1 mg/L)	6573	Ran-specific GTPase-activating protein	A0A210Q6H5	GO:0005622, GO:0046907	Down
B[a]P (100 µg/L) + C60 (1 mg/L)	29159	Uncharacterized protein	K1R543		Down

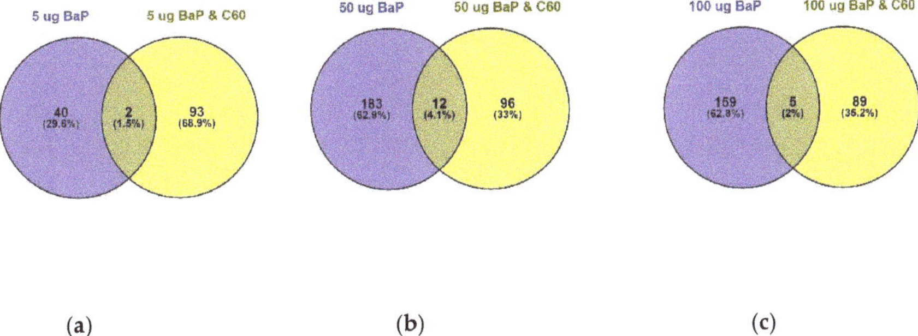

(a)   (b)   (c)

Figure 6. Venn diagram visualising the overlap between 5 µg/L (**a**), 50 µg/L (**b**) and 100 µg/L (**c**) of B[a]P with a mixture solution containing the same B[a]P concentrations in addition to 1 mg/L of C_{60} following 24 h exposure. Overlap is based on $p < 0.05$ and FDR set at 1%.

3.4.2. GO Functional Enrichment

Gene ontologies were directly annotated using a custom annotation database derived from UNIPROTKB (bivalvia) with enrichment carried out using GOfuncR. This provides a controlled vocabulary to describe gene product characteristics in three independent ontologies viz. biological process, molecular function and cellular components. Based on the R package GOfuncR, 31, 35 and 23 GO nodes were found enriched at a threshold of $p < 0.05$ (Family wise error rate (FWER) correction) following treatment with B[a]P, C_{60} or co-mixtures (5–100 µg/L B[a]P and 1 mg/L C_{60}). The top GO terms are listed in Table 3 (threshold set FWER = 0.01), while the full list separated by treatment and concentration can be found in supplemental material (Spreadsheet S1). Irrespective of treatment, biological process records the majority of enriched terms. The ability of a stress organism to adjust its cellular processes via transcriptional and subsequently proteomic processes allows it where possible to minimise cellular damage, which may lead to organism death. GO analysis revealed 30 enriched proteins following B[a]P exposure, 42 following C_{60} exposure and 31 in the mixture exposure. The response of *M. galloprovincialis* to B[a]P is characterised by a predominant enrichment of Biological processes (67% or 20 GO's) with the majority of these occurring at 100 µg/L. When compared to the mixture model at the same concentration, seven terms are absent in the mixture model compared to the single exposure viz. DNA metabolic processes (GO:0006259), DNA repair (GO:0006281), Cellular response to DNA damage stimulus (GO:0006974), cellular response to stress (GO:0033554), metabolic processes (GO:0008152), cellular metabolic processes (GO:0044237) primary metabolic processes (GO:0044238) and organic substance metabolic processes (GO:0071704). The absence of these enriched terms at the highest mixture concentration of B[a]P and C_{60} in association with the reduction in differentially expressed proteins (when compared to single exposure and 50 µg/L) suggest an antagonistic interaction between the two common contaminants. This may be explained by known properties of the chemicals. nC_{60} is an exceptional free radical scavenger [101,102], while B[a]P has been shown to produce free radicals under a variety of conditions [103]. B[a]P contributes approximately 50% of the total carcinogenic potential of the PAH group [104]. Transcriptomic alterations related to B[a]P are likely to be related to genotoxic mechanisms in addition to other biological processes such as mitochondrial activities and immune response as outlined previously [33]. In contrast, Zhang et al. demonstrated that aqueous C_{60} aggregates induced apoptosis of macrophage by changing the mitochondrial membrane potential [105]. As predicted by the literature, enriched GO terms following single nC_{60} exposure are predominantly related to changes to the membrane-enclosed, organelle and intracellular lumen, while mixed exposure resulted in enrichment of mitochondrial components (viz. matrix, ribosome and protein complex). This enrichment of organelle cellular components correlates with enrichment of the ribosome KEGG pathway (ko03010, 35 proteins at 1 mg/L C_{60}), suggesting an increase in the production of newly synthesised organelle proteins which must find its way from site of production in the cytosol to the organelle where it functions. It was not feasible to quantify changes in cellular components in the digestive gland during this study; however, we can postulate from prior studies that observed changes may be linked to changes in the mitochondria. Mitochondria are essential eukaryotic organelles required for a range of metabolic, signalling and development processes. Using fullerenol, a polyhydroxylated fullerene derivative, Yang et al. demonstrated significant changes to isolated mitochondria via mitochondrial swelling, collapse of membrane potential, decreased of membrane fluidity and alterations to the ultrastructure [106]. The increase in protein production via the ribosome at the highest concentration may reflect the activation of a repair mechanism for damage to this structure. In a recent review, the main negative molecular and cellular responses associated with carbon nanotube (CNTs) in mammals were associated with oxidative stress which can promote inflammation, mitochondrial oxidation and activation of apoptosis [107]. Additionally, Zhang et al. reported on a loss in mitochondrial membrane potential in a mouse in vitro model, in association with increase in cellular ROS suggesting mitochondria associated apoptosis [105]. In a typical aquatic NP exposure, uptake is followed by localisation into the endosomes, lysosomes and digestive associated cells as well as the lumen of digestive tubules [22,27,108]. This NP exposure response can be followed by disruption or modification to mitochondrial activity [30]. Although the current study would support the hypothesis of mitochondrial damage/repair, further work will need to be carried out to verify.

Table 3. Subset of enriched Gene Ontology (GO) terms with an family wise error (FWER) threshold of 1% (or 0.01) following B[a]P (5–100 μg/L), C_{60} (0.01–1 mg/L) and a mixture of B[a]P (5–100 μg/L) and C_{60} (1 mg/L) treatments. Cellular component and biological processes are abbreviated to CC and BP, respectively.

Treatment	Ontology	GO-ID	GO-ID Name	FWER
B[a]P (100 μg/L)	BP	GO:0006139	Nucleobase-containing compound metabolic process	0.01
B[a]P (100 μg/L)	BP	GO:0006725	Cellular aromatic compound metabolic process	0.01
B[a]P (100 μg/L)	BP	GO:0034641	Cellular nitrogen compound metabolic process	0.01
B[a]P (100 μg/L)	BP	GO:0046483	Heterocycle metabolic process	0.01
B[a]P (100 μg/L)	BP	GO:0090304	Nucleic acid metabolic process	0.01
B[a]P (100 μg/L)	BP	GO:1901360	Organic cyclic compound metabolic process	0.01
C_{60} (0.01 mg/L)	BP	GO:0000226	Microtubule cytoskeleton organization	0.01
C_{60} (0.1 mg/L)	CC	GO:0031974	Membrane-enclosed lumen	0.01
C_{60} (0.1 mg/L)	CC	GO:0043233	Organelle lumen	0.01
C_{60} (0.1 mg/L)	CC	GO:0070013	Intracellular organelle lumen	0.01

3.4.3. KEGG Pathway Enrichment

To further analyse the identified proteins per treatment, KEGG pathway analysis was performed. Using the bioconductor package clusterProfiler, protein sequences were assigned to DEPs ($p < 0.05$) and submitted to GhostKoala to obtained KEGG Orthology numbers (KO). In general, 52–56% of entries were successfully annotated with approximately 92% of annotations associated with the mollusca taxonomy. Variation between enrichment was described per treatment and concentration as follows:

B[a]P: at 5 μg/L exposure, 52 enriched processes were identified and include ribosome processes (26 genes), thermogenesis (19 genes), protein processing in endoplasmic reticulum (13 genes) and mTOR signalling pathway (nine genes). At 50 μg/L exposure, 38 pathways were enriched and ribosome (26 genes), protein processing in the endoplasmic reticulum (17 genes) and phagosome (13 genes). Finally, at 100 μg/L, 26 enriched processes were identified including ribosome (26 genes), RNA transport (16 genes), protein processing in the endoplasmic reticulum (16 genes), biosynthesis of amino acids (16 genes) and endocytosis (15 genes). The mTOR signalling pathway was not enriched at either 50 or 100 μg/L.

The majority of enriched pathways identified can be grouped under genetic information processing, cellular processes, environmental information processing and metabolism. The top enriched pathways identified per concentration were plotted to identify commonalties and differences between differing concentrations of B[a]P (Figure 7a) based on genes identified in that pathway. Interestingly, unique pathways appear to be activated dependent on exposure concentration, with only the ribosome pathway consistently present and enriched at all concentrations potentially indicating the high degree of translation which may be occurring as a consequence of PAH exposure.

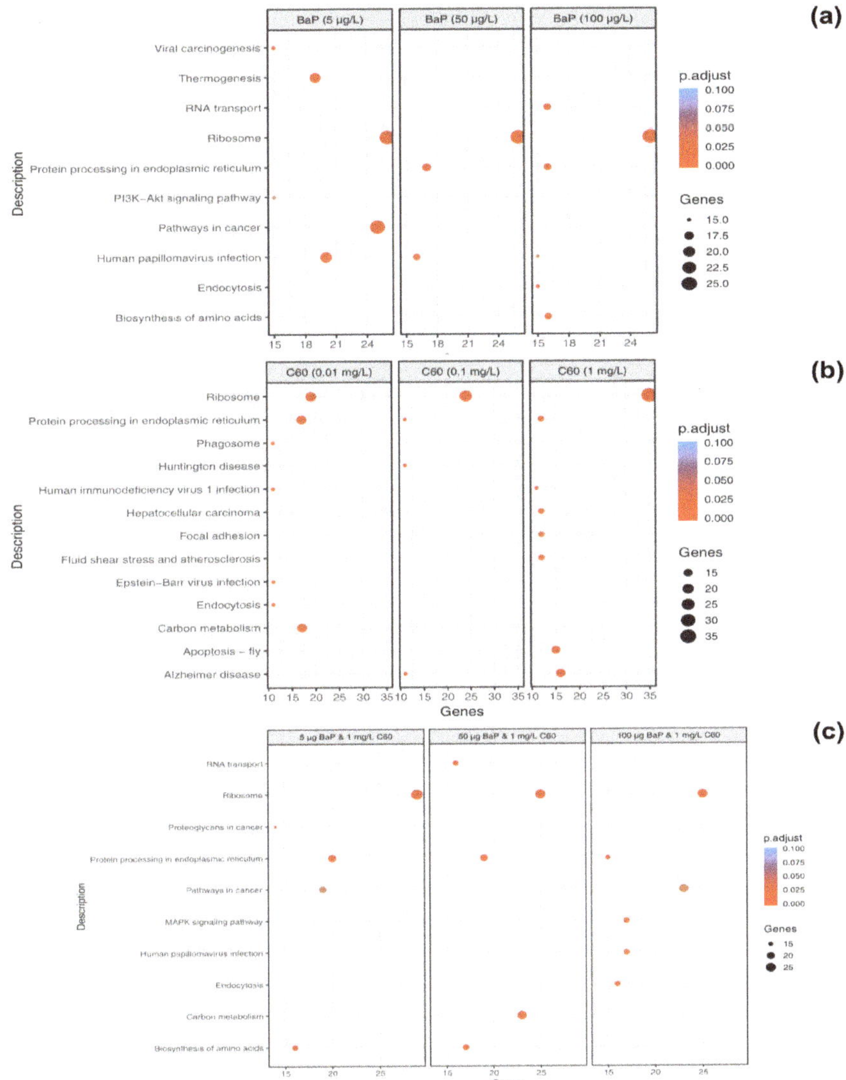

Figure 7. Dotplot of enriched KEGG pathways for differentially expressed genes (DEGs) ($p < 0.05$) that were common between concentrations of B[a]P (**a**), C_{60} (**b**) and a mixture of 5, 50 and 100 µg/L with 1 mg/L C_{60} (**c**). Along the *x*-axis, genes represent the number of genes identified as enriched in this particular pathway. The size and colour of each dot represents the gene number and adjustment *p* based on FDR correction.

C_{60}: at 0.01 mg/L exposure, 33 enriched pathways were identified while 12 enriched pathways were identified at 0.1 mg/L exposure and 35 enriched pathways identified at 1 mg/L exposure ($p < 0.05$, FDR = 5%). The top enriched pathways were illustrated in Figure 7b, with an absence of enrichment of certain pathways dependent on treatment concentration. For example, thermogenesis was only enriched at the highest concentration of 1 mg/L with 12 genes identified in the pathway. The ribosome is the top enriched pathway at all concentrations of C_{60} with 19 genes enriched at 0.01 mg/L exposure, 24 genes enriched at 0.1 mg/L exposure and 35 genes enriched at the highest concentration of 1 mg/L. This is closely followed by protein processing in endoplasmic reticulum, which is broadly comparable in terms of genes between 0.01 mg/L (17 genes), 0.1 mg/L (11 genes) and 1 mg/L (16 genes, Figure 8) exposure. The enriched pathways can be broadly grouped into predominantly genetic information processing, metabolism and cellular processes.

Mixtures: Under mixture scenario, C_{60} at a constant concentration of 1 mg/L was mixed with 5, 50 and 100 µg/L of B[a]P resulting in 50, 38 and 54 enriched pathways, respectively. At the lower mixture concentration of 5 µg/L B[a]P and C_{60}, the top three enriched descriptive terms were related to the ribosome (29 genes), protein processing in endoplasmic reticulum (20 genes) and pathways in cancer (23 genes). At 50 µg/L B[a]P and C_{60}, the top three enriched descriptive terms were related to the ribosome (23 genes), carbon metabolism (23 genes) and protein processing in endoplasmic reticulum (19 genes).

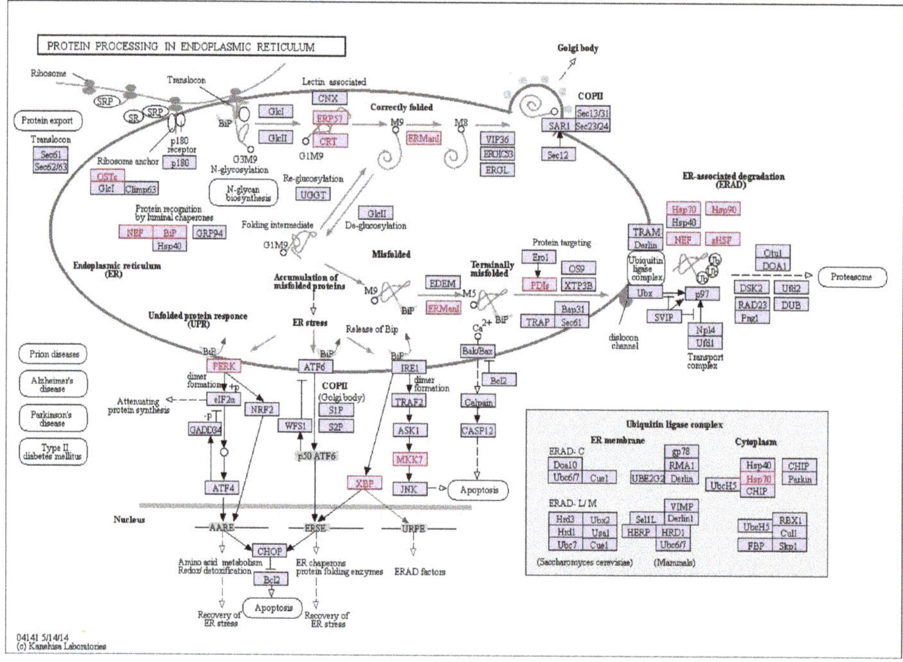

Figure 8. Interaction network of differentially expressed genes in the digestive gland of *M. galloprovincialis* involved in protein processing in the endoplasmic reticulum during exposure to 1 mg/L *n*C_{60}. Genes which are differentially expressed during exposure are highlighted in red.

Finally, at 100 µg/L B[a]P and C_{60}, the top three enriched descriptive terms were related to the ribosome (25 genes), pathways in cancer (23 genes) and mitogen-activated protein kinase signalling pathway (MAPK) signalling pathway (17 genes). Key genes consistently identified in the protein processing in the endoplasmic reticulum (irrespective of treatment) include *Hsp70*, *Hsp90*, *TRAP*, *PDIs* and *OSTs*. At the highest concentration of B[a]P and C_{60}, genes identified in pathways in cancer include *GSTs*, *CASP3* and *Wnt*. The top pathways based on quantity of genes present in the pathway were presented in Figure 7 with clear trends towards an absence of enrichment in certain pathways based on mixture concentration, e.g., MAPK signalling, which is only present at the top exposure concentration combination.

KEGG pathway analysis can provide physiological pathway information for various experiments with prior studies using it to aid in identification of mode of action of environmental contaminants [77]. In the current study, irrespective of exposure conditions or concentrations, the top enriched pathway identified using KEGG was the Ribosome with 19–39 genes identified in the pathway dependent on treatment and concentration. This was followed by protein processing in the endoplasmic reticulum and carbon metabolism. The ribosome is a large complex molecule made of RNA and proteins that perform the essential task of protein synthesis in the cell. They also serve as the initiation point for several translation-associated functions including protein folding and degradation of defective or nonstop mRNAs. Previous studies have demonstrated a change in regulation of genes which encode ribosomal protein subunits following B[a]P exposure, with the suggestion that mRNA directed protein synthesis is reduced in mussels exposed to higher B[a]P loads [33]. Additionally, *M. galloprovincialis* has been shown to response to B[a]P exposure via changes in abundance of proteins related to synthesis and degradation, energy supply (via ATP) and structural proteins [77]. Proteomic results for B[a]P exposure to digestive gland tissue are in agreement with prior studies and support the observed trends identified using transcriptomic methodologies. In the second most enriched pathway (viz. protein processing in endoplasmic reticulum), three heat shock proteins viz. *HSP70, HSP90 and HSP40* and other molecular chaperones were identified dependent on exposure conditions. This is not surprising given that many Heat Shock Proteins (*HSPs)* function as molecular chaperones to protect damaged proteins from aggregation, unfold protein aggregates or refold damaged proteins or target them for efficient removal [109]. These proteins regulate cell response to oxidative stress with *HSP70* strongly upregulated by heat stress and toxic chemicals. *HSP70* plays several essential roles in cellular protein metabolism [110,111] while *HSP40* facilitates cellular recovery from adverse effects of damaged or misfolded proteins (proteotoxic stress). Changes in HSPs, in addition to up/down regulation of *HSP40, HSP70* and *HSP90* have typically been reported in response to thermal stress in bivalves [95,112,113] and other environmental contaminants such as B[a]P [33]. In general, the consistent enrichment of genes involved in the endoplasmic-reticulum associated protein degradation (ERAD) pathway suggest that aqueous fullerene exposure targets the cellular pathway involved in targeting misfolding proteins for ubiquitination (post-translational modification) and subsequent degradation by proteasomes (protein degrading complex, breaks peptide bonds). It is interesting to note the overlap between organismal response to fullerene exposure and that of organismal response to thermal stress. Observed enrichment pathways in the current study viz protein processing in endoplasmic reticulum, apoptosis, ubiquitin mediated proteolysis, endocytosis, spliceosome, and MAPK signalling pathway have been observed as differentially enriched in oysters as a response to thermal stress [112].

3.5. Notes

The lack of consensus regarding C_{60} toxicity may be partly due to limited studies which incorporate both a physiological and ecological approach. As a consequence, little is still known about NP bioavailability, mode of uptake, ingestion rates and actual internal concentrations related to ADME [27]. Generally, the greater the water solubility of fullerene aggregates (through e.g., stirring, surface modifications, sonication), the less the toxicity associated with the exposure [88]. Gomes et al. highlight that, while mussels represent a target for environmental exposure to nanoparticles, exposure duration may significantly contribute to NPs' mediated toxicity [114]. As such, it is possible that the lack of differentially expressed proteins identified in this study is a factor of limited exposure duration. Limited exposure duration in the region of days or hours is common in the literature, and it would be of interest to explore long term exposure to NPs to look at the long-term impact and adaptation of mussels in the marine environment. Species specific responses to C_{60} are abundant in the literature and it would be remiss to not discuss how our results align with other marine invertebrates. Exposure to ROS can cause a range of reversible and irreversible modifications of protein amino acid side-chains which has been reviewed by Ghezzi and Bonetto [115]. Within the field of aquatic ecotoxicology, the toxic impact and potential mechanisms of single contaminant exposures have been extensively studied via laboratory experiments (in vivo, in vitro and in silico) and field monitoring. However, harder to predict is the effects of mixtures of pollutants in the environment. Biological damage observed cannot simply be linked to the actual environmental condition as mixtures of contaminants are known to exist in the aquatic ecosystem. This is further complicated with respect to nanomaterials due to their inherent properties which can amplify or negate the toxic effects of other compounds [75]. Complicated interactions may occur which make interpretation complex. For example, proteomic analysis of *Mytilus galloprovincialis* revealed that single Cu and B[a]P exposure in addition to a combination of the two generate different protein profiles with a non-additive profile [77]. Differences in mixture response compared to single exposure are likely to be related to individual chemical properties and toxicity mechanisms of B[a]P and C_{60}, as has been noted in B[a]P co-exposed with various metals [116]. C_{60} concentration was kept constant with increasing concentrations of B[a]P in an experimental design that has been previously carried out using algae and crustacean species [22]. This may reflect limited proteome changes at the exposure concentrations, with concentrations of C_{60} in the range of 10–500 ppb have been reported to be 10 fold below the no observable adverse effect level (NOAEL) [117,118]. At 1 mg/L, an increase in Glutathione S-Transferase (GST) activity in the digestive gland has been reported [108]. C_{60} is known to bind to minor grooves of double stranded DNA and trigger unwinding and disruption of the DNA helix [100]. C_{60} adsorbs onto cell-membrane P-glycoprotein through hydrophobic interactions, but the stability and secondary structure of the protein are barely affected [119]. P-glycoprotein is present in *Mytilus galloprovincialis* [120]. C_{60} and its derivatives are known to impact DNA and RNA in terms of stability, replication and reactivity in addition to structural stabilisation [121,122]. In a recent study, Canesi et al. determined that C_{60} fullerene exposure to *Mytilus galloprovincialis* hemocytes did not induce significant cytoxicity, and instead stimulated immune and inflammatory parameters such as lysozyme release, oxidative burst and nitric oxide (NO) production [10]. Nanomaterial suspensions can induce inflammatory processes in bivalve hemocytes akin to those observed in vertebrate cells [10]. Results from mammalian studies suggest that C_{60} fullerene exposure results predominantly in inflammatory responses [123].

4. Conclusions

This study has confirmed our hypothesis of an interaction between B[a]P and C_{60}, two ubiquitous environmental contaminants. We demonstrated for the first time an apparent antagonistic relationship at the genotoxic and the proteome expression level, which is not visible at lower exposure concentrations. This response is not explained by expected strong sorption of B[a]P on C_{60} as no difference in bioaccumulation was noted, but rather by the free radical scavenger propriety of C_{60}. No Trojan horse effects were observed for uptake or toxicity of the co-contaminants B[a]P in interaction with C_{60}. Proteome profile is dependent on concentration and treatment. The exposure to the three conditions had overlap and common mechanisms of response irrespective of differences in mode of action. The provided list of condition specific differentially expressed proteins and enriched pathways (Spreadsheet S1) may represent a step towards definitively identifying mode of action of these compounds in bivalves when combined with other OMICs based approaches. It should be noted that the antagonistic proteome response observed in the current study between B[a]P and C_{60} is based on a single concentration of the fullerene and as such represents a general overview of toxicological behaviour. It is possible that that this antagonistic interaction will change when another dose range is selected [90]. Gomes et al. previously highlighted that, while mussels represent a target for environmental exposure to nanoparticles, exposure duration may significantly contribute to NPs' mediated toxicity [114]. As such, further work must be carried out to explore mixture effects at different concentrations and over differing exposure duration.

Supplementary Materials: The following are available online at http://www.mdpi.com/2079-4991/9/7/987/s1, Figure S1: (a) bright-field TEM and (b) intensity-weighted particle size distribution of nC60 aggregates present in mussel-exposed seawater, Figure S2: (a) bright-field TEM and (b) point EDX spectroscopy analysis of $Er_3N@C_{80}$, Figure S3: Dark-field STEM and EDX spectroscopy mapping analysis of $Er_3N@C_{80}$, confirming the necessity for spectroscopy to confirm the presence of labelled fullerenes, using the characteristic X-rays emitted from Er upon electron irradiation, Figure S4: (a,c,e) dark-field STEM and (b,d,f) corresponding point EDX spectroscopy analysis of cross-sections of mussel digestive gland exposed to $Er_3N@C_{80}$, Figure S5: Volcano plots representing the differentially expressed proteins with exposure to B[a]P, C_{60} or a mixture of the two (5–50–100 µg/L B[a]P, 1 mg/L C_{60}), Table S1: The influence of benzo[a]pyrene (B[a]P) of the hydrodynamic diameter (d_H) of nC_{60} in mussel-exposed seawater as determined by DLS, Table S2: The concentration of B[a]P in seawater at T0, day 1 and day 3, Table S3: The concentration of nC_{60} in seawater at T0, day 1 and day 3, Spreadsheet S1: Full list of DEPs, enriched Gene Ontology (GO) terms and KEGG pathways, R script S1: R script used for proteomics analysis.

Author Contributions: Conceptualization, A.N.J.; methodology, A.B., G.A.R., L.M.L., V.S., and Y.A.; formal analysis, A.B., L.M.L., and G.A.R.; investigation, A.B., L.M.L., G.A.R., Y.A., N.J.W., F.A., V.M.A., and V.S.; resources, A.N.J.; data curation, A.B., L.M.L.,V.S., and G.A.R.; writing—original draft preparation, A.B., L.M.L., and G.A.R.; writing—review and editing, A.B., L.M.L., G.A.R., Y.A., N.J.W., F.A., M.N.M., V.M.A., V.S., A.N.K., J.W.R., and A.N.J.; visualization, A.B. and L.M.L.; supervision, A.N.J.; project administration, A.N.J.; funding acquisition, A.N.J.

Funding: This study is mainly supported by Natural Environment Research Council (NERC), UK (Grant No. NE/L006782/1; PI: ANJ). Additional Support from the Engineering and Physical Sciences Research Council (EPSRC) (Grant No. EP/L022494/1) and the University of Nottingham is acknowledged. Work at King's College London was further supported by the National Institute for Health Research Health Protection Research Unit (NIHR HPRU) in the Health Impact of Environmental Hazards at King's College London in partnership with Public Health England (PHE) and Imperial College London.

Acknowledgments: The authors would like to thank Alessandro La Torre and Julie Watts for technical assistance and discussion. The authors would like to express their sincere thanks to Jian-Wen Qiu, Hong Kong Baptist University, for sharing the information related to mussel proteome. The views expressed in this article are those of the authors and not necessarily those of the National Health Service, the National Institute for Health Research, the Department of Health and Social Care or Public Health England.

Conflicts of Interest: The authors declare no conflict of interest.

References and Notes

1. Bergin, I.L.; Witzmann, F.A. Nanoparticle toxicity by the gastrointestinal route: Evidence and knowledge gaps. *Int. J. Biomed. Nanosci. Nanotechnol.* **2013**, *3*, 163. [CrossRef] [PubMed]

2. Cheng, X.; Kan, A.T.; Tomson, M.B. Naphthalene adsorption and desorption from aqueous C_{60} fullerene. *J. Chem. Eng. Data* **2004**, *49*, 675–683. [CrossRef]

3. Georgakilas, V.; Perman, J.A.; Tucek, J.; Zboril, R. Broad family of carbon nanoallotropes: Classification, chemistry, and applications of fullerenes, carbon dots, nanotubes, graphene, nanodiamonds, and combined superstructures. *Chem. Rev.* **2015**, *115*, 4744–4822. [CrossRef] [PubMed]

4. Andrievsky, G.; Klochkov, V.; Karyakina, E.; Mchedlov-Petrossyan, N. Studies of aqueous colloidal solutions of fullerene C_{60} by electron microscopy. *Chem. Phys. Lett.* **1999**, *300*, 392–396. [CrossRef]

5. Chen, Z.; Westerhoff, P.; Herckes, P. Quantification of C_{60} fullerene concentrations in water. *Environ. Toxicol. Chem.* **2008**, *27*, 1852. [CrossRef] [PubMed]

6. Henry, T.B.; Petersen, E.J.; Compton, R.N. Aqueous fullerene aggregates (nC_{60}) generate minimal reactive oxygen species and are of low toxicity in fish: A revision of previous reports. *Curr. Opin. Biotechnol.* **2011**, *22*, 533–537. [CrossRef] [PubMed]

7. Markovic, Z.; Trajkovic, V. Biomedical potential of the reactive oxygen species generation and quenching by fullerenes (C_{60}). *Biomaterials* **2008**, *29*, 3561–3573. [CrossRef]

8. Blickley, T.M.; McClellan-Green, P. Toxicity of aqueous fullerene in adult and larval fundulus heteroclitus. *Environ. Toxicol. Chem.* **2008**, *27*, 1964. [CrossRef]

9. Della Torre, C.; Maggioni, D.; Ghilardi, A.; Parolini, M.; Santo, N.; Landi, C.; Madaschi, L.; Magni, S.; Tasselli, S.; Ascagni, M.; et al. The interactions of fullerene C_{60} and benzo(a)pyrene influence their bioavailability and toxicity to zebrafish embryos. *Environ. Pollut.* **2018**, *241*, 999–1008. [CrossRef]

10. Canesi, L.; Ciacci, C.; Vallotto, D.; Gallo, G.; Marcomini, A.; Pojana, G. In vitro effects of suspensions of selected nanoparticles (C_{60} fullerene, TiO_2, SiO_2) on *Mytilus* hemocytes. *Aquat. Toxicol.* **2010**, *96*, 151–158. [CrossRef]

11. Freixa, A.; Acuña, V.; Gutierrez, M.; Sanchís, J.; Santos, L.H.M.L.M.; Rodriguez-Mozaz, S.; Farré, M.; Barceló, D.; Sabater, S. Fullerenes influence the toxicity of organic micro-contaminants to river biofilms. *Front. Microbiol.* **2018**, *9*, 1–12. [CrossRef] [PubMed]

12. Yang, J.L.; Li, Y.F.; Guo, X.P.; Liang, X.; Xu, Y.F.; Ding, D.W.; Bao, W.Y.; Dobretsov, S. The effect of carbon nanotubes and titanium dioxide incorporated in PDMS on biofilm community composition and subsequent mussel plantigrade settlement. *Biofouling* **2016**, *32*, 763–777. [CrossRef] [PubMed]

13. Lehto, M.; Karilainen, T.; Róg, T.; Cramariuc, O.; Vanhala, E.; Tornaeus, J.; Taberman, H.; Jänis, J.; Alenius, H.; Vattulainen, I.; et al. Co-Exposure with fullerene may strengthen health effects of organic industrial chemicals. *PLoS ONE* **2014**, *9*, e114490. [CrossRef] [PubMed]

14. Park, J.-W.; Henry, T.B.; Ard, S.; Menn, F.-M.; Compton, R.N.; Sayler, G.S. The association between $nC_{(60)}$ and 17α-ethinylestradiol (EE2) decreases EE2 bioavailability in zebrafish and alters nanoaggregate characteristics. *Nanotoxicology* **2011**, *5*, 406–416. [CrossRef] [PubMed]

15. Kim, K.T.; Jang, M.H.; Kim, J.Y.; Kim, S.D. Effect of preparation methods on toxicity of fullerene water suspensions to Japanese medaka embryos. *Sci. Total Environ.* **2010**, *408*, 5606–5612. [CrossRef] [PubMed]

16. Nielsen, G.D.; Roursgaard, M.; Jensen, K.A.; Poulsen, S.S.; Larsen, S.T. In vivo biology and toxicology of fullerenes and their derivatives. *Basic Clin. Pharmacol. Toxicol.* **2008**, *103*, 197–208. [CrossRef] [PubMed]

17. Fiorito, S.; Serafino, A.; Andreola, F.; Bernier, P. Effects of fullerenes and single-wall carbon nanotubes on murine and human macrophages. *Carbon N. Y.* **2006**, *44*, 1100–1105. [CrossRef]

18. Oberdörster, G.; Oberdörster, E.; Oberdörster, J. Nanotoxicology: An emerging discipline evolving from studies of ultrafine particles. *Environ. Health Perspect.* **2005**, *113*, 823–839. [CrossRef] [PubMed]

19. Kennedy, A.J.; Hull, M.S.; Steevens, J.A.; Dontsova, K.M.; Chappell, M.A.; Gunter, J.C.; Weiss, C.A. Factors influencing the partitioning and toxicity of nanotubes in the aquatic environment. *Environ. Toxicol. Chem.* **2008**, *27*, 1932–1941. [CrossRef]

20. Trpkovic, A.; Todorovic-Markovic, B.; Trajkovic, V. Toxicity of pristine versus functionalized fullerenes: Mechanisms of cell damage and the role of oxidative stress. *Arch. Toxicol.* **2012**, *86*, 1809–1827. [CrossRef]

21. Zhu, S.; Oberdörster, E.; Haasch, M.L. Toxicity of an engineered nanoparticle (fullerene, C_{60}) in two aquatic species, Daphnia and fathead minnow. *Mar. Environ. Res.* **2006**, *62*, S5. [CrossRef] [PubMed]

22. Baun, A.; Sørensen, S.N.; Rasmussen, R.F.; Hartmann, N.B.; Koch, C.B. Toxicity and bioaccumulation of xenobiotic organic compounds in the presence of aqueous suspensions of aggregates of nano-C_{60}. *Aquat. Toxicol.* **2008**, *86*, 379–387. [CrossRef] [PubMed]

23. Henry, T.B.; Wileman, S.J.; Boran, H.; Sutton, P. Association of Hg^{2+} with aqueous $(C_{60})n$ aggregates facilitates increased bioavailability of Hg^{2+} in zebrafish (*Danio rerio*). *Environ. Sci. Technol.* **2013**, *47*, 9997–10004. [CrossRef] [PubMed]

24. Velzeboer, I.; Kwadijk, C.J.A.F.; Koelmans, A.A. Strong sorption of PCBs to nanoplastics, microplastics, carbon nanotubes, and fullerenes. *Environ. Sci. Technol.* **2014**, *48*, 4869–4876. [CrossRef] [PubMed]

25. Farkas, J.; Bergum, S.; Nilsen, E.W.; Olsen, A.J.; Salaberria, I.; Ciesielski, T.M.; Baczek, T.; Konieczna, L.; Salvenmoser, W.; Jenssen, B.M. The impact of TiO_2 nanoparticles on uptake and toxicity of benzo(a)pyrene in the blue mussel (*Mytilus edulis*). *Sci. Total Environ.* **2015**, *511*, 469–476. [CrossRef] [PubMed]

26. Holmstrup, M.; Bindesbøl, A.M.; Oostingh, G.J.; Duschl, A.; Scheil, V.; Köhler, H.R.; Loureiro, S.; Soares, A.M.V.M.; Ferreira, A.L.G.; Kienle, C.; et al. Interactions between effects of environmental chemicals and natural stressors: A review. *Sci. Total Environ.* **2010**, *408*, 3746–3762. [CrossRef] [PubMed]

27. Canesi, L.; Ciacci, C.; Fabbri, R.; Marcomini, A.; Pojana, G.; Gallo, G. Bivalve molluscs as a unique target group for nanoparticle toxicity. *Mar. Environ. Res.* **2012**, *76*, 16–21. [CrossRef] [PubMed]

28. de Lafontaine, Y.; Gagné, F.; Blaise, C.; Costan, G.; Gagnon, P.; Chan, H.M. Biomarkers in zebra mussels (Dreissena polymorpha) for the assessment and monitoring of water quality of the St Lawrence River (Canada). *Aquat. Toxicol.* **2000**, *50*, 51–71. [CrossRef]

29. Hu, M.; Lin, D.; Shang, Y.; Hu, Y.; Lu, W.; Huang, X.; Ning, K.; Chen, Y.; Wang, Y. CO_2-induced pH reduction increases physiological toxicity of nano-TiO_2 in the mussel *Mytilus coruscus*. *Sci. Rep.* **2017**, *7*, 1–11. [CrossRef] [PubMed]

30. Gomes, T.; Pereira, C.G.; Cardoso, C.; Bebianno, M.J. Differential protein expression in mussels *Mytilus galloprovincialis* exposed to nano and ionic Ag. *Aquat. Toxicol.* **2013**, *136–137*, 79–90. [CrossRef] [PubMed]

31. D'Agata, A.; Fasulo, S.; Dallas, L.J.; Fisher, A.S.; Maisano, M.; Readman, J.W.; Jha, A.N. Enhanced toxicity of "bulk" titanium dioxide compared to "fresh" and "aged" nano-TiO$_2$ in marine mussels (*Mytilus galloprovincialis*). *Nanotoxicology* **2014**, *8*, 549–558. [CrossRef] [PubMed]

32. Dallas, L.J.; Bean, T.P.; Turner, A.; Lyons, B.P.; Jha, A.N. Oxidative DNA damage may not mediate Ni-induced genotoxicity in marine mussels: Assessment of genotoxic biomarkers and transcriptional responses of key stress genes. *Mutat. Res.* **2013**, *754*, 22–31. [CrossRef] [PubMed]

33. Banni, M.; Sforzini, S.; Arlt, V.M.; Barranger, A.; Dallas, L.J.; Oliveri, C.; Aminot, Y.; Pacchioni, B.; Millino, C.; Lanfranchi, G.; et al. Assessing the impact of benzo[a]pyrene on marine mussels: Application of a novel targeted low density microarray complementing classical biomarker responses. *PLoS ONE* **2017**, *12*, e0178460. [CrossRef] [PubMed]

34. Sanchís, J.; Aminot, Y.; Abad, E.; Jha, A.N.; Readman, J.W.; Farré, M. Transformation of C$_{60}$ fullerene aggregates suspended and weathered under realistic environmental conditions. *Carbon* **2018**, *128*, 54–62. [CrossRef]

35. Wiśniewski, J.R.; Zougman, A.; Nagaraj, N.; Mann, M. Universal sample preparation method for proteome analysis. *Nat. Methods* **2009**, *6*, 359–362. [CrossRef] [PubMed]

36. Rappsilber, J.; Ishihama, Y.; Mann, M. Stop and go extraction tips for matrix-assisted laser desorption/ionization, nanoelectrospray, and LC/MS sample pretreatment in proteomics. *Anal. Chem.* **2003**, *75*, 663–670. [CrossRef] [PubMed]

37. Sequiera, G.L.; Sareen, N.; Sharma, V.; Surendran, A.; Abu-El-Rub, E.; Ravandi, A.; Dhingra, S. High throughput screening reveals no significant changes in protein synthesis, processing, and degradation machinery during passaging of mesenchymal stem cells. *Can. J. Physiol. Pharmacol.* **2018**, 1–8. [CrossRef] [PubMed]

38. *R: A Language and Environment for Statistical Computing*, version 3.5.0; R Foundation for Statistical Computing: Vienna, Austria, 2013.

39. Kessner, D.; Chambers, M.; Burke, R.; Agus, D.; Mallick, P. ProteoWizard: Open source software for rapid proteomics tools development. *Bioinformatics* **2008**, *24*, 2534–2536. [CrossRef]

40. Pedersen, T.L. MSGFplus: An interface between R and MS-GF+, 2017.

41. Kim, S.; Pevzner, P.A. MS-GF+ makes progress towards a universal database search tool for proteomics. *Nat. Commun.* **2014**, *5*, 1–10. [CrossRef]

42. Ding, Q.; Zhang, J. seqRFLP: Simulation and visualization of restriction enzyme cutting pattern from DNA sequences, 2012.

43. Levitsky, L.I.; Ivanov, M.V.; Lobas, A.A.; Gorshkov, M.V. Unbiased false discovery rate estimation for shotgun proteomics based on the target-decoy approach. *J. Proteome Res.* **2017**, *16*, 393–397. [CrossRef]

44. Deutsch, E.W.; Csordas, A.; Sun, Z.; Jarnuczak, A.; Perez-Riverol, Y.; Ternent, T.; Campbell, D.S.; Bernal-Llinares, M.; Okuda, S.; Kawano, S.; et al. The ProteomeXchange consortium in 2017: Supporting the cultural change in proteomics public data deposition. *Nucleic Acids Res.* **2017**, *45*, D1100–D1106. [CrossRef] [PubMed]

45. Perez-Riverol, Y.; Csordas, A.; Bai, J.; Bernal-Llinares, M.; Hewapathirana, S.; Kundu, D.J.; Inuganti, A.; Griss, J.; Mayer, G.; Eisenacher, M.; et al. The PRIDE database and related tools and resources in 2019: Improving support for quantification data. *Nucleic Acids Res.* **2019**, *47*, D442–D450. [CrossRef] [PubMed]

46. Fu, X.; Gharib, S.A.; Green, P.S.; Aitken, M.L.; Frazer, D.A.; Park, D.R.; Vaisar, T.; Heinecke, J.W. Spectral index for assessment of differential protein expression in shotgun proteomics. *J. Proteome Res.* **2008**, *7*, 845–854. [CrossRef] [PubMed]

47. Pursiheimo, A.; Vehmas, A.P.; Afzal, S.; Suomi, T.; Chand, T.; Strauss, L.; Poutanen, M.; Rokka, A.; Corthals, G.L.; Elo, L.L. Optimization of statistical methods impact on quantitative proteomics data. *J. Proteome Res.* **2015**, *14*, 4118–4126. [CrossRef] [PubMed]

48. Karpievitch, Y.V.; Dabney, A.R.; Smith, R.D. Normalization and missing value imputation for label-free LC-MS analysis. *BMC Bioinform.* **2012**, *13*, S5. [CrossRef]

49. Välikangas, T.; Suomi, T.; Elo, L.L. A systematic evaluation of normalization methods in quantitative label-free proteomics. *Brief. Bioinform.* **2018**, *19*, 1–11. [CrossRef]

50. Wei, R.; Wang, J.; Su, M.; Jia, E.; Chen, S.; Chen, T.; Ni, Y. Missing value imputation approach for mass spectrometry-based metabolomics data. *Sci. Rep.* **2018**, *8*, 663. [CrossRef]

51. Lazar, C.; Gatto, L.; Ferro, M.; Bruley, C.; Burger, T. Accounting for the multiple natures of missing values in label-free quantitative proteomics data sets to compare imputation strategies. *J. Proteome Res.* **2016**, *15*, 1116–1125. [CrossRef]

52. Gregori, J.; Sanchez, A.; Villanueva, J. msmsEDA and msmsTests: R/Bioconductor packages for spectral count label-free proteomics data analysis, 2016.

53. Yu, G.; Wang, L.-G.; Han, Y.; He, Q.-Y. clusterProfiler: An R Package for comparing biological themes among gene clusters. *Omics J. Integr. Biol.* **2012**, *16*, 284–287. [CrossRef]

54. Kanehisa, M.; Sato, Y.; Morishima, K. BlastKOALA and GhostKOALA: KEGG Tools for Functional Characterization of Genome and Metagenome Sequences. *J. Mol. Biol.* **2016**, *428*, 726–731. [CrossRef]

55. Oliveros, J. VENNY. An interactive tool for comparing lists with Venn's diagrams, 2007.

56. Helbock, H.J.; Beckman, K.B.; Shigenaga, M.K.; Walter, P.B.; Woodall, A.A.; Yeo, H.C.; Ames, B.N. DNA oxidation matters: The HPLC-electrochemical detection assay of 8-oxo-deoxyguanosine and 8-oxo-guanine. *Proc. Natl. Acad. Sci. USA* **1998**, *95*, 288–293. [CrossRef] [PubMed]

57. Akcha, F.; Burgeot, T.; Budzinski, H.; Pfohl-Leszkowicz, A.; Narbonne, J. Induction and elimination of bulky benzo[a]pyrene-related DNA adducts and 8-oxodGuo in mussels *Mytilus galloprovincialis* exposed in vivo to B[a]P-contaminated feed. *Mar. Ecol. Prog. Ser.* **2000**, *205*, 195–206. [CrossRef]

58. Barranger, A.; Heude-Berthelin, C.; Rouxel, J.; Adeline, B.; Benabdelmouna, A.; Burgeot, T.; Akcha, F. Parental exposure to the herbicide diuron results in oxidative DNA damage to germinal cells of the Pacific oyster *Crassostrea gigas*. *Comp. Biochem. Physiol. Part C Toxicol. Pharmacol.* **2016**, *180*, 23–30. [CrossRef] [PubMed]

59. Phillips, D.H.; Arlt, V.M. 32P-Postlabeling analysis of DNA adducts. In *Molecular Toxicology Protocols*; Keohavong, P., Grant, S.G., Eds.; Humana Press: Totowa, NJ, USA, 2014; pp. 127–138; ISBN 978-1-62703-739-6.

60. Reed, L.; Mrizova, I.; Barta, F.; Indra, R.; Moserova, M.; Kopka, K.; Schmeiser, H.H.; Wolf, C.R.; Henderson, C.J.; Stiborova, M.; et al. Cytochrome b_5 impacts on cytochrome P450-mediated metabolism of benzo[a]pyrene and its DNA adduct formation: Studies in hepatic cytochrome b_5/P450 reductase null (HBRN) mice. *Arch. Toxicol.* **2018**, *92*, 1625–1638. [CrossRef] [PubMed]

61. Kucab, J.E.; van Steeg, H.; Luijten, M.; Schmeiser, H.H.; White, P.A.; Phillips, D.H.; Arlt, V.M. *TP53* mutations induced by BPDE in Xpa-WT and Xpa-Null human *TP53* knock-in (Hupki) mouse embryo fibroblasts. *Mutat. Res. Fundam. Mol. Mech. Mutagen.* **2015**, *773*, 48–62. [CrossRef]

62. *R: A Language and Environment for Statistical Computing*, version 3.6.0; R Foundation for Statistical Computing: Vienna, Austria, 2019.

63. David, R.; Ebbels, T.; Gooderham, N. Synergistic and antagonistic mutation responses of human MCL-5 cells to mixtures of Benzo[a]pyrene and 2-amino-1-methyl-6-phenylimidazo[4,5-b]pyridine: Dose-related variation in the joint effects of common dietary carcinogens. *Environ. Health Perspect.* **2016**, *124*, 88–96. [CrossRef]

64. Katsifis, S.P.; Kinney, P.L.; Hosselet, S.; Burns, F.J.; Christie, N.T. Interaction of nickel with mutagens in the induction of sister chromatid exchanges in human lymphocytes. *Mutat. Res. Mutagen. Relat. Subj.* **1996**, *359*, 7–15. [CrossRef]

65. Schlesinger, R.B.; Zelikoff, J.T.; Chen, L.C.; Kinney, P.L. Assessment of toxicologic interactions resulting from acute inhalation exposure to sulfuric acid and ozone mixtures. *Toxicol. Appl. Pharmacol.* **1992**, *115*, 183–190. [CrossRef]

66. Fitzpatrick, P.J.; Krag, T.O.B.; Højrup, P.; Sheehan, D. Characterization of a glutathione S-transferase and a related glutathione-binding protein from gill of the blue mussel, *Mytilus edulis*. *Biochem. J.* **2015**, *305*, 145–150. [CrossRef]

67. Di, Y.; Aminot, Y.; Schroeder, D.C.; Readman, J.W.; Jha, A.N. Integrated biological responses and tissue-specific expression of *p53* and *ras* genes in marine mussels following exposure to benzo(α)pyrene and C_{60} fullerenes, either alone or in combination. *Mutagenesis* **2016**, *32*, 77–90. [CrossRef]

68. Gomes, T.; Pereira, C.G.; Cardoso, Ć.; Sousa, V.S.; Teixeira, M.R.; Pinheiro, J.P.; Bebianno, M.J. Effects of silver nanoparticles exposure in the mussel *Mytilus galloprovincialis*. *Mar. Environ. Res.* **2014**, *101*, 208–214. [CrossRef] [PubMed]

69. Tedesco, S.; Doyle, H.; Redmond, G.; Sheehan, D. Gold nanoparticles and oxidative stress in *Mytilus edulis*. *Mar. Environ. Res.* **2008**, *66*, 131–133. [CrossRef] [PubMed]

70. Kahru, A.; Dubourguier, H.C. From ecotoxicology to nanoecotoxicology. *Toxicology* **2010**, *269*, 105–119. [CrossRef] [PubMed]

71. Isakovic, A.; Markovic, Z.; Todorovic-Marcovic, B.; Nikolic, N.; Vranjes-Djuric, S.; Mirkovic, M.; Dramicanin, M.; Harhaji, L.; Raicevic, N.; Nikolic, Z.; et al. Distinct cytotoxic mechanisms of pristine versus hydroxylated fullerene. *Toxicol. Sci.* **2006**, *91*, 173–183. [CrossRef] [PubMed]

72. Du, C.; Zhang, B.; He, Y.; Hu, C.; Ng, Q.X.; Zhang, H.; Ong, C.N.; Lin, Z. Biological effect of aqueous C_{60} aggregates on *Scenedesmus obliquus* revealed by transcriptomics and non-targeted metabolomics. *J. Hazard. Mater.* **2017**, *324*, 221–229. [CrossRef] [PubMed]

73. Kuznetsova, G.P.; Larina, O.V.; Petushkova, N.A.; Kisrieva, Y.S.; Samenkova, N.F.; Trifonova, O.P.; Karuzina, I.I.; Ipatova, O.M.; Zolotaryov, K.V.; Romashova, Y.A.; et al. Effects of fullerene C_{60} on proteomic profile of *danio rerio* fish embryos. *Bull. Exp. Biol. Med.* **2014**, *156*, 694–698. [CrossRef] [PubMed]

74. Levi, N.; Hantgan, R.R.; Lively, M.O.; Carroll, D.L.; Prasad, G.L. C_{60}-Fullerenes: Detection of intracellular photoluminescence and lack of cytotoxic effects. *J. Nanobiotechnol.* **2006**, *4*, 1–11. [CrossRef] [PubMed]

75. Ferreira, J.L.R.; Lonné, M.N.; França, T.A.; Maximilla, N.R.; Lugokenski, T.H.; Costa, P.G.; Fillmann, G.; Antunes Soares, F.A.; de la Torre, F.R.; Monserrat, J.M. Co-exposure of the organic nanomaterial fullerene C_{60} with benzo[a]pyrene in *Danio rerio* (zebrafish) hepatocytes: Evidence of toxicological interactions. *Aquat. Toxicol.* **2014**, *147*, 76–83. [CrossRef]

76. Song, Q.; Chen, H.; Li, Y.; Zhou, H.; Han, Q.; Diao, X. Toxicological effects of benzo(a)pyrene, DDT and their mixture on the green mussel *Perna viridis* revealed by proteomic and metabolomic approaches. *Chemosphere* **2016**, *144*, 214–224. [CrossRef] [PubMed]

77. Maria, V.L.; Gomes, T.; Barreira, L.; Bebianno, M.J. Impact of benzo(a)pyrene, Cu and their mixture on the proteomic response of *Mytilus galloprovincialis*. *Aquat. Toxicol.* **2013**, *144–145*, 284–295. [CrossRef]

78. Linard, E.N.; Apul, O.G.; Karanfil, T.; Van Den Hurk, P.; Klaine, S.J. Bioavailability of Carbon Nanomaterial-Adsorbed Polycyclic Aromatic Hydrocarbons to *Pimphales promelas*: Influence of Adsorbate Molecular Size and Configuration. *Environ. Sci. Technol.* **2017**, *51*, 9288–9296. [CrossRef] [PubMed]

79. Santín, G.; Eljarrat, E.; Barceló, D. Bioavailability of classical and novel flame retardants: Effect of fullerene presence. *Sci. Total Environ.* **2016**, *565*, 299–305. [CrossRef] [PubMed]

80. Sanchís, J.; Olmos, M.; Vincent, P.; Farré, M.; Barceló, D. New Insights on the Influence of Organic Co-Contaminants on the Aquatic Toxicology of Carbon Nanomaterials. *Environ. Sci. Technol.* **2016**, *50*, 961–969. [CrossRef] [PubMed]

81. Su, Y.; Yan, X.; Pu, Y.; Xiao, F.; Wang, D.; Yang, M. Risks of single-walled carbon nanotubes acting as contaminants-carriers: Potential release of phenanthrene in Japanese medaka (*Oryzias latipes*). *Environ. Sci. Technol.* **2013**, *47*, 4704–4710. [CrossRef] [PubMed]

82. Della Torre, C.; Parolini, M.; Del Giacco, L.; Ghilardi, A.; Ascagni, M.; Santo, N.; Maggioni, D.; Magni, S.; Madaschi, L.; Prosperi, L.; et al. Adsorption of B(α)P on carbon nanopowder affects accumulation and toxicity in zebrafish (*Danio rerio*) embryos. *Environ. Sci. Nano* **2017**, *4*. [CrossRef]

83. Hüffer, T.; Kah, M.; Hofmann, T.; Schmidt, T.C. How redox conditions and irradiation affect sorption of PAHs by dispersed fullerenes (nC_{60}). *Environ. Sci. Technol.* **2013**, *47*, 6935–6942. [CrossRef]

84. Xia, X.; Chen, X.; Zhao, X.; Chen, H.; Shen, M. Effects of carbon nanotubes, chars, and ash on bioaccumulation of perfluorochemicals by *Chironomus plumosus* larvae in sediment. *Environ. Sci. Technol.* **2012**, *46*, 12467–12475. [CrossRef]

85. Al-Subiai, S.N.; Arlt, V.M.; Frickers, P.E.; Readman, J.W.; Stolpe, B.; Lead, J.R.; Moody, A.J.; Jha, A.N. Merging nano-genotoxicology with eco-genotoxicology: An integrated approach to determine interactive genotoxic and sub-lethal toxic effects of C_{60} fullerenes and fluoranthene in marine mussels, *Mytilus* sp. *Mutat. Res. Genet. Toxicol. Environ. Mutagen.* **2012**, *745*, 92–103. [CrossRef]

86. Sanchís, J.; Llorca, M.; Olmos, M.; Schirinzi, G.F.; Bosch-Orea, C.; Abad, E.; Barceló, D.; Farré, M. Metabolic responses of *Mytilus galloprovincialis* to fullerenes in mesocosm exposure experiments. *Environ. Sci. Technol.* **2018**, *52*, 1002–1013. [CrossRef]

87. Baird, W.M.; Hooven, L.A.; Mahadevan, B. Carcinogenic polycyclic aromatic hydrocarbon-DNA adducts and mechanism of action. *Environ. Mol. Mutagen.* **2005**, *45*, 106–114. [CrossRef]

88. Johnston, H.J.; Hutchison, G.R.; Christensen, F.M.; Aschberger, K.; Stone, V. The biological mechanisms and physicochemical characteristics responsible for driving fullerene toxicity. *Toxicol. Sci.* **2009**, *114*, 162–182. [CrossRef] [PubMed]

89. Rondags, A.; Yuen, W.Y.; Jonkman, M.F.; Horváth, B. Fullerene C_{60} with cytoprotective and cytotoxic potential: Prospects as a novel treatment agent in Dermatology? *Exp. Dermatol.* **2017**, *26*, 220–224. [CrossRef] [PubMed]

90. Groten, J.P.; Feron, V.J.; Suhnel, J. Toxicology of simple and complex mixtures. *Trends Pharmacol. Sci.* **2001**, *22*, 316–322. [CrossRef]

91. Slattery, M.; Ankisetty, S.; Corrales, J.; Marsh-Hunkin, K.E.; Gochfeld, D.J.; Willett, K.L.; Rimoldi, J.M. Marine proteomics: A critical assessment of an emerging technology. *J. Nat. Prod.* **2012**, *75*, 1833–1837. [CrossRef] [PubMed]

92. Campos, A.; Danielsson, G.; Farinha, A.P.; Kuruvilla, J.; Warholm, P.; Cristobal, S. Shotgun proteomics to unravel marine mussel (*Mytilus edulis*) response to long-term exposure to low salinity and propranolol in a Baltic Sea microcosm. *J. Proteomics* **2016**, *137*, 97–106. [CrossRef] [PubMed]

93. Pales Espinosa, E.; Koller, A.; Allam, B. Proteomic characterization of mucosal secretions in the eastern oyster, *Crassostrea virginica*. *J. Proteomics* **2016**, *132*, 63–76. [CrossRef]

94. Campos, A.; Apraiz, I.; da Fonseca, R.R.; Cristobal, S. Shotgun analysis of the marine mussel *Mytilus edulis* hemolymph proteome and mapping the innate immunity elements. *Proteomics* **2015**, *15*, 4021–4029. [CrossRef] [PubMed]

95. Fields, P.A.; Zuzow, M.J.; Tomanek, L. Proteomic responses of blue mussel (*Mytilus*) congeners to temperature acclimation. *J. Exp. Biol.* **2012**, *215*, 1106–1116. [CrossRef]

96. Franco-Martínez, L.; Martínez-Subiela, S.; Escribano, D.; Schlosser, S.; Nöbauer, K.; Razzazi-Fazeli, E.; Romero, D.; Cerón, J.J.; Tvarijonaviciute, A. Alterations in haemolymph proteome of *Mytilus galloprovincialis* mussel after an induced injury. *Fish. Shellfish Immunol.* **2018**, *75*, 41–47. [CrossRef]

97. Murgarella, M.; Puiu, D.; Novoa, B.; Figueras, A.; Posada, D.; Canchaya, C. A first insight into the genome of the filter-feeder mussel *Mytilus galloprovincialis*. *PLoS ONE* **2016**, *11*, e0151561. [CrossRef]

98. De Rosa, C.T.; El-Masri, H.A.; Pohl, H.; Cibulas, W.; Mumtaz, M.M. Implications of chemical mixtures in public health practice. *J. Toxicol. Environ. Health Part B* **2004**, *7*, 339–350. [CrossRef] [PubMed]

99. Rey-Salgueiro, L.; Martínez-Carballo, E.; Cid, A.; Simal-Gándara, J. Determination of kinetic bioconcentration in mussels after short term exposure to polycyclic aromatic hydrocarbons. *Heliyon* **2017**, *3*, e00231. [CrossRef] [PubMed]

100. Xu, J.; Corry, D.; Patton, D.; Liu, J.; Jackson, S. F-Actin plaque formation as a transitional membrane microstructure which plays a crucial role in cell-cell reconnections of rat hepatic cells after isolation. *J. Interdiscip. Histopathol.* **2012**, *1*, 50. [CrossRef]

101. Anilkumar, P.; Lu, F.; Cao, L.; G. Luo, P.; Liu, J.-H.; Sahu, S.; Tackett II, K.N.; Wang, Y.; Sun, Y.-P. Fullerenes for applications in biology and medicine. *Curr. Med. Chem.* **2011**, *18*, 2045–2059. [CrossRef] [PubMed]

102. Bakry, R.; Vallant, R.M.; Najam-ul-Haq, M.; Rainer, M.; Szabo, Z.; Huck, C.W.; Bonn, G.K. Medicinal applications of fullerenes. *Int. J. Nanomed.* **2007**, *2*, 639–649.

103. Sullivan, P.D. Free radicals of benzo(a)pyrene and derivatives. *Environ. Health Perspect.* **1985**, *64*, 283–295. [CrossRef] [PubMed]

104. Souza, T.; Jennen, D.; van Delft, J.; van Herwijnen, M.; Kyrtoupolos, S.; Kleinjans, J. New insights into BaP-induced toxicity: Role of major metabolites in transcriptomics and contribution to hepatocarcinogenesis. *Arch. Toxicol.* **2016**, *90*, 1449–1458. [CrossRef]

105. Zhang, B.; Bian, W.; Pal, A.; He, Y. Macrophage apoptosis induced by aqueous C_{60} aggregates changing the mitochondrial membrane potential. *Environ. Toxicol. Pharmacol.* **2015**, *39*, 237–246. [CrossRef]

106. Yang, L.Y.; Gao, J.L.; Gao, T.; Dong, P.; Ma, L.; Jiang, F.L.; Liu, Y. Toxicity of polyhydroxylated fullerene to mitochondria. *J. Hazard. Mater.* **2016**, *301*, 119–126. [CrossRef]

107. Costa, P.M.; Bourgognon, M.; Wang, J.T.W.; Al-Jamal, K.T. Functionalized carbon nanotubes: From intracellular uptake and cell-related toxicity to systemic brain delivery. *J. Control. Release* **2016**, *241*, 200–219. [CrossRef]

108. Canesi, L.; Fabbri, R.; Gallo, G.; Vallotto, D.; Marcomini, A.; Pojana, G. Biomarkers in *Mytilus galloprovincialis* exposed to suspensions of selected nanoparticles (Nano carbon black, C60 fullerene, Nano-TiO_2, Nano-SiO_2). *Aquat. Toxicol.* **2010**, *100*, 168–177. [CrossRef] [PubMed]

109. Verghese, J.; Abrams, J.; Wang, Y.; Morano, K.A. Biology of the Heat Shock Response and Protein Chaperones: Budding Yeast (*Saccharomyces cerevisiae*) as a Model System. *Microbiol. Mol. Biol. Rev.* **2012**, *76*, 115–158. [CrossRef] [PubMed]

110. Ryan, M.T.; Pfanner, N. Hsp70 proteins in protein translocation. *Adv. Protein Chem.* **2001**, *59*, 223–242. [CrossRef] [PubMed]

111. Pratt, W.B.; Toft, D.O. Regulation of signaling protein function and trafficking by the hsp90/hsp70-based chaperone machinery. *Exp. Biol. Med.* **2003**, *228*, 111–133. [CrossRef] [PubMed]

112. Li, J.; Zhang, Y.; Mao, F.; Tong, Y.; Liu, Y.; Zhang, Y.; Yu, Z. Characterization and identification of differentially expressed genes involved in thermal adaptation of the Hong Kong Oyster *Crassostrea hongkongensis* by digital gene expression profiling. *Front. Mar. Sci.* **2017**, *4*, 1–12. [CrossRef]

113. Negri, A.; Oliveri, C.; Sforzini, S.; Mignione, F.; Viarengo, A.; Banni, M. Transcriptional response of the Mussel *Mytilus galloprovincialis* (Lam.) following exposure to heat stress and copper. *PLoS ONE* **2013**, *8*, e66802. [CrossRef] [PubMed]

114. Gomes, T.; Pinheiro, J.P.; Cancio, I.; Pereira, C.G.; Cardoso, C.; Bebianno, M.J. Effects of copper nanoparticles exposure in the mussel *Mytilus galloprovincialis*. *Environ. Sci. Technol.* **2011**, *45*, 9356–9362. [CrossRef]

115. Ghezzi, P.; Bonetto, V. Redox proteomics: Identification of oxidatively modified proteins. *Proteomics* **2003**, *3*, 1145–1153. [CrossRef]

116. Chen, S.; Qu, M.; Ding, J.; Zhang, Y.; Wang, Y.; Di, Y. BaP-metals co-exposure induced tissue-specific antioxidant defense in marine mussels *Mytilus coruscus*. *Chemosphere* **2018**, *205*, 286–296. [CrossRef]

117. Kamat, J.P.; Devasagayam, T.P. a.; Priyadarsini, K.I.; Mohan, H.; Mittal, J.P. Oxidative damage induced by the fullerene C_{60} on photosensitization in rat liver microsomes. *Chem. Biol. Interact.* **1998**, *114*, 145–159. [CrossRef]

118. Kamat, J.P.; Devasagayam, T.P.A.; Priyadarsini, K.I.; Mohan, H. Reactive oxygen species mediated membrane damage induced by fullerene derivatives and its possible biological implications. *Toxicology* **2000**, *155*, 55–61. [CrossRef]

119. Xu, X.; Li, R.; Ma, M.; Wang, X.; Wang, Y.; Zou, H. Multidrug resistance protein P-glycoprotein does not recognize nanoparticle C_{60}: Experiment and modeling. *Soft Matter* **2012**, *8*, 2915–2923. [CrossRef]

120. Smital, T.; Sauerborn, R.; Hackenberger, B.K. Inducibility of the P-glycoprotein transport activity in the marine mussel *Mytilus galloprovincialis* and the freshwater mussel *Dreissena polymorpha*. *Aquat. Toxicol.* **2003**, *65*, 443–465. [CrossRef]

121. Xu, X.; Wang, X.; Li, Y.; Wang, Y.; Yang, L. A large-scale association study for nanoparticle C60 uncovers mechanisms of nanotoxicity disrupting the native conformations of DNA/RNA. *Nucleic Acids Res.* **2012**, *40*, 7622–7632. [CrossRef] [PubMed]

122. An, H.; Jin, B. DNA exposure to buckminsterfullerene (C_{60}): Toward DNA stability, reactivity, and replication. *Environ. Sci. Technol.* **2011**, *45*, 6608–6616. [CrossRef] [PubMed]

123. Park, E.J.; Roh, J.; Kim, Y.; Park, K. Induction of inflammatory responses by carbon fullerene (C_{60}) in cultured RAW264.7 cells and in intraperitoneally injected mice. *Toxicol. Res.* **2013**, *26*, 267–273. [CrossRef] [PubMed]

Article

In Vitro Effects of Titanium Dioxide Nanoparticles (TiO_2NPs) on Cadmium Chloride ($CdCl_2$) Genotoxicity in Human Sperm Cells

Marianna Santonastaso [1], Filomena Mottola [2], Concetta Iovine [2], Fulvio Cesaroni [3], Nicola Colacurci [1] and Lucia Rocco [2,*]

[1] Department of Woman, Child and General and Special Surgery, University of Campania "Luigi Vanvitelli", 80138 Napoli, Italy; marianna.santonastaso@unicampania.it (M.S.); Nicola.colacurci@unicampania.it (N.C.)

[2] Department of Environmental, Biological and Pharmaceutical Sciences and Technologies, University of Campania "Luigi Vanvitelli", 81100 Caserta, Italy; filomena.mottola@unicampania.it (F.M.); concetta.iovine@unicampania.it (C.I.)

[3] PMA Center of Cassinate, 03043 Cassino, Italy; fulviocesaroni@hotmail.it

* Correspondence: lucia.rocco@unicampania.it

Received: 30 April 2020; Accepted: 3 June 2020; Published: 5 June 2020

Abstract: The environmental release of titanium dioxide nanoparticles (TiO_2NPs) associated with their intensive use has been reported to have a genotoxic effect on male fertility. TiO_2NP is able to bind and transport environmental pollutants, such as cadmium (Cd), modifying their availability and/or toxicity. The aim of this work is to assess the in vitro effect of TiO_2NPs and cadmium interaction in human sperm cells. Semen parameters, apoptotic cells, sperm DNA fragmentation, genomic stability and oxidative stress were investigated after sperm incubation in cadmium alone and in combination with TiO_2NPs at different times (15, 30, 45 and 90 min). Our results showed that cadmium reduced sperm DNA integrity, and increased sperm DNA fragmentation and oxidative stress. The genotoxicity induced by TiO_2NPs-cadmium co-exposure was lower compared to single cadmium exposure, suggesting an interaction of the substances to modulate their reactivity. The Quantitative Structure-Activity Relationship (QSAR) computational method showed that the interaction between TiO_2NPs and cadmium leads to the formation of a sandwich-like structure, with cadmium in the middle, which results in the inhibition of its genotoxicity by TiO_2NPs in human sperm cells.

Keywords: sperm DNA damage; titanium dioxide nanoparticles; cadmium; oxidative stress; male infertility

---

1. Introduction

Titanium dioxide is a colorless, crystalline and poorly soluble powder. The small dimensions of the crystals are responsible for their particular physical–chemical characteristics and enhanced reactivity, unlike other solid materials and larger particles with the same chemical composition [1,2]. Titanium dioxide used in the form of nanoparticles (TiO_2NPs) showed different properties such as robust oxidation, biocompatibility and photocatalysis. Therefore, TiO_2NPs are used in a wide range of applications, including pharmaceuticals, cosmetics, paints, medicine and engineering. However, there are some concerns about the possible biological effects associated with their use [3]. In fact, the increased use of TiO_2NPs in industry and in daily applications (domestic, cosmetic, food) has attracted growing interest because, to date, we cannot yet accurately predict and control the impact on health due to their release into the environment. It is known that TiO_2NPs induce in vivo and in vitro genotoxicity and cytotoxicity on several experimental models, by altering the genome stability, increasing the apoptosis and decreasing the cell viability in different vertebrates [4–7].

TiO$_2$NPs negatively influence male fertility, as they lead to a reduced sperm quality and daily sperm production, reduced weight of the testes and histopathological testicular changes. A review on the reproductive and developmental toxicity of nanomaterials indicates that the studies are generally performed in adult or pre-pubertal/pubertal rats or mice [8]. Recently, it has been shown that TiO$_2$NPs have adverse effects on human reproduction by inducing DNA sperm damage. Human ejaculated spermatozoa treated with TiO$_2$NPs showed a loss of DNA integrity, probably due to the production of intracellular reactive oxygen species (ROS) [9].

In addition to toxicity and genotoxicity caused by their inherent and unique properties [10,11], TiO$_2$NPs were also demonstrated to interact with pollutants, either organic or heavy metals, modulating bioaccumulation and toxic responses in co-exposed organisms, and modifying their fate, behavior, bioavailability and toxicity for the ecosystem and human health [12,13]. In fact, TiO$_2$NPs were able to phagocytize and carry other pollutants and/or drugs, hence skipping the natural cellular defenses, through the mechanism known as the "Trojan Horse effect" [14–16]. However, the interaction between TiO$_2$NPs and co-existing contaminants in the environment remains unclear, with conflicting results. On human amniocytes in vitro, TiO$_2$NPs increased the genotoxicity of lincomycin through a loss of DNA integrity, apoptosis and DNA damage [17]. In zebrafish larvae, *Daphnia magna* and carp, TiO$_2$NPs enhanced lead (Pb), copper (Cu), arsenic (As) (III), zinc (Zn) and cadmium (Cd) bioaccumulation and toxicity [18–22], while, in algae (*Chlamydomonas reinhardtii* and *Microcystis aeruginosa*) and amphipods (*Gammarus fossarum*), TiO$_2$NPs reduced the bioavailability and toxicity of Cd and Cu [23–25]. TiO$_2$NPs enhanced the Cd and nickel (Ni) reproductive and developmental toxicity in *Caenorhabditis elegans* in a dose-dependent manner [26]. Among heavy metal, Cd is the most widespread in industrial applications, ranked as the seventh most toxic heavy metal, with a specific toxicological profile (ATSDR 2012) that describes its adverse effects for living organisms and human health [27]. Due to its application in fertilizers, battery, pigments and plastics [28], Cd may enter the natural environment and impact human health and the environment [29]. Cd and its compounds were classified as type I human carcinogens in 1993 by the International Agency for Research on Cancer, IARC [30]. Cd is reported as toxic for organs, such as the kidney, liver and stomach, causing respiratory and bone disease, as well as neurological disorders [31–33]. Finally, Cd exerts negative effects on human reproduction. In fact, Cd concentration in the human seminal plasma is closely related to working conditions, food and cigarette smoke, with a reduced fertility being observed in highly exposed patients [34]. In vitro studies have shown that Cd affects sperm motility and the sperm's ability to reach and penetrate into the oocyte [35–38]. Furthermore, rats exposed to Cd showed a reduced testicular volume sperm concentration and testosterone concentration in the Leydig cells, as well as an increased follicle-stimulating hormone (FSH) concentration in the serum [39].

Cd-induced damage depends on the dose, duration of exposure, type of contact, as well as the interaction with other materials and/or nanomaterials. The effects of TiO$_2$NPs and Cd^{2+} co-exposure have been investigated in plants and aquatic species [6,40–46]. A recent study showed that a non-cytotoxic concentration of TiO$_2$NPs enhanced the toxicological potential of Cd^{2+} in human liver (HepG2) and human breast cancer (MCF-7) cells [47]. However, the overall results are conflicting, underlining that the influence of TiO$_2$NPs on Cd^{2+} accumulation and toxicity varies according to the species, tissues, culture media and physical–chemical behavior of particles in exposure media. Moreover, their influence on reproductive health is still scarcely investigated.

This study aimed to evaluate the genotoxicity of Cd and to investigate the combined effects of TiO$_2$NPs and cadmium chloride (CdCl$_2$) on human ejaculated sperm cells in vitro. In this study, we attempted to determine whether TiO$_2$NPs could modify the possible Cd genotoxic responses. To achieve our goal, we investigated cytotoxicity, genotoxicity, oxidative stress and apoptosis in human sperm cells after TiO$_2$NPs and CdCl$_2$ co-exposure. To our knowledge, this is the first study to evaluate the responses of human sperm on CdCl$_2$ and TiO$_2$NPs co-exposure. It provided new insights into the TiO$_2$NPs' interaction with heavy metal and clarified the potential reproductive health risk of manufactured nanoparticles as carriers of contaminants.

2. Material and Methods

2.1. Chemicals

Nano-powder of TiO_2NPs (Aeroxide) was supplied by Evonik Degussa (Essen, Germany; Lot. 614061098). Aeroxide is a mixture of 75% rutile and 25% anatase forms with an average primary particle size of 21 nm and declared purity of 99.9%. $CdCl_2$ (CAS number 10108-64-2, 99.999% purity) was acquired from Sigma-Aldrich (St. Louis, Missouri, USA). $CdCl_2$ was dissolved in water to a stock concentration of 100 mM. Benzene (CAS number 71-43-2, 99.8 purity) was provided from Sigma-Aldrich (St. Louis, MO, USA).

2.2. Preparation and Characterization of TiO_2NPs

TiO_2NPs stock solutions (10.0 mg/L) were prepared by dispersing the NPs in Eagle's Minimum Essential Medium (MEM) with sonication (40 kHz frequency, Dr. Hielscher UP 200S, Teltow, Germany) and were dosed according to Santonastaso and collaborators [9]. Briefly, we acquired UV-Vis spectra of TiO_2NPs in the range of 200–600 nm using a Shimadzu UV-1700 double-beam spectrophotometer. UV spectra of 1 μg/L TiO_2NPs-enriched culture medium at 15, 45 and 90 min showed a secondary peak at longer wavelengths probably due to the formation of agglomerates (Table 1). The primary particle diameter and shape were determined by Zeiss-LIBRA120 (Carl Zeiss Oberkochen, Germany) Transmission Electron Microscope (TEM). TiO_2NPs' TEM images showed a size distribution ranging approximately from 20 to 60 nm, with a partly irregular and semispherical shape, and agglomeration occurred with a diameter in the range of 400 ± 52 nm (Figure 1).

Table 1. Dispersion of tested 1 μg/L TiO_2NPs-enriched culture media at 0, 15, 30, 45 and 90 min calculated on the TiO_2NPs calibration curve, obtained by plotting the absorbance at the maximum wavelength (325 nm) vs. the different sonicated standard solutions' dose levels.

Time (min)	nTiO$_2$NPs [1 μg/L]
0	1.01 ± 0.01
15	0.91 ± 0.05
30	0.87 ± 0.02
45	0.69 ± 0.01
90	0.27 ± 0.03

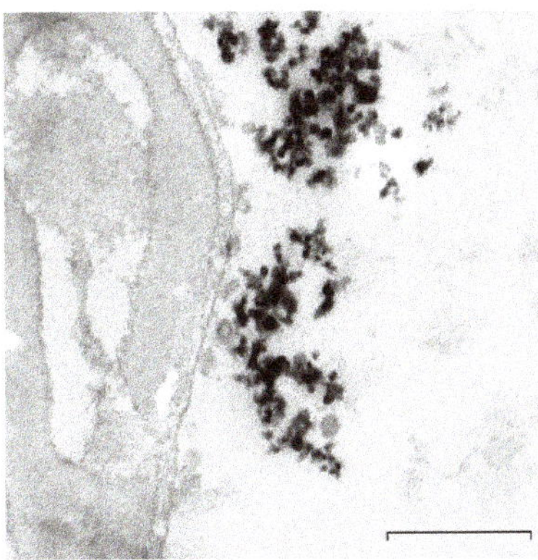

Figure 1. TEM micrograph showing the aggregation pattern of TiO_2 nanoparticles (bar 0.5 μm).

2.3. Sample Collection, Evaluation and Exposure Procedure

Semen samples were obtained by masturbation from 125 men between 25 and 39 years old, underwent routine semen analysis and were examined in our Reproduction Biology Laboratory (University of Campania L. Vanvitelli). Patients were informed about the purpose of the study and they signed written informed consent. Subjects on any medication or antioxidant supplementation were not included. All ejaculates presenting normal semen parameters with a seminal white blood cell count $<0.5 \times 10^6$ /mL, which was less than the pathologic concentration, were used in the study (Table 2) [48]. After liquefaction at room temperature for 30 min, the semen volume, pH, sperm concentration, motility, morphology and viability were determined according to the World Health Organization (WHO) guidelines (2010) [48]. The percentage of morphologically abnormal spermatozoa was evaluated by Test-simplets® pre-stained slides (Origio, Cooper Surgical, Inc., Trumbull, CT, USA). Sperm vitality was assessed by the eosin–nigrosine staining. The ejaculates were purified by discontinuous density gradient centrifugation. The sperm preparation was done using a 45%–90% double density gradient (SPERM GRAD™; Vitrolife, Göteborg, Sweden) in order to obtain a sufficient number of selected spermatozoa to perform the experiments.

Table 2. Parameters of semen fluid selected for the study. Sperm parameters were expressed as mean ± SD.

Sperm Parameters	Mean± SD
Semen volume (mL)	3.5 ± 0.5
Ph	6.9 ± 0.3
Sperm concentration (*10^6 sperm/mL)	60.4 ± 15.6
Vitality (%)	70.8 ± 5.8
Motility (%)	
Progressive	69.0 ± 15.7
Non-progressive	20.0 ± 5.5
Immobile	11 ± 9
Normal morphology (%)	15 ± 6.5

Each purified sample was divided into four aliquots (1×10^6 sperm/mL). One aliquot was exposed to 10 μM of $CdCl_2$, another aliquot was co-exposed to 1 μg/L of TiO_2NPs and 10 μM of $CdCl_2$; one aliquot was treated with 0.4 μL/mL benzene as positive control [49], while an untreated aliquot was used as negative control each time. TiO_2NPs and $CdCl_2$ concentrations were selected based on previous in vitro studies [9,36]. The exposition was evaluated after 15, 30, 45 and 90 min (min). Incubation was performed in Eagle's Minimum Essential Medium (MEM) at 36.5 °C. After exposure, the samples were centrifuged for 5 min at 1500 rpm, the supernatant was removed, the pellet was re-suspended in 500 μL bicarbonate buffer and the semen parameters were evaluated.

2.4. Comet Assay

The Comet assay and the relative statistical analyses were performed according to Santonastaso and collaborators [9]. Briefly, sperm cells were mixed with low melting point (LMP, 0.7%) agarose and were included into normal melting point (NMP, 21%) agarose layers on slides for 30 min. Then, another LMP layer was added. The slides were treated with lysis buffer (NaCl 2.5 M, Na_2EDTA 0.1 M, Tris-Base 0.4 M, Triton X-100 1%, dimethyl sulfoxide (DMSO) 10%, pH 10) and were enzymatically digested with proteinase K (0.5 μg/L). After washing with a neutralizing solution, the slides were incubated in alkaline buffer (NaOH 10N, EDTA 200 mM, pH 12.1) for 10 min and then exposed to electrophoresis (25 V, 300 mA, 20 min). After fixing in cold methanol and staining with ethidium bromide, the slides were observed by using a fluorescence microscope with 60X magnification (Nikon Eclipse E-600). The images were acquired by means of the "OpenComet" software [50]. The Comet assay was performed in triplicate. The parameter that was considered was the percentage of damaged DNA in the comet tail (% Tail DNA) (Figure 2).

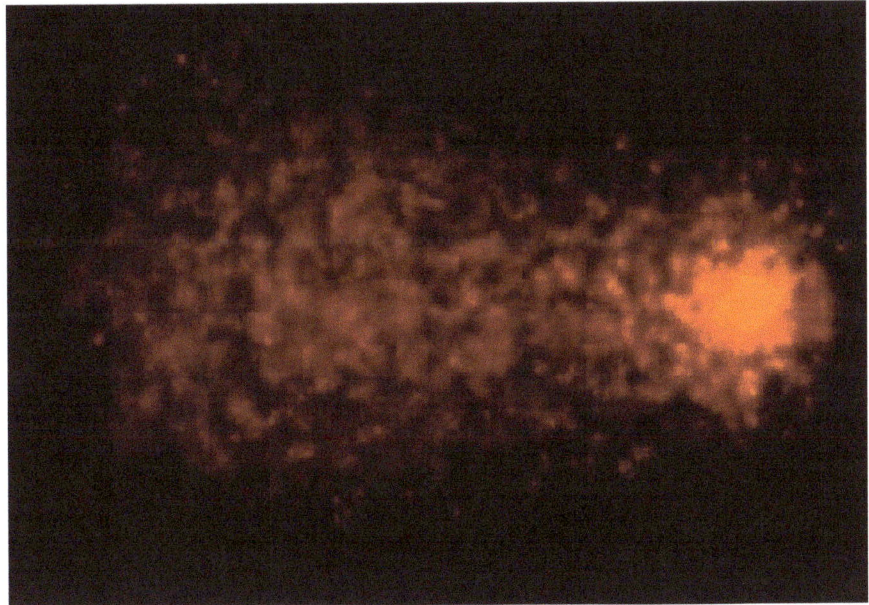

Figure 2. Comet tail DNA in human sperm cell analyzed using "OpenComet" software.

2.5. TUNEL Test

The TUNEL test was performed according to Santonastaso and collaborators [9] using the In-Situ Cell Death Detection Kit (Roche Diagnostics, Basel, Switzerland). The sperm samples were put on glass slides, fixed in 4% paraformaldehyde for 1 h and air-dried. After 2 min incubation in a permeabilizing solution, the slides were washed in bicarbonate buffer and air-dried. Then, 5 μL of terminal deoxy nucleotidyl transferase enzyme solution and 45 μL of label solution were placed on each slide. After 1 h incubation in a humid chamber at 37 °C, the slides were stained in 4′,6-diamidino-2-phenylindole (DAPI) solution for 5 min, and 100 μL of 1,4 diazobicyclo (2,2,2) octane (DABCO) solution (20×) was added to each slide. The TUNEL test was performed in triplicate. The slides were analyzed by using a fluorescence microscope (Nikon Eclipse E-600) equipped with BP 330–380 nm and LP 420 nm filters. The percentage of sperm with fragmented DNA was referred to as the percentage of TUNEL-positive sperm.

2.6. Genomic DNA Extraction, RAPD-PCR Technique and Genomic Template Stability

Sperm cells' DNA was extracted and purified from 200 μL/sample, using the High pure PCR template preparation Kit (ROCHE Diagnostics®). The RAPD method is a PCR-based technique that amplifies random DNA fragments with the use of single short primers under low annealing conditions. The PCR amplifications were performed in 25 μL of reaction mixture containing 40 ng of DNA, 5 pmoL/μL of primer D11 (5′-d[GTCCCGACGA]-3′) and Taq DNA recombinant polymerase (Roche Diagnostics, Basel, Switzerland), including nucleotides (dNTPs 0.4 mM), magnesium chloride and DNA polymerase. D11 primer was selected to yield amplification products with a reasonable number of bands [9]. The PCR program consisted of an initial denaturation at 94 °C for 2 min, then 45 cycles, each of them including DNA denaturation at 95 °C for 1 min, annealing at 36 °C for 1 min and extension at 72 °C for 1 min. The reaction products were analyzed by an electrophoretic run on 2% agarose gel and gel staining with 10× ethidium bromide. RAPD-PCR patterns were acquired by ChemiDoc Gel Imaging System (Bio-Rad, Hercules, CA, USA). The change in the number of the bands and the variation in their intensity are associated with alterations of genetic material [51].

The polymorphic pattern generated by the RAPD-PCR profiles allowed the calculation of the Genomic Template Stability (GTS, %) as follows: GTS% = (1 − a/n) × 100, where a is the average number of polymorphic bands detected in each exposed sample, and n is the number of total bands in the non-treated cells. The template genomic stability of the control was set to 100% [52].

2.7. DCF Assay

The intracellular sperm levels of ROS were measured by DCF assay with a 2,7-dichlorodihydrofluorescein diacetate ($DCFH_2$-DA) probe, as described in Santonastaso and collaborators [9]. Briefly, 13 µm $DCFH_2$-DA was added to 150 µL sperm suspension in MEM. After 30 min incubation at 37 °C in the dark, the cell suspension was washed with bicarbonate buffer and counterstained with 4′,6-Diamidine-2′-phenylindole dihydrochloride (DAPI) solution for 5 min. Then, the sperm cells were transferred to a glass slide and observed by using a fluorescent microscope (Nikon Eclipse E-600) equipped with BP 330–380 nm and LP 420 nm filters. Intracellular ROS were visually scored comparing the control and samples and were measured as the percentage of sperm cells exhibiting a response (green cells) on the total sperm cells (Figure 3). The DCF assay was performed in triplicate.

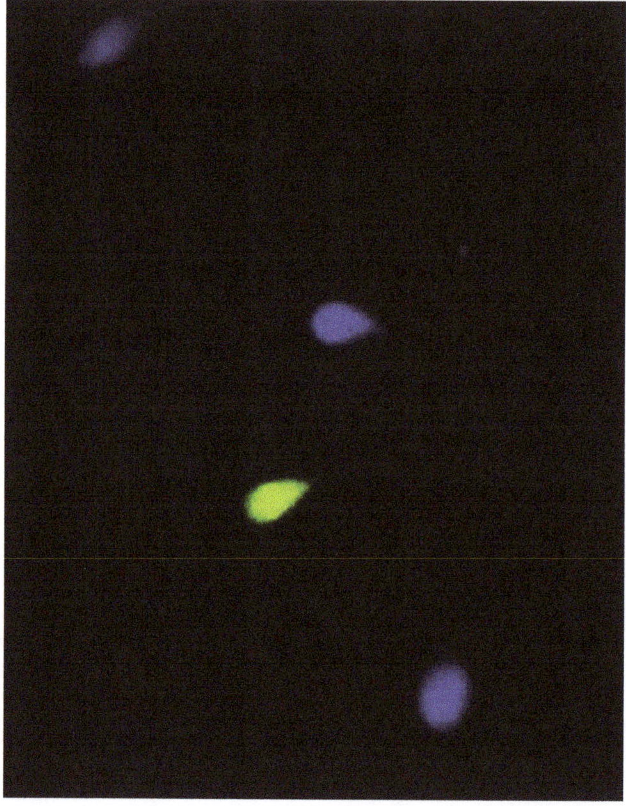

Figure 3. Intracellular ROS (green cell) in human sperm cell analyzed by fluorescent microscopy using $DCFH_2$-DA probe.

2.8. Statistical Analysis

All sperm parameters and the experimental data were expressed as the mean ± standard deviation (SD). Differences in the percentage of DNA damage, GTS%, DFI% and intracellular ROS% among the experimental groups were compared using the unpaired Student's *t*-test by GraphPad Prism 6. The effect was considered significant if *p*-value ≤ 0.01 with respect to the negative control.

3. Results

3.1. Sperm Motility Is Reduced after 90 min TiO_2NPs and $CdCl_2$ Co-Exposure

$CdCl_2$ exposure and TiO_2NPs-$CdCl_2$ co-exposure did not cause a statistically significant reduction in vitality. $CdCl_2$ treatment induced a statistically significant reduction of motility (progressive and non-progressive) after 30 min, while TiO_2NPs-$CdCl_2$ co-exposure reduced sperm motility after 90 min (*p*-value ≤ 0.01) (Table 3).

Table 3. Sperm vitality and motility after $CdCl_2$ exposure and TiO_2NPs co-exposure. The values were expressed as mean ± SD. * p ≤ 0.01.

Substances Concentration	Exposure Minutes	Vitality (%)	Motility (P + NP) (%)
$CdCl_2$ 10 μM μg/L	15	71 ± 4.5	81 ± 5.5
	30	65 ± 2.5	69 ± 8.0 *
	45	65 ± 5.6	57 ± 5.8 *
	90	61 ± 4.5	55 ± 7.5 *
$CdCl_2$ 10 μM + TiO_2NPs 1 μg/L	15	72 ± 3.2	78 ± 6.9
	30	70 ± 2.7	74 ± 8.5
	45	69 ± 4.1	70 ± 8.7
	90	65 ± 4.5	58 ± 5.5 *

3.2. TiO_2NPs and $CdCl_2$ Co-Exposure Causes Time-Dependent Loss of Sperm DNA Integrity

The results of the Comet assay showed that TiO_2NPs-$CdCl_2$ co-exposure induced a time-dependent loss of human sperm DNA integrity with statistically significant values (*p*-value ≤ 0.01) after 30 min. Furthermore, $CdCl_2$ exposure already reduced the sperm DNA integrity after 15 min (Figure 4).

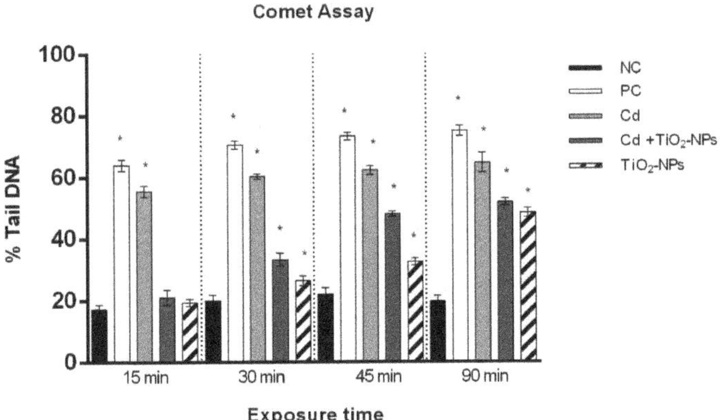

Figure 4. Percentage of DNA in the comet tail (ordinate) in human sperm after different exposure times (abscissa) to $CdCl_2$ and $CdCl_2$ + TiO_2NPs co-exposure. The black bars are negative controls (NC); the white bars are positive controls (PC) (benzene 0.4 μL/mL); the light grey bars are 10 μM $CdCl_2$-treated sperm (Cd); the dark grey bars are 10 μM $CdCl_2$ + 1 μg/L TiO_2NPs-treated sperm (Cd + TiO_2NPs); the striped bars are 1 μg/L TiO_2NPs-treated sperm. * p ≤ 0.01.

3.3. TiO$_2$NPs and CdCl$_2$ Combined Exposure Induces Sperm DNA Fragmentation

The data from the TUNEL test displayed a statistically significant (*p*-value ≤ 0.01) increase of sperm DNA fragmentation starting from 30 min of TiO$_2$NPs and CdCl$_2$ co-exposure, whereas the CdCl$_2$ single exposure induced sperm DNA fragmentation starting as early as after 15 min. The DNA Fragmentation Index (DFI) corresponding to the cut-off value (26%) associated to male infertility [53] was not exceeded after the TiO$_2$NPs and CdCl$_2$ combined exposure (Figure 5).

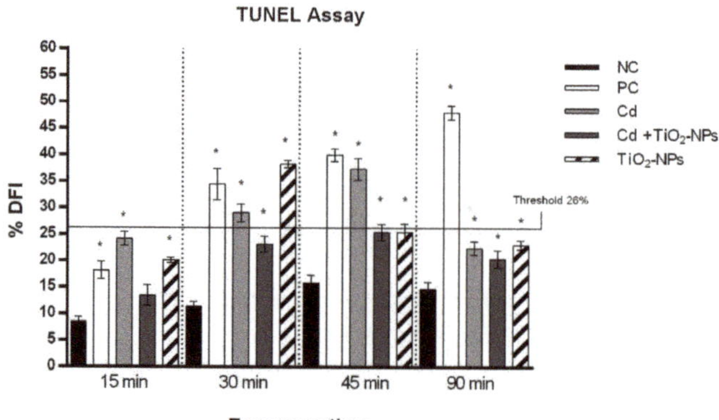

Figure 5. Percentage of DNA fragmentation index (ordinate) in human sperm after different exposure times (abscissa) to CdCl$_2$ and CdCl$_2$ + TiO$_2$NPs co-exposure. The black bars are negative controls (NC); the white bars are positive controls (PC) (benzene 0.4 µL/mL); the light grey bars are 10 µM CdCl$_2$-treated sperm (Cd); the dark grey bars are 10 µM CdCl$_2$ + 1 µg/L TiO$_2$NPs-treated sperm (Cd + TiO$_2$NPs); the striped bars are 1 µg/L TiO$_2$NPs-treated sperm. * *p* ≤ 0.01.

3.4. TiO$_2$NPs in Combination with CdCl$_2$ Produce Intracellular Oxidative Stress in Sperm Cells

A statistically significant increase (*p*-value ≤ 0.01) of intracellular ROS was observed in sperm cells exposed to CdCl$_2$ alone starting from 15 min and in combination with TiO$_2$NPs starting from 30 min with respect to the control. The increasing intracellular oxidative stress was time-dependent after co-exposure to CdCl$_2$ and TiO$_2$NPs (Figure 6).

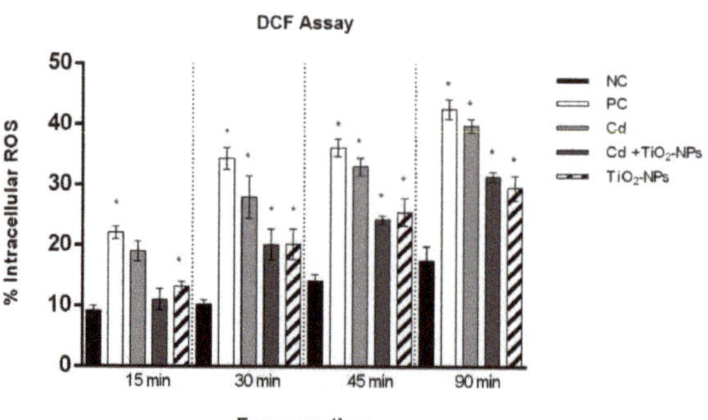

Figure 6. Percentage of intracellular ROS (ordinate) in human sperm after different exposure times (abscissa) to CdCl$_2$ and CdCl$_2$ + TiO$_2$NPs co-exposure. The black bars are negative controls (NC); the white bars are positive controls (PC) (benzene 0.4 µL/mL); the light grey bars are 10 µM CdCl$_2$-treated sperm (Cd); the dark grey bars are 10 µM CdCl$_2$ + 1 µg/L TiO$_2$NPs-treated sperm (Cd + TiO$_2$NPs); the striped bars are 1 µg/L TiO$_2$NPs-treated sperm. * *p* ≤ 0.01.

3.5. TiO$_2$NPs and CdCl$_2$ Co-Exposure Generates Sperm DNA Polymorphic Profiles Alterations

The RAPD-PCR analysis showed a variation of polymorphic profiles of sperm DNA exposed to CdCl$_2$ alone and in combination with TiO$_2$NPs compared to non-treated sperm DNA. CdCl$_2$ treatment induced the appearance of one band and the disappearance of two bands with respect to the control after 15 min, while after 30 of exposure mins three news bands appeared and one band disappeared. After 45 exposure mins we evidenced the appearance of three news bands and the disappearance of two bands with respect to the control. The exposure to CdCl$_2$ for 90 mins induced the prevalence of bands' disappearance when compared to the control. The electrophoretic patterns relative to the TiO$_2$NPs-CdCl$_2$ co-exposure showed the prevalence of bands' appearance when compared to non-treated sperm cells (Table 4).

Table 4. Molecular sizes (bp) of appeared and disappeared bands after amplification with primer D11 in human sperm DNA exposed to CdCl$_2$ and CdCl$_2$ with TiO$_2$NPs co-exposure. * Control bands are at: 190, 270, 450, 500, 510, 560, 650, 850, 910, 950, 1000 and 2000 bp.

Substances Concentration	Exposure Minutes	Gained Bands	Lost Bands *
CdCl$_2$ 10 μM μg/L	15	210	560, 850
	30	300, 350, 900	560
	45	210, 700, 900	270, 850
	90	600	270, 450, 500, 560, 850
CdCl$_2$ 10 μM + TiO$_2$NPs 1 μg/L	15	700	-
	30	210, 350, 900	-
	45	210, 530, 900	850
	90	300, 400, 900	850

3.6. TiO$_2$NPs in Combination with CdCl$_2$ Reduce Sperm Genome Stability

The human sperm genome stability (GTS%) significantly decreased in the CdCl$_2$-exposed sperms in a time-dependent manner. The genome stability of co-exposed sperms was statistically reduced after 30 min (Figure 7).

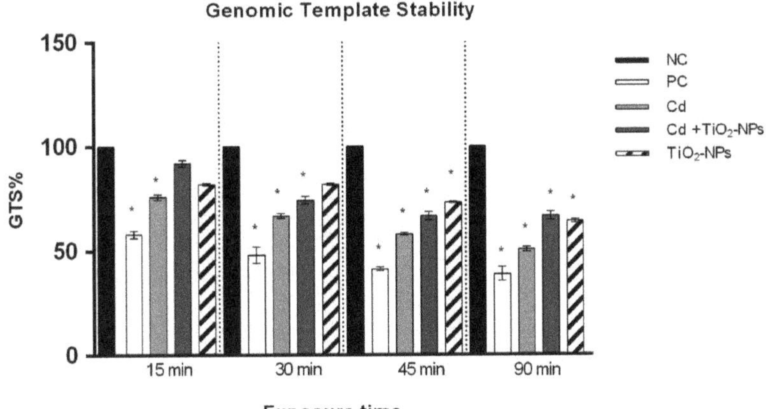

Figure 7. Changes in the percentage of Genome Template Stability in human sperm DNA (ordinate) after different exposure times (abscissa) to CdCl$_2$ and CdCl$_2$ + TiO$_2$NPs co-exposure, as evidenced by the RAPD-PCR technique. The black bars are the negative controls (NC); the white bars are the positive controls (PC) (benzene 0.4 μL/mL); the light grey bars are 10 μM CdCl$_2$-treated sperm (Cd); the dark grey bars are 10 μM CdCl$_2$ + 1 μg/L TiO$_2$NPs-treated sperm (Cd + TiO$_2$NPs); the striped bars are 1 μg/L TiO$_2$NPs-treated sperm. * $p \leq 0.01$.

4. Discussion

The chemical–physical characteristics of TiO_2NPs make them capable of absorbing and transporting numerous compounds through biological barriers, including Cd, with a mechanism known as the "Trojan Horse" effect. The transport can take place either by simple diffusion from the caveola systems or by endocytosis with transport mediated by ABC family proteins [14].

The effects of NPs' and heavy metals' co-exposure on the living organism are still unclear. Data on heavy metal and TiO_2NPs genotoxicity and cytotoxicity are controversial, especially because these interactions are species-specific, often tissue-specific and related to physical–chemical features of the co-exposure medium [22,54,55]. The aim of our work was to examine in vitro the genotoxic responses induced by Cd alone and in association with TiO_2NPs in human sperm cells at different times (15, 30, 45 and 90 min).

Scarce amounts of data are available regarding TiO_2NPs' effects on reproduction/fertility. In adult male Wistar rats, TiO_2NPs' daily oral exposure (50 mg/kg body weight (BW)/day) caused significant time-dependent adverse effects such as decreased testis and prostate weight, disrupted hormone profile (i.e., significant decreased serum testosterone level and increased serum estradiol, Luteinizing hormone (LH) and Follicle stimulating hormone (FSH) levels), impaired spermatogenesis, lipid peroxidation and inflammation in testicular tissues. Moreover, effects on semen parameters were also reported: normal sperm counts decreased from 88% (control) to 68% after 21 days of exposure [56]. Song and collaborators [57] examined the testes and sperm quality in male mice after an oral 5–10 nm TiO_2NP exposure to of 0, 10, 50 and 100 mg/kg BW for 28 days. TiO_2NPs exposure caused sperm malformations, a sperm cell micronucleus rate and levels of markers indicating cell damage in the testes, a further reduction in the germ cell number and spherospermia, interstitial glands, malalignment and vacuolization in spermatogenic cells at the two highest dose levels (50 and 100 mg/kg BW). The superoxide dismutase (SOD) activity significantly decreased at the highest dose level (100 mg/kg BW) and the malondialdehyde significantly increased at the two highest dose levels (50 and 100 mg/kg BW), both of which are markers indicating cell damage in testis, although the weights of the testicles and epididymis were not affected. Conversely, no effect on male reproductive parameters (weight of reproductive organs, daily sperm production and plasma testosterone levels) was reported for adult male mice after weekly TiO_2NP intratracheal instillation for seven weeks [58].

Our previous study showed that TiO_2NPs cause a loss of sperm DNA integrity and a statistically significant increase in DNA fragmentation and DNA strand breaks, inducing apoptosis [9]. In males, apoptosis plays a physiological role by maintaining an appropriate germ cell to Sertoli cell ratio, removing defective germ cells and controlling sperm production [59]; however, elevated apoptosis levels can damage the spermatozoa. Our results showed that TiO_2NPs were genotoxic on human sperm cells in vitro, significantly affecting the reproductive potential.

Furthermore, exposure to Cd induced apoptosis in rats and frog testes, rat Leydig cells, trophoblast cells of rat placenta and granulosa cells of chicken ovarian follicles [60–64]. This heavy metal, commonly present in the environment in the form of $CdCl_2$, is able to alter sperms' motility and their capacity to reach and penetrate into the oocyte by altering the sperm enzyme acetylcholine transferase and the oxygen uptake [35].

Sperms' acute exposure to Cd may impair the sperm fertilization potential in vitro; in fact, exposure to $CdCl_2$ results in a decreased progressive and hyperactivated sperm motility, as well as increased caspase activation, which suggests the triggering of an apoptotic pathway [36]. In vitro Cd-exposed murine spermatozoa exhibit an altered sperm fertilization potential, producing a lower number of pronuclei than controls during in vitro fertilization [37]. A severe reduction of sperm motility and kinematic parameters was also shown in a rat model exposed to Cd in vivo [38].

Our results showed that exposure to $CdCl_2$ caused a reduction of human sperm motility in a time-dependent manner and a decrease of the genome template stability, which was associated with an increased level of DNA strand breaks, apoptosis and oxidative stress. These results confirmed the genotoxic potential of Cd though the induction of an oxidative microenvironment [65]. Our results showed that the presence of TiO$_2$NPs can reduce the cytotoxicity and genotoxicity associated with single Cd exposure as sperm showed a higher motility, and a lower induction of DNA strand breaks as well as of the apoptotic pathway after co-exposure highlighted by the Comet assay. We also estimated human sperm DNA fragmentation by TUNEL test. The sperm DNA fragmentation index (DFI) is considered a valuable early marker of the presence and harmful effects of pollution [66]. A pathological parameter (26%) was correlated with the inability to fertilize the egg cell, and was thus associated with poliabortivity [67]. Cd and TiO$_2$NPs co-exposure induced a lower sperm DNA fragmentation than that observed by single Cd exposure for all exposure times, and the pathological value of DFI was never exceeded; otherwise, Cd caused an increase beyond the pathological value after 30 and 45 exposure mins. Sperm DNA fragmentation decreased with an increasing exposure time, suggesting that sperms could undergo increasing damage of the genetic material in the first exposure times (15, 30 and 45 min), but there was a decrease of the sperm DNA fragmentation at 90 min, thus probably implying that sperms activated the cellular repair mechanisms, as evidenced by the Comet assay, at an exposure time longer than 45 min.

Alterations in RAPDs' profiles have allowed for the detection of genomic instability as different molecular events (genomic rearrangements, point mutations, deletions and insertions) of sperm treated with respect to the sperm negative control. In fact, the appearance of a new band is related to point mutations and/or rearrangements, while the disappearance of the band is attributable to DNA adducts and breaks in the double helix [68]. The RAPD-PCR analysis was able to detect DNA changes not necessarily related to apoptotic processes. We observed that CdCl$_2$ treatment induced both the appearance of new bands and disappearance of bands in comparison with the control, while CdCl$_2$-TiO$_2$NPs co-exposure resulted in the prevalence of new bands' appearance with respect to the non-treated sperm cells as well as those induced by treatment with 1 µg/L TiO$_2$NPs, as previously reported by Santonastaso and collaborators [9]. Thanks to RAPDs' profile alterations, we evaluated the sperm genomic template stability percentage (%GTS). The decrease of %GTS depends on the time exposure to single Cd and Cd-TiO$_2$NPs. The decrease of %GTS is the first molecular response to a toxicant and has been demonstrated as being directly related to the extent of DNA damage and/or to the efficiency of DNA repair and replication [7,69]. Our study showed a statistically significant decrease in genomic stability in sperm exposed to Cd already after 15 min, which reached deep values after 90 exposure minutes, while co-exposure significantly impacted genomic stability after only 30 min.

The changes in RAPDs' profiles highlighted in sperm cells exposed to Cd alone and in association with TiO$_2$NPs could be related to oxidative DNA damage through ROS direct damage on genetic material [70,71]. To clarify this assumption, we assessed intracellular ROS production in both experimental conditions. Co-exposure resulted in lower intracellular ROS levels when compared to single Cd exposure, indicating that ROS detrimental effects could be the main mechanism for Cd and TiO$_2$NPs' genotoxicity.

The results showed that sperm DNA damage probably due to ROS action induced by Cd-TiO$_2$NPs co-exposure is lower than in the case of Cd single exposure, and we may speculate that the Cd genotoxic potential was inhibited and/or masked by TiO$_2$NPs; thus, no "Trojan Horse" effect was demonstrated in human spermatozoa in vitro. Unfortunately, evidence on transport, absorption, agglomeration and reactivity in the synergy between nanoparticles and Cd is still scarce and contradictory. Contrary to our results, other studies reported that TiO$_2$NPs potentiated Cd-induced pro-oxidants generation (ROS and lipid peroxidation), antioxidants depletion (glutathione level and glutathione reductase, SOD and catalase enzymes) and apoptosis (by altering the gene expression of p53, bax and bcl-2), along with the alteration of the mitochondrial membrane potential in HepG2 and MCF-7 cells [72]. Xia and collaborators [73] reported that TiO$_2$NPs may promote Cd-induced cellular oxidative stress in human embryo kidney 293T (HEK293T) cells, as indicated by the changes in the SOD activity and ROS concentration. Cd and TiO$_2$NPs exert additive or synergistic effects on HEK293T cells' oxidative damage, and these effects vary with different proportions and concentrations of CdCl$_2$ and TiO$_2$NPs in the mixture. TiO$_2$NPs assume different behaviors depending on features of the exposure medium and duration of exposure. In salt water, TiO_2NPs begin to form aggregates such as sedimentations after about 6 h; it was also highlighted that the presence of $CdCl_2$ increases the aggregation between the nanoparticles and their sedimentation [12]. However, when the size of TiO_2NPs is around 23.8 nm, they do not form visible and appreciable aggregates, independently of the duration of exposure in fresh water [7].

Nanoparticles show a strong tendency to form agglomerates in solution due to their high surface area [74]. The agglomerate (or cluster) is defined as a compound formed by the secondary aggregates (which are joined by weak chemical bonds), which can be broken through manipulation. The state of dispersion of a particulate system describes the relative number of primary particles (aggregated) present in a suspension medium. The general view is that the degree and type of agglomerates formed may influence the toxicity of NPs. The aggregate can also be defined as a compound of the secondary primary particles (which are joined by means of strong chemical bonds), which behaves as a single unit [75]. Our results by TEM evidenced that TiO_2NP agglomerates in the range of 400 ± 52 nm do not penetrate into the human sperm cells. They exercise a pro-oxidant effect on the polyunsaturated fatty acids of the sperm membrane, resulting in intracellular ROS generation and the induction of different DNA damage degrees, depending on the endpoints investigated in the sperm exposure medium (MEM) [9,15,46,76]. In this study, we demonstrated that sperm genotoxicity induced by TiO_2NPs and $CdCl_2$ co-exposure significantly decreased when compared to that induced by Cd. Therefore, we can speculate that TiO_2NPs and Cd could form complexes, whose size does not allow them to penetrate into the sperm cells. Moreover, the degree and the type of aggregates could influence TiO_2NPs' and Cd's sperm genotoxicity in MEM. In industrial wastewaters, the presence of TiO_2NPs and Cd^{2+} results in the formation of a ternary surface complex with arsenic that inhibits Cd release into the aqueous phase and that, hence, facilitates the immobilization of the heavy metal [77]. As shown in the Quantitative Structure-Activity Relationship (QSAR) computational method [77], the interaction between TiO_2NPs and Cd can result in the formation of a "sandwich structure", where Cd, placed at the center, is completely masked by the TiO_2NPs, which are located at the outer surfaces of the structures. It can be concluded that TiO_2NPs–$CdCl_2$ co-exposure could lead to the formation of aggregates with a reduced genotoxic activity due to their sandwich organization [46].

Based on our knowledge, this is the first study reporting the genotoxic effects of the combined exposure to TiO_2NPs and $CdCl_2$ of human sperm cells in vitro. The association of the two tested contaminants leads to a reduction in the single molecules' genotoxic effects, probably due to the formation of aggregates, as predicted by the in silico analysis. Although these results could be interpreted as positive in relation to health risks, we must underline that our data are limited to an in vitro system; hence, we cannot draw any conclusion on a systemic impact of the co-exposure. Therefore, considering the widespread existence of these contaminants in the environment, further studies are necessary in order to clarify the pathways that are responsible for TiO_2NPs and $CdCl_2$ genotoxicity and to better clarify their potential interaction.

Author Contributions: Data curation, M.S.; Funding acquisition, L.R.; Methodology, F.M., C.I., F.C. and N.C.; Project administration, L.R.; Writing—original draft, M.S.; Writing—review & editing, L.R. All authors have read and agreed to the published version of the manuscript.

Funding: This study was supported by ICURE project, POR CAMPANIA FESR 2014/2020.

Acknowledgments: The authors would like to thank Angela Marano, Marcella Cammarota and Severina Pacifico for their contribute in some steps of the research. In addition, we thank Renata Finelli for her assistance in revising the English of this manuscript.

Conflicts of Interest: The authors declare no conflict of interest.

References

1. Johnston, H.J.; Hutchison, G.R.; Christensen, F.M.; Peters, S.; Hankin, S.; Stone, V. Identification of the mechanisms that drive the toxicity of TiO 2particulates: The contribution of physicochemical characteristics. *Part. Fibre Toxicol.* **2009**, *6*, 1–27. [CrossRef] [PubMed]
2. Lehner, R.; Wang, X.; Marsch, S.; Hunziker, P. Intelligent nanomaterials for medicine: Carrier platforms and targeting strategies in the context of clinical application. *Nanomed. Nanotechnol. Biol. Med.* **2013**, *9*, 742–757. [CrossRef] [PubMed]
3. Oberdörster, G.; Oberdörster, E.; Oberdörster, J. Nanotoxicology: An emerging discipline evolving from studies of ultrafine particles. *Environ. Health Perspect.* **2005**, *113*, 823–839. [CrossRef] [PubMed]
4. Chen, T.; Yan, J.; Li, Y. Genotoxicity of titanium dioxide nanoparticles. *J. Food Drug Anal.* **2014**, *22*, 95–104. [CrossRef]
5. Demir, E.; Akça, H.; Turna, F.; Aksakal, S.; Burgucu, D.; Kaya, B.; Tokgün, O.; Vales, G.; Creus, A.; Marcos, R. Genotoxic and cell-transforming effects of titanium dioxide nanoparticles. *Environ. Res.* **2015**, *136*, 300–308. [CrossRef] [PubMed]
6. Nigro, M.; Bernardeschi, M.; Costagliola, D.; Della Torre, C.; Frenzilli, G.; Guidi, P.; Lucchesi, P.; Mottola, F.; Santonastaso, M.; Scarcelli, V.; et al. n-TiO$_2$ and CdCl$_2$ co-exposure to titanium dioxide nanoparticles and cadmium: Genomic, DNA and chromosomal damage evaluation in the marine fish European sea bass (*Dicentrarchus labrax*). *Aquat. Toxicol.* **2015**, *168*, 72–77. [CrossRef]
7. Rocco, L.; Santonastaso, M.; Mottola, F.; Costagliola, D.; Suero, T.; Pacifico, S.; Stingo, V. Genotoxicity assessment of TiO$_2$ nanoparticles in the teleost *Danio rerio*. *Ecotoxicol. Environ. Saf.* **2015**, *113*, 223–230. [CrossRef] [PubMed]
8. Larsen, P.B.; Mørck, T.A.; Andersen, D.N.; Hougaard, K.S. A Critical Review of Studies on the Reproductive and Developmental Toxicity of Nanomaterials Month 20xx. Available online: https://euon.echa.europa.eu/documents/23168237/24095696/critical_review_of_studies_on_reproductive_and_developmental_toxicity_of_nanomaterials_en.pdf/c83f78ef-7136-ef4b-268c-c5d9b7bf1fea (accessed on 4 June 2020).
9. Santonastaso, M.; Mottola, F.; Colacurci, N.; Iovine, C.; Pacifico, S.; Cammarota, M.; Cesaroni, F.; Rocco, L. In vitro genotoxic effects of titanium dioxide nanoparticles (n-TiO$_2$) in human sperm cells. *Mol. Reprod. Dev.* **2019**, *86*, 1369–1377. [CrossRef]
10. D'Agata, A.; Fasulo, S.; Dallas, L.J.; Fisher, A.S.; Maisano, M.; Readman, J.W.; Jha, A.N. Enhanced toxicity of "bulk" titanium dioxide compared to "fresh" and "aged" nano-TiO$_2$ in marine mussels (*Mytilus galloprovincialis*). *Nanotoxicology* **2014**, *8*, 549–558. [CrossRef] [PubMed]
11. Frenzilli, G.; Bernardeschi, M.; Guidi, P.; Scarcelli, V.; Lucchesi, P.; Marsili, L.; Fossi, M.C.; Brunelli, A.; Pojana, G.; Marcomini, A.; et al. Effects of in vitro exposure to titanium dioxide on DNA integrity of bottlenose dolphin (*Tursiops truncatus*) fibroblasts and leukocytes. *Mar. Environ. Res.* **2014**, *100*, 68–73. [CrossRef] [PubMed]
12. Della Torre, C.; Buonocore, F.; Frenzilli, G.; Corsolini, S.; Brunelli, A.; Guidi, P.; Kocan, A.; Mariottini, M.; Mottola, F.; Nigro, M.; et al. Influence of titanium dioxide nanoparticles on 2,3,7,8-tetrachlorodibenzo-p-dioxin bioconcentration and toxicity in the marine fish European sea bass (*Dicentrarchus labrax*). *Environ. Pollut.* **2015**, *196*, 185–193. [CrossRef] [PubMed]
13. Deng, R.; Lin, D.; Zhu, L.; Majumdar, S.; White, J.C.; Gardea-Torresdey, J.L.; Xing, B. Nanoparticle interactions with co-existing contaminants: Joint toxicity, bioaccumulation and risk. *Nanotoxicology* **2017**, *11*, 591–612. [CrossRef] [PubMed]
14. Nowack, B.; Bucheli, T.D. Occurrence, behavior and effects of nanoparticles in the environment. *Environ. Pollut.* **2007**, *150*, 5–22. [CrossRef] [PubMed]
15. Canesi, L.; Frenzilli, G.; Balbi, T.; Bernardeschi, M.; Ciacci, C.; Corsolini, S.; Della Torre, C.; Fabbri, R.; Faleri, C.; Focardi, S.; et al. Interactive effects of n-TiO$_2$ and 2,3,7,8-TCDD on the marine bivalve *Mytilus galloprovincialis*. *Aquat. Toxicol.* **2014**, *153*, 53–65. [CrossRef] [PubMed]
16. Sendra, M.; Pintado-Herrera, M.G.; Aguirre-Martínez, G.V.; Moreno-Garrido, I.; Martin-Díaz, L.M.; Lara-Martín, P.A.; Blasco, J. Are the TiO$_2$ NPs a "Trojan horse" for personal care products (PCPs) in the clam Ruditapes philippinarum? *Chemosphere* **2017**, *185*, 192–204. [CrossRef] [PubMed]

17. Mottola, F.; Iovine, C.; Santonastaso, M.; Romeo, M.L.; Pacifico, S.; Cobellis, L.; Rocco, L. NPs-TiO$_2$ and lincomycin coexposure induces DNA damage in cultured human amniotic cells. *Nanomaterials* **2019**, *9*, 1511. [CrossRef] [PubMed]

18. Fan, W.H.; Cui, M.; Shi, Z.W.; Cheng Tan, X.P.Y. Enhanced oxidative stress and physiological damage in daphnia magna by copper in the presence of Nano-TiO$_2$. *J. Nanomater.* **2012**, 1–7. [CrossRef]

19. Miao, W.; Zhu, B.; Xiao, X.; Li, Y.; Dirbaba, N.B.; Zhou, B.; Wu, H. Effects of titanium dioxide nanoparticles on lead bioconcentration and toxicity on thyroid endocrine system and neuronal development in zebrafish larvae. *Aquat. Toxicol.* **2015**, *161*, 117–126. [CrossRef]

20. Sun, H.; Zhang, X.; Zhang, Z.; Chen, Y.; Crittenden, J.C. Influence of titanium dioxide nanoparticles on speciation and bioavailability of arsenite. *Environ. Pollut.* **2009**, *157*, 1165–1170. [CrossRef]

21. Tan, C.; Fan, W.H.; Wang, W.X. Role of titanium dioxide nanoparticles in the elevated uptake and retention of cadmium and zinc in *Daphnia magna*. *Environ. Sci. Technol.* **2012**, *46*, 469–476. [CrossRef]

22. Tan, C.; Wang, W.X. Modification of metal bioaccumulation and toxicity in *Daphnia magna* by titanium dioxide nanoparticles. *Environ. Pollut.* **2014**, *186*, 36–42. [CrossRef] [PubMed]

23. Chen, J.; Qian, Y.; Li, H.; Cheng, Y.; Zhao, M. The reduced bioavailability of copper by nano-TiO$_2$ attenuates the toxicity to *Microcystis aeruginosa*. *Environ. Sci. Pollut. Res.* **2015**, *22*, 12407–12414. [CrossRef] [PubMed]

24. Rosenfeldt, R.R.; Seitz, F.; Zubrod, J.P.; Feckler, A.; Merkel, T.; Lüderwald, S.; Bundschuh, R.; Schulz, R.; Bundschuh, M. Does the presence of titanium dioxide nanoparticles reduce copper toxicity? A factorial approach with the benthic amphipod *Gammarus fossarum*. *Aquat. Toxicol.* **2015**, *165*, 154–159. [CrossRef] [PubMed]

25. Yang, W.W.; Li, Y.; Miao, A.J.; Yang, L.Y. Cd^{2+} toxicity as affected by bare TiO$_2$ nanoparticles and their bulk counterpart. *Ecotoxicol. Environ. Saf.* **2012**, *85*, 44–51. [CrossRef] [PubMed]

26. Wang, J.; Dai, H.; Nie, Y.; Wang, M.; Yang, Z.; Cheng, L.; Liu, Y.; Chen, S.; Zhao, G.; Wu, L.; et al. TiO$_2$ nanoparticles enhance bioaccumulation and toxicity of heavy metals in *Caenorhabditis elegans* via modification of local concentrations during the sedimentation process. *Ecotoxicol. Environ. Saf.* **2018**, *162*, 160–169. [CrossRef] [PubMed]

27. Waalkes, M.P.; Coogan, T.P.; Barter, R.A. Toxicological principles of metal carcinogenesis with special emphasis on cadmium. *Crit. Rev. Toxicol.* **1992**, *22*, 175–201. [CrossRef] [PubMed]

28. Capaldo, A.; Gay, F.; Scudiero, R.; Trinchella, F.; Caputo, I.; Lepretti, M.; Marabotti, A.; Esposito, C.; Laforgia, V. Histological changes, apoptosis and metallothionein levels in *Triturus carnifex* (Amphibia, Urodela) exposed to environmental cadmium concentrations. *Aquat. Toxicol.* **2016**, *173*, 63–73. [CrossRef]

29. Cuypers, A.; Plusquin, M.; Remans, T.; Jozefczak, M.; Keunen, E.; Gielen, H.; Opdenakker, K.; Nair, A.R.; Munters, E.; Artois, T.J.; et al. Cadmium stress: An oxidative challenge. *Biometals* **2010**, *23*, 927–940. [CrossRef]

30. International Agency for Research on Cancer (IARC). Summaries & Evaluations: Cadmium and Cadmium Compounds (Group 1). Available online: http://www.inchem.org/documents/iarc/vol58/mono58-2.html (accessed on 4 June 2020).

31. Nordberg, G.F. Cadmium and health in the 21st Century—Historical remarks and trends for the future. *Proc. Biometals* **2004**, *17*, 485–489. [CrossRef]

32. Moitra, S.; Blanc, P.D.; Sahu, S. Adverse respiratory effects associated with *Cadmium exposure* in small-scale jewellery workshops in India. *Thorax* **2013**, *68*, 565–570. [CrossRef]

33. Pourahmad, J.; O'Brien, P.J. A comparison of hepatocyte cytotoxic mechanisms for Cu^{2+} and Cd^{2+}. *Toxicology* **2000**, *143*, 263–273. [CrossRef]

34. Benoff, S.; Jacob, A.; Hurley, I.R. Male infertility and environmental exposure to lead and cadmium. *Hum. Reprod. Update* **2000**, *6*, 107–121. [CrossRef] [PubMed]

35. Alabi, N.S.; Whanger, P.D.; Wu, A.S.H. Interactive effects of organic and inorganic selenium with cadmium and mercury on spermatozoal oxygen consumption and motility in vitro1. *Biol. Reprod.* **1985**, *33*, 911–919. [CrossRef] [PubMed]

36. Marchiani, S.; Tamburrino, L.; Farnetani, G.; Muratori, M.; Vignozzi, L.; Baldi, E. Acute effects on human sperm exposed in vitro to cadmium chloride and diisobutyl phthalate. *Reproduction* **2019**, *158*, 281–290. [CrossRef] [PubMed]

37. Zhao, L.L.; Ru, Y.F.; Liu, M.; Tang, J.N.; Zheng, J.F.; Wu, B.; Gu, Y.H.; Shi, H.J. Reproductive effects of cadmium on sperm function and early embryonic development in vitro. *PLoS ONE* **2017**, *12*, e0186727. [CrossRef] [PubMed]

38. Hardneck, F.; Israel, G.; Pool, E.; Maree, L. Quantitative assessment of heavy metal effects on sperm function using computer-aided sperm analysis and cytotoxicity assays. *Andrologia* **2018**, *50*, 1–9. [CrossRef] [PubMed]

39. Laskey, J.W.; Berman, E.; Ferrell, J.M. The use of cultured ovarian fragments to assess toxicant alterations in steroidogenesis in the sprague-dawley rat. *Reprod. Toxicol.* **1995**, *9*, 131–141. [CrossRef]

40. Hu, X.; Chen, Q.; Jiang, L.; Yu, Z.; Jiang, D.; Yin, D. Combined effects of titanium dioxide and humic acid on the bioaccumulation of cadmium in Zebrafish. *Environ. Pollut.* **2011**, *159*, 1151–1158. [CrossRef] [PubMed]

41. Ji, Y.; Zhou, Y.; Ma, C.; Feng, Y.; Hao, Y.; Rui, Y.; Wu, W.; Gui, X.; Le, V.N.; Han, Y.; et al. Jointed toxicity of TiO$_2$ NPs and Cd to rice seedlings: NPs alleviated Cd toxicity and Cd promoted NPs uptake. *Plant Physiol. Biochem.* **2017**, *110*, 82–93. [CrossRef]

42. Manesh, R.R.; Grassi, G.; Bergami, E.; Marques-Santos, L.F.; Faleri, C.; Liberatori, G.; Corsi, I. Co-exposure to titanium dioxide nanoparticles does not affect cadmium toxicity in radish seeds (Raphanus sativus). *Ecotoxicol. Environ. Saf.* **2018**, *148*, 359–366. [CrossRef] [PubMed]

43. Hartmann, N.B.; Legros, S.; Von der Kammer, F.; Hofmann, T.; Baun, A. The potential of TiO$_2$ nanoparticles as carriers for cadmium uptake in *Lumbriculus variegatus* and *Daphnia magna*. *Aquat. Toxicol.* **2012**, *118–119*, 1–8. [CrossRef] [PubMed]

44. Hartmann, N.B.; Engelbrekt, C.; Zhang, J.; Ulstrup, J.; Kusk, K.O.; Baun, A. The challenges of testing metal and metal oxide nanoparticles in algal bioassays: Titanium dioxide and gold nanoparticles as case studies. *Nanotoxicology* **2013**, *7*, 1082–1094. [CrossRef] [PubMed]

45. Yang, W.W.; Miao, A.J.; Yang, L.Y. Cd^{2+} toxicity to a green alga *Chlamydomonas reinhardtii* as influenced by its adsorption on TiO$_2$ engineered nanoparticles. *PLoS ONE* **2012**, *7*, e32300.

46. Rocco, L.; Santonastaso, M.; Nigro, M.; Mottola, F.; Costagliola, D.; Bernardeschi, M.; Guidi, P.; Lucchesi, P.; Scarcelli, V.; Corsi, I.; et al. Genomic and chromosomal damage in the marine mussel *Mytilus galloprovincialis*: Effects of the combined exposure to titanium dioxide nanoparticles and cadmium chloride. *Mar. Environ. Res.* **2015**, *111*, 144–148. [CrossRef] [PubMed]

47. Ahamed, M.; Akhtar, M.J.; Alaizeri, Z.A.M.; Alhadlaq, H.A. TiO$_2$ nanoparticles potentiated the cytotoxicity, oxidative stress and apoptosis response of cadmium in two different human cells. *Environ. Sci. Pollut. Res.* **2020**, *27*, 10425–10435. [CrossRef] [PubMed]

48. WHO laboratory manual for the Examination and processing of human semen. *World Health Organ.* **2010**, *1*, 10–17.

49. Rocco, L.; Peluso, C.; Cesaroni, F.; Morra, N. Genomic damage in human sperm cells exposed in vitro to environmental pollutants. *J. Environ. Anal. Toxicol.* **2012**, *2*, 1–7. [CrossRef]

50. Gyori, B.M.; Venkatachalam, G.; Thiagarajan, P.S.; Hsu, D.; Clement, M.V. OpenComet: An automated tool for comet assay image analysis. *Redox Biol.* **2014**, *2*, 457–465. [CrossRef]

51. Noel, S.; Rath, S.K. Randomly amplified polymorphic DNA as a tool for genotoxicity: An assessment. *Toxicol. Ind. Health* **2006**, *22*, 267–275. [CrossRef]

52. Rocco, L.; Valentino, I.V.; Scapigliati, G.; Stingo, V. RAPD-PCR analysis for molecular characterization and genotoxic studies of a new marine fish cell line derived from *Dicentrarchus labrax*. *Cytotechnology* **2014**, *66*, 383–393. [CrossRef]

53. Chenlo, P.H.; Curi, S.M.; Pugliese, M.N.; Ariagno, J.I.; Sardi-Segovia, M.; Furlan, M.J.; Repetto, H.E.; Zeitler, E.; Cohen, M.; Mendeluk, G.R. Fragmentation of sperm DNA using the TUNEL method. *Actas Urológicas Españolas* **2014**, *38*, 608–612. [CrossRef] [PubMed]

54. Zheng, W.-W.; Ge Song, Q.W.; Liu, S.; Zhu, X.; Deng, S.; Zhong, A.; Tan, Y. Ying Tan Sperm DNA damage has a negative effect on early embryonic development following in vitro fertilization. *Asian J. Androl.* **2018**, *20*, 75–79. [CrossRef] [PubMed]

55. Mansouri, B.; Maleki, A.; Johari, S.A.; Shahmoradi, B.; Mohammadi, E.; Shahsavari, S.; Davari, B. Copper bioaccumulation and depuration in common carp (*Cyprinus carpio*) following Co-exposure to TiO$_2$ and CuO nanoparticles. *Arch. Environ. Contam. Toxicol.* **2016**, *71*, 541–552. [CrossRef] [PubMed]

56. Shahin, N.N.; Mohamed, M.M. Nano-sized titanium dioxide toxicity in rat prostate and testis: Possible ameliorative effect of morin. *Toxicol. Appl. Pharmacol.* **2017**, *334*, 129–141. [CrossRef] [PubMed]

57. Song, G.; Lin, L.; Liu, L.; Wang, K.; Ding, Y.; Niu, Q.; Mu, L.; Wang, H.; Shen, H.; Guo, S. Toxic effects of anatase titanium dioxidenanoparticles on spermatogenesis and testiclesin male mice. *Polish J. Environ. Stud.* **2017**, *26*, 2739–2745. [CrossRef]

58. Lauvås, A.J.; Skovmand, A.; Poulsen, M.S.; Kyjovska, Z.O.; Roursgaard, M.; Goericke-Pesch, S.; Vogel, U.; Hougaard, K.S. Airway exposure to TiO$_2$ nanoparticles and quartz and effects on sperm counts and testosterone levels in male mice. *Reprod. Toxicol.* **2019**, *90*, 134–140. [CrossRef]

59. Elmore, S. Apoptosis: A review of programmed cell death. *Toxicol. Pathol.* **2007**, *35*, 495–516. [CrossRef]

60. Erboga, M.; Kanter, M. Effect of cadmium on trophoblast cell proliferation and apoptosis in different gestation periods of rat placenta. *Biol. Trace Elem. Res.* **2016**, *169*, 285–293. [CrossRef]

61. Jia, Y.; Lin, J.; Mi, Y.; Zhang, C. Quercetin attenuates cadmium-induced oxidative damage and apoptosis in granulosa cells from chicken ovarian follicles. *Reprod. Toxicol.* **2011**, *31*, 477–485. [CrossRef]

62. Sönmez, M.F.; Tascioglu, S. Protective effects of grape seed extract on cadmium-induced testicular damage, apoptosis, and endothelial nitric oxide synthases expression in rats. *Toxicol. Ind. Health* **2016**, *32*, 1486–1494. [CrossRef]

63. Zhang, H.; Cai, C.; Shi, C.; Cao, H.; Han, Z.; Jia, X. Cadmium-induced oxidative stress and apoptosis in the testes of frog *Rana limnocharis*. *Aquat. Toxicol.* **2012**, *122–123*, 67–74. [CrossRef] [PubMed]

64. Zou, L.; Su, L.; Sun, Y.; Han, A.; Chang, X.; Zhu, A.; Liu, F.; Li, J.; Sun, Y. Nickel sulfate induced apoptosis via activating ROS-dependent mitochondria and endoplasmic reticulum stress pathways in rat Leydig cells. *Environ. Toxicol.* **2017**, *32*, 1918–1926. [CrossRef] [PubMed]

65. Thévenod, F. Cadmium and cellular signaling cascades: To be or not to be? *Toxicol. Appl. Pharmacol.* **2009**, *238*, 221–239. [CrossRef] [PubMed]

66. Bosco, L.; Notari, T.; Ruvolo, G.; Roccheri, M.C.; Martino, C.; Chiappetta, R.; Carone, D.; Lo Bosco, G.; Carrillo, L.; Raimondo, S.; et al. Sperm DNA fragmentation: An early and reliable marker of air pollution. *Environ. Toxicol. Pharmacol.* **2018**, *58*, 243–249. [CrossRef]

67. Payne, J.F.; Raburn, D.J.; Couchman, G.M.; Price, T.M.; Jamison, M.G.; Walmer, D.K. Redefining the relationship between sperm deoxyribonucleic acid fragmentation as measured by the sperm chromatin structure assay and outcomes of assisted reproductive techniques. *Fertil. Steril.* **2005**, *84*, 356–364. [CrossRef]

68. Liu, W.; Li, P.J.; Qi, X.M.; Zhou, Q.X.; Zheng, L.; Sun, T.H.; Yang, Y.S. DNA changes in barley (*Hordeum vulgare*) seedlings induced by cadmium pollution using RAPD analysis. *Chemosphere* **2005**, *61*, 158–167. [CrossRef]

69. Atienzar, F.A.; Jha, A.N. The random amplified polymorphic DNA (RAPD) assay and related techniques applied to genotoxicity and carcinogenesis studies: A critical review. *Mutat. Res. Rev. Mutat. Res.* **2006**, *613*, 76–102. [CrossRef]

70. Liu, W.; Yang, Y.S.; Zhou, Q.; Xie, L.; Li, P.; Sun, T. Impact assessment of cadmium contamination on rice (*Oryza sativa* L.) seedlings at molecular and population levels using multiple biomarkers. *Chemosphere* **2007**, *67*, 1155–1163. [CrossRef]

71. Long, T.C.; Saleh, N.; Tilton, R.D.; Lowry, G.V.; Veronesi, B. Titanium dioxide (P25) produces reactive oxygen species in immortalized brain microglia (BV2): Implications for nanoparticle neurotoxicity. *Environ. Sci. Technol.* **2006**, *40*, 4346–4352. [CrossRef]

72. Alahmar, A. Role of oxidative stress in male infertility: An updated review. *J. Hum. Reprod. Sci.* **2019**, *12*, 4–18. [CrossRef]

73. Xia, B.; Chen, J.; Zhou, Y. Cellular oxidative damage of HEK293T cells induced by combination of CdCl$_2$ and nano-TiO$_2$. *J. Huazhong Univ. Sci. Technol. Med. Sci.* **2011**, *31*, 290–294. [CrossRef]

74. Mohanraj, V.J.; Chen, Y. Nanoparticles—A review. *Trop. J. Pharm. Res.* **2006**, *5*, 561–573. [CrossRef]

75. Sager, T.M.; Porter, D.W.; Robinson, V.A.; Lindsley, W.G.; Schwegler-Berry, D.E.; Castranova, V. Improved method to disperse nanoparticles for in vitro and in vivo investigation of toxicity. *Nanotoxicology* **2007**, *1*, 118–129. [CrossRef]

76. Leisegang, K.; Henkel, R.; Agarwal, A. Redox regulation of fertility in aging male and the role of antioxidants: A savior or stressor. *Curr. Pharm. Des.* **2016**, *23*, 4438–4450. [CrossRef] [PubMed]

77. Muster, W.; Breidenbach, A.; Fischer, H.; Kirchner, S.; Müller, L.; Pähler, A. Computational toxicology in drug development. *Drug Discov. Today* **2008**, *13*, 303–310. [CrossRef] [PubMed]

Review

Evaluation of Ecotoxicology Assessment Methods of Nanomaterials and Their Effects

Bianca-Vanesa Boros and Vasile Ostafe *

Advanced Environmental Research Laboratories, Department of Biology—Chemistry, West University of Timisoara, Oituz 4, 300086 Timisoara, Romania; bianca.boros91@e-uvt.ro
* Correspondence: vasile.ostafe@e-uvt.ro

Received: 28 February 2020; Accepted: 20 March 2020; Published: 26 March 2020

Abstract: This paper describes the ecotoxicological effects of nanomaterials (NMs) as well as their testing methods. Standard ecotoxicity testing methods are applicable to nanomaterials as well but require some adaptation. We have taken into account methods that meet several conditions. They must be properly researched by a minimum of ten scientific articles where adaptation of the method to the NMs is also presented; use organisms suitable for simple and rapid ecotoxicity testing (SSRET); have a test period shorter than 30 days; require no special equipment; have low costs and have the possibility of optimization for high-throughput screening. From the standard assays described in guidelines developed by organizations such as Organization for Economic Cooperation and Development and United States Environmental Protection Agency, which meet the required conditions, we selected as methods adaptable for NMs, some methods based on algae, duckweed, amphipods, daphnids, chironomids, terrestrial plants, nematodes and earthworms. By analyzing the effects of NMs on a wide range of organisms, it has been observed that these effects can be of several categories, such as behavioral, morphological, cellular, molecular or genetic effects. By comparing the EC_{50} values of some NMs it has been observed that such values are available mainly for aquatic ecotoxicity, with the most sensitive test being the algae assay. The most toxic NMs overall were the silver NMs.

Keywords: nanomaterials; organisms suitable for simple and rapid ecotoxicity testing SSRET; ecotoxicological test batteries; ecotoxicology

---

1. Introduction

Both the study of nanomaterials and that of ecotoxicology are constantly increasing, as evidenced by the increasing number of scientific articles from Web of Science Core Collection (Web of Science, Clarivate Analytics) during 2010–2019 (Figure 1a,b). Due to the wide range of applications of nanomaterials, as well as the increased importance of protecting the environment and analyzing the ecotoxicity of potential pollutants, investigating the ecotoxicological effects of nanomaterials is essential. This is also underlined by the increasing number of scientific articles on the ecotoxicity of nanomaterials from Web of Science Core Collection (Web of Science, Clarivate Analytics) in 2010–2019 (Figure 1c). With the emergence of the need to assess the ecotoxicity of nanomaterials, the need arises to adapt standard ecotoxicity tests for testing nanomaterials.

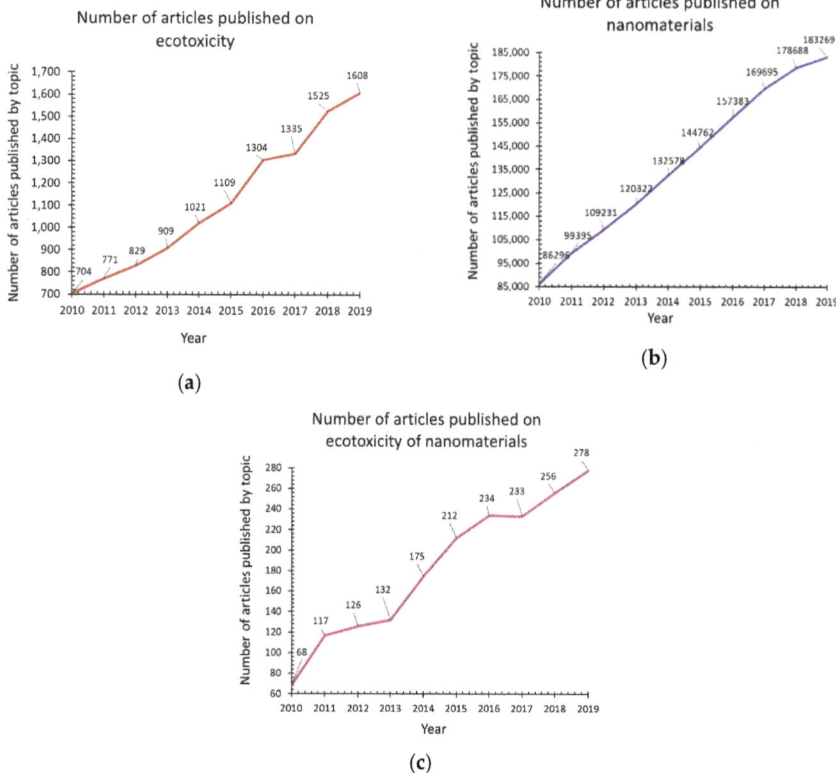

Figure 1. Bibliometric analysis of research in 2010–2019 from Web of Science Core Collection (Web of Science, Clarivate Analytics) on (**a**) ecotoxicity (search strategy: topic search (TS)=(Ecotox*) (containing "ecotox" in the topic of the articles)); (**b**) nanomaterials (search strategy: TS=(Nano*)) and (**c**) ecotoxicity of nanomaterials (search strategy: TS=(Ecotox* AND Nano*)).

In the present work, a series of bibliometric investigations were carried out to observe and analyze the state of the research on the ecotoxicity, nanomaterials and the ecotoxicity of the nanomaterials (Table 1.). These studies have highlighted the importance of research on the ecotoxicity of nanomaterials, the adaptation of the standard methods for testing the ecotoxicity of nanomaterials, and also highlighted the need to study the ecotoxicity in case of less studied nanomaterials. All bibliometric data were obtained from Web of Science Core Collection (Web of Science, Clarivate Analytics) in January 2020.

Table 1. Summary of bibliometric analysis of research on ecotoxicity of nanomaterials (NMs) from Web of Science Core Collection (Web of Science, Clarivate Analytics) presented in this review.

Ecotoxicity	Nanomaterials	Ecotoxicity of NMs	Ecotoxicity of NMs by Type
1. Ecotoxicity by topic (Figure 1a)	4. Nanomaterials by topic (Figure 1b)	6. NMs vs. ecotoxicity of NMs by topic (Figure 1c and Figure 15)	7. Ecotoxicity of types of NMs by topic (Figure 17)
2. Ecotoxicity by title vs. by topic (Figure 2)	5. Types of NMs by topic (Figure 5)		8. Categories of NMs and ecotoxicity of NMs by topic (Figure 16)
3. Aquatic vs. terrestrial ecotoxicity by topic (Figure 3)			

In this paper we address both the adaptation of the standard ecotoxicity testing methods for nanomaterials, as well as the mechanisms of the ecotoxicological effects of these nanomaterials described as their biological and biochemical effects.

1.1. Ecotoxicity

The term "ecotoxicology" was first used by Ernst Haeckel in 1866 in the sense of the "economy of nature" and the "science of all interactions between organisms and their environment". In the following years the science of ecotoxicology was developed as the study, using the tools of toxicology, of the effects of radiation and chemical pollutants from the biosphere on organisms from the environment [1,2]. In 1969, René Truhaut defined the term "ecotoxicology" as "the branch of toxicology concerned with the study of toxic effects, caused by natural or synthetic pollutants, to the constituents of ecosystems, animal, vegetable and microbial, in an integral context" [2].

The ecotoxicological research is under rapid development due to the pollution of the environment caused by the rapid industrial development and it is speeded up by severe industrial accidents. Since ecotoxicology became an important part in ecological and environmental risk assessment, ecotoxicity assessment policies were developed [2].

The environment hazard assessment framework uses ecotoxicity tests which are tools used to answer questions about the dangers of chemical substances that may be released into the environment [3].

The percentage of scientific articles by year in the Web of Science Core Collection (Web of Science, Clarivate Analytics) from the total of articles in 2010–2019 is increasing (Figure 2). This highlights the increasing importance of ecotoxicological testing in all areas and the necessity of adapting these assays to the new types of materials.

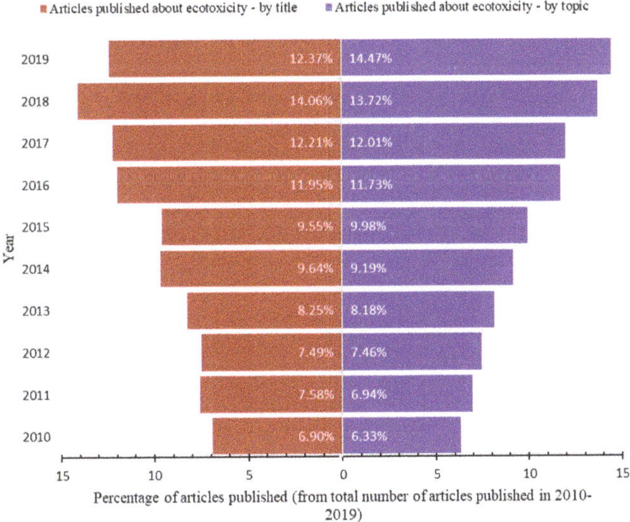

Figure 2. Bibliometric analysis of research on ecotoxicity in 2010–2019 from Web of Science Core Collection (Web of Science, Clarivate Analytics), the search strategy used to retrieve the data being: title search (TI)=(Ecotox*) (containing "ecotox" in the title of the articles) and TS=(Ecotox*) (containing "ecotox" in the topic of the articles).

The ecotoxicological assessment developed as aquatic and terrestrial, the terrestrial assessment lagging the aquatic one [2]. Although the number of scientific articles published both in the subject of aquatic ecotoxicity and in terrestrial ecotoxicity is increasing, the ratio between the two types remains relatively constant even during the period 2010–2019, the aquatic ecotoxicity being more researched than the terrestrial one (Figure 3). This highlights the necessity to further develop the study on terrestrial ecotoxicity to reach the level of the aquatic one, or at least to get close to its level.

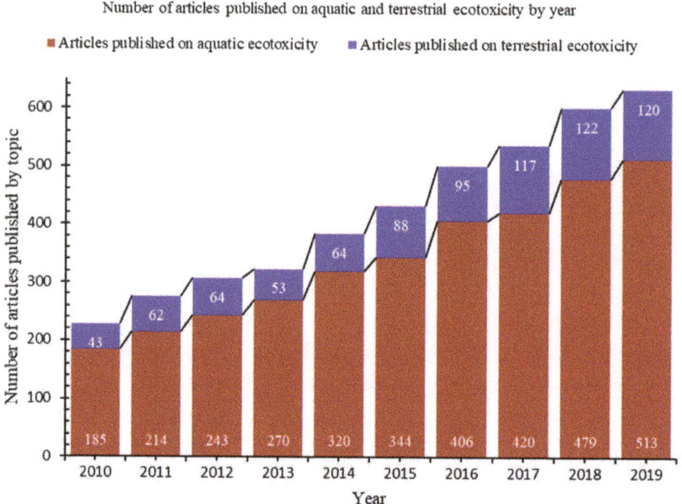

Figure 3. Bibliometric analysis of research on aquatic and terrestrial ecotoxicity in 2010–2019 from Web of Science Core Collection (Web of Science, Clarivate Analytics), the search strategy used to retrieve the data being: TS=(aquat* AND ecotox*) (containing both "aquat" and "ecotox" in the topic of the articles) and TS=(terrestr* AND ecotox*).

The ultimate concern of ecotoxicology is the establishment of the consequences of the effects at population level and on whole ecosystems, but in practice much work is done at individual level [4].

Standardized tests under the forms of guidelines were formulated by different organizations, such as Organization for Economic Cooperation and Development (OECD), United States Environmental Protection Agency (EPA), International Organization for Standardization (ISO), Government of Canada (GC) or American Society for Testing and Materials (ASTM). The standardized assay can be conducted on:

- aquatic organisms

    - algae
    - plants
    - invertebrates
    - vertebrates

- terrestrial organisms

    - algae
    - plants
    - invertebrates
    - vertebrates

These assays use a wide range of organisms depending on the type of test needed. A list of these organisms is provided in Tables 2 and 3.

Table 2. Algae, plants and invertebrates used in standard ecotoxicity assays.

LE [1]	OT [2]	Organism Category	Common Name of Group	Species	Standard Guidelines
Aquatic	Plants	Algae	Cyanobacteria	*Anabaena flos-aquae* *Synechococcus leopoliensis* *Microcystis aeruginosa*	OECD 201; EPA 850.4500; 850.4550; ISO 8692; 10253; 11044; ASTM E1218-04; GC EPS1/RM/25;
			Diatoms	*Navicula pelliculosa* *Skeletonema costatum* *Thalassiosira pseudonana* *Phaeodactylum tricornutum*	
			Green algae	*Raphidocelis subcapitata* *(Pseudokirchneriella* *subcapitata; Selenastrum* *capricornutum)* *Desmodesmus subspicatus* *(Scenedesmus subspicatus)* *Dunaliella tertiolecta*	
			Red algae	*Ceramium tenuicorne*	ISO 10710;
		Angiosperms	Duckweed	*Lemna minor* *Lemna gibba* *Spirodela polyrhiza*	OECD 221; EPA 850.4400; ISO 20079; 20227; ASTM E1415-91; GC EPS1/RM/37;
			Great Manna grass	*Glyceria maxima*	OECD 239;
			Watermilfoil	*Myriophyllum spicatum* *Myriophyllum aquaticum*	OECD 238; 239; ISO 16191;
	Animals	Snails	Mud snail	*Potamopyrgus antipodarum*	OECD 242;
			Pond snail	*Lymnaea stagnalis*	OECD 243;
		Bivalves	Clams	*Mercenaria mercenaria*	EPA 850.1025; 850.1055; 850.1710; ISO 17244; ASTM E2455-06; E724-98;
			Mussel	*Mytilus edulis* *Mytilus galloprovincialis*	
			Oyster	*Crassostrea virginica* *Crassostrea gigas*	
		Oligochaetes	Blackworms	*Lumbriculus variegatus*	OECD 225; 315;
			Tubificid worms	*Tubifex tubifex* *Branchiura sowerbyi*	
		Crustaceans	Amphipods	*Gammarus fasciatus* *Gammarus pseudolimnaeus* *Gammarus lacustris* *Ampelisca abdita* *Eohaustorius estuaries* *Rhepoxynius abronius* *Leptocheirus plumulosus* *Hyalella azteca*	EPA 850.1020; 850.1735; 850.1740; ISO 16712; GC EPS1/RM/26; EPS1/RM/33; EPS1/RM/35;
			Copepods	*Acartia tonsa* *Nitocra spinipes*	ISO 14669; 16778; 18220; ASTM E2317-04;
			Mysids	*Americamysis bahia* *(Mysidopsis bahia)*	EPA 850.1035; ASTM E1191-03a; E1463-92;
			Ostracods	*Heterocypris incongruens*	ISO 14371;

Table 2. *Cont.*

LE [1]	OT [2]	Organism Category	Common Name of Group	Species	Standard Guidelines
Aquatic	Animals	Crustaceans	Penaeids	*Farfantepenaeus aztecus (Penaeus aztecus)* *Farfantepenaeus duorarum (Penaeus duorarum)* *Litopenaeus setiferus (Penaeus setiferus)*	EPA 850.1045;
			Water fleas	*Daphnia magna* *Daphnia pulex* *Ceriodaphnia dubia*	OECD 202; 211; EPA 850.1010; 850.1300; ISO 6341; 10706; 20665; ASTM E1193-97; E1295-01; GC EPS1/RM/11; EPS1/RM/14; EPS1/RM/21;
		Insects	Chironomids	*Chironomus riparius* *Chironomus yohimatsui* *Chironomus tenans* *Chironomus dilutus*	OECD 218; 219; 233; 235; EPA 850.1735; GC EPS1/RM/32;
Terrestrial	Plants	Angiosperms	Monocots	*Allium cepa* *Avena sativa* *Hordeum vulgare* *Lolium perenne* *Oryza sativa* *Secale cereal* *Sorghum bicolor* *Triticum aestivum* *Zea mays*	OECD 208; 227; EPA 850.4100; 850.4150; 850.4230; 850.4300; 850.4800; ISO 11269-1; 11269-2; 17126; 18763; 21479; 22030; ASTM E1963-09; GC EPS1/RM/45; EPS1/RM/56;
			Dicots	*Beta vulgaris* *Brassica campestris var. chinensis* *Brassica napus* *Brassica oleracea var. capitate* *Brassica rapa* *Cucumis sativus* *Daucus carota* *Fagopyrum esculentum* *Glycine max (Glycine. soja)* *Gossypium sp.* *Helianthus annuus* *Lactuca sativa* *Lepidium sativum* *Linum usitatissimum* *Lotus corniculatus* *Phaseolus aureus* *Phaseolus vulgaris* *Pisum sativum* *Raphanus sativus* *Sinapis alba* *Solanum lycopersicum (Lycopersicon esculentum)* *Trifolium pratense* *Trigonella foenum-graecum* *Vicia sativa*	

Table 2. *Cont.*

LE [1]	OT [2]	Organism Category	Common Name of Group	Species	Standard Guidelines
Terrestrial	Animals	Snails	Helicidae	*Helix aspersa aspersa*	ISO 15952;
		Nematodes		*Caenorhabditis elegans*	ISO 10872; ASTM E2172-01;
		Oligochaetes	Earthworms	*Eisenia foetida* *Eisenia andrei*	OECD 207; 222; 317; EPA 850.3100; ISO 11268-1; 11268-2; 11268-3; 17512-1; 23611-1; ASTM E1676-12; GC EPS1/RM/43;
			Potworms	*Enchytraeus buchholzi* *Enchytraeus albidus* *Enchytraeus crypticus* *Enchytraeus luxuriosus*	OECD 220; 222; 317; ISO 16387; 23611-3; ASTM E1676-12;
		Mites		*Hypoaspis aculeifer (Geolaelaps aculeifer)*	OECD 226;
		Springtails		*Folsomia candida* *Folsomia fimetaria*	OECD 232; ISO 11267; 17512-2; GC EPS1/RM/47;
		Insects	Beetles	*Oxythyrea funesta*	ISO 20963;
			Bumble bees	*Bombus sp.*	OECD 246; 247;
			Honeybees	*Apis mellifera*	OECD 213; 214; 237; 245; EPA 850.3020; 850.3030; 850.3040;
			Leaf cutter bees	*Megachile rotundata*	EPA 850.3040;
			Sweat bees	*Nomia melanderi*	EPA 850.3040;
			Flies	*Musca autumnalis* *Scathophaga stercoraria*	OECD 228;

[1] LE = life environment; [2] OT = organism type

In Europe, there is a widespread tendency to reduce as much as possible the use of assays that involve the use of vertebrate organisms. Draft reports from EU REACH Implementation Project 3.3 recommend the use of vertebrates only if necessary. For example, in the aquatic ecotoxicity assessment, the sequence of testing involves the use of simple organisms for first test, and assays on fish only if necessary [3].

In conformity with this trend, in the present review, we used the term "SSRET organisms" for organisms suitable for simple and rapid ecotoxicity testing. We considered as SSRET organisms those organisms that are superior not from an evolutionary, taxonomical or complexity point of view, but from the point of view of suitability for ecotoxicity assessment. This term was assigned to species of plants (algae and vascular plants) and animals (only invertebrates) that are suitable due to their mechanisms of reproduction, size, culture method and also the lack of a proper ethical regulation, including only species that are not on the International Union for Conservation of Nature (IUCN) Red List [5] or protected by law. Only ecotoxicity assays that use SSRET organisms were studied because these give advantages like short culture period, are cheaper to maintain, need smaller laboratory space, etc. The organisms used in standard guidelines for ecotoxicity testing also include SSRET organisms from various categories such as those described in Figure 4.

Table 3. Vertebrates used in standard ecotoxicity assays.

Organism Type	Life Environment	Common Name	Standard Guidelines
Vertebrates	Aquatic	fish	OECD 203; 204; 210; 212; 215; 229; 230; 234; 236; 240; 305; EPA 850.1075; 850.1400; 850.1730; ISO 10229; 22082; 7346-1; 7346-2; 7346-3; 12890; 15088; 23893-1; 23893-2; 23893-3; ASTM E1192-97; E729-96; E1241-05; E1711-12; GC EPS1/RM/09; EPS1/RM/10; EPS1/RM/13; EPS1/RM/22; EPS1/RM/28;
		amphibians	OECD 231; 241; ISO 21427-1; ASTM E1192-97; E2591-07; E729-96;
	Terrestrial	birds	OECD 205; 206; 223; EPA 850.2100; 850.2200; 850.2300; ASTM E857-05;
		mammals	EPA 850.2400; ASTM E1163-10; E1619-11;

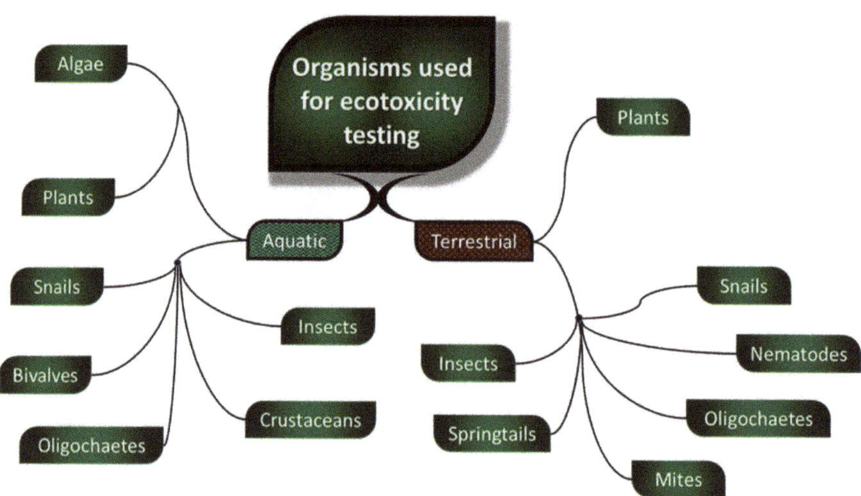

Figure 4. Categories of organisms that are used in standard guidelines for ecotoxicity assessment, several of the mentioned organisms being suitable for simple and rapid ecotoxicity testing (SSRET).

The assays that use vertebrates and other unsuitable animal species, including fish, amphibians, reptiles, birds and mammals, that are subjected to ethical regulations and that are difficult to use in ecotoxicity assays due to their larger size, longer reproduction cycles and difficult culture methods, were excluded from our study. The assays conducted on vertebrates, even though they are needed in some cases, present disadvantages like ethical problems, expensive culture methods, long testing periods.

1.2. Nanomaterials

"Nanomaterial" (NM) is a term that usually describes materials that have external dimensions or internal structures measured in nanoscale. The materials exhibit additional or different unique properties, than the same chemical substance in non-nano form (bulk material or pristine chemical), such as optical, magnetic, conductive, mechanical and often biological properties [6]. The definition of nanomaterials (NMs) according to the International Organization for Standardization (ISO) is: a nanomaterial is a material with any external dimension or internal structure or surface structure in the nanoscale (1–100 nm) [3].

In this study the term "nanomaterial" is addressed to any chemical or mixture resulted after a chemical reaction or a physical process and which have at least one dimension in nanorange. Anthropogenic bionanomolecules are named bionanomaterials (BNMs) and natural biomacromolecules (polymers such as proteins, polysaccharides, nucleic acids), which naturally possess at least one nanodimension, are not named nanomaterials but pristine bionanomolecules (PBNMs).

To properly understand and appreciate the diversity of NMs, some form of classification is required:

- by dimension

  - zero-dimensional (0D)
  - one-dimensional (1D)
  - two-dimensional (2D)
  - three-dimensional (3D)

- by composition

  - carbon based NMs
  - organic based NMs
  - inorganic-based NMs

    - metal-based NMs
    - metal oxide NMs

  - composite-based NMs [6]

Nanomaterials with all dimensions (x, y, z) less than 100 nm and length equal to width are zero-dimensional nanoparticles (0D-NPs). This class of NMs includes mostly spherical materials, but cubes and polygonal shapes can also be found. Types of 0D-NPs are: molecules, particles, atomic clusters and grains, for example silver and gold nanoparticles and quantum dots [6,7].

Nanomaterials with two dimensions (x, y) less than 100 nm and one dimension beyond the nanoscale (> 100 nm) are 1D-NMs. Examples of 1D-NMs are nanotubes, nanowires, nanofilaments, nanofibers, nanorods, nanowhiskers, etc. [6,7].

The materials that have three arbitrary dimensions beyond the nanoscale (> 100 nm) but have a nanocrystalline structure or involve the presence of features at the nanoscale are 3D-NMs. Examples of 3D-NMs include bulk materials composed of individual blocks, such as skeletons of fibers and nanotubes, honeycombs, composites of layers, fullerites, etc. [6,7].

Nanomaterials can be classified into two main categories: natural NMs, which existed in the environment long before the nanotechnology era started, and anthropogenic NMs. Airborne nanocrystals of sea salts, soil colloids, fullerenes, carbon nanotubes, biogenic magnetite etc. are some examples of natural NMs. Anthropogenic NMs can be further divided into two categories: incidental and engineered/manufactured. Incidental NMs are produced unintentionally in manmade processes (e.g., carbon nanotubes, carbon black and fullerenes, platinum and rhodium-containing nanoparticles from combustion byproducts). Engineered/manufactured NMs are materials that are produced intentionally due to their nanospecific properties [8].

The largest percentage of scientific articles on nanomaterials from the Web of Science Core Collection (Web of Science, Clarivate Analytics) in 2010–2019 is observed in the case of carbon nanotubes, followed by fullerene-based nanomaterials and those from gold and silver. The least investigated are the nanomaterials from polyhydroxyalkanoates, polylactic acid, alginate and cerium (Figure 5).

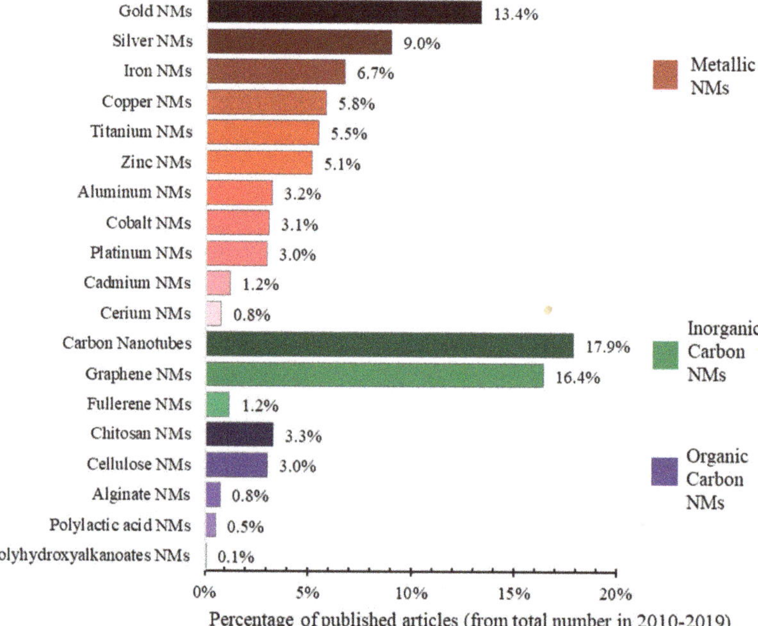

Figure 5. Bibliometric analysis of research on various types of nanomaterials in 2010–2019 from Web of Science Core Collection (Web of Science, Clarivate Analytics), the search strategy used to retrieve the data being: TS=(Nano* AND polyhydroxyalk*), TS=(Nano* AND polylact*), TS=(Nano* AND algin*), TS=(Nano* AND cellulos*), TS=(Nano* AND Chitosan*) TS=(Nano* AND fuller*), TS=(Nano* AND graphen*), TS=(Nanotub* AND carbon*), TS=(Nano* AND cerium*), TS=(Nano* AND cadmium*), TS=(Nano* AND platinum*), TS=(Nano* AND cobalt*), TS=(Nano* AND aluminum*), TS=(Nano* AND zinc*), TS=(Nano* AND titanium*), TS=(Nano* AND copper*), TS=(Nano* AND iron*), TS=(Nano* AND silver*) and TS=(Nano* AND gold*).

Nanotechnology is an interdisciplinary field with enormous potential in the fields of clean energy, medicine, chemistry, physics, nano-electronics, agriculture, astronomy, and environmental remediation. The physical, chemical, biological, catalytic and optical properties of materials become different from their bulk counterparts when their size becomes nanoscale (1–100 nm) [9].

2. Ecotoxicity Assessment of Nanomaterials on SSRET Organisms

In recent years the manufacturing, production and use of engineered nanomaterials (ENMs) in a wide range of products has increased. Releases of ENMs may occur during the use of nano-enabled consumer and industrial products either by nonintentional releases, like weathering of products containing NMs, intentional releases, like NMs used for environmental remediation, or accidental releases, like accidental spills during production, transportation or disposal [10].

There is a trend to include the ecotoxicity tests as part of the safe-by-design concept of creation, fabrication, utilization and disposal of ENMs.

Ecotoxicity assays can be either acute (short-term) or chronic (long-term). Acute assays are the most common measurement of ecotoxicological effects and are frequently used first to evaluate the survival of the organisms. Chronic tests are used when results from short-term tests combined with large safety factors suggest that there may be risks to the environment. These assays evaluate the sublethal effects on organism growth or reproduction [3].

The assessment of ecotoxicological effects of previously untested substances, as in the case of nanomaterials, is a challenging task. NMs are very rarely tested as potential pollutants, in contrast with their large diffusion. The main difficulties in assessing toxicity are a consequence of their colloidal nature and dynamics, as systems in which smaller or larger aggregates can form in poorly predictable ways, making it difficult to measure shape, size and concentration in the final sample. A basic requirement for good practice in toxicology laboratories is the use of approved or certified standards, which seems far from fulfillment in the case of NMs. The organization of the dispersed nanophase, in all naturally occurring systems like water, soil, air and their combinations, depends equally from the physical chemistry of the ENMs and from that of the environment, as well as from the modalities of suspension [11].

The assays selected for the ecotoxicity test battery for the assessment of the ecotoxicological effects of nanomaterials must be suitable for NMs because these materials present specific properties, different from their bulk materials, which may not be compatible with the assessment methods from the standard guidelines. There are several aspects in which NMs could interfere with the testing methods such as:

- agglomeration in test media during the procedures of test suspensions preparations;
- agglomeration, dissolution, and association with dissolved chemical species and colloidal/particulate matter already present in the natural waters;
- sedimentation due to agglomeration in the water column;
- chemical transformation processes such as sulphidation, hydroxylation;
- adhesion/deposition of nanomaterial onto soil minerals [10].

To our knowledge, there are not yet published articles or other documents stating that one or more of the standardized ecotoxicity assays are not suitable for NMs. Furthermore, the conclusion of the OECD Working Party on Manufactured Nanomaterials (WPMN) was that, in principle, the test guidelines are considered to be suitable also for the testing of NMs, although some adaptations were found to be necessary [12].

Since the assessment of ecotoxicity of NMs is imperative, the already developed and standardized assays must be sorted and selected, in order to create an efficient and adapted ecotoxicity test battery for NMs. Thus, we considered that an assay can be used to test NMs if the following conditions (Figure 6) are met:

- minimum 10 articles have already been published with the assay applied and adapted to NMs;
- only SSRET organisms are used;
- assessment period not longer than 30 days;
- no special apparatus or training needed;
- low testing costs;
- high-throughput methods (if possible).

Even if the majority of the standard tests could be adaptable, we only considered those that have already been published in minimum 10 scientific articles in which the assay was applied and adapted to NMs. We chose this rule in order to ensure that the theoretical adaptability is proven by research results.

Assays that have testing periods longer than 30 days were excluded because the ecotoxicity test battery's purpose is to make the testing easier and more relevant. Also, the companies that produce NMs or products that contain NMs, want to test their materials as fast as they can to ensure their profit, thus they cannot wait long periods of time for the results.

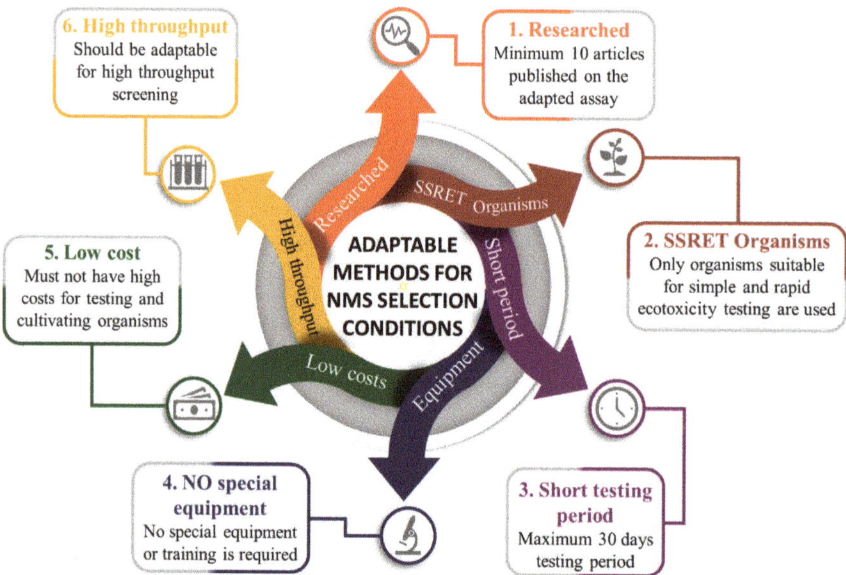

Figure 6. Condition for the selection of standard methods for ecotoxicity assessment that can be adapted for nanomaterial.

The use of special apparatus and/or special training increases testing time by forcing the companies to search for laboratories that have these specific apparatus and staff. These kinds of assays can also increase the cost of testing since special reagents might be needed. The cost of the assessment is important too, since the profit of a company that produce NMs or products that contain NMs depends on the cost of testing.

An ideal ecotoxicity test battery should be easy and fast to apply, with specific tests for the selected material, and, in this case NMs, give relevant data for the material's safety for the environment. From this point of view, high-throughput assays would be the best choice since these minimize the time of the test, by conducting several tests at once, and reducing the quantities of reagents needed, thus lowering the costs too. High-throughput screening (HTS) is an alternative technique to the "classical" method for scientific experimentation that allows a researcher to quickly conduct several tests concomitantly, by using, for example, a microtiter plate instead of test tubes.

An ideal ecotoxicity test should be predictive (the effect of NMs should be predictable outside the range of tested concentration) and transferable (the effect of NMs should be the same or very similar on non-tested organisms).

From the tests listed in Table 4, which contains assays with testing time less than 30 days, only the tests that are compatible or adaptable for nanomaterials are selected and described. If a test is widely used for nanomaterials (minimum 10 articles published) but the majority of the described adaptations are only for the NM suspension preparation, it is still described if at least a few test adaptations are mentioned. Also, the tests that cannot be made without special apparatus or training were eliminated from the descriptions. The tests that were eliminated are listed below:

- the watermilfoil assays (sediment free toxicity assay and water-sediment toxicity assay) were eliminated due to the restricted number of published articles; in the published articles there are presented only the modification for the preparation of NM suspension, but not for the assay itself;
- the snail assays (on mud snail and on pond snail) were eliminated due to the restricted number of published articles, because bioconcentration was primarily assessed which takes longer than

30 days and because it involves the use of special apparatus like a flame atomic absorption spectrometer or MC-ICP-MS (Multicollector-Inductively Coupled Plasma Mass Spectrometer);

- the bivalve, aquatic oligochaetes, mysid, penaeid, enchytraeid, mite, springtail and the bumblebee assays were eliminated due to the restricted number of published articles; the published articles present only the modification for the preparation of NM suspension, but not for the assay itself;
- the fly assay was eliminated because only a few articles were found for ecotoxicity assessment of NMs; even if there are numerous ecotoxicity articles for NMs using as test organism *Drosophila sp.*, these assays were not included in the description due to a lack of standardization.

Table 4. Analysis of ecotoxicity assays that correspond to the criteria described above. For simplification of table only tests less than 30 days are showed.

LE [1]	OT [2]	Organism	Test	NMAP [3]	SAT [4]	TD [5]
Aquatic	Plants	Algae	Growth inhibition	✓	✗	72 h
			Toxicity	✓	✗	96 h
		Duckweed	Growth inhibition	✓	✗	7 days
			Toxicity	✓	✗	
		Watermilfoil	Sediment free toxicity	✗	✗	14 days
			Water-sediment toxicity	✗	✗	
	Animals	Mud snails	Reproduction	✗	✗	28 days
		Pond snails	Reproduction	✗	✗	28 days
		Bivalves	Acute toxicity	✗	✗	48–96 h
		Oligochaetes	Sediment-water toxicity	✗	✗	28 days
		Amphipods	Acute toxicity	✓	✗	96 h
			Spiked whole sediment toxicity	✓	✗	10 days
		Mysids	Acute toxicity	✗	✗	96 h
		Penaeids	Acute toxicity	✗	✗	96 h
		Water fleas	Acute immobilization	✓	✗	48 h
			Acute toxicity	✓	✗	48 h
			Chronic toxicity	✓	✗	21 days
			Reproduction	✓	✗	21 days
		Chironomids	Acute immobilization	✓	✗	48 h
			Sediment-water toxicity with spiked sediment	✓	✗	20–28 days
			Sediment-water toxicity with spiked water	✓	✗	20–28 days
			Spiked whole sediment toxicity	✓	✗	10 days
Terrestrial	Plants	Angiosperms	Early seedling growth toxicity	✓	✗	14 days
			Seedling emergence/Seedling growth	✓	✗	14–21 days
			Vegetative vigor	✓	✗	21–28 days
	Animals	Nematodes	Toxicity	✓	✗	96 h
		Earthworms	Acute toxicity (contact and soil)	✓	✗	14 days
			Subchronic toxicity	✓	✗	28 days
		Mites	Reproduction	✗	✗	14 days
		Springtails	Reproduction	✗	✗	21–28 days
		Bumblebees	Acute contact toxicity	✗	✗	48–96 h
			Acute oral toxicity	✗	✗	48–96 h
		Honeybees	Acute contact toxicity	✗	✗	48–96 h
			Acute oral toxicity	✗	✗	48–96 h
			Chronic oral toxicity	✗	✗	10 days
			Larval toxicity	✗	✗	7 days
			Toxicity of residue on foliage	✗	✗	24 h
		Flies	Developmental toxicity	✗	✗	18–23 days

[1] LE = life environment; [2] OT = organism type; [3] NMAP = nanomaterial adaptation possibility; [4] SAT = special apparatus and training; [5] TD = test duration.

Only eight standard assays were selected as being adaptable for NMs, five aquatic assays and three terrestrial ones: the algae, duckweed, amphipods, daphnids, chironomids, terrestrial plants, nematodes and earthworm assay (Figure 7).

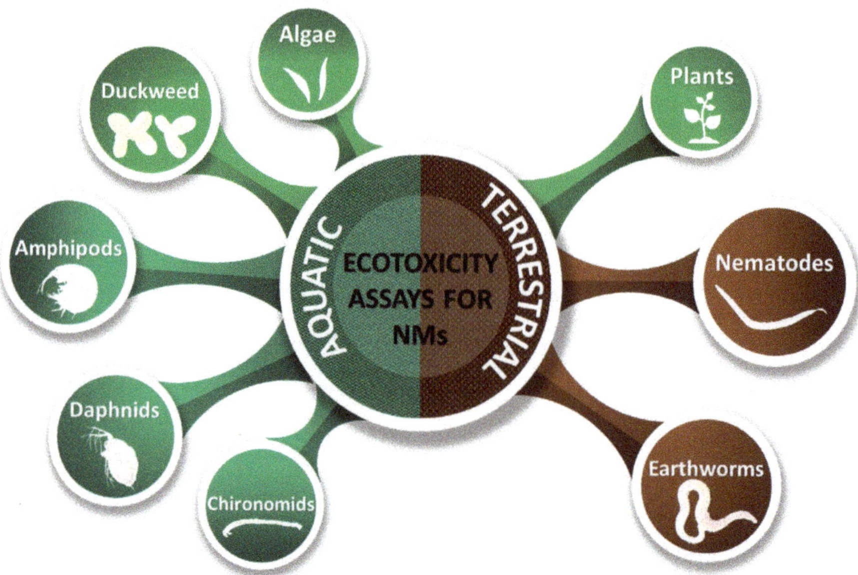

Figure 7. Standard assays selected to be adaptable for NMs.

3. Description of Ecotoxicity Assays Suitable for Nanomaterials and the Necessary Adaptations

3.1. General Adaptations and Considerations

Multiple factors need to be taken into consideration when bioassays involving exposure to suspended NMs are interpreted, such as: the relevance and the appropriateness for assessing the tested NMs; the accuracy of the representation of the exposure, e.g., whether the frequency of characterization measurements sufficient to capture changes in exposure during the bioassay; the consistency of the exposure, such as stable concentration, agglomeration and dissolution; whether maintaining a consistent exposure is possible in the bioassay method-specific test system; whether nanospecific bioassay acceptability criteria, e.g., sufficiently consistent exposure concentration with respect to agglomeration and dissolution, are met; and whether the characterization and monitoring data during the bioassay are amenable to expressing data by an alternative dose metric [13].

3.1.1. Physico-Chemical Adaptations and Considerations

A wide range of dispersing techniques is used (application of solvents, dispersion or stabilizing agents, bath or probe sonication, stirring, etc.) and the methods also vary with respect to applied concentrations, time, etc. The NMs could be significantly altered by the preparation method, the used solvents could interact producing toxic byproducts and the properties of the dispersions depend on the dispersing method, and thus the ecotoxic effects of the NMs could be influenced and the comparability between tests could be hampered [14].

As described by previous OECD documents for metal toxicity testing, such as algae testing, it is important to exclude metal chelators such as ethylenediaminetetraacetic acid (EDTA) for NMs where dissolved metal ions may impact the toxicity (e.g., ZnO ang Ag NPs). The interactions between the chelators and the NM surface may influence the NM behaviors and transformations [13].

3.1.2. Biological Adaptations and Considerations

It is important to ensure that only the portion of the contaminant that has actually accumulated is considered when bioaccumulation, bioconcentration, bioamplification and trophic transfer factor (TTF) are assessed in ecotoxicity studies. Only contaminants that are absorbed inside the organism must be considered. The contaminants should not be considered if these are on the surface of the organism or if these are ingested but can be eliminated by simple excretion. Thus, NMs that are trapped inside the digestive tract and prevent normal exchanges such as nutrient uptake, which can result in physiological impairments, should be considered as accumulated. To correctly consider a NM accumulated in the digestive tract, the organisms should be rinsed and allowed to depurate (i.e., they should be placed in a non-contaminated environment long enough for complete excretion to occur) so that only the absorbed contaminant is left in the organism. Absorption of NMs by the organism can result from several processes such as direct entrance in the cell (e.g., intestinal, skin or gill cells) or crossing of epithelial barriers through intercellular junctions to enter the blood circulation and then be distributed to the whole organism, without entering the cells [15].

3.2. Aquatic Plants Used for Ecotoxicity Testing of Nanomaterials

3.2.1. Algae

These organisms are used in ecotoxicity assessment through growth inhibition and toxicity assays. These assays are regulated by several standard guidelines created by organizations like OECD [16], EPA [17,18] and ISO [19,20].

The purpose of these assays is the determination of the ecotoxic effects of a substance on the growth of freshwater microalgae and/or cyanobacteria (Figure 8). Test organisms that are in the exponential growth phase are exposed to the test substances over a period of 72 hours (during which effects over several generations can be assessed), which can be extended to 96 hours for the assessment of toxicity by obtaining population growth data such as cell density [16,17].

- Adaptations

  - an EDTA-free version of the OECD medium for *Raphidocelis subcapitata* is recommended for metallic NMs [12];
  - by supplying iron as $FeSO_4$ instead of $FeCl_3$, the OECD medium permits a better growth for the algae. With this medium the amount of phosphorus is also increased, as mono and dibasic salts [12];
  - a shacking procedure is recommended during the tests [21];
  - the determination of algal biomass by cell counting with a hemocytometer is very laborious, has a large variance and it may not reflect the true biomass if there are changes in the mean size of the cells as a response to a toxicant or other condition. The interference caused by NMs could be avoided by the determination of biomass by fluorometry followed by *Chl a* extraction [12];
  - the determination of in vivo chlorophyll content could be realized in microtiter plates but NMs might interfere with the measurements [21];
  - the in vivo determination of fluorescence was found to be an unstable parameter; it depends on the prior light exposure of the culture, since different results are obtained by the repeated measurement of chlorophyll on the same sample [12];
  - the most reliable quantification method of algal biomass for testing the effects of NMs has been found to be the measurement of fluorescence of chlorophyll extracts. However the background fluorescence of the NMs should be reduced as much as possible [21].

- Uptake

  - The uptake of NMs in bacteria or algae can't take place without the attraction of the NMs to a biofilm or their absorption to a substrate. The attachment of NMs to the cell membrane is assumed to take place via electrostatic attraction, but there are other forces that could be involved such as random collision. If no uptake is observed in microalgae it is possible that the concentration of NMs was too low for attachment by collision to occur or that the electrostatic attraction was too weak for adsorption to have taken place [22].
  - Released metal ions from metallic NMs may diffuse across cell membranes. For some cyanobacteria (e.g., *Anabaena flos-aquae*), however, by the production of extracellular polysaccharides the internalization of NMs could be avoided due to the electrostatic interaction of these substances with NMs, which are entrapped outside the cell [22].

- Advantages

  - short testing period of 72–96 hours;
  - high-throughput assay microplates can be used;
  - simple culture method;
  - simple quantification method: chlorophyll extraction;
  - one of the most sensitive ecotoxicological assays [23].

- Disadvantages

  - fluorescence reader is necessary for optimal chlorophyll quantification;
  - NM–algae aggregates may form: heteroaggregation of NM-algae was reported [21];
  - NMs may cause shading by scattering the light from reaching algal cells and thereby reduce their growth rate, rather than or in addition to any toxic effect. Even if the algae can temporarily overcome the shading from distant NMs, the adhesion of NMs to the algal cells can result in permanent shading and limitation of nutrient availability [24];
  - if optical density measurements are used, there could be interferences with the NMs [23].

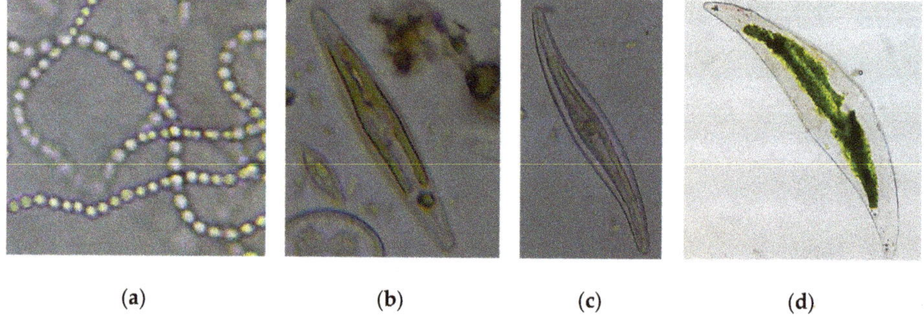

| (a) | (b) | (c) | (d) |

Figure 8. Examples of algae that can be used for ecotoxicity assessment: (**a**) cyanobacteria; (**b**) diatoms and (**c,d**) green algae.

3.2.2. Duckweed

These organisms (Figure 9) are used in ecotoxicity assessment through growth inhibition and toxicity assays. These assays are regulated by several standard guidelines formulated by organizations like OECD [25], EPA [26] and ISO [27,28].

Figure 9. Example of duckweed species recommended for ecotoxicity assessment by standard guidelines: *Lemna minor* (common duckweed).

Exponentially growing duckweed cultures are grown as monocultures in test substances of different concentrations over a period of seven days. The quantification of substance-related effects on vegetative growth can be based on the assessment of different variables such as frond number, which is the primary variable of measurement, and at least one other (e.g., total frond area, fresh or dry weight, chlorophyll content). The endpoint of this assays is percent inhibition in average specific growth rate and 50 percent inhibition (IC_{50}) [25].

- Adaptations

  - the careful stirring twice a day of the growth medium is recommended to minimize the sedimentation of NMs [29];
  - it is recommended to wash the plant samples using EDTA-Na_2 (0.02 M, five times) and distilled water (five times) to avoid the possible attachment of metallic NMs to the samples [30].

- Advantages

  - simple cultivation method;
  - can be grown indefinitely as genetically homogeneous clonal colonies due to their predominantly vegetative reproduction [31];
  - ready contact with substances dissolved in the culture medium is ensured by their high surface-to-volume ratio and lack of cuticle on their surface in contact with water [31];
  - some species of duckweed have a wide pH tolerance such as *Spirodela polyrhiza* [32].

- Disadvantages

  - medium testing period: seven days;
  - relatively large laboratory space necessary for testing of multiple experiments at once;
  - the actual environmental conditions, under which the test organisms live in nature, are not accurately reflected by the standard experimental conditions employed in any standardized duckweed toxicity test [31].

3.3. Aquatic Invertebrates Used for Ecotoxicity Testing of Nanomaterials

3.3.1. Amphipods

These organisms (Figure 10) are used in ecotoxicity assessment through acute toxicity and spiked whole sediment toxicity assays. These assays are regulated by several standard guidelines formulated by organizations like EPA [33–35] and ISO [36].

Figure 10. Example of amphipod species that can be used for ecotoxicity assessment: *Gammarus sp.*

In the acute toxicity assay, young gammarid amphipods are exposed to the test substance and to appropriate controls for 96 hours. Observations are made regarding the survival of the organisms and other toxic effects. The relationship between aqueous concentrations of the test substance and mortality of gammarids over the full concentration–response curve is determined by this assay [33].

The spiked whole sediment toxicity assay involves the monitoring of amphipods during the test for sediment avoidance and other toxic effects observable without disturbing the sediment. The survival and growth are determined at the termination of the test [34,35].

- Adaptations

  - only NM suspension preparation adaptations are specified.
- Advantages

  - short testing period for the acute toxicity assay: 96 hours.
- Disadvantages

  - relatively large laboratory space necessary for testing of multiple experiments at once.

3.3.2. Daphnia

These organisms (Figure 11) are used in ecotoxicity assessment through an acute toxicity (immobilization) assay or a reproduction assay. Both assays are regulated by several standard guidelines formulated by organizations like OECD [37,38], EPA [39,40] and ISO [41–43].

The acute toxicity assay involves the use of young daphnids (with age less than 24 hours at the start of the test) and their exposure to the test substance at a range of concentrations for a period of 48 hours. The endpoint of this assay is immobilization (lack of motility) of daphnids, which is recorded at 24 hours and at 48 hours and compared with control values. The half maximal effective concentration at 48 hours is calculated by analyzing the obtained results [37].

In the reproduction assay young female daphnids (aged less than 24 hours) are exposed to the test substance (at a range of concentrations) added to water. The assessment of the effect of chemicals on the reproductive output of daphnids is the primary objective of this assay. At the end of the test, after 21 days, the total number of living descendants produced is determined. Other ways can be also used for the expression of the reproductive output of the parent animals but these should be reported in addition to the total number of living offspring produced at the end of the test [38].

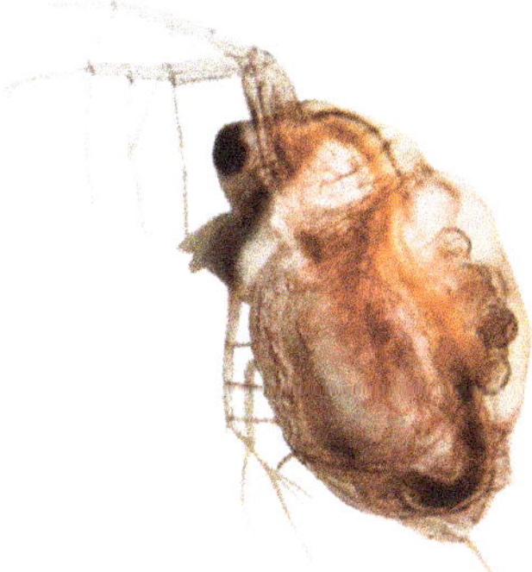

Figure 11. Example of daphnia species recommended for ecotoxicity assessment by standard guidelines: *Daphnia magna*.

NMs or NM-algae associations could be consumed by *Daphnia magna*. The water with the NMs is funneled towards the daphnia's mouth by its feeding appendages. In the gut lumen, the NMs are accumulated after being compacted by the undigested food and other materials. The accumulation of NMs in the gut lumen and their condensation into a tightly packed form was also observed in marine copepods such as *Tigriopus japonicus* [22].

- Adaptations

  - the use of a medium with a very low ionic strength, under which daphnids can grow and reproduce normally, and which has a pH value where more stable dispersions can be obtained is recommended [12];
  - NMs can be absorbed on the exoskeleton, cuticle and antenna of crustaceans like daphnids. This influences the mobility, molting and swimming velocity of the tested crustaceans. Thus, the inclusion of both lethality and immobilization assays is recommended. The use of only immobility as endpoint, such as the OECD assay, may be problematic in cases where immobility reflects physical impairment rather than toxicity [24];
  - to prevent the mechanical impairment of daphnids by absorbed NMs, a mesh could be inserted at the bottom of the test vessel to prevent the contact of the daphnids with larger clusters of NMs that accumulate at the bottom of the beaker [24];
  - greater water hardness is used for *Daphnia magna* growth and reproduction assays which leads to a greater agglomeration rate of NMs for charge-stabilized NMs, resulting in less consistent exposure. Thus, using daphnid species that are adapted for softer waters is recommended (such as *Daphnia pulex*) [13];
  - to increase the contact between the daphnids and the NMs the use of shallow test vessels or semi-static and flow-through systems is recommended [21];
  - the feeding of daphnids during the reproduction assay is not recommended. It was observed that the outcome of the test is food quantity dependent, as the addition of higher food levels resulted in higher animal survival, growth and reproduction compared to tests with lower food levels. Also, the uptake of NMs is influenced by the presence of food. This may influence the chronic effects found in long-term exposure tests. Furthermore, the presence of food (algae) in daphnia reproduction test may affect the observed toxicity, due to the interaction of NMs with algal exudates which affect the bioavailability of the NMs [24].

- Advantages

  - ○ short testing period for acute toxicity assay: 48 hours;
  - ○ daphnia are particle-feeding organisms, a relevant model for NMs [23];
  - ○ daphnids are one of the most sensitive organisms used in ecotoxicity assessment of chemicals [37].

- Disadvantages

  - ○ long testing period for reproduction and chronic assays: 21 days;
  - ○ high sample volume is required (50 mL per concentration) [23];
  - ○ test medium may induce agglomeration and sedimentation of NMs [44];
  - ○ NMs may affect the movement of the daphnids by provoking physical effects on their surface [21].

3.3.3. Chironomids

These organisms (Figure 12) are used in ecotoxicity assessment through a sediment water toxicity assay or a sediment water life cycle toxicity assay, using either spiked sediment or spiked water or an acute immobilization assay. These assays are regulated by several standard guidelines formulated by organizations like OECD [45–48] and EPA [34].

Figure 12. Example of chironomid that can be used for ecotoxicity assessment.

The sediment water toxicity assay with spiked sediment involves the exposure of first instar chironomid larvae to a concentration range of the test chemical in sediment–water systems, where the test substance is spiked into the sediment. The first instar larvae are subsequently introduced into test beakers with stabilized sediment and water concentrations. At the end of the test the emergence and development rate of chironomids are measured. If required, after ten days, the larval survival and weight can be also measured [47].

The sediment water toxicity assay with spiked water involves the exposure of first instar chironomid larvae to a concentration range of the test chemical in sediment–water systems, where the test substance is spiked into the water. At the end of the test the emergence and development rate of chironomids are measured [48].

The sediment water life cycle toxicity assay with spiked water or sediment involves the exposure of first instar chironomid larvae to a concentration range of the test substance in a sediment-water system. The first instar larvae (first generation) are placed into test beakers that contain the test substance spiked into the sediment or water. At the end of the test the chironomid emergence, time to emergence and sex ratio of the fully emerged and living midges are assessed. To facilitate the swarming, mating and oviposition of the emerged adults, these are transferred to breeding cages and the number of egg ropes produced is assessed. The second generation first instar larvae obtained from these egg ropes are then placed into freshly prepared test beakers to determine their viability through an assessment of their emergence, time to emergence and the sex ratio of the fully emerged and living midges [46].

The acute immobilization assay involves the exposure of first instar *Chironomus sp.* larvae to a range of concentrations of the test substance in water-only vessels for a period of 48 hours. The immobilization of the chironomids is recorded both at 24 hours and at the end of the test, at 48 hours [45].

- Adaptations

    ○ an additional parameter is recommended for the assessment of the toxic effects. This parameter is represented by morphological malfunctions revealed by the analysis of the mouthpart structures [49].

- Advantages

    ○ short testing period for acute tests: 48 hours;
    ○ chironomids have a widespread distribution. The aquatic sediment environment is strongly influenced by them through processes such as sediment ingestion, digestion, resuspension, excretion, and secretion, bioirrigation and bioturbation [50].

- Disadvantages

    ○ long testing periods for the sediment–water system tests: 10–28 days.

3.4. Terrestrial Plants Used for Ecotoxicity Testing of Nanomaterials

These organisms (Figure 13) are used in ecotoxicity assessment through an early seedling growth toxicity, a seedling emergence and seedling growth assay and a vegetative vigor assay. These assays are regulated by several standard guidelines formulated by organizations like OECD [51,52], EPA [53–55] and ISO [56–60].

(a) (b) (c)

(d) (e) (f)

Figure 13. Examples of terrestrial plant species recommended for ecotoxicity assessment by standard guidelines: (**a**) *Cucumis sativus* (cucumber); (**b**) *Trifolium pratense* (red clover); (**c**) *Lotus corniculatus* (common birdsfoot trefoil); (**d**) *Phaseolus vulgaris* (common bean); (**e**) *Pisum sativum* (pea) and (**f**) *Helianthus annuus* (sunflower).

The effect of a test substance applied to the roots or the leaves of several terrestrial plant species is assessed thorough the early seedling growth toxicity assay. After germination of seeds planted in the potting containers, seedlings are thinned (by pinching the stem at the support medium surface) to the 10 most uniform seedlings per pot to which the test substances are applied. The application of the test substances can produce either a foliar exposure (by exposing the plants in a fumigation chamber to gas or by spraying or dusting the foliage) or a root exposure (by nutrient solution or by sorption to support media). Seedlings emerging after this time are also pinched off at the surface. After 14 days, the plants are harvested, and their growth is analyzed by parameters such as observed phytotoxicity, length or dry weigh of the shoot or the root, seedling survival, length and weight of whole plants [55].

The effects of a test substance dosed to soil are assessed on seedling emergence and early growth of higher plants. The effects of the test substance on seeds are assessed after 14–21 days, after 50% emergence of the seedlings in the control group. Visual assessment of seedling emergence, dry or fresh shoot weight or height and visible detrimental effects on different parts of the plant are the measured endpoint of this assays, which are compared to those of untreated control plants [51].

Following deposition of the test substance on the leaves and above-ground portions of plants, the potential effects of those test substances are assessed through the vegetative vigor assay. The test substance is sprayed on plants, grown from seed to the 2–4 leaf stage, and their leaf surfaces at different concentrations. At various intervals through 21–28 days from treatment, the plants are evaluated for effects on vigor and growth against untreated control. Shoot dry or fresh weight and height, and visible detrimental effects on different parts of the plant are the assessed endpoints of this test [52].

- Adaptations

  - to avoid NMs precipitation, which are poorly soluble in water, and to distribute NMs evenly, the use of plant agar tests is recommended. In the preparation of the agar solutions, to avoid the possible precipitation of NMs, the test plates were immediately hardened after pouring in a freezer [61].

- Advantages

  - plants are critical to the function of ecosystems and the integrity of the food supply, thus plants being an essential component of the environment [62];
  - NMs are of an extensive interest for application on plants for uses in agriculture and horticulture [63].

- Disadvantages

  - long testing periods: 14–28 days;
  - the analysis of NM content of plants may imply the use of special apparatus like ICP-MS [61];
  - special techniques like the scanning electron microscopy–cathodoluminescence technique may be used for the analysis of NM content of agar paste [61];
  - the biological effects determined could depend on the species selected for the assay [64].

3.5. Terrestrial Invertebrates Used for Ecotoxicity Testing of Nanomaterials

3.5.1. Nematodes

These organisms (Figure 14) are used in ecotoxicity assessment through a soil and sediment toxicity test on growth, fertility and reproduction of *Caenorhabditis elegans*. This assay is regulated by several standard guidelines formulated by organizations like ISO [65] and ASTM [66].

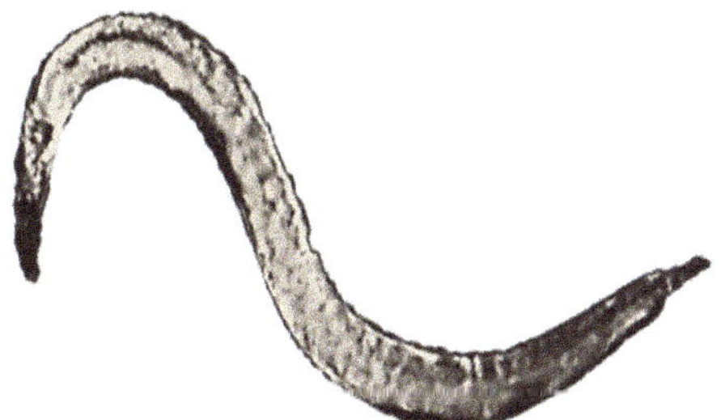

Figure 14. Example of nematode that can be used for ecotoxicity assessment.

The soil and sediment toxicity test on growth, fertility and reproduction of *Caenorhabditis elegans* involves the maintenance of nematode stock cultures grown on agar medium. For the aqueous assay, the stock solutions are transferred to 12-well microplates, while for the soil and sediment assays, the dry material is placed into test vessels and moistened with medium (to achieve 40% water content for the artificial control sediment and the soil samples and original water content for natural sediment samples). *Escherichia coli* suspended in medium is then mixed into the aqueous, sediment, and soil samples as food supply. At the end of the test, after 96 hours, the worms are heat-killed and then mixed with an aqueous solution of rose Bengal to stain them for easier counting [67].

- Adaptations

  - the NM suspension can be prepared by sonication [68];
  - during the test, the testing plates can be shaken [68].

- Advantages

  - short testing period: 96 hours;
  - can be applied in microplates, thus it is a high throughput assay [68];
  - a major change in the abundance of soil invertebrates such as nematodes, which are key organisms in soil, could have serious adverse effects on the entire terrestrial system [66];
  - due to its ability to grow and reproduce in both soil and aqueous environments, *Caenorhabditis elegans* is a well suited organism for toxicity assessment [69].

- Disadvantages

  - NMs can aggregate and may settle on the bottom of test vessels. This could potentially increase the local concentration and exposure to the NMs and could also change the exposure from NMs to NM aggregates [69];
  - the reproductive output could be decreased by the shacking of the test vessels [69];

3.5.2. Earthworms

These organisms (Figure 15) are used in ecotoxicity assessment through an acute toxicity assay, and a subchronic toxicity assay. These assays are regulated by several standard guidelines formulated by organizations like OECD [70], EPA [71] and ISO [72].

Figure 15. Example of earthworm that can be used for ecotoxicity assessment.

The simple contact acute toxicity test involves the exposure of earthworms to test substances on moist filter paper to identify potentially toxic chemicals to earthworms in soil. This test gives reproducible results with the recommended species and it is easy to perform [70].

Data more representative of natural exposure of earthworms to chemicals can be obtained using the artificial soil acute toxicity test. In this test earthworms are kept in samples of a precisely defined artificial soil to which a range of concentrations of the test substance has been applied. Mortality is assessed seven and 14 days after application [70].

The subchronic toxicity assay involves the placement of acclimated worms in test chambers that contain artificial soil with the test substance. The earthworms ingest the soil mixture and at the end of the test, after 28 days, their mortality and other effects are assessed [71].

- Adaptations

  - the reduction of organic matter content of the standard OECD soil is recommended, due to the reduction of bioavailability of NMs by artificial soil (the NMs are absorbed to soil organic matter) [12];
  - it is recommended to dose the NMs by directly adding the dry nanopowder to the soil, because it was observed that NMs agglomerate in water at the concentration needed for the dosing of the soil [73];
  - to ensure a continuous exposure even if the worms would attempt to escape into the added food mixture, the food should also be dosed with NMs [73].

- Advantages

  - *Eisenia fetida* is a recommended species for ecotoxicity assays as it can be easily cultured in the laboratory [74];
  - earthworms represent 60–80% of the total soil biomass and have a wide range distribution of soil, thus being an ideal organism for use in ecotoxicity assays, as these are also sensitive organisms that are readily available. Furthermore, historical data are available for their use in ecotoxicity assessment [74].

- Disadvantages

  - long testing period: 14–28 days.

4. Ecotoxicity of Nanomaterials

Nanotechnology is slowly becoming an essential part of daily life in different forms, such as pharmaceuticals, food packaging, biosensors, cosmetics etc., and thus it might present an unintended risk to human health and the environment [9].

The environment could be polluted intentionally or accidentally by large quantities of NMs due to the increasing presence of these materials in commercial products. It is crucial that the potential environmental and health impacts of NMs are assessed and the environment is protected to ensure a sustainable nanotechnology industry [75].

Both the percentage of scientific articles on nanomaterials and the percentage of scientific articles on the ecotoxicity of nanomaterials from the Web of Science Core Collection (Web of Science, Clarivate Analytics) are increasing during 2010–2019 (Figure 16). This highlights directly the importance of testing the ecotoxicity of nanomaterials, but also, indirectly, the importance of adapting standard ecotoxicity tests to NMs.

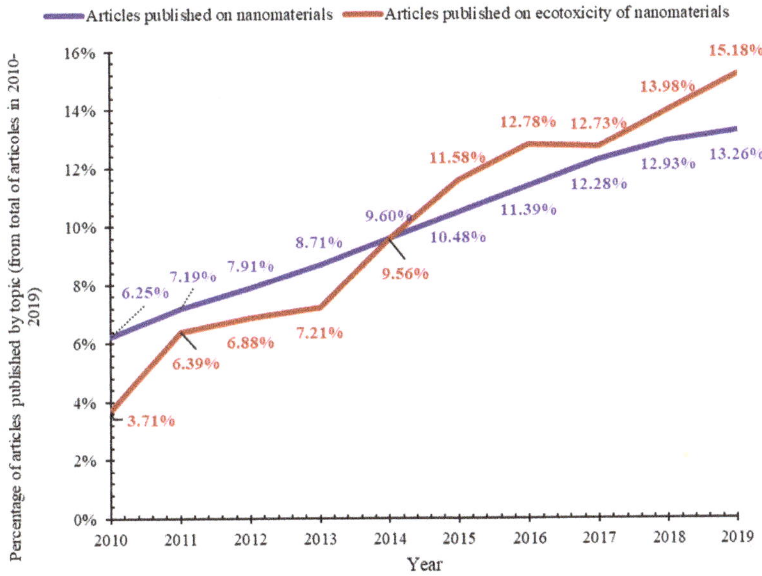

Figure 16. Bibliometric analysis of research on nanomaterials and their ecotoxicity from 2010 to 2019 from Web of Science Core Collection (Web of Science, Clarivate Analytics). The search strategy used to retrieve the data was TS=(Nano*) and TS=(Ecotox* AND Nano*).

Both the highest percentage of scientific articles on nanomaterials, as well as the highest percentage of scientific articles on the ecotoxicity of nanomaterials from the Web of Science Core Collection (Web of Science, Clarivate Analytics) can be observed in the case of metallic nanomaterials, followed by those of inorganic carbon. Only in the case of metallic nanomaterials, the percentage of scientific articles on the ecotoxicity of the nanomaterials was higher than the percentage of scientific articles on nanomaterials, percentages relative to the total scientific articles from 2010–2019 from the same topic (Figure 17). This highlights the necessity of scientific research on the ecotoxicity of organic carbon nanomaterials.

The highest percentages of scientific articles on the ecotoxicity of different types of nanomaterials during 2010–2019 from the Web of Science Core Collection (Web of Science, Clarivate Analytics) can be observed in the case of titanium nanomaterials, carbon nanotubes and zinc nanomaterials. The lowest percentages can be observed in the case of nanomaterials of polyhydroxyalkanoates, polylactic acid, alginate, aluminum, platinum, chitosan and cobalt (Figure 18). This highlights the necessity of scientific research on the ecotoxicity of the least researched nanomaterials.

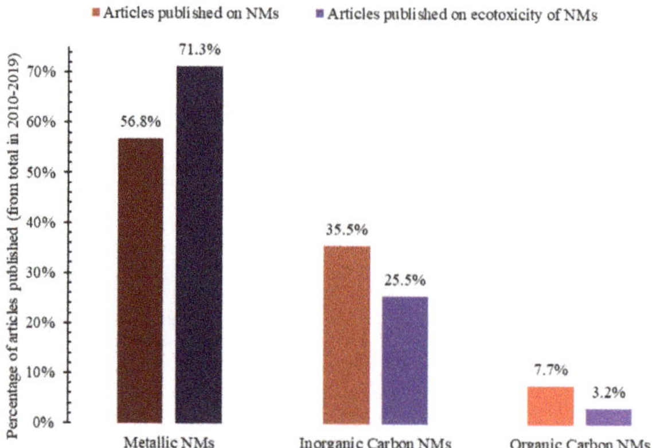

Figure 17. Bibliometric analysis of research on nanomaterials and their ecotoxicity by nanomaterial category in 2010–2019 from Web of Science Core Collection (Web of Science, Clarivate Analytics), the search strategy used to retrieve the data being: TS=(Nano* AND polyhydroxyalk*) AND TS=(Ecotox*), TS=(Nano* AND polylact*) AND TS=(Ecotox*), TS=(Nano* AND algin*) AND TS=(Ecotox*), TS=(Nano* AND aluminum*) AND TS=(Ecotox*), TS=(Nano* AND platinum*) AND TS=(Ecotox*), TS=(Nano* AND Chitosan*) AND TS=(Ecotox*), TS=(Nano* AND cobalt*) AND TS=(Ecotox*), TS=(Nano* AND cellulos*) AND TS=(Ecotox*), TS=(Nano* AND cerium*) AND TS=(Ecotox*), TS=(Nano* AND cadmium*) AND TS=(Ecotox*), TS=(Nano* AND graphen*) AND TS=(Ecotox*), TS=(Nano* AND fuller*) AND TS=(Ecotox*), TS=(Nano* AND silver*) AND TS=(Ecotox*), TS=(Nano* AND iron*) AND TS=(Ecotox*), TS=(Nano* AND gold*) AND TS=(Ecotox*), TS=(Nano* AND copper*) AND TS=(Ecotox*), TS=(Nano* AND zinc*) AND TS=(Ecotox*), TS=(Nanotub* AND carbon*) AND TS=(Ecotox*) and TS=(Nano* AND titanium*) AND TS=(Ecotox*).

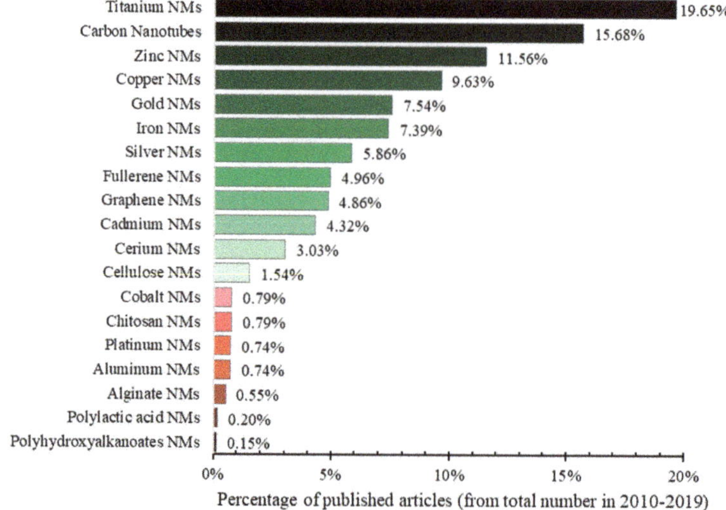

Figure 18. Bibliometric analysis of research on ecotoxicity of various types of nanomaterials in 2010–2019 from Web of Science Core Collection (Web of Science, Clarivate Analytics), the search strategy used to retrieve the data being: sum of the codes for NMs from each category (metallic NMs: aluminum, cadmium, cerium, cobalt, copper, gold, iron, platinum, silver, titanium and zinc; inorganic carbon NMs: carbon nanotubes, fullerene and graphene; organic carbon NMs: alginate, cellulose, chitosan, polyhydroxyalkanoates and polylactic acid).

The conducting of risk assessment or the establishment of environmental quality standards and guidelines is difficult due to the poorly understood toxic effects of NMs, especially on wildlife [3].

The use of NMs depends on testing their safety prior to their application due to their harmful effects, even if nanotechnology also offers potential advantages and promises. The major hazards of toxicity of NMs lie in their chemistry, composition, surface properties, size, shape, solubility, nondegradable properties, routes of exposure, interactions with biological molecules, bioavailability, tissue distribution and property to accumulate in routes of entry, tissues and cells [76].

- Surface of NMs

    - control the distribution of materials in tissue;
    - NMs undergo adsorption on macromolecules of the tested organism;
    - ionic crystal NMs are observed to be accumulated in cytoplasm or body fluid through circulation.

- Size of NMs

    - controls the distribution and penetration of tissue by NMs;
    - reduction in size to nanoscale level leads to an increase of surface-to-volume ratio, thereby increasing the number of chemical molecules on surface, leading to an increase in intrinsic toxicity;
    - the NMs size can control the dose–response relationship in relation to its solubility and toxicity.

- Shape of NMs

    - plays an important role in determining the toxic nature of NMs, as high aspect ratio NMs (with only one or two dimensions in nanoscale) like nanofibers (that are longer than 10–20 μm and thinner than 3 μm) may remain in the pleural cavity, causing lung inflammation and even cancer (for example asbestos and carbon nanotubes).

- Aggregation of NMs

    - NMs tend to form aggregates;
    - the size of aggregates/agglomerates influences the residence time and reduces the potential for a NM to be inhaled;
    - the aggregation/agglomeration is controlled by external environment like air and dispersion media;
    - NMs may undergo disaggregation and disagglomeration within respiratory system, thereby penetrating lung cells [76].

The ecotoxicity of NMs is usually expressed by different concentration values such as EC_{50} (half maximal effective concentration), IC_{50} (half maximal inhibitory concentration), LOEC (lowest observed effect concentration), NOEC (no observed effect concentration), etc. To emphasize the ecotoxicity of these nanomaterials, biological and biochemical effects are presented on both SSRET organisms and vertebrates and other unsuitable (non SSRET) organisms, if data is available. Biological effects describe both anatomical and morphological aspects, as well as aspects related to body functions, such as locomotion and digestion. Biochemical aspects describe the mechanisms of action of these materials, such as interaction with metabolic pathways or certain molecules in the body, such as enzymes.

4.1. Ecotoxicity of Bionanomolecules

Both natural (pristine bionanomolecules) and engineered (bionanomaterials) bionanomolecules have a wide range of applications, which increases their likelihood of reaching the environment. Thus, as potential pollutants, it is necessary to test their ecotoxicity as well as to elucidate the mechanisms by which these nanomaterials affect the organisms.

Due to their applications in medicine and pharmacology, polymeric bionanomolecules, such as chitosan, alginate, poly lactic acid, etc., have been in focus in recent years. These polymeric BNMs can be synthetized by different methods, such as solvent evaporation, nanoprecipitation, interfacial polymerization and controlled gelation [77].

Although these bionanomolecules are mostly biocompatible and biodegradable, they can still have some ecotoxicological effects on some organisms. This is because biocompatibility tests are performed at lower concentrations, which certainly do not cause toxic effects. But the purpose of ecotoxicity tests is to determine the concentrations at which toxic effects occur, thus testing concentrations much higher than those that would be used in vivo or that would occur in the environment in normal conditions.

4.1.1. Nanochitosan

Chitosan, obtained by the deacetylation of chitin, is a polysaccharide that has two monomer units, N-acetyl-D-glucosamine and D-glucosamine (Figure 19). The many special properties of chitosan are due to its cationic nature and functional groups like amine and hydroxyl [78].

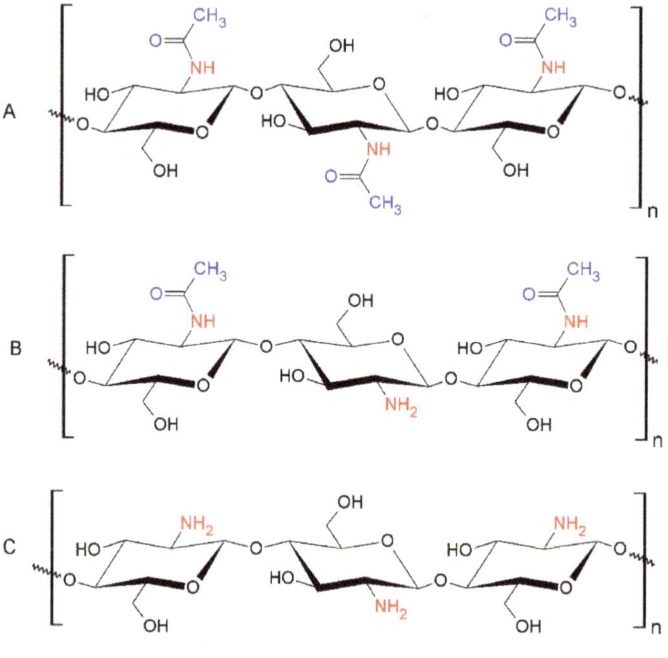

Figure 19. Structure of chitosan pristine bionanomolecules.

In nature, chitin, the source of chitosan, exists as long and straight microfibrils, that have an indeterminate length and a diameter of 2.8 nm [79]. These molecules (chitin and chitosan) may be considered as pristine bionanomolecules, as they have a dimension in nanoscale.

Chitosan bionanomaterials can be fabricated using different processes such as ionic gelation, covalent crosslinking or self-assembly. The most common forms of chitosan BNMs are nanogels, micelles, nanofibers, nanospheres and nanoparticles [80].

Chitosan BNMs and PBNMs are used as drug carriers, via various routes of administration such as oral, nasal, ocular or intravenous. The drugs these nano-chitosans could deliver could be variate, such as genes, proteins or antibiotics [81]. Another potential application is the encapsulation of vitamins, probiotics, phytochemicals, flavors, enzymes etc. [78]. In medicine, the most common applications are in tissue engineering, wound healing, cancer diagnosis, etc. [82].

The number of published articles aimed at determining the ecotoxicological effects of chitosan bionanomolecules is very low. Most published articles are in the field of medicine, chitosan BNMs being used for bone tissue regeneration and in wound healing.

The scarce data on ecotoxic effects show that chitosan BNMs have an antifungal effect on species like *Fusarium oxysporum*. The antifungal effects [83] include induction of morphological changes such as abnormal shapes of hyphae and large vesicles in mycelium, and induction of irreversible membrane damage, as well as mycelium surface damage, appearing as ruptures or holes. Cell disintegration was also observed due to the increase of membrane permeability (Figure 20). Chitosan BNMs also increased the reagent oxygen species (ROS) production in *F. oxysporum*, that can increase the oxidative stress, even cause the release of cytochrome C, leading to apoptosis (Figure 21) [84].

Chitosan–silver composite NPs were tested on *Danio rerio* and *Paratelphusa hydrodromous*. The predation of *D. rerio* larvae was not inhibited, it actually increased with approximately 19%. The superoxide dismutase (SOD) activity from contaminated hepatopancreas tissue of fresh water crab, *P. hydrodromous*, was stimulated. A possible explanation might be the enhancement of pre-existing enzyme or the synthesis of new enzymes [85].

There is a knowledge gap in the ecotoxicity of chitosan based BNMs, as available research is focused on applications, especially in medicine. There is data on the ecotoxicity of NMs capped with chitosan, but the data is also scarce. The effects of chitosan capped poly(ε-caprolactone) (CS-PCL) NPs were tested on *Daphnia similis*. No acute toxicity was observed for the CS-PCL NPs, but these could enhance the retention time of the NPs in the gut of the organism, prolonging the exposure to substances that would be loaded into these NPs, which could potentially be toxic [86].

Chitosan capped Ag NPs, at concentrations above 5 mg/L, induced cell division in the root meristem of *Allium cepa*. An increase in the total number of cells in prophase also occurred, which could be associated with a violation of the supramolecular structure of the chromosomes. Other observed effects were the presence of lagging chromosomes and the induction of polyploidy [87].

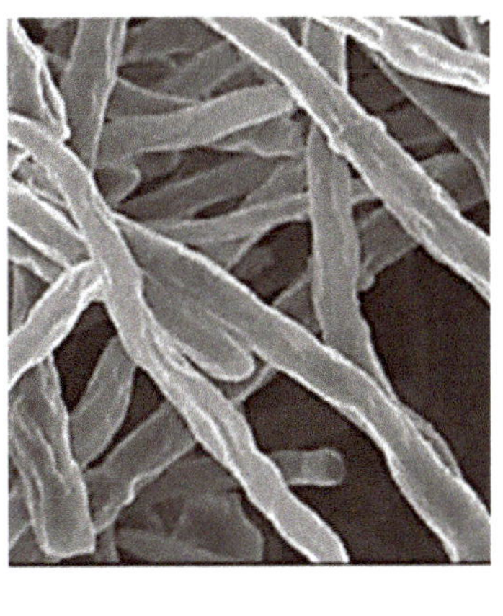

(a)

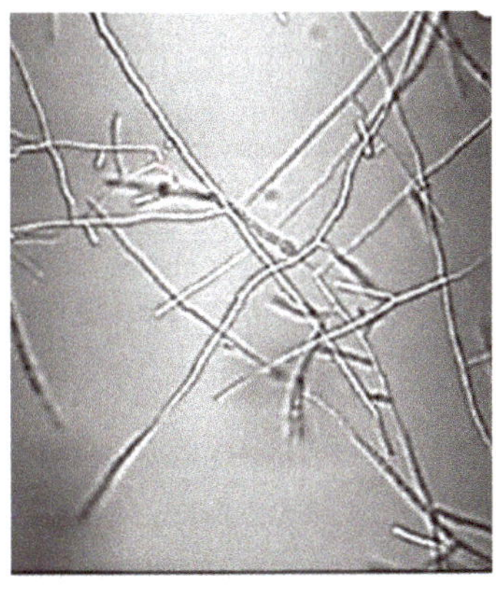

(b)

Figure 20. *Cont.*

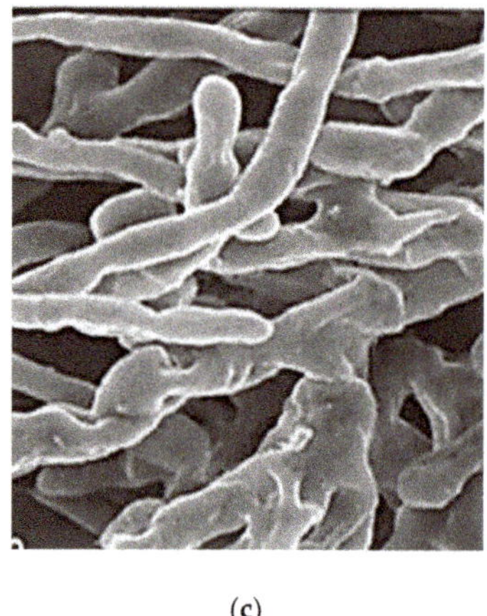 

(c)   (d)

Figure 20. Comparison of ultrastructural and cell permeability changes in *F. oxysporum* mycelium upon chitosan bionanomaterial (BNM) exposure: (**a**) and (**c**) show ultrastructural changes determined by SEM analysis of *F. oxysporum* mycelium of control (**a**) and chitosan BNMs (400 μg/mL) (**c**); (**b**) and (**d**) show cell permeability analysis data by propidium iodide assay; (**b**) influx of control; (**d**) influx of chitosan BNMs treatment (modified after [84]).

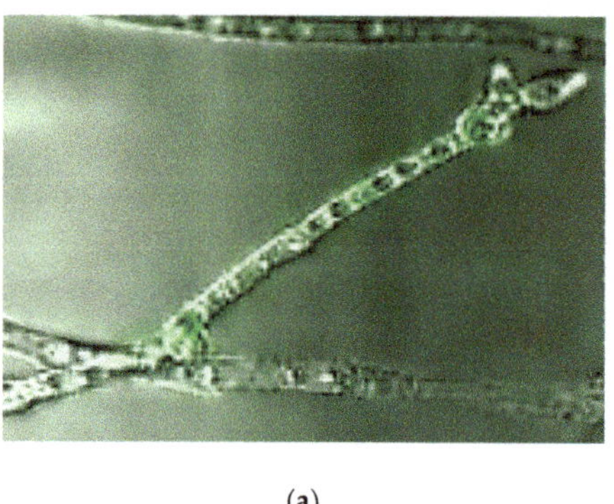 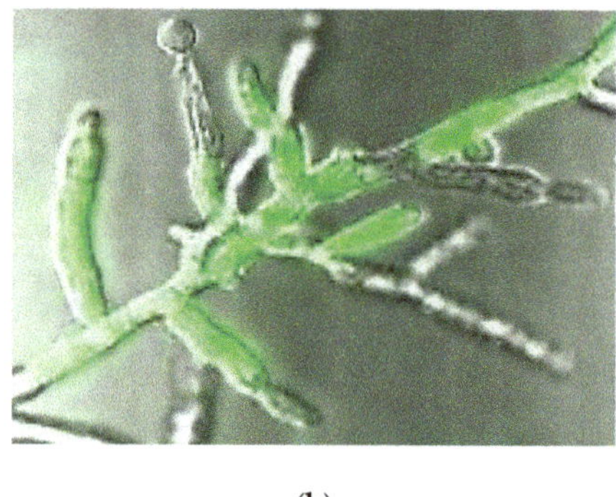

(a)   (b)

Figure 21. Confocal laser scanning microscopy (CLSM) image of *F. oxysporum* mycelium showing ROS production level upon exposure of chitosan BNMs. (**a**) control; (**b**) chitosan BNMs (400 μg/mL) (modified after [84]).

The effects of PBNMs of chitosan on plants include chromatin alterations and DNA damage, accumulation of hydrogen peroxide, synthesis of abscisic acid, increase in cytosolic Ca^{2+}, oxidative burst, etc. [88]. In worms (*Tubifex tubifex*), a weight loss was recorded after treatment. The levels of metallothionein were significantly higher than in control. Glutathione, glutathione-S-transferase, glutathione-reductase and catalase activities were also increased for the chitosan PBNM treatment [89]

4.1.2. Nanoalginic acid

Alginic acid is an anionic polysaccharide composed of two repeating units, forming a linear polymer: *D*-mannuronate and *L*-guluronate (Figure 22). It is a linear binary copolymer with homopolymeric blocks of either *D*-mannuronate, either *L*-guluronate, interspersed with alternating monomers. Due to its low toxicity, biocompatibility, biodegradability and mild gelation at addition of cation (ex. Ca^{2+}), it is extensively investigated and used for many biomedical applications [77,90].

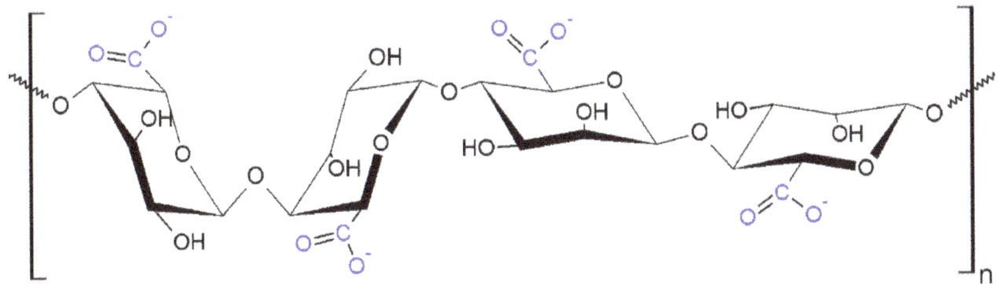

Figure 22. Structure of alginic acid pristine bionanomolecules.

There is a lack of information regarding the ecotoxicity of both BNMs and PBNMs of alginate in the literature. The research topics mainly covered are the use of alginate in medicine [91–94], and its use in wastewater remediation [95–98].

There are articles that study alginate as PBNMs used in the reduction of ecotoxicity of other NMs or as dispersant materials. For example, alginate was studied as a dispersant in ecotoxicity of TiO_2 NMs on embryos of zebrafish (*Danio rerio*). Alginate, with a z-average diameter of 178 nm, reduced the amount of TiO_2 stuck to the glass vials, thus reducing the loss of the tested NMs, not by neutralizing the surface charge of nano-TiO_2 (both TiO_2 and alginate have negative charges), but through steric hindrance [99]. Alginate PBNMs were also tested for the reduction of ecotoxicity of TiO_2 NMs on *Artemia franciscana* and *Phaeodactylum tricornutum*. The growth of *P. tricornutum* was not significantly affected by the presence of alginate in either the negative or positive control (the reference toxicant was ($K_2Cr_2O_7$)) tests. The immobilization after 24 or 48 h of *A. franciscana* was not significantly affected by the presence of alginate in the negative control, but it made a significant difference in the positive control, where, at 24 h, alginate reduced the ecotoxicological effects of the reference toxicant ($CuSO_4 * 5H_2O$). Thus, alginate reduced the toxicity of $CuSO_4 * 5H_2O$, but not of $K_2Cr_2O_7$, acting as a confounding factor. Alginate also reduced the bioavailability of TiO_2 NMs in the *P. tricornutum* assay, a possible explanation being a capping or coating effect of the alginate. In the *A. franciscana* assay, no significant differences were observed between the TiO_2 NMs with and without alginate [100].

One of the few data available on the ecotoxicity of alginate BNMs assesses the effects of chitosan–alginate nanoparticles on bullfrog (*Lithobates catesbeianus*) tadpoles, but not as its principal objective. The BNMs ecotoxicity is compared to clomazone pestanal®in its free form and associated with the chitosan–alginate nanoparticles. In all exposed groups (clomazone, BNMs and clomazone-BNMs) there was a significant increase of melanomacrophage centers, triggering a hepatic response in the tadpoles. Both groups exposed to BNMs presented hepatic sinusoids full of erythrocytes and abundant melanomacrophage centers which reflect that the BNMs might be recognized as toxin by the tadpole organism [101].

The ecotoxicity of alginate PBNM hydrogels was tested on two microalgae (*Halamphora coffeaeformis* and *Cylindrotheca closterium*) in correlation with the jellifying agent used: calcium chloride, copper sulfate or zinc acetate. All three metal ions caused an approximately 85% inhibition of algae adhesion, while the growth reduction was the highest for Cu^{2+}, followed by Zn^{2+} [102].

The ecotoxicity of alginate PBNMS is indirectly assessed, not being the scope of the articles. For example, in the case of rainbow trout, the nitrogen digestibility and dry matter content of the feces were reduced, while the protein content, visceral and liver weights and the mortality were not influenced by alginate as PBNMs [103]. In cats, the intraperitoneal injection of alginate caused proteinuria and hematuria, the kidney tubules being occluded by erythrocytes. Both intravenous and intraperitoneal administration caused renal tubular damage, necrosis and fragmentation of the liver cells [104]. PBNMs of sodium alginate did not cause ecotoxic effects on *Ceriodaphnia cornuta*, as no mortality was observed in an indirect assessment [105].

4.1.3. Nanocellulose

Cellulose is the most abundant natural polymer found in nature and it is mainly extracted from wood pulp [106,107]. It is a linear carbohydrate polymer, its monomeric units being β-D-glucopyranose molecules with β (1→4) covalent linkage (Figure 23). The length of the biopolymer chains present variations with the origin and treatment of the raw material. For example, in case of wood pulp, there are 300–1700 anhydroglucose units (AGUs), while plant fibers such as cotton and bacterial cellulose have 800–1000 AGUs. The numerous applications of cellulose and its derivatives, such as coatings, films, membranes, pharmaceuticals, etc., are due to its distinct fiber morphology. Elementary fibrils (1.5–3.5 nm lateral dimensions (LDs)), microfibrils (10–30 nm LDs) and microfibrillar bands (100 nm LDs) define the morphological hierarchy of cellulose [107].

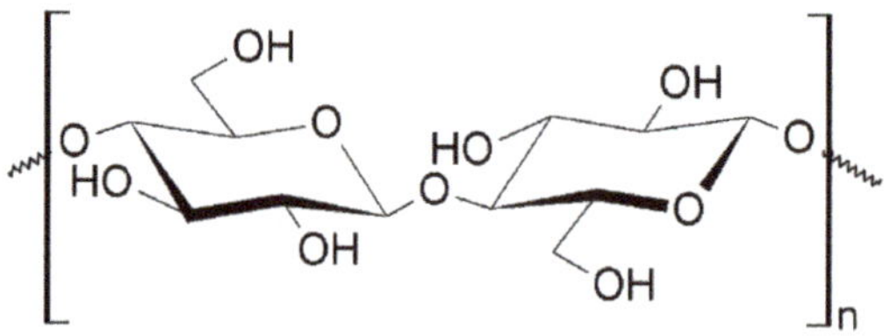

Figure 23. Structure of cellulose pristine bionanomolecules.

Recently, considerable interest has been focused on finding new material applications for cellulose such as development of cellulose nanocrystals. Nanocrystalline cellulose can readily be obtained by subjecting native cellulose to strong acid hydrolysis, but the most common method of preparing cellulose nanomaterials is aqueous and solvent solution casting [106].

Due to their excellent mechanical properties, good biocompatibility and its renewable nature, cellulose BNMs are used in the fields of biomedical engineering and material science [108]. For example cellulose BNMs, such as cellulose nanocrystals, nanowhiskers [109] and nanofibers, are of interest for use in reinforcing fillers instead of cellulose fibers [106].

Since cellulose BNMs and PBNMs are biocompatible and biodegradable, their ecotoxicological effects are questionable. For example, cellulose BNMs caused no decrease of nematode (*C. elegans*) number [110]; caused mechanical inhibition of mobility in *D. magna* neonates [111]; had LC_{50} values higher than 1 g/L for *D. magna* and *O. mykiss*, and 0.3 g/L for *C. dubia* [112]; induced relatively low mortality or any other developmental impairment towards embryonic zebrafish [113]; etc. Cellulose in PBNM form showed no toxic effects toward *Pseudomonas putida*, *S. capricornutum*, *D. magna* and *D. rerio* [114]; showed toxicity to *Toxoptera graminum* [115]; etc.

4.1.4. Nano polyhydroxyalkanoates (PHA)

Polyhydroxyalkanoates (PHA) are linear polyesters of hydroxyalkanoates (Figure 24), with a molecular weight in the range of 50–1000 kDa. All monomers units are in the $D(-)$ configuration due to the biosynthetic enzymes that are stereospecific. These polymers are synthesized by gram negative and positive bacteria from at least 75 different genera, and are accumulated intracellularly, under conditions of nutrient stress, to levels as high as 90% of the cells' dry weight, to act as carbon and energy reserve [116].

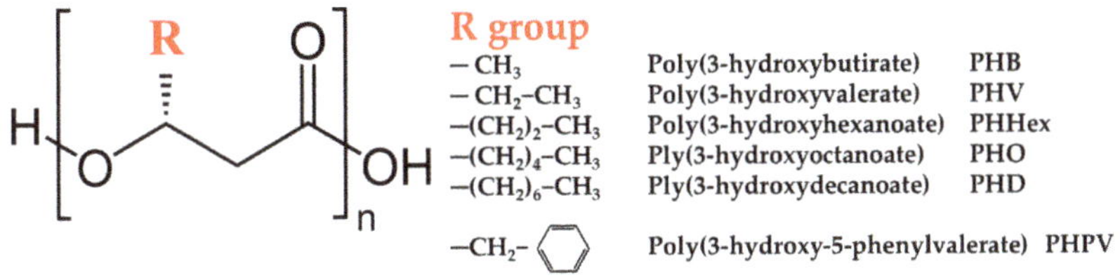

Figure 24. Structure of polyhydroxyalkanoate (PHA) pristine bionanomolecules.

PHA are biocompatible, nontoxic and biodegradable, which promotes their applications as plastic replacement, in packaging, as chiral precursors for chemical synthesis of optically active compounds, as biodegradable carriers for medicine, drugs, herbicides and insecticides, as osteosynthetic materials for bone growth stimulations, etc. [116].

PHA nanomaterials are developed by bacterial direct synthesis, phase separation, as nanobiocomposites, etc., for different applications such as in medicine. For example, PHA nanofibrous matrices are developed for use as materials in cell growth supporting [117], or nanobeads of PHA produced for use as biomaterial in various applications in medicine [118]. PHA nanogranules have applications in fields such as drug delivery, bioseparation, enzyme immobilization, protein purification and vaccines [119].

There is a knowledge gap in the ecotoxicity of PHAs as both bionanomaterials and pristine bionanomolecules, highlighted by Hauser et al. in their environmental hazard assessment of polymeric and inorganic nanomaterials used in drug delivery [120]. The data available shows that PHA nanofibers are able to support the growth of rat neural stem cells [121]. Poly(3-hydroxybutirate) (P3HB) granules showed a decrease in concentration after dispersion in different organs in rats [122]. The degradation of PHAs in mammals happened gradually in six months or more, not causing weight loss in mice [123].

4.1.5. Nano polylactic acid (PLA)

Polylactic acid (PLA) is an aliphatic polyester with its monomer being exclusively lactic acid. It is a biocompatible, bioresorbable and biodegradable polymer with high strength and thermoplastic properties [124,125]. Applications of PLA are as packaging materials [125], biomedical materials for implants, sutures, screws and plates, and in textile applications [124]. PLA is also used in tissue engineering as fixation device materials [126] and in drug delivery [127].

PLA can be produced from renewable resources, its monomer being mainly synthesized by bacteria from the genus *Lactobacillus*, through the Embden–Meyerhof pathway, predominantly from glucose (Figure 25). Some species synthesize the *L*(+)-isomer of lactic acid, such as *L. amylophilus* and *L. salivarius*, while other species, like *L. acidophilus* and *L. jensenii*, yield the *D* isomer [125]. The synthesis of PLA implies several steps, starting with the production of its monomer and ending with the polymerization of lactic acid. The polymerization can follow three main routes: 1. condensation polymerization; 2. azeotropic dehydrative condensation and 3. ring-opening polymerization [124].

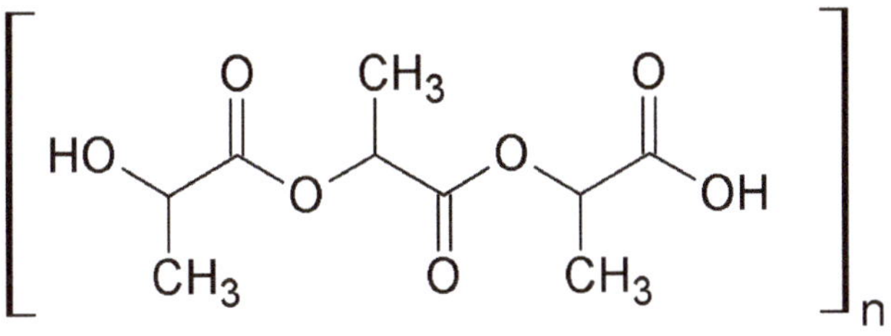

Figure 25. Structure of poly (*L*-lactic acid) (PLA) pristine bionanomolecules.

Nanoparticles of PLA are used as delivery systems in medicine due to their low toxicity and hydrolytic degradability [124].

Hauser et al. highlighted the lack of information available on the ecotoxicity of bionanomolecules of polylactic acid [120]. The little available information reveals that PLA caused a slight increase in mitochondrial activity in mouse fibroblasts with a significant decrease of DNA synthesis [128], but it didn't cause an acute or chronic inflammatory response in rats, with no polymer rejection by the organism [129].

4.2. Ecotoxicity of Carbon-Based Nanomaterials

Carbon-based NMs are mainly composed of carbon. The most common forms are hollow spheres and ellipsoids, referred to as fullerenes, or cylinders, known as carbon nanotubes (CNTs) [6].

4.2.1. Carbon Nanotubes

The most studied carbon-based NMs are carbon nanotubes. These are built from sp^2-hybridized carbon atoms assembled via very short *s*-bonds as aromatic rings. The rings are assembled according to a planar periodic lattice looking like a single-atom-thick hexagonal pavement [130].

CNTs have a wide range of properties, such as morphological (length, diameter, bundling) and structural (number of walls, metallic or semiconducting electrical behavior). CNTs, short or entangled, may be readily taken up by cells in a passive way, such as passive diffusion or by piercing the cell membrane, or in an active way through mechanisms like endocytosis or phagocytosis [130].

A production capacity that exceeds several thousand tons per year reflects a worldwide commercial interest in carbon nanotubes. The use of the CNTs extends over various application areas such as coatings and films (e.g., paints that reduce biofouling of ship hulls, thin-film heaters for the defrosting of windows), composite materials (e.g., electrically conductive fillers or flame-retardant additives in plastics), microelectronics (e.g., CNT thin-film transistors), energy storage (e.g., lithium ion batteries for notebook computers and mobile phones), environment (e.g., water purification) and biotechnology (e.g., biosensors, medical devices) [131].

The investigation of the potential impact of CNTs on the different environmental compartments (soil, water) is very important because CNT-containing materials may not be properly disposed in the absence of specific regulation. Initial studies on a single environment compartment revealed that biological species, such as crustaceans, worms, amphibian larvae, all interact with CNTs, which transit, without visible harm, through the gastrointestinal pathway. In most cases, at low concentration, less than 10 mg/L in aquatic studies, no significant effect is observed. The actual CNT concentrations in the environment are expected to be few orders of magnitude below this value of 10 mg/L, since only available data are based on calculations, upon hypotheses on the transfer between air, soil, and water. However, some toxicity is generally noticed at higher concentrations, which seems in most cases possible to correlate with 'mechanical' effects, such as perturbation of the digestion or the respiration related more to the presence of large amount of foreign material in the body [130].

The dispersion of CNTs can be directly modified by organisms. Protozoan cells, such as *Tetrahymena thermophila* and *Stylonychia mutilus*, that ingest multiwalled or single-walled carbon nanotubes (MWCNTs or SWCNTs), without any discrimination of bacterial food, excrete these as sedimented granules in micron size. Impaired ingestion of bacteria by phagocytosis (bacterivory) and impaired regulation of bacterial growth can occur both for parental cells and to the two daughter cells during cell division. Thus, the ingested CNT may affect protozoan food intake, and could be transferred between generations and move up the food chain [132].

In algae, the toxic effects of CNTs are due to direct contact with the surface. Thus, CNT shading and formation of algae–CNT agglomerates can inhibit algal growth, as studies on *Pseudokirchneriella subcapitata* and *Chlorella vulgaris* suggest [132].

The bioaccumulation of environmental contaminants, such as hydrophobic organic contaminants (HOC), can be influenced by the presence of CNTs. The bioaccumulation of HOC was significantly reduced in the presence of SWCNTs in *Streblospio benedicti*, a deposit/suspension feeding polychaete, while in the deposit-feeding meiobentic copepod *Amphiascus tenuiremis* the HOC bioaccumulation was less affected [132].

The food processing of *Daphnia magna* was affected by the CNTs aggregated in the gut in the presence of food that contributed to the toxicity of the CNTs, which were, however, not able to cross the gut lumen. The presence of food reduced the elimination time of MWCNTs from a day to a few hours. The lipid coating of SWCNTs (that increase its water-solubility) was removed by the digestive system of *D. magna* making the CNTs less water soluble and more prone to sedimentation [132]. At high concentrations, MWCNTs adhered to the external surface of daphnids, being absorbed, and together with the ingestion of these materials, it was significant enough to cause sinking of daphnids to the bottom of the test vessels by the prevention of mobility through the water column. At low concentrations of MWCNTs, only ingestion was observed as shown in (Figure 26) [75].

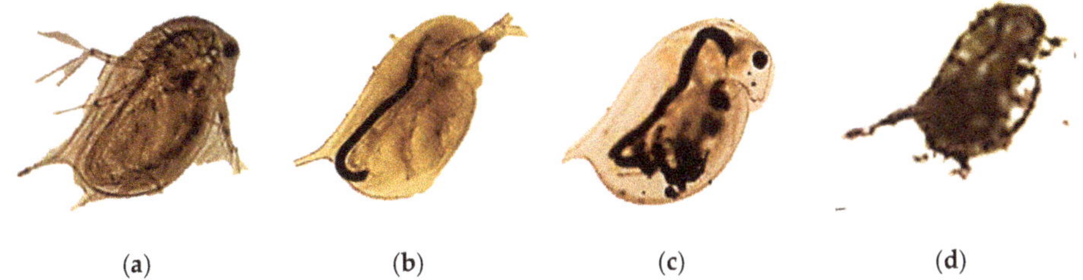

(a)   (b)   (c)   (d)

Figure 26. The uptake and adsorption of multi-wall carbon nanotubes (MWCNTs) by *Daphnia magna* after 48 h of exposure. (**a**) control; (**b**) 5 mg/L MWCNTs; (**c**) 50 mg/L MWCNTs and (**d**) 100 mg/L MWCNTs. (Modified after [75]).

The bioaccumulation of HOC or perfluorochemicals (PFC) in benthic larvae of *Chironomus plumosus* was reduced by the addition of MWCNTs [132].

In *Arabidopsis sp.* plants translation is affected by MWCNTs. The consequences are increased colonization of bacteria in infections, stress indication and root hair development inhibition [76].

The bioaccumulation of pyrene in terrestrial oligochaete *Eisenia foetida* was reduced by the presence of SWCNTs and MWCNTs in high concentrations, due to the increase of polycyclic aromatic hydrocarbons (PAH) elimination and of uptake by CNTs [132].

4.2.2. Fullerenes

Fullerenes are a key topic nowadays in nanotechnology and industrial research due to their excellent unique properties, such as high symmetry. Fullerenes have a hexagonal ground state with sp^2 bonding. The most symmetric molecule, with the largest number of symmetry operations is Buckminster C60 fullerene [6].

The use of the fullerenes extends over various application areas such as electronics (e.g., electrodes, solar cells) [133] or biology and medicine (e.g., antioxidants, antiviral agents, drug and gene delivery) [134]. The probability of environmental pollution with fullerenes increases due to the increased production and commercial applications. Thus, there is a considerable interest in the effects and behavior of carbon fullerenes in the environment [135].

The growth of the algae *P. subcapitata* was not inhibited by C60 fullerene, the EC_{50} being more than 90 mg/L, while fullerol even had a beneficial effect at 1mg/L, thus being a ROS scavenger [22].

C60 fullerene causes clubbing and tentacle retraction in chronic exposure in *Hydra attenuata* [11].

In mussels, such as *Mytilus edulis*, the accumulation of C60 fullerene took place in the digestive glands, followed by gills. At 0.1 and 1mg/L concentrations, several histopathological abnormalities were observed, such as necrosis in digestive tubules, hypoplasia in frontal and lateral cilia and atrophy in adductor muscle myocytes [22].

The heart rate of *D. magna* was increased by exposure to C60, along with the amplification of stereotypical movements in swimming and feeding and reduction of reproductive rates [11]. Acute exposure to C60 caused modulation of vertical position of daphnids over time and reduction of their swimming velocity, while chronic exposure caused cellular damage to the alimentary canal, delayed molting and inhibited reproduction [22].

In the nematode *Caenorhabditis elegans* hydroxylated fullerenes are able to induce apoptosis [136].

In higher organisms, such as fish and mammals, fullerenes can cross the membrane of eukaryote cells, as well as the blood–brain barrier, and can accumulate lysosomes and mitochondria [11].

4.2.3. Graphene

Graphene is formed from strongly sp^2-bonded carbon atoms that are arranged in a planar monolayer that forms a two-dimensional honeycomb lattice [137].

Graphene is used in various area such as field effects transistors, sensors, transparent conductive films, clean energy devices, etc. [138].

In *Euglena gracilis* graphene oxide (GO) induced growth inhibition, decrease of chlorophyll a content, but not of chlorophyll *b* and carotenoids. GO induced oxidative stress as the activities of catalase (CAT) and superoxide dismutase (SOD) were increased in comparison with control. There were no evident damages in the ultrastructure of the protozoa, however the cells were clearly covered with a layer of GO [139].

Fifty percent growth inhibition was induced by GO in Raphidocelis subcapitata, along with oxidative stress and membrane damage. GO also decreased the autofluorescence intensity, due to oxidative stress and/or a shading effect due to the agglomeration of GO [140].

No toxicity was observed by the exposure of *Artemia salina* to graphene at maximum concentrations, however the microscopical analysis showed the presence of pristine graphene monolayer flakes (PGMF) and graphene nanopowder grade C1 (GNC1) in the gut of the crustaceans (Figure 27). An altered pattern of oxidative stress biomarkers was observed at a 48-hour exposure. PGMF and GNC1 induced an increase in catalase and glutathione peroxidase activities, as well as an increase in the level of lipid peroxidation of membranes [141].

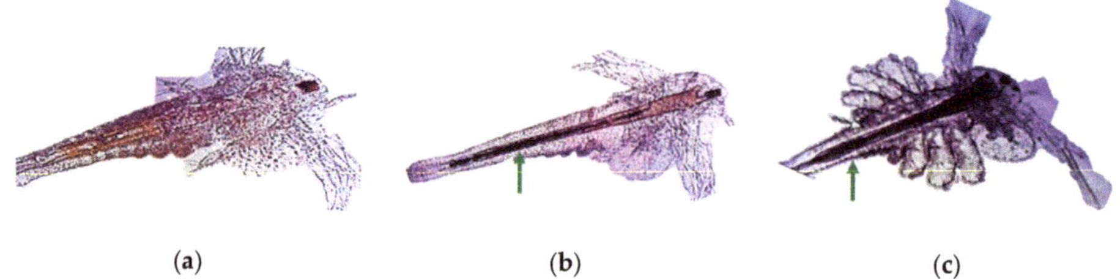

(a)   (b)   (c)

Figure 27. Light microscope images of (a) control, (b) pristine graphene monolayer flake (PGMF) and (c) graphene nanopowder grade C1 (GNC1-)treated *Artemia salina* (24 h exposure, 1.25 mg/L. Arrows indicate the presence of PGMF or GNC1 in the gut. (Modified after [141]).

After a 96-hour co-exposure to Cu and GO, the roots of duckweeds were covered with GO fragments (Figure 28). GO significantly decreased the nutrient contents of *L. minor* only at concentrations of at least 5 mg/L [86].

(a)

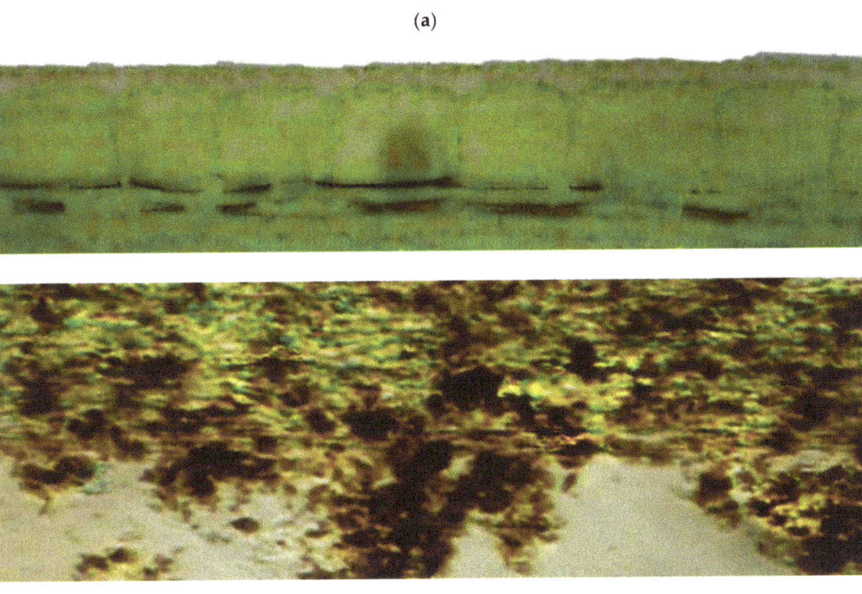

(b)

Figure 28. Roots of *Lemna minor* in (**a**) control and (**b**) treatment with 5 mg/L graphene oxide (GO). (Modified after [86]).

4.3. Ecotoxicity of Metallic Nanomaterials

The exposure of plants to metallic nanomaterials causes nanotoxicity at physiological level such as root length inhibition, biomass decrease, altered transpiration rate, and plant developmental delays. The NMs that enter plant tissue cause the disruption of chlorophyll synthesis in leaves. Evidence of genotoxicity of metallic NMs to plants is provided by the analysis of the mitotic index, chromosomal aberrations, and micronuclei induction [62].

4.3.1. Aluminum Nanomaterials

Aluminum based NMs have several industrial applications, such as absorbents, abrasives, desiccants etc., due to their excellent dielectric and abrasive properties. These highly adsorptive materials are, in contrast, chemically active and potentially hazardous to the environment [142].

It was observed that Al_2O_3 NMs decreased the viability of algal populations at short-term exposure, while at a long term exposure a gradual recovery was observed [15]. The exposure of aluminum NMs to *Scenedesmus obliquus* caused oxidative stress by altering the SOD activity and the concentrations of glutathione and malondialdehyde [76].

The toxicity of Al_2O_3 NMs to *D. magna* is dose-dependent and it is higher than its bulk material's. The potential ecotoxicity and environmental health effect of these NMs cannot be neglected as these are ingested by the daphnids [75].

The toxicity of aluminum NMs to plants is species dependent as these had no obvious effect on cucumber, significantly retarded root elongation of ryegrass and lettuce and promoted the root growth of radish and rape [143].

Ni/γ-Al_2O_3 (NiNC) nanoceramics were analyzed by the AMPHITOX assay on the larvae of the *Rhinella arenarum* toad. The sublethal effects were mainly hyperkinesia and reduced swimming movements, an expression of behavioral alteration, as well as collapsed cavities, edema and axial flexure. The contents of Al and Ni were higher in heads than in tails, and these elements were found in the oral disc [142].

4.3.2. Cerium Nanomaterials

Cerium oxide nanomaterials (CeO_2 NMs) have a wide range of applications in engineering and biomedical manufacturing industries and have the ability to act as a redox catalyst, thus it may be able to both induce or alleviate oxidative stress in organisms [144].

The potential toxicity of ceria NMs was assessed on the unicellular alga, *Chlamydomonas reinhardtii*, by unbiased transcriptomics and metabolomics approaches to provide insight into molecular toxicity pathways. Although the ceria NMs were internalized in *C. reinhardtii* into intracellular vesicles, no significant toxicity was observed on the algal growth at any concentrations. The only effects, such as downregulation of photosynthesis and carbon fixation with associated effects on energy metabolism, were observed at ceria NM-concentrations higher than environmental levels [145].

Daphnia similis and *Daphnia pulex* were used for the assessment of toxicity of CeO_2 NMs. *D. similis* was 350 times more sensitive to the ceria NMs than *D. pulex* in acute ecotoxicity assessment. The NMs were absorbed on both species, but less strongly on *D. pulex*, and for both species the swimming velocities (SV) were differently and significantly affected. The different toxicity for the two species can be explained by the differences in morphology, such as the presence of reliefs on the cuticle and a longer distal spine in *D. similis* acting as traps for the CeO_2 aggregates (Figure 29). *D. similis* also has double swimming velocity than *D. pulex*, thus it can collide with twice as much NMs [146].

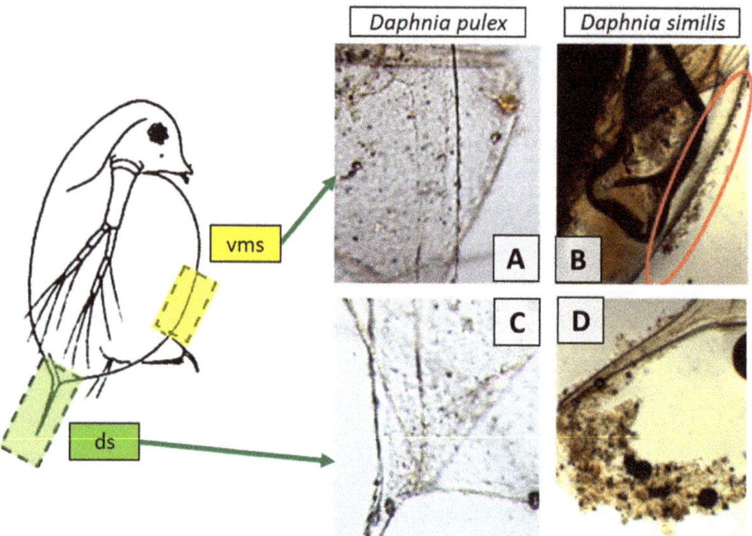

Figure 29. Representative image of distal spine (ds) and ventral margin of the shield (vms) in *Daphnia pulex* and *D. simillis* exposed to 10 mg/L of CeO2 NPs for 48 h. Note the accumulation of particles onto the cuticle of *D. simillis*. (Modified after [146]).

The toxic effect on the nematode *C. elegans* was dose-dependent growth inhibition. Mildly altered growth was observed for some metal and oxidative stress-sensitive mutant nematode strains in comparison with the wild-type. *C. elegans* ingested CeO_2 NMs but these materials were not detected inside the nematode cells. The aggregation of NMs around bacterial food and/or inside the gut tract may cause, at least in part, the growth inhibition, due to the inhibition of feeding caused by these aggregates (Figure 30) [144].

(**a**)   (**b**)

Figure 30. Representative dark field image of (**a**) control wild-type nematode (green arrow indicates the light scattered by tissue aggregates from nematode extraintestinal tissues) and of (**b**) a wild-type nematode exposed to 12.5 mg/L CeO2 NPs for 24 h (green arrow indicates ingested CeO2 NPs). (Modified after [144]).

4.3.3. Cadmium Nanomaterials

Cadmium based quantum dots (QDs), also called colloidal semiconductor nanocrystals, have several uses in biomedical applications, such as fluorescent biosensors (used in the detection of proteins, nucleic acids, etc.) and in bio-imaging (cellular or in vivo targeting and imaging) [147].

The effects of cadmium telluride quantum dots (CdTe QDs) were assessed on a microbial food chain, composed of Escherichia coli as prey and *Paramecium caudatum* as predator. It was observed that the QDs caused the loss of the bacterivory potential of paramecium, including an ∼12 h delay in doubling time. When paramecium was exposed to the QDs (25 mg/L at 24 h), these NMs were bioaccumulated, as shown by the fluorescence based stoichiometric analysis (Figure 31) [148].

L-Cysteine-capped CdS NMs expressed a high uptake in the roots of *Spirodela polyrhiza*. This was confirmed by epifluorescence microscopy where the presence of NMs was observed inside the root tissues (as particles with different sizes in intracellular spaces), the NMs aggregates appearing as optically dense signals under fluorescence (Figure 32). The entrance of NMs into roots is done through intercellular plasmodesmata, capillary forces, osmotic pressure, pores in cell walls, or via the highly regulated symplastic route [149].

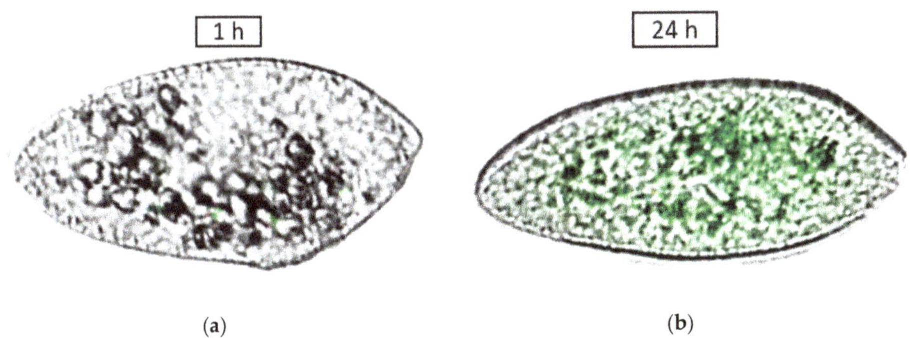

(**a**)   (**b**)

Figure 31. *Cont.*

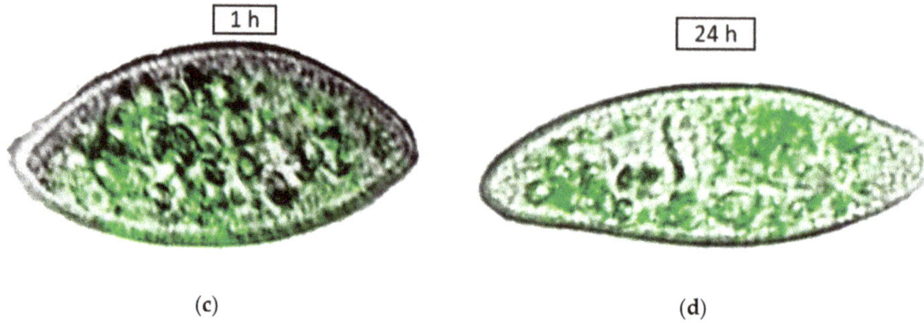

(c)   (d)

Figure 31. Fluorescence-based measurement of accumulation of cadmium telluride quantum dots (CT QDs) in *Paramecium* cells. Superimposed (fluorescence and bright field) images of (**a**) and (**b**) control *Paramecium* (exposed with *E. coli*) at (**a**) 1 and (**b**) 24 h; and (**c**) and (**d**) of treated *Paramecium* (*E. coli* + QDs) at (**c**) 1 h and (**d**) 24 h. (Modified after [148]).

(a)

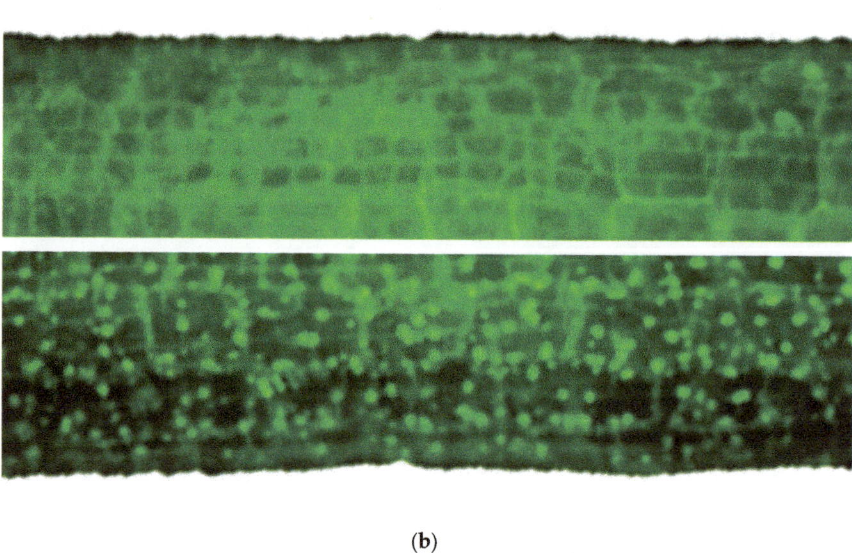

(b)

Figure 32. Fluorescence microscopic images of roots of (**a**) control and (**b**) treated *Spirodela polyrhiza* plant after four days of exposure to *L*-Cysteine-capped CdS NPs.(B) indicates the remarkable presence of nanoparticle aggregates inside the root tissues (modified after [149]).

Cadmium telluride quantum dots (CdTe QDs) caused abnormal foraging behavior (related to the altered function of the motor neurons) in *C. elegans*, at long-term early onset exposure. Thus, a decrease in fluorescence of the motor neurons cell bodies was observed, indicating an alteration in their development [136]. The main route of exposure of nematodes to QDs was determined, by fluorescence microscopy, to be through the digestive tracts [150].

4.3.4. Cobalt Nanomaterials

Cobalt NMs are of great interest in both life-sciences and industry, due to their wide range of applications, such as in lithium-ion batteries, gas sensors or medicine [151].

Decline in the growth rate and reduction in biomass concentration of two cyanobacteria, *Microcystis* and *Oscillatoria*, was observed at exposure to Co NMs. Other observed effects were the reduction of carotenoid, protein and carbohydrates contents and the decrease of SOD activity with increase of NMs concentration in the microalgae [152].

Allium cepa was used to investigate the effects of cobalt oxide NMs as an indicator organism. The observed phytotoxic effect at root level was the inhibition of root elongation due to the massive adsorption of NMs into the root system [153].

4.3.5. Copper Nanomaterials

Due to their potential applications in diverse fields, such as biomedicine, electronics, and optics, copper NMs have been the focus of intensive study [154].

When present in high concentrations, copper is supposed to be highly toxic in aquatic systems, causing irreversible damage [9]. Due to their antimicrobial and biocidal properties, copper oxide (CuO) NMs are frequently used. These NMs may represent an important source of contamination in the aquatic environment, due to their application in antifouling paints used on boats and immersed structures [155].

The dissolution and adsorption of CuO NMs onto cell walls, of the prokaryotic alga *Microcystis aeruginosa*, was observed to be enhanced by dissolved organic carbon (DOC). The cell walls were crossed by the NMs through the cell wall pores and the cell plasma membrane was crossed via endocytosis, thus the NMs reaching the thylakoids and granules [22]. In *L. minor*, the CuO NMs alter the activity of antioxidative enzymes such as guaiacol peroxidase, glutathione reductase and ascorbate peroxidase, increasing necrosis and bleaching [76].

A mesocosm that modeled tidal cycles was used for the assessment of CuO NMs toxicity on the worm *Hediste diversicolor* and the clam *Scrobicularia plana*. In both organisms, the observed effects were oxidative stress defense system responses (affected oxidative stress markers were: glutathione *S*-transferase (GST) and catalase (CAT)) and induction of genotoxicity (comet assay was used to asses DNA damage) [15].

The effects of Cu NMs were assessed on cowpea, *Vigna unguiculata*; specifically, how NMs affect the ascorbate peroxidase (APX), catalase (CAT), superoxide dismutase (SOD), glutathione reductase (GR) activities, lipid peroxidation, Cu uptake and bioaccumulation in roots, leaves and seeds [76].

4.3.6. Gold Nanomaterials

Due to their functions in medicine and therapeutics, in electronics, catalysis, cosmetic, and food industries, gold NMs are of special interest [156].

The exposure of marine bivalves to Au NMs was assessed. In *Scrobicularia plana*, Au NMs formed aggregates and gold was accumulated in the soft tissues of the clams. Biochemical effects of NMs were metallothionein induction, increase in catalase, superoxide dismutase and glutathione S-transferase activities (indicating defense against oxidative stress), while a behavioral effect was the impairment of the burrowing behavior [156]. In *Mytilus edulis*, Au NMs enhanced stress parameters in digestive glands, mantle and hematocytes, paradoxically protecting from the oxidative stress due to menadione [11].

The vegetative uptake of gold NMs was assessed using as a model poplar plants *Populus deltoides* × *nigra*. The Au NMs were observed in the cytoplasm and various organelles of root and leaf cells, and these accumulated in the plasmodesma of the phloem complex in root cells suggesting that the transport between cells and translocation throughout the whole plant were done with ease, inferring the potential for entry and transfer in food webs [157].

4.3.7. Iron Nanomaterials

Iron NMs have a wide range of applications such as magnetic, electrical, catalytic and biomedical (e.g., MRI contrast enhancer) [158]. These NMs are also suitable for immobilization and degradation of soil contaminants due to their high specific surface area and high reactivity. Thus, their use in soil clean-up purposes could cause potential hazards to soil organisms and macrophytes [64].

Severe negative effects of nanosized zero valent iron (nZVI) were observed on *Heterocypris incongruens*, an ostracod, and on *Folsomia candida*, a collembolan. The effects were observed after seven days and prolonged exposure led to the oxidation of nZVI, reducing its toxicity [75].

A test battery, composed from algae (*Pseudokirchneriella subcapitata, Chlamydomonas sp.*), crustaceans (Daphnia magna), plants (*Raphanus sativus, Lolium multiflorum*) and worms (*Eisenia fetida, Lumbriculus variegatus*), was used for the assessment of the effects of nZVI. The testing of the iron NMs was difficult due to their turbidity, aggregation and sedimentation behavior in aqueous media, but nZVI proved to be toxic. The observed effects for plants were the inhibition of root elongation in *Raphanus sativus* and *Lolium multiflorum* [97].

The effects of iron-based NMs were tested on three plant species: *Lepidium sativum, Sinapis alba* and *Sorghum saccharatum*. Microscopy images show that the NMs aggregated on the surface of the plants, visible as black spots, sometimes even forming a coating. The accumulation of iron NMs inside the tissue was observed in longitudinal sections of roots (Figure 33). However, the NMs did not enter the palisade cells or the xylem, as shown in transverse sections [64].

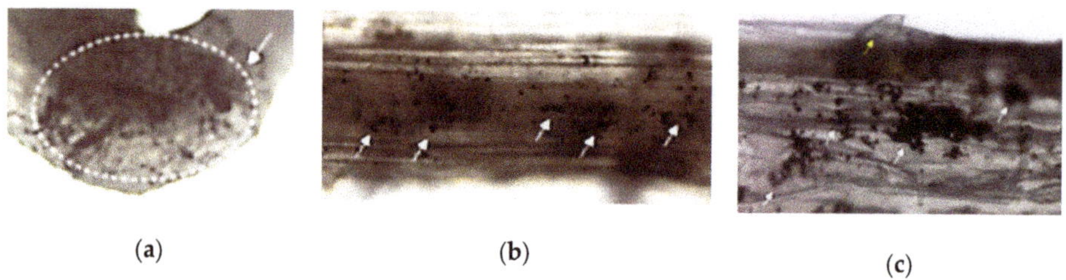

<table>
<tr><td>(a)</td><td>(b)</td><td>(c)</td></tr>
</table>

Figure 33. Bright-field micrographs illustrating (**a**) root apex of *Lepidium sativum* and (**b**) and (**c**) root longitudinal sections of (**b**) *Sinapis alba* and (**c**) *Sorghum saccharatum* treated with 992 mg/L of *n*Fe for 72 h at 25 °C. White arrows indicate *n*Fe aggregates, while yellow one shows root hairs. (modified after [64]).

4.3.8. Platinum Nanomaterials

Platinum NMs have a wide range of applications in fields such as CO oxidation, hydrogen or methanol fuel cells, electrochemical oxidation of ethanol or formic acid, oxygen reduction and glucose detection [159].

Antioxidant effect of Pt NMs coated with polyvinylpyrrolidone was observed in larvae of *C. elegans* (L4 development stage). The observed effects were counteraction of induction of oxidative stress by paraquat (an intracellular free radical-generating compound) and prolongation of life span of wild-type and short-living mutant mev-1 worms [11]. The life span of wild-type N2 nematodes was also extended, regardless of thermotolerance or dietary restriction. The NMs reduced the accumulation of lipofuscin (an endogenous autofluorescent marker that increases in concentration with oxidative stress) and ROS induced by paraquat. The effects of Pt NMs were compared to EUK-8, a superoxide dismutase (SOD)/catalase mimetic. The results showed similar results for the two tested substances, suggesting that Pt NMs are a superoxide dismutase (SOD)/catalase mimetic [160].

4.3.9. Silver Nanomaterials

The major applications of Ag NMs include their use as optical sensors, catalysts, in engineering, optics, electronics, and, most importantly in the biomedical field, as a bactericidal and therapeutic agent [161].

The effects of Ag NPs were tested on two microalgal species: *Dunaliella tertiolecta* and *Chlorella vulgaris*. The observed effects were depletion of chlorophyll content, inhibition of photosystem II (PSII) electron transport, membrane damage via lipid peroxidation possibly via ROS-mediated processes [22].

In adult *Mytilus edulis*, poly(allyl)amine (PAAm)-capped silver NPs caused the formation of a donut shaped microstructures on the nacreous layer of the bivalve, which can be explained by the disturbance of the shell calcification process. In the oyster *Crassostrea virginica*, Ag NPs caused the inhibition of embryo development and the destabilization of the lysosomal membrane of hepatopancreas cells of adults [22].

In *D. magna*, silver NPs accumulated in the gut, under the carapace, in the brood chamber and on the antennae and body surface, affecting their swimming behavior [11,22].

The effects of silver NMs on the nematode *C. elegans* were neurotoxicity, reduction of velocity, flex, amplitude, and wavelength of the body bend of exposed worms and reduction of survival and reproductivity [136].

4.3.10. Titanium Nanomaterials

Titanium dioxide NMs are frequently used in the production of paper, plastics, paints, cosmetics and welding rod coating material [162].

The effects of TiO_2 NMs on the algae *P. subcapitata* were light shading, interference on nutrient uptake through adsorption onto the cell surfaces and production of ROS which cause lipid peroxidation of cell membrane, leading to leaching of DNA from the algal cells. In *Chlamydomonas reinhardtii*, TiO_2 NMs caused up-regulation of genes associated with antioxidant activities, such as superoxide dismutase (SOD), glutathione peroxidase (GPX), catalase (CAT) and plastid terminal oxidase (ptox2), but these did not alter the transcriptions of genes associated with photosynthesis and carotenoid biosynthesis [22].

In *D. magna* TiO_2 NMs caused the reduction of brood size and body length, disruption of digestive enzymes, such as amylase and esterase, affecting nutrient assimilation and energy allocation. The increase of GST, GPX, and CAT activities demonstrate the induction of oxidative stress by titanium NMs [22].

In *C. elegans*, it was found that proteins involved in oxidative stress protection and metal elimination, such as SOD isoforms, metallothioneins and heat shock proteins, have a key role in resistance to TiO_2 NPs. These NPs led to a substantial decrease in both head thrash or body bend in nematode mutants (SOD-2, SOD-3, metallothionein-2 and heat shock protein-16.48) compared to the wild type [136].

The symbiosis of *Rhizobium leguminosarum* bv. *viciae* 3841 on garden peas *Pisum sativum* was affected by TiO_2 NMs by delaying root nodule formation and the onset of nitrogen fixation [76].

4.3.11. Zinc Nanomaterials

Zinc oxide NMs have been extensively used in products like coatings, paints and sunscreens due to their chemical stability and strong adsorption ability. These also have a high inherent risk of water contamination, being able to reach high concentrations in surface waters posing significant threat to aquatic ecosystems [163].

The ZnO NMs at lower concentration cause toxic effects, such as decrease of cell viability (Figure 34) due to compromised membrane integrity, on algae mainly due to the Zn ions. However, at higher concentrations, the growth of algae less sensitive to Zn ions, such as *Phaeodactylum tricornutum*, was inhibited, as the contact between algal cells and NM particles increased [22,163].

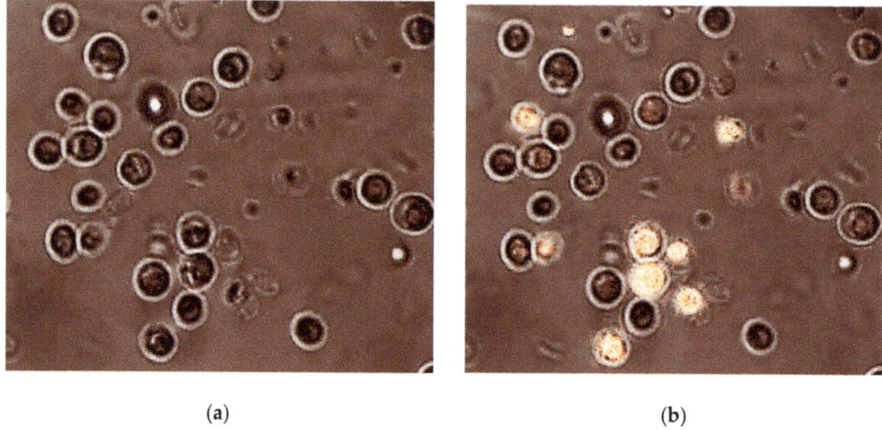

(a)   (b)

Figure 34. Confocal images of (**a**) untreated cell with intact membrane, preventing entry of PI (propidium iodide) dye, resulting in unstained cells and (**b**) cells treated with 300 mg/L ZnO NPs (zinc oxide nanoparticles) for 72 h showing PI stained cells due to compromised membrane integrity. (Modified after [163]).

The feeding and defecating rates of the snail *Lymnaea stagnalis* were suppressed by ZnO NMs, 86% of the Zn being retained inside the organism. The reproduction and survival of copepods was also altered by ZnO NMs, along with the impairment of their movement [15,22].

Genotoxicity and cytotoxicity were the effects of exposure of *Allium cepa* to ZnO NPs. The mitotic index was inhibited in a concentration dependent way, indicating a mitodepressive effect of the NPs, which may prevent several cells from entering the prophase and blocking the mitotic cycle during interphase inhibiting DNA/protein synthesis (Figure 35). The presence of ZnO NP deposits inside the cell matrix of *A. cepa* is shown in microscopy images confirming their internalization and agglomeration [143].

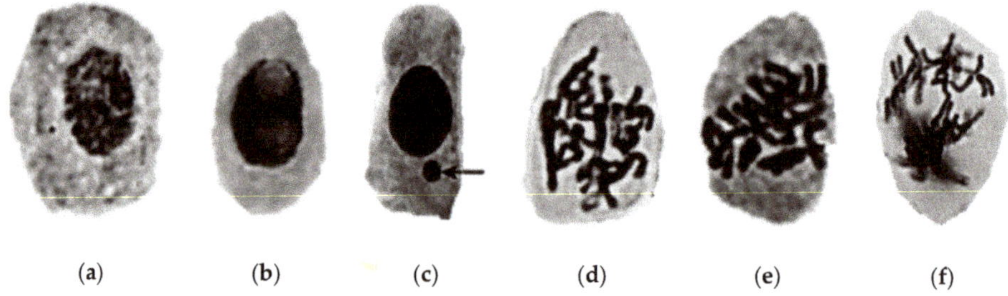

(a)   (b)   (c)   (d)   (e)   (f)

Figure 35. Chromosomal aberrations observed in *Allium cepa* meristematic cells exposed to ZnO NPs. (**a**) Normal cell in prophase; (**b**) binucleated cells at early telophase; (**c**) prophase nucleus with micronucleus; (**d**) disturbed metaphase; (**e**) disturbed anaphase and (**f**) multipolar anaphase. (Modified after [143]).

4.4. Comparison of Ecotoxicity of Described NMs Based on Their Half Maximal Effective Concentration Values

In order to compare the ecotoxicity of NMs and to classify these NMs into toxicity classes, from published data, the half maximal effective concentrations (EC_{50}) were identified and analyzed. The concentration values were taken into account only for the ecotoxicity tests selected as being adaptable for nanomaterials. In order to be able to compare the concentrations of different types of nanomaterials, one model species was selected for each test, because the concentrations show interspecific variations. The selected model species were:

- Aquatic tests

    - algae assay—*Raphidocelis subcapitata*
    - duckweed assay—*Lemna minor*
    - daphnid assay—*Daphnia magna*

- Terrestrial tests

    - plant assay—*Allium cepa*
    - nematode assay—Caenorhabditis elegans
    - earthworm assay—*Eisenia foetida*

In the ecotoxicity comparison tables (Table 6 and Table 9), for both the aquatic and terrestrial environments, only nanomaterials that had EC_{50} values for at least one of the tests are presented, even if these represent one of the mentioned special cases. The low number of nanomaterials described in the comparison tables is due to the fact that some, although they are studied from an ecotoxicological point of view, either do not present the results in the form of EC_{50}, or they are realized on species other than those chosen as a model.

Where EC_{50} values were presented for multiple NM types based on the same substance, which might differ in shape and/or size, the mean value of all types of tested NMs was included in the comparison table. The average EC_{50} values were calculated as a mean of the concentration of each aquatic and terrestrial assay in order to compare the NMs, but only for those NMs that had available data for all assays, for the rest of NMs "not applicable" (N/A) was entered in the table. The data that was not an exact value but was represented as greater or smaller than a value, the value +1 and –1, respectively, was taken into calculation.

4.4.1. Comparison of Ecotoxicity of NMs in the Aquatic Environment

The aquatic ecotoxicity of substances during short term exposure (acute toxicity), for all organisms types (including algae, plants, invertebrates and fish) is divided into five toxicity categories (Table 5), according to both U.S. Environmental Protection Agency [164] and United Nations [165].

Table 5. Toxicity categories for aquatic ecotoxicity.

Categories According to U.S. EPA [164]		Categories According to U.N. [165]		EC_{50} (mg/L)
Very highly toxic	(VHT)	Acute 1.1	(A1.1)	< 0.1
Highly toxic	(HT)	Acute 1.2	(A1.2)	0.1–1
Moderately toxic	(MT)	Acute 2	(A2)	> 1–10
Slightly toxic	(ST)	Acute 3	(A3)	> 10–100
Practically nontoxic	(PNT)	Acute 4	(A4)	> 100

The EC_{50} values for aquatic ecotoxicity assays suitable for NMs of some nanomaterials were compared (Table 6).

Only cadmium, copper, iron, silver and titanium-based NMs had available data regarding their EC_{50} values for all the selected assays and model test species. By comparing the average EC_{50} values for all assays, silver NMs are the most toxic (VHT/A1.1), followed by the highly toxic (HT/A1.2) copper NMs, the moderately toxic (MT/A2) cadmium-based NMs and the slightly toxic (ST/A3) iron and titanium NMs. The order of their toxicity is: Ag NMs > Cu NMs > Cd NMs > Ti NMs > Fe NMs.

The most sensitive aquatic assay for NMs is the algae assay (according to data for *Raphidocelis subcapitata*), while the least sensitive is the duckweed assay (according to data for *Lemna minor*) as shown by the average EC_{50} values per assay.

Table 6. Comparison of EC_{50} values for some NMs and classification in aquatic toxicity categories. The classification into toxicity categories is color-coded as follows: values that are in the VHT/A1.1 category have dark red shading, in HT/A1.2 have red shading, in MT/A2 have orange shading, in ST/A3 have yellow shading and in PNT/A4 have green shading.

NMs Based on:	EC_{50} Value (mg/L) for Aquatic Organisms			Tox. Cat. for Mean EC_{50} for All Assays
	Algae 72 h Test	Duckweed 168 h (7 d) Test	Daphnid 48 h Test	
Carbon NT	29.9 [166]	NDA	>100 [23]	N/A
Fullerene	NDA	NDA	11 [167]	N/A
Graphene	20 [140]	NDA	20 [167]	N/A
Cerium	NDA	NDA	52.42 ** [168]	N/A
Cadmium	3.5 * [169]	0,45 [170]	0.33 * [169]	1.427
Copper	0.7 [23]	0,84 [171]	0.9 [23]	0.813
Gold	0.048 [172]	NDA	>30 [23]	N/A
Iron	0.07 [173]	>100 [173]	43.41 [173]	48.16
Platinum	NDA	0.213 [174]	0.444 [174]	N/A
Silver	0.003 [23]	0,03 [175]	0.003 [23]	0.012
Titanium	6.8 [23]	>90 [176]	29.5 [176]	42.433
Zinc	0.14 [23]	NDA	1.87 [23]	N/A
AVERAGE EC_{50}	6.795	32.255	24.323	

NDA=no data available; N/A not applicable; * if there was no data on EC_{50}, the value of other concentrations, such as IC_{50} or median lethal concentration (LC_{50}), were entered in the table; ** if there was no data regarding EC_{50} at the standard test time interval, the value of the concentrations at other time intervals were entered in the table;

The EC_{50} values for aquatic ecotoxicity assays suitable for NMs of some nanomaterials were classified into the five toxicity categories for the aquatic environment (Table 7).

Table 7. Classification of NMs into the five aquatic toxicity categories based on their EC_{50} values.

Toxicity Category	EC_{50} (mg/L)	Algae Assay	Duckweed Assay	Daphnid Assay
Very highly toxic	< 0.1	Ag, Au, Fe NMs	Ag NMs	Ag NMs
Highly toxic	0.1–1	Zn, Cu NMs	Pt, Cd, Cu NMs	Cd, Pt, Cu NMs
Moderately toxic	> 1–10	Cd, Ti NMs	-	Zn NMs
Slightly toxic	> 10–100	Graphene, Carbon NT	Ti NMs	Fullerene, graphene, Ti, Au, Fe, Ce NMs
Practically nontoxic	> 100	-	Fe NMs	Carbon NT

For algae, the most toxic NMs were those based on silver, gold and iron, and the least toxic were the carbon nanotubes and graphene. The most toxic NMs to duckweed were silver and platinum NMs, iron NMs being the least toxic. For daphnids, the highest toxicity was observed for silver NMs, followed by cadmium and platinum NMs, the least toxic being cerium NMs and carbon nanotubes.

4.4.2. Comparison of Ecotoxicity of NMs in the Terrestrial Environment

The terrestrial ecotoxicity of substances during short term exposure (acute toxicity), for both plants and soil dwelling invertebrates, is divided into three toxicity categories (Table 8), according to [177].

Table 8. Toxicity categories for terrestrial ecotoxicity.

Toxicity Categories	EC_{50} (mg/kg Soil Dry Weight)
Very toxic (VT)/Acute 1 (A1)	≤ 10
Toxic (T)/Acute 2 (A2)	> 10 – ≤ 100
Harmful (H)/Acute 3 (A3)	> 100 – ≤ 1000

The EC$_{50}$ values for terrestrial ecotoxicity assays suitable for NMs of some nanomaterials were compared (Table 9).

Table 9. Comparison of EC$_{50}$ values for some NMs and classification in terrestrial toxicity categories. The classification into toxicity categories is color-coded as follows: values that are in the VT/A1 category have red shading, in T/A2 have orange shading and in H/A3 have yellow shading.

NMs Based on:	EC$_{50}$ Value (mg/kg Soil Dry Weight) for Terrestrial Organisms			Tox. Cat. for Mean EC$_{50}$ for All Assays
	Plants 72 h Test	Nematodes 24 h Test	Earthworms 28 d Test	
Cerium	NDA	NDA	294.6 [178]	N/A
Copper	NDA	NDA	197 [179]	N/A
Silver	12.973 [180]	2.553 [181]	31 ** [182]	15.508
Titanium	NDA	18 [183]	NDA	N/A
Zinc	NDA	NDA	179 *** [184]	N/A

NDA=no data available; N/A not applicable; * if there was no data on EC$_{50}$, the value of other concentrations, such as IC$_{50}$ or LC$_{50}$, were entered in the table; ** if there was no data regarding EC$_{50}$ at the standard test time interval, the value of the concentrations at other time intervals were entered in the table; *** if there was no data regarding EC$_{50}$ for the selected model species, but there were values for species of the same genus, these values were entered in the table.

Only silver NMs had available data regarding their EC$_{50}$ values on all the selected assays and model test species. The silver NMs were the most toxic towards nematodes, followed by plants and earthworms. By considering the earthworm assay, where all NMs that were taken into account had EC$_{50}$ values, the order of toxicity for the NMs is: Ag NMs > Zn NMs > Cu NMs > Ce NMs.

The most sensitive terrestrial assay, based on the scarce available data, was the nematode assay (according to data for *Caenorhabditis elegans*), followed by the plant assay (according to data for *Allium cepa*) and the earthworm assay (according to data for *Eisenia foetida*).

The EC$_{50}$ values for terrestrial ecotoxicity assays suitable for NMs of some nanomaterials were classified into the three toxicity categories for the terrestrial environment (Table 10).

Table 10. Classification of NMs into the three terrestrial toxicity categories based on their EC$_{50}$ values.

Toxicity Categories.	EC$_{50}$ (mg/kg Soil Dw)	Plant Assay	Nematode Assay	Earthworm Assay
Very toxic	≤ 10	-	Ag NMs	-
Toxic	> 10 – ≤ 100	Ag NMs	Ti NMs	Ag NMs
Harmful	> 100 – ≤ 1000	-	-	Zn, Cu, Ce NMs

For nematodes the most toxic NM was based on silver, and the least toxic on titanium. For earthworms, the most toxic NM was based on silver as well, and the least toxic on cerium.

By comparing EC$_{50}$ values of silver nanomaterials obtained in aquatic and terrestrial ecotoxicity tests, it can be observed that aquatic tests are more sensitive than terrestrial ones.

5. Final Remarks

Bibliometric investigations revealed that the research on ecotoxicity, nanomaterials and the ecotoxicity of nanomaterials increased during the period 2010–2019. This highlights the need for further research into the ecotoxicity of nanomaterials.

The ecotoxicological effects of a potentially polluting substances such as nanomaterials can be analyzed by applying ecotoxicity tests. Such tests have been developed over time, creating standard test guides.

Although these standards use a wide range of organisms, from the simplest, such as algae and invertebrates, to the most complex, such as higher plants and vertebrates, only the use of simple organisms is recommended in the first phase, and of the most complex only in case of need.

This trend has been applied by us as well, only SSRET (suitable for simple and rapid ecotoxicity testing) organisms being considered and included in our study regarding the testing of nanomaterial ecotoxicity.

Due to the special properties of nanomaterials with respect to their bulk material and their wide range of applications, it is essential to test the ecotoxicity of NMs. The use of standard tests for testing the ecotoxicity of nanomaterials is possible but requires some adaptation. We considered as adaptive tests only those that use only SSRET organisms, are suitable for nanomaterials (i.e., were applied and adapted to NMs in at least ten scientific articles), have a shorter duration than 30 days, do not require special equipment or training and have low costs. Also, the possibility of these tests being high throughput is an advantage.

The tests considered to be adaptable for nanomaterials are the algal, duckweed, amphipod, daphnid and chironomid tests in the aquatic environment, and the terrestrial plants, nematodes and earthworm tests in the terrestrial environment.

The adaptations of these tests to nanomaterials include the methods of preparing the material solutions, as well as different methods of preparing the culture media, quantifying the number of organisms, etc.

Analyzing the different effects of nanomaterials on SSRET and other organisms, it can be observed that these effects are complex and are of different categories (Figure 36). NMs can physically affect organisms, for example by blocking the digestive tract or by attaching to the surface of the body. Also, they may have effects at the cellular level, by breaking the plasma membrane, at the molecular level, by inducing the production of different enzymes, or at the genetic level, by altering the chromosomes. The behavior of some organisms might also be affected, by inducing abnormal feeding or swimming behaviors.

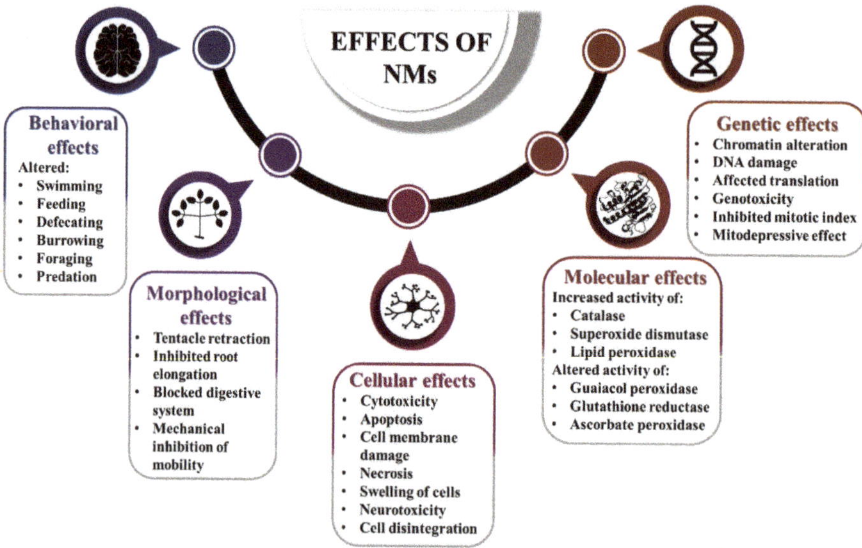

Figure 36. Summary of effects of different categories that NMs can have on different types of organism. The described effects can occur in all or only a few organism types.

Comparing the EC_{50} values of some nanomaterials obtained in aquatic and terrestrial ecotoxicity tests, that were selected as conforming for nanomaterials and are conducted on selected model species, it can be observed that such values are available mainly for aquatic ecotoxicity. It can also be observed that from the three selected aquatic tests, the algae, duckweed and daphnids tests, the most sensitive was the algae test, while the least sensitive was the duckweed test. From the selected terrestrial tests, the plant, nematode and earthworm tests, the most sensitive was the nematode test, the least sensitive being the earthworm test.

It can also be observed that in aquatic tests only for nanomaterials based on cadmium, copper, iron, silver and titanium EC_{50} values were described for all three selected tests. The toxicity order of these four NMs was Ag > Cu > Cd > Ti > Fe. The most toxic NMs to algae were silver, gold and iron, to duckweed and daphnids silver, while the least toxic to algae were graphene and carbon NTs, to duckweed iron, and to daphnids carbon NT.

In the terrestrial tests only silver NMs had EC_{50} values for all the selected assays. Comparing the concentration for the earthworm assay, the toxicity order of NMs was possible: Ag > Zn > Cu > Ce.

In conclusion, the assessment of the ecotoxicity of nanomaterials and its mechanism are essential, as well as the adaptation of the standard ecotoxicity testing methods for nanomaterials. There are still difficulties regarding the testing of NMs, but these could be resolved by further research on this subject. It is clear that the effects of nanomaterials on different types of organisms are complex, thus these must be further analyzed.

Author Contributions: Conceptualization V.O. and B.-V.B.; writing—original draft preparation B.-V.B.; writing—review and editing V.O. and B.-V.B. All authors have read and agreed to the published version of the manuscript.

Funding: This work was supported by the ERA-Net COFASP Project "Polymeric NanoBioMaterials for drug delivery: developing and implementation of safe-by-design concept enabling safe healthcare solutions" (project no. PN3-P3-285/14/2017 program H2020, financed by the Romanian National Authority for Scientific Research UEFISCDI).

Acknowledgments: The authors acknowledge the financial support of the project PNIII-28-PFE BID.

Conflicts of Interest: The authors declare no conflict of interest.

References

1. Sandermann, H. Molecular ecotoxicology of plants. *Trends Plant Sci.* **2004**, *9*, 406–413. [CrossRef] [PubMed]
2. Kahru, A.; Dubourguier, H.C. From ecotoxicology to nanoecotoxicology. *Toxicology* **2010**, *269*, 105–119. [CrossRef] [PubMed]
3. Crane, M.; Handy, R.D.; Garrod, J.; Owen, R. Ecotoxicity test methods and environmental hazard assessment for engineered nanoparticles. *Ecotoxicology (Lond. Engl.)* **2008**, *17*, 421–437. [CrossRef] [PubMed]
4. Walker, C.H.; Greig-Smith, P.W.; Crossland, N.O.; Brown, R. Ecotoxicology. In *Animals and Alternatives in Toxicology: Present Status and Future Prospects*; Balls, M., Bridges, J., Southee, J., Eds.; Macmillan Education UK: London, UK, 1991; pp. 223–251. [CrossRef]
5. International Union for Conservation of Nature. The IUCN Red List of Threatened Species. Version 2019-3. Available online: http://www.iucnredlist.org (accessed on 28 January 2020).
6. Sudha, P.N.; Sangeetha, K.; Vijayalakshmi, K.; Barhoum, A. Chapter 12—Nanomaterials history, classification, unique properties, production and market. In *Emerging Applications of Nanoparticles and Architecture Nanostructures*; Barhoum, A., Makhlouf, A.S.H., Eds.; Elsevier: Amsterdam, The Netherlands, 2018; pp. 341–384. [CrossRef]
7. Saleh, T.A.; Gupta, V.K. Chapter 4—Synthesis, classification, and properties of nanomaterials. In *Nanomaterial and Polymer Membranes*; Saleh, T.A., Gupta, V.K., Eds.; Elsevier: Amsterdam, The Netherlands, 2016; pp. 83–133. [CrossRef]
8. Dhawan, A.; Sharma, V. Toxicity assessment of nanomaterials: Methods and challenges. *Anal. Bioanal. Chem.* **2010**, *398*, 589–605. [CrossRef]
9. Naqvi, S.; Kumar, V.; Gopinath, P. Chapter 11—Nanomaterial toxicity: A challenge to end users. In *Applications of Nanomaterials*; Mohan Bhagyaraj, S., Oluwafemi, O.S., Kalarikkal, N., Thomas, S., Eds.; Woodhead Publishing: Cambridge, UK, 2018; pp. 315–343. [CrossRef]
10. Baun, A.; Sayre, P.; Steinhauser, K.G.; Rose, J. Regulatory relevant and reliable methods and data for determining the environmental fate of manufactured nanomaterials. *NanoImpact* **2017**, *8*, 1–10. [CrossRef]

11. Cattaneo, A.G.; Gornati, R.; Chiriva-Internati, M.; Bernardini, G. Ecotoxicology of nanomaterials: The role of invertebrate testing. *ISJ-Invert. Surviv. J.* **2009**, *6*, 78–97.

12. Hund-Rinke, K.; Baun, A.; Cupi, D.; Fernandes, T.F.; Handy, R.; Kinross, J.H.; Navas, J.M.; Peijnenburg, W.; Schlich, K.; Shaw, B.J.; et al. Regulatory ecotoxicity testing of nanomaterials—Proposed modifications of OECD test guidelines based on laboratory experience with silver and titanium dioxide nanoparticles. *Nanotoxicology* **2016**, *10*, 1442–1447. [CrossRef]

13. Petersen, E.J.; Diamond, S.A.; Kennedy, A.J.; Goss, G.G.; Ho, K.; Lead, J.; Hanna, S.K.; Hartmann, N.B.; Hund-Rinke, K.; Mader, B.; et al. Adapting OECD aquatic toxicity tests for use with manufactured nanomaterials: Key issues and consensus recommendations. *Environ. Sci. Technol.* **2015**, *49*, 9532–9547. [CrossRef]

14. Organisation for Economic Co-operation and Development. *ENV/JM/MONO1: Ecotoxicology and Environmental Fate of Manufactured Nanomaterials: Test Guidelines. Environment Directorate Joint Meeting of the Chemicals Committee and the Working Party on Chemicals, Pesticides and Biotechnology*; OECD: Paris, France, 2014.

15. Bour, A.; Mouchet, F.; Silvestre, J.; Gauthier, L.; Pinelli, E. Environmentally relevant approaches to assess nanoparticles ecotoxicity: A review. *J. Hazard. Mater.* **2015**, *283*, 764–777. [CrossRef]

16. Organisation for Economic Co-operation and Development. *Freshwater Alga and Cyanobacteria, Growth Inhibition Test. OECD Guidelines for the Testing of Chemicals.* 2; Test No. 201; OECD: Paris, France, 2011; Available online: https://www.oecd-ilibrary.org/environment/oecd-guidelines-for-the-testing-of-chemicals-section-2-effects-on-biotic-systems_20745761 (accessed on 13 November 2019).

17. U.S. Environmental Protection Agency. *Algal Toxicity*; OCSPP 850.4500; EPA: Washington, DC, USA, 2012. Available online: www.epa.gov (accessed on 12 November 2019).

18. U.S. Environmental Protection Agency. *Cyanobacteria (Anabaena flos-aquae) Toxicity*; OCSPP 850.4550; EPA: Washington, DC, USA, 2012. Available online: www.epa.gov (accessed on 12 November 2019).

19. International Organization for Standardization. *Fresh Water Algal Growth Inhibition Test with Unicellular Green Algae*; ISO Standard No. 8692; ISO: Geneva, Switzerland, 2012; Available online: www.iso.org (accessed on 3 December 2019).

20. International Organization for Standardization. *Marine Algal Growth Inhibition Test with Skeletonema sp. and Phaeodactylum Tricornutum*; ISO Standard No. 10253; ISO: Geneva, Switzerland, 2016; Available online: www.iso.org (accessed on 3 December 2019).

21. Organisation for Economic Co-operation and Development. *ENV/JM/MONO(1). Ecotoxicology and Environmental Fate of Manufactured Nanomaterials: Test Guidelines*; OECD: Paris, France, 2014.

22. Wong, S.W.Y.; Leung, K.M.Y.; Djurisic, A.B. A Comprehensive review on the aquatic toxicity of engineered nanomaterials. *Rev. Nanosci. Nanotechnol.* **2013**, *2*, 79–105. [CrossRef]

23. Bondarenko, O.M.; Heinlaan, M.; Sihtmae, M.; Ivask, A.; Kurvet, I.; Joonas, E.; Jemec, A.; Mannerstrom, M.; Heinonen, T.; Rekulapelly, R.; et al. Multilaboratory evaluation of 15 bioassays for (eco)toxicity screening and hazard ranking of engineered nanomaterials: FP7 project NANOVALID. *Nanotoxicology* **2016**, *10*, 1229–1242. [CrossRef] [PubMed]

24. Hjorth, R.; Skjolding, L.M.; Sørensen, S.N.; Baun, A. Regulatory adequacy of aquatic ecotoxicity testing of nanomaterials. *NanoImpact* **2017**, *8*, 28–37. [CrossRef]

25. Organisation for Economic Co-operation and Development: Lemna sp. *Growth Inhibition Test; OECD Guidelines for the Testing of Chemicals.* 2; Test No. 221; OECD: Paris, France, 2006. [CrossRef]

26. U.S. Environmental Protection Agency. *Aquatic Plant Toxicity Test Using Lemna spp*; OCSPP 850.4400; EPA: Washington, DC, USA, 2012. Available online: www.epa.gov (accessed on 12 November 2019).

27. International Organization for Standardization. *Determination of the Toxic Effect of Water Constituents and Waste Water on Duckweed (Lemna Minor)—Duckweed Growth Inhibition Test*; ISO Standard No. 20079; ISO: Geneva, Switzerland, 2005; Available online: www.iso.org (accessed on 3 December 2019).

28. International Organization for Standardization. *Determination of the Growth Inhibition Effects of Waste Waters, Natural Waters and Chemicals on the Duckweed Spirodela Polyrhiza—Method Using a Stock Culture Independent Microbiotest*; ISO Standard No. 20227; ISO: Geneva, Switzerland, 2017; Available online: www.iso.org (accessed on 3 December 2019).

29. Dolenc Koce, J. Effects of exposure to nano and bulk sized TiO_2 and CuO in *Lemna minor*. *Plant Physiol. Biochem.* **2017**, *119*, 43–49. [CrossRef] [PubMed]

30. Yue, L.; Zhao, J.; Yu, X.; Lv, K.; Wang, Z.; Xing, B. Interaction of CuO nanoparticles with duckweed (*Lemna minor*. L): Uptake, distribution and ROS production sites. *Environ. Pollut.* **2018**, *243*, 543–552. [CrossRef]

31. Ziegler, P.; Sree, K.S.; Appenroth, K.J. Duckweeds for water remediation and toxicity testing. *Toxicol. Environ. Chem.* **2016**, *98*, 1127–1154. [CrossRef]

32. Sallenave, R.; Fomin, A. Some advantages of the duckweed test to assess the toxicity of environmental samples. *Acta Hydrochim. Hydrobiol.* **1997**, *25*, 135–140. [CrossRef]

33. U.S. Environmental Protection Agency. *Gammarid Amphipod Acute Toxicity Test*; OCSPP 850.1020; EPA: Washington, DC, USA, 2016. Available online: www.epa.gov (accessed on 12 November 2019).

34. U.S. Environmental Protection Agency. *Spiked Whole Sediment 10-Day Toxicity Test, Freshwater Invertebrates*; OCSPP 850.1735; EPA: Washington, DC, USA, 2016. Available online: www.epa.gov (accessed on 12 November 2019).

35. U.S. Environmental Protection Agency. *Spiked Whole Sediment 10-Day Toxicity Test, Saltwater Invertebrates*; OCSPP 850.1740; EPA: Washington, DC, USA, 2016. Available online: www.epa.gov (accessed on 18 November 2019).

36. International Organization for Standardization. *Determination of Acute Toxicity of Marine or Estuarine Sediment to Amphipods*; ISO Standard No. 16712; ISO: Geneva, Switzerland, 2005; Available online: www.iso.org (accessed on 3 December 2019).

37. Organisation for Economic Co-operation and Development. *Daphnia sp. Acute Immobilisation Test*; OECD Guidelines for the Testing of Chemicals. 2; Test No. 202; OECD: Paris, France, 2004. [CrossRef]

38. Organisation for Economic Co-operation and Development. *Daphnia Magna Reproduction Test*; OECD Guidelines for the Testing of Chemicals. 2; Test No. 211; OECD: Paris, France, 2012. [CrossRef]

39. U.S. Environmental Protection Agency. *Aquatic Invertebrate Acute Toxicity Test, Freshwater Daphnids*; OCSPP 850.1010; EPA: Washington, DC, USA, 2016. Available online: www.epa.gov (accessed on 18 November 2019).

40. U.S. Environmental Protection Agency. *Daphnid Chronic Toxicity Test*; OCSPP 850.1300; EPA: Washington, DC, USA, 2016. Available online: www.epa.gov (accessed on 18 November 2019).

41. International Organization for Standardization. *Determination of the Inhibition of the Mobility of Daphnia Magna Straus (Cladocera, Crustacea)—Acute Toxicity Test*; ISO Standard No. 6341; ISO: Geneva, Switzerland, 2012; Available online: www.iso.org (accessed on 3 December 2019).

42. International Organization for Standardization. *Determination of Chronic Toxicity to Ceriodaphnia Dubia*; ISO Standard No. 20665; ISO: Geneva, Switzerland, 2008; Available online: www.iso.org (accessed on 3 December 2019).

43. International Organization for Standardization. *Determination of Long Term Toxicity of Substances to Daphnia Magna Straus (Cladocera, Crustacea)*; ISO Standard No. 10706; ISO: Geneva, Switzerland, 2000; Available online: www.iso.org (accessed on 3 December 2019).

44. Wiench, K.; Wohlleben, W.; Hisgen, V.; Radke, K.; Salinas, E.; Zok, S.; Landsiedel, R. Acute and chronic effects of nano- and non-nano-scale TiO_2 and ZnO particles on mobility and reproduction of the freshwater invertebrate *Daphnia magna*. *Chemosphere* **2009**, *76*, 1356–1365. [CrossRef]

45. Organisation for Economic Co-operation and Development. *Chironomus sp., Acute Immobilisation Test*; OECD Guidelines for the Testing of Chemicals. 2; Test No. 235; OECD: Paris, France, 2011. [CrossRef]

46. Organisation for Economic Co-operation and Development. *Sediment-Water Chironomid Life-Cycle Toxicity Test Using Spiked Water or Spiked Sediment*; OECD Guidelines for the Testing of Chemicals. 2; Test No. 233; OECD: Paris, France, 2010. [CrossRef]

47. Organisation for Economic Co-operation and Development. *Sediment-Water Chironomid Toxicity Using Spiked Sediment*; OECD Guidelines for the Testing of Chemicals. 2; Test No. 218; OECD: Paris, France, 2004. [CrossRef]

48. Organisation for Economic Co-operation and Development. *Sediment-Water Chironomid Toxicity Using Spiked Water*; OECD Guidelines for the Testing of Chemicals. 2; Test No. 219; OECD: Paris, France, 2004. [CrossRef]

49. Tomilina, I.I.; Gremyachikh, V.A.; Grebenyuk, L.P.; Klevleeva, T.R. The effect of zinc oxide nano- and microparticles and zinc ions on freshwater organisms of different trophic levels. *Inland Water Biol.* **2014**, *7*, 88–96. [CrossRef]

50. Oberholster, P.J.; Musee, N.; Botha, A.M.; Chelule, P.K.; Focke, W.W.; Ashton, P.J. Assessment of the effect of nanomaterials on sediment-dwelling invertebrate *Chironomus tentans* larvae. *Ecotoxicol. Environ. Saf.* **2011**, *74*, 416–423. [CrossRef] [PubMed]

51. Organisation for Economic Co-operation and Development. *Terrestrial Plant Test: Seedling Emergence and Seedling Growth Test; OECD Guidelines for the Testing of Chemicals.* 2; Test No. 208; OECD: Paris, France, 2006. [CrossRef]

52. Organisation for Economic Co-operation and Development. *Terrestrial Plant Test: Vegetative Vigour Test; OECD Guidelines for the Testing of Chemicals.* 2; Test No. 227; OECD: Paris, France, 2006. [CrossRef]

53. U.S. Environmental Protection Agency. *Seedling Emergence and Seedling Growth;* OCSPP 850.4100; EPA: Washington, DC, USA, 2012. Available online: www.epa.gov (accessed on 12 November 2019).

54. U.S. Environmental Protection Agency. *Vegetative Vigor;* OCSPP 850.4150; EPA: Washington, DC, USA, 2012. Available online: www.epa.gov (accessed on 18 November 2019).

55. U.S. Environmental Protection Agency. *Early Seedling Growth Toxicity Test;* OCSPP 850.4230; EPA: Washington, DC, USA, 2012. Available online: www.epa.gov (accessed on 12 November 2019).

56. International Organization for Standardization. *Determination of the Effects of Pollutants on Soil Flora—Part 1: Method for the Measurement of Inhibition of Root Growth;* ISO Standard No. 11269-1; ISO: Geneva, Switzerland, 2012; Available online: www.iso.org (accessed on 3 December 2019).

57. International Organization for Standardization. *Determination of the Effects of Pollutants on Soil Flora—Part 2: Effects of Contaminated Soil on the Emergence and Early Growth of Higher Plants;* ISO Standard No. 11269-2; ISO: Geneva, Switzerland, 2012; Available online: www.iso.org (accessed on 9 December 2019).

58. International Organization for Standardization. *Determination of the Effects of Pollutants on Soil Flora—Screening Test for Emergence of Lettuce Seedlings (Lactuca sativa L.);* ISO Standard No. 17126; ISO: Geneva, Switzerland, 2005; Available online: www.iso.org (accessed on 9 December 2019).

59. International Organization for Standardization. *Determination of the Toxic Effects of Pollutants on Germination and Early Growth of Higher Plants;* ISO Standard No. 18763; ISO: Geneva, Switzerland, 2016; Available online: www.iso.org (accessed on 9 December 2019).

60. International Organization for Standardization ISO-22030. *Biological Methods—Chronic Toxicity in Higher Plants;* ISO: Geneva, Switzerland, 2005; Available online: www.iso.org (accessed on 9 December 2019).

61. Lee, W.M.; An, Y.J.; Yoon, H.; Kweon, H.S. Toxicity and bioavailability of copper nanoparticles to the terrestrial plants mung bean (*Phaseolus radiatus*) and wheat (*Triticum aestivum*): Plant agar test for water-insoluble nanoparticles. *Environ. Toxicol. Chem.* **2008**, *27*, 1915–1921. [CrossRef] [PubMed]

62. Ma, C.; White, J.C.; Dhankher, O.P.; Xing, B. Metal-based nanotoxicity and detoxification pathways in higher plants. *Environ. Sci. Technol.* **2015**, *49*, 7109–7122. [CrossRef] [PubMed]

63. Khodakovskaya, M.; Dervishi, E.; Mahmood, M.; Xu, Y.; Li, Z.; Watanabe, F.; Biris, A.S. Carbon nanotubes are able to penetrate plant seed coat and dramatically affect seed germination and plant growth. *ACS Nano* **2009**, *3*, 3221–3227. [CrossRef] [PubMed]

64. Libralato, G.; Costa Devoti, A.; Zanella, M.; Sabbioni, E.; Mičetić, I.; Manodori, L.; Pigozzo, A.; Manenti, S.; Groppi, F.; Volpi Ghirardini, A. Phytotoxicity of ionic, micro- and nano-sized iron in three plant species. *Ecotoxicol. Environ. Saf.* **2016**, *123*, 81–88. [CrossRef] [PubMed]

65. International Organization for Standardization. *Determination of the Toxic Effect of Sediment and Soil Samples on Growth, Fertility and Reproduction of Caenorhabditis Elegans (Nematoda);* ISO Standard No. 10872; ISO: Geneva, Switzerland, 2010; Available online: www.iso.org (accessed on 9 December 2019).

66. American Society for Testing and Materials-. *Standard Guide for Conducting Laboratory Soil Toxicity Tests with the Nematode Caenorhabditis Elegans;* Standard No. E2172-01; ASTM: West Conshohocken, PA, USA, 2014.

67. Hoss, S.; Ahlf, W.; Bergtold, M.; Bluebaum-Gronau, E.; Brinke, M.; Donnevert, G.; Menzel, R.; Mohlenkamp, C.; Ratte, H.T.; Traunspurger, W.; et al. Interlaboratory comparison of a standardized toxicity test using the nematode *Caenorhabditis elegans* (ISO 10872). *Environ. Toxicol. Chem.* **2012**, *31*, 1525–1535. [CrossRef]

68. Jung, S.K.; Qu, X.; Aleman-Meza, B.; Wang, T.; Riepe, C.; Liu, Z.; Li, Q.; Zhong, W. Multi-endpoint, high-throughput study of nanomaterial toxicity in *Caenorhabditis elegans*. *Environ. Sci. Technol.* **2015**, *49*, 2477–2485. [CrossRef]

69. Hanna, S.K.; Cooksey, G.A.; Dong, S.; Nelson, B.C.; Mao, L.; Elliott, J.T.; Petersen, E.J. Feasibility of using a standardized *Caenorhabditis elegans* toxicity test to assess nanomaterial toxicity. *Environ. Sci. Nano* **2016**, *3*, 1080–1089. [CrossRef]

70. Organisation for Economic Co-operation and Development. *Earthworm, Acute Toxicity Tests; OECD Guidelines for the Testing of Chemicals.* 2; Test No. 207; OECD: Paris, France, 1984. [CrossRef]

71. U.S. Environmental Protection Agency. *Earthworm Subchronic Toxicity Test*; OCSPP 850.3100; EPA: Washington, DC, USA, 2012. Available online: www.epa.gov (accessed on 18 November 2019).

72. International Organization for Standardization. *Effects of Pollutants on Earthworms—Part 1: Determination of Acute Toxicity to Eisenia Fetida/Eisenia Andrei*; ISO Standard No. 11268-1; ISO: Geneva, Switzerland, 2012; Available online: www.iso.org (accessed on 9 December 2019).

73. Hooper, H.L.; Jurkschat, K.; Morgan, A.J.; Bailey, J.; Lawlor, A.J.; Spurgeon, D.J.; Svendsen, C. Comparative chronic toxicity of nanoparticulate and ionic zinc to the earthworm *Eisenia veneta* in a soil matrix. *Environ. Int.* **2011**, *37*, 1111–1117. [CrossRef]

74. Hu, C.W.; Li, M.; Cui, Y.B.; Li, D.S.; Chen, J.; Yang, L.Y. Toxicological effects of TiO$_2$ and ZnO nanoparticles in soil on earthworm *Eisenia fetida*. *Soil Biol. Biochem.* **2010**, *42*, 586–591. [CrossRef]

75. Zhu, X.; Zhu, L.; Chen, Y.; Tian, S. Acute toxicities of six manufactured nanomaterial suspensions to *Daphnia magna*. *J. Nanopart. Res.* **2009**, *11*, 67–75. [CrossRef]

76. Ghosh, S. *Nanomaterials Safety: Toxicity and Health Hazards*; Walter de Gruyter GmbH & Co KG: Berlin, Germany, 2018; p. 288.

77. Daemi, H.; Barikani, M. Synthesis and characterization of calcium alginate nanoparticles, sodium homopolymannuronate salt and its calcium nanoparticles. *Sci. Iran.* **2012**, *19*, 2023–2028. [CrossRef]

78. Zhao, L.-M.; Shi, L.-E.; Zhang, Z.-L.; Chen, J.-M.; Shi, D.-D.; Yang, J.; Tang, Z.-X. Preparation and application of chitosan nanoparticles and nanofibers. *Braz. J. Chem. Eng.* **2011**, *28*, 353–362. [CrossRef]

79. Andersen, S.O. Chapter 64—Cuticle. In *Encyclopedia of Insects*, 2nd ed.; Resh, V.H., Cardé, R.T., Eds.; Academic Press: San Diego, CA, USA, 2009; pp. 245–246. [CrossRef]

80. Li, J.; Cai, C.; Li, J.; Li, J.; Li, J.; Sun, T.; Wang, L.; Wu, H.; Yu, G. Chitosan-based nanomaterials for drug delivery. *Molecules* **2018**, *23*, 2661. [CrossRef] [PubMed]

81. Wang, J.J.; Zeng, Z.W.; Xiao, R.Z.; Xie, T.; Zhou, G.L.; Zhan, X.R.; Wang, S.L. Recent advances of chitosan nanoparticles as drug carriers. *Int. J. Nanomed.* **2011**, *6*, 765–774. [CrossRef]

82. Jayakumar, R.; Menon, D.; Manzoor, K.; Nair, S.V.; Tamura, H. Biomedical applications of chitin and chitosan based nanomaterials—A short review. *Carbohydr. Polym.* **2010**, *82*, 227–232. [CrossRef]

83. Matica, M.A.; Aachmann, F.L.; Tøndervik, A.; Sletta, H.; Ostafe, V. Chitosan as a wound dressing starting material: Antimicrobial properties and mode of action. *Int. J. Mol. Sci.* **2019**, *20*, 5889. [CrossRef]

84. Dananjaya, S.H.S.; Erandani, W.K.C.U.; Kim, C.-H.; Nikapitiya, C.; Lee, J.; De Zoysa, M. Comparative study on antifungal activities of chitosan nanoparticles and chitosan silver nano composites against *Fusarium oxysporum* species complex. *Int. J. Biol. Macromol.* **2017**, *105*, 478–488. [CrossRef]

85. Murugan, K.; Anitha, J.; Dinesh, D.; Suresh, U.; Rajaganesh, R.; Chandramohan, B.; Subramaniam, J.; Paulpandi, M.; Vadivalagan, C.; Amuthavalli, P.; et al. Fabrication of nano-mosquitocides using chitosan from crab shells: Impact on non-target organisms in the aquatic environment. *Ecotoxicol. Environ. Saf.* **2016**, *132*, 318–328. [CrossRef]

86. Hu, C.; Liu, L.; Li, X.; Xu, Y.; Ge, Z.; Zhao, Y. Effect of graphene oxide on copper stress in *Lemna minor* L.: Evaluating growth, biochemical responses, and nutrient uptake. *J. Hazard. Mater.* **2018**, *341*, 168–176. [CrossRef]

87. Souza, P.M.S.; Corroqué, N.A.; Morales, A.R.; Marin-Morales, M.A.; Mei, L.H.I. PLA and organoclays nanocomposites: Degradation process and evaluation of ecotoxicity using *Allium cepa* as test organism. *J. Polym. Environ.* **2013**, *21*, 1052–1063. [CrossRef]

88. Hadwiger, L.A. Multiple effects of chitosan on plant systems: Solid science or hype. *Plant Sci.* **2013**, *208*, 42–49. [CrossRef]

89. Mosleh, Y.Y.; Paris-Palacios, S.; Ahmed, M.T.; Mahmoud, F.M.; Osman, M.A.; Biagianti-Risbourg, S. Effects of chitosan on oxidative stress and metallothioneins in aquatic worm *Tubifex tubifex* (Oligochaeta, Tubificidae). *Chemosphere* **2007**, *67*, 167–175. [CrossRef] [PubMed]

90. Lee, K.Y.; Mooney, D.J. Alginate: Properties and biomedical applications. *Prog. Polym. Sci.* **2012**, *37*, 106–126. [CrossRef] [PubMed]

91. Sowjanya, J.A.; Singh, J.; Mohita, T.; Sarvanan, S.; Moorthi, A.; Srinivasan, N.; Selvamurugan, N. Biocomposite scaffolds containing chitosan/alginate/nano-silica for bone tissue engineering. *Colloids Surf. B Biointerfaces* **2013**, *109*, 294–300. [CrossRef] [PubMed]

92. Zhang, S.M.; Cui, F.Z.; Liao, S.S.; Zhu, Y.; Han, L. Synthesis and biocompatibility of porous nano-hydroxyapatite/collagen/alginate composite. *J. Mater. Sci. Mater. Med.* **2003**, *14*, 641–645. [CrossRef]

93. Kim, H.L.; Jung, G.Y.; Yoon, J.H.; Han, J.S.; Park, Y.J.; Kim, D.G.; Zhang, M.; Kim, D.J. Preparation and characterization of nano-sized hydroxyapatite/alginate/chitosan composite scaffolds for bone tissue engineering. *Mater. Sci. Eng. C* **2015**, *54*, 20–25. [CrossRef]

94. Varaprasad, K.; Raghavendra, G.M.; Jayaramudu, T.; Seo, J. Nano zinc oxide–sodium alginate antibacterial cellulose fibres. *Carbohydr. Polym.* **2016**, *135*, 349–355. [CrossRef]

95. Fourest, E.; Volesky, B. Contribution of sulfonate groups and alginate to heavy metal biosorption by the dry biomass of *Sargassum fluitans. Environ. Sci. Technol.* **1996**, *30*, 277–282. [CrossRef]

96. Mata, Y.N.; Blázquez, M.L.; Ballester, A.; González, F.; Muñoz, J.A. Biosorption of cadmium, lead and copper with calcium alginate xerogels and immobilized *Fucus vesiculosus. J. Hazard. Mater.* **2009**, *163*, 555–562. [CrossRef]

97. Fourest, E.; Volesky, B. Alginate properties and heavy metal biosorption by marine algae. *Appl. Biochem. Biotechnol.* **1997**, *67*, 215–226. [CrossRef]

98. Lau, P.S.; Tam, N.F.Y.; Wong, Y.S. Wastewater nutrients (N and P) removal by carrageenan and alginate immobilized *Chlorella vulgaris. Environ. Technol.* **1997**, *18*, 945–951. [CrossRef]

99. Shih, Y.J.; Su, C.C.; Chen, C.W.; Dong, C.D.; Liu, W.S.; Huang, C.P. Adsorption characteristics of nano-TiO_2 onto zebrafish embryos and its impacts on egg hatching. *Chemosphere* **2016**, *154*, 109–117. [CrossRef]

100. Callegaro, S.; Minetto, D.; Pojana, G.; Bilanicová, D.; Libralato, G.; Volpi Ghirardini, A.; Hasellöv, M.; Marcomini, A. Effects of alginate on stability and ecotoxicity of nano-TiO_2 in artificial seawater. *Ecotoxicol. Environ. Saf.* **2015**, *117*, 107–114. [CrossRef]

101. Oliveira, C.R.d.; Fraceto, L.F.; Rizzi, G.M.; Salla, R.F.; Abdalla, F.C.; Costa, M.J.; Silva-Zacarin, E.C.M. Hepatic effects of the clomazone herbicide in both its free form and associated with chitosan-alginate nanoparticles in bullfrog tadpoles. *Chemosphere* **2016**, *149*, 304–313. [CrossRef] [PubMed]

102. Abi Nassif, L.; Rioual, S.; Trepos, R.; Fauchon, M.; Farah, W.; Hellio, C.; Abboud, M.; Lescop, B. Development of alginate hydrogels active against adhesion of microalgae. *Mater. Lett.* **2019**, *239*, 180–183. [CrossRef]

103. Storebakken, T. Binders in fish feeds: I. Effect of alginate and guar gum on growth, digestibility, feed intake and passage through the gastrointestinal tract of rainbow trout. *Aquaculture* **1985**, *47*, 11–26. [CrossRef]

104. Chenoweth, M.B. The Toxicity of sodium alginate in cats. *Ann. Surg.* **1948**, *127*, 1173–1181. [CrossRef] [PubMed]

105. Vijayakumar, S.; Malaikozhundan, B.; Ramasamy, P.; Vaseeharan, B. Assessment of biopolymer stabilized silver nanoparticle for their ecotoxicity on *Ceriodaphnia cornuta* and antibiofilm activity. *J. Environ. Chem. Eng.* **2016**, *4*, 2076–2083. [CrossRef]

106. Savadekar, N.R.; Mhaske, S.T. Synthesis of nano cellulose fibers and effect on thermoplastics starch based films. *Carbohydr. Polym.* **2012**, *89*, 146–151. [CrossRef] [PubMed]

107. Klemm, D.; Heublein, B.; Fink, H.-P.; Bohn, A. Cellulose: Fascinating biopolymer and sustainable raw material. *Angew. Chem. Int. Ed.* **2005**, *44*, 3358–3393. [CrossRef]

108. Abitbol, T.; Rivkin, A.; Cao, Y.; Nevo, Y.; Abraham, E.; Ben-Shalom, T.; Lapidot, S.; Shoseyov, O. Nanocellulose, a tiny fiber with huge applications. *Curr. Opin. Biotechnol.* **2016**, *39*, 76–88. [CrossRef]

109. Mincea, M.; Negrulescu, A.; Ostafe, V. Preparation, modification, and applications of chitin nanowhiskers: A review. *Rev. Adv. Mater. Sci.* **2012**, *30*, 225–242.

110. Pitkänen, M.; Kangas, H.; Laitinen, O.; Sneck, A.; Lahtinen, P.; Peresin, M.S.; Niinimäki, J. Characteristics and safety of nano-sized cellulose fibrils. *Cellulose* **2014**, *21*, 3871–3886. [CrossRef]

111. Vartiainen, J.; Pöhler, T.; Sirola, K.; Pylkkänen, L.; Alenius, H.; Hokkinen, J.; Tapper, U.; Lahtinen, P.; Kapanen, A.; Putkisto, K.; et al. Health and environmental safety aspects of friction grinding and spray drying of microfibrillated cellulose. *Cellulose* **2011**, *18*, 775–786. [CrossRef]

112. Kovacs, T.; Naish, V.; O'Connor, B.; Blaise, C.; Gagné, F.; Hall, L.; Trudeau, V.; Martel, P. An ecotoxicological characterization of nanocrystalline cellulose (NCC). *Nanotoxicology* **2010**, *4*, 255–270. [CrossRef] [PubMed]

113. Harper, B.J.; Clendaniel, A.; Sinche, F.; Way, D.; Hughes, M.; Schardt, J.; Simonsen, J.; Stefaniak, A.B.; Harper, S.L. Impacts of chemical modification on the toxicity of diverse nanocellulose materials to developing zebrafish. *Cellulose* **2016**, *23*, 1763–1775. [CrossRef]

114. VanGinkel, C.G.; Gayton, S. The biodegradability and nontoxicity of carboxymethyl cellulose (DS 0.7) and intermediates. *Environ. Toxicol. Chem.* **1996**, *15*, 270–274. [CrossRef]

115. Chada, H.L. Toxicity of cellulose acetate and vinyl plastic cages to barley plants and greenbugs1. *J. Econ. Entomol.* **1962**, *55*, 970–972. [CrossRef]

116. Reddy, C.S.K.; Ghai, R.; Kalia, V.C. Polyhydroxyalkanoates: An overview. *Bioresour. Technol.* **2003**, *87*, 137–146. [CrossRef]

117. Li, X.T.; Zhang, Y.; Chen, G.Q. Nanofibrous polyhydroxyalkanoate matrices as cell growth supporting materials. *Biomaterials* **2008**, *29*, 3720–3728. [CrossRef]

118. Dinjaski, N.; Prieto, M.A. Smart polyhydroxyalkanoate nanobeads by protein based functionalization. *Nanomed. Nanotechnol. Biol. Med.* **2015**, *11*, 885–899. [CrossRef]

119. Ran, G.; Tan, D.; Dai, W.; Zhu, X.; Zhao, J.; Ma, Q.; Lu, X. Immobilization of alkaline polygalacturonate lyase from *Bacillus subtilis* on the surface of bacterial polyhydroxyalkanoate nano-granules. *Appl. Microbiol. Biotechnol.* **2017**, *101*, 3247–3258. [CrossRef] [PubMed]

120. Hauser, M.; Li, G.; Nowack, B. Environmental hazard assessment for polymeric and inorganic nanobiomaterials used in drug delivery. *J. Nanobiotechnol.* **2019**, *17*, 56. [CrossRef] [PubMed]

121. Xu, X.Y.; Li, X.T.; Peng, S.W.; Xiao, J.F.; Liu, C.; Fang, G.; Chen, K.C.; Chen, G.Q. The behaviour of neural stem cells on polyhydroxyalkanoate nanofiber scaffolds. *Biomaterials* **2010**, *31*, 3967–3975. [CrossRef] [PubMed]

122. Saito, T.; Tomita, K.; Juni, K.; Ooba, K. In vivo and In Vitro degradation of poly(3-hydroxybutyrate) in pat. *Biomaterials* **1991**, *12*, 309–312. [CrossRef]

123. Akaraonye, E.; Keshavarz, T.; Roy, I. Production of polyhydroxyalkanoates: The future green materials of choice. *J. Chem. Technol. Biotechnol.* **2010**, *85*, 732–743. [CrossRef]

124. Avérous, L. Chapter 21—Polylactic acid: Synthesis, properties and applications. In *Monomers, Polymers and Composites from Renewable Resources*; Belgacem, M.N., Gandini, A., Eds.; Elsevier: Amsterdam, The Netherlands, 2008; pp. 433–450. [CrossRef]

125. Hartmann, M.H. High Molecular Weight Polylactic Acid Polymers. In *Biopolymers from Renewable Resources*; Kaplan, D.L., Ed.; Springer: Berlin/Heidelberg, Germany, 1998; pp. 367–411. [CrossRef]

126. Lopes, M.S.; Jardini, A.L.; Filho, R.M. Poly (lactic acid) production for tissue engineering applications. *Procedia Eng.* **2012**, *42*, 1402–1413. [CrossRef]

127. Smith, A.; Hunneyball, L.M. Evaluation of poly(lactic acid) as a biodegradable drug delivery system for parenteral administration. *Int. J. Pharm.* **1986**, *30*, 215–220. [CrossRef]

128. Ignatius, A.A.; Claes, L.E. In vitro biocompatibility of bioresorbable polymers: Poly(L, DL-lactide) and poly(L-lactide-co-glycolide). *Biomaterials* **1996**, *17*, 831–839. [CrossRef]

129. Bos, R.R.M.; Rozema, F.B.; Boering, G.; Nijenhius, A.J.; Pennings, A.J.; Verwey, A.B.; Nieuwenhuis, P.; Jansen, H.W.B. Degradation of and tissue reaction to biodegradable poly(L-lactide) for use as internal fixation of fractures: A study in rats. *Biomaterials* **1991**, *12*, 32–36. [CrossRef]

130. Monthioux, M.; Flahaut, E.; Laurent, C.; Escoffier, W.; Raquet, B.; Bacsa, W.; Puech, P.; Machado, B.; Serp, P. Properties of carbon nanotubes. In *Handbook of Nanomaterials Properties*; Bhushan, B., Luo, D., Schricker, S.R., Sigmund, W., Zauscher, S., Eds.; Springer: Berlin/Heidelberg, Germany, 2014; pp. 1–49. [CrossRef]

131. De Volder, M.F.L.; Tawfick, S.H.; Baughman, R.H.; Hart, A.J. Carbon nanotubes: Present and future commercial applications. *Science* **2013**, *339*, 535–539. [CrossRef]

132. Jackson, P.; Jacobsen, N.R.; Baun, A.; Birkedal, R.; Kühnel, D.; Jensen, K.A.; Vogel, U.; Wallin, H. Bioaccumulation and ecotoxicity of carbon nanotubes. *Chem. Cent. J.* **2013**, *7*, 154. [CrossRef] [PubMed]

133. Imahori, H.; Umeyama, T. Chapter 13 Fullerene-modified electrodes and solar cells. In *Fullerenes: Principles and Applications (2)*; Langa, F., Nierengarten, J.F., Eds.; Royal Society of Chemistry: London, UK, 2011. [CrossRef]

134. Bianco, A.; Da Ros, T. Chapter 14 Biological applications of fullerenes. In *Fullerenes: Principles and Applications (2)*; Langa, F., Nierengarten, J.F., Eds.; Royal Society of Chemistry: London, UK, 2011. [CrossRef]

135. Tao, X.; Fortner, J.D.; Zhang, B.; He, Y.; Chen, Y.; Hughes, J.B. Effects of aqueous stable fullerene nanocrystals (nC60) on *Daphnia magna*: Evaluation of sub-lethal reproductive responses and accumulation. *Chemosphere* **2009**, *77*, 1482–1487. [CrossRef]

136. Mendonça, M.C.P.; Rizoli, C.; Ávila, D.S.; Amorim, M.J.B.; de Jesus, M.B. nanomaterials in the environment: perspectives on *in vivo* terrestrial toxicity testing. *Front. Environ. Sci.* **2017**, *5*. [CrossRef]

137. Cao, C.; Wu, X.; Xi, X.; Filleter, T.; Sun, Y. Mechanical characterization of graphene. In *Handbook of Nanomaterials Properties*; Bhushan, B., Luo, D., Schricker, S.R., Sigmund, W., Zauscher, S., Eds.; Springer: Berlin/Heidelberg, Germany, 2014; pp. 121–135. [CrossRef]

138. Zhu, Y.; Murali, S.; Cai, W.; Li, X.; Suk, J.W.; Potts, J.R.; Ruoff, R.S. Graphene and graphene oxide: synthesis, properties, and applications. *Adv. Mater.* **2010**, *22*, 3906–3924. [CrossRef]

139. Hu, C.; Wang, Q.; Zhao, H.; Wang, L.; Guo, S.; Li, X. Ecotoxicological effects of graphene oxide on the protozoan *Euglena gracilis*. *Chemosphere* **2015**, *128*, 184–190. [CrossRef] [PubMed]

140. Nogueira, P.F.M.; Nakabayashi, D.; Zucolotto, V. The effects of graphene oxide on green algae *Raphidocelis subcapitata*. *Aquat. Toxicol.* **2015**, *166*, 29–35. [CrossRef]

141. Pretti, C.; Oliva, M.; Pietro, R.D.; Monni, G.; Cevasco, G.; Chiellini, F.; Pomelli, C.; Chiappe, C. Ecotoxicity of pristine graphene to marine organisms. *Ecotoxicol. Environ. Saf.* **2014**, *101*, 138–145. [CrossRef]

142. Svartz, G.; Papa, M.; Gosatti, M.; Jordán, M.; Soldati, A.; Samter, P.; Guraya, M.M.; Pérez Coll, C.; Perez Catán, S. Monitoring the ecotoxicity of γ-Al$_2$O$_3$ and Ni/γ-Al$_2$O$_3$ nanomaterials by means of a battery of bioassays. *Ecotoxicol. Environ. Saf.* **2017**, *144*, 200–207. [CrossRef]

143. Kumari, M.; Khan, S.S.; Pakrashi, S.; Mukherjee, A.; Chandrasekaran, N. Cytogenetic and genotoxic effects of zinc oxide nanoparticles on root cells of *Allium cepa*. *J. Hazard. Mater.* **2011**, *190*, 613–621. [CrossRef]

144. Arnold, M.C.; Badireddy, A.R.; Wiesner, M.R.; Di Giulio, R.T.; Meyer, J.N. Cerium oxide nanoparticles are more toxic than equimolar bulk cerium oxide in *Caenorhabditis elegans*. *Arch. Environ. Contam. Toxicol.* **2013**, *65*, 224–233. [CrossRef]

145. Taylor, N.S.; Merrifield, R.; Williams, T.D.; Chipman, J.K.; Lead, J.R.; Viant, M.R. Molecular toxicity of cerium oxide nanoparticles to the freshwater alga *Chlamydomonas reinhardtii* is associated with supra-environmental exposure concentrations. *Nanotoxicology* **2016**, *10*, 32–41. [CrossRef]

146. Artells, E.; Issartel, J.; Auffan, M.; Borschneck, D.; Thill, A.; Tella, M.; Brousset, L.; Rose, J.; Bottero, J.-Y.; Thiéry, A. Exposure to cerium dioxide nanoparticles differently affect swimming performance and survival in two daphnid species. *PLoS ONE* **2013**, *8*, e71260. [CrossRef]

147. Li, J.; Zhu, J.J. Quantum dots for fluorescent biosensing and bio-imaging applications. *Analyst* **2013**, *138*, 2506–2515. [CrossRef] [PubMed]

148. Gupta, G.S.; Kumar, A.; Senapati, V.A.; Pandey, A.K.; Shanker, R.; Dhawan, A. Laboratory scale microbial food chain to study bioaccumulation, biomagnification, and ecotoxicity of cadmium telluride quantum dots. *Environ. Sci. Technol.* **2017**, *51*, 1695–1706. [CrossRef] [PubMed]

149. Khataee, A.; Movafeghi, A.; Nazari, F.; Vafaei, F.; Dadpour, M.R.; Hanifehpour, Y.; Joo, S.W. The toxic effects of l-Cysteine-capped cadmium sulfide nanoparticles on the aquatic plant *Spirodela polyrrhiza*. *J. Nanoparticle Res.* **2014**, *16*, 2774. [CrossRef]

150. Contreras, E.Q.; Cho, M.; Zhu, H.; Puppala, H.L.; Escalera, G.; Zhong, W.; Colvin, V.L. Toxicity of quantum dots and cadmium salt to *Caenorhabditis elegans* after multigenerational exposure. *Environ. Sci. Technol.* **2013**, *47*, 1148–1154. [CrossRef]

151. Petrarca, C.; Perrone, A.; Verna, N.; Verginelli, F.; Ponti, J.; Sabbioni, E.; Di Giampaolo, L.; Dadorante, V.; Schiavone, C.; Boscolo, P.; et al. Cobalt nano-particles modulate cytokine in vitro release by human mononuclear cells mimicking autoimmune disease. *Int. J. Immunopathol. Pharm.* **2006**, *19*, 11–14.

152. Anusha, L.; Devi, C.S.; Sibi, G. Inhibition effects of cobalt nano particles against fresh water algal blooms caused by *Microcystis* and *Oscillatoria*. *Am. J. Appl. Sci. Res.* **2017**, *3*, 26–32. [CrossRef]

153. Ghodake, G.; Seo, Y.D.; Lee, D.S. Hazardous phytotoxic nature of cobalt and zinc oxide nanoparticles assessed using *Allium cepa*. *J. Hazard. Mater.* **2011**, *186*, 952–955. [CrossRef]

154. Rubilar, O.; Rai, M.; Tortella, G.; Diez, M.C.; Seabra, A.B.; Durán, N. Biogenic nanoparticles: Copper, copper oxides, copper sulphides, complex copper nanostructures and their applications. *Biotechnol. Lett.* **2013**, *35*, 1365–1375. [CrossRef]

155. Perreault, F.; Oukarroum, A.; Melegari, S.P.; Matias, W.G.; Popovic, R. Polymer coating of copper oxide nanoparticles increases nanoparticles uptake and toxicity in the green alga *Chlamydomonas reinhardtii*. *Chemosphere* **2012**, *87*, 1388–1394. [CrossRef] [PubMed]

156. Pan, J.F.; Buffet, P.E.; Poirier, L.; Amiard-Triquet, C.; Gilliland, D.; Joubert, Y.; Pilet, P.; Guibbolini, M.; Risso de Faverney, C.; Roméo, M.; et al. Size dependent bioaccumulation and ecotoxicity of gold nanoparticles in an endobenthic invertebrate: The Tellinid clam *Scrobicularia plana*. *Environ. Pollut.* **2012**, *168*, 37–43. [CrossRef]

157. Zhai, G.; Walters, K.S.; Peate, D.W.; Alvarez, P.J.J.; Schnoor, J.L. Transport of gold nanoparticles through plasmodesmata and precipitation of gold ions in woody poplar. *Environ. Sci. Technol. Lett.* **2014**, *1*, 146–151. [CrossRef] [PubMed]

158. Huber, D.L. Synthesis, properties, and applications of iron nanoparticles. *Small* **2005**, *1*, 482–501. [CrossRef] [PubMed]

159. Chen, A.; Holt-Hindle, P. platinum-based nanostructured materials: synthesis, properties, and applications. *Chem. Rev.* **2010**, *110*, 3767–3804. [CrossRef]

160. Kim, J.; Takahashi, M.; Shimizu, T.; Shirasawa, T.; Kajita, M.; Kanayama, A.; Miyamoto, Y. Effects of a potent antioxidant, platinum nanoparticle, on the lifespan of *Caenorhabditis elegans*. *Mech. Ageing Dev.* **2008**, *129*, 322–331. [CrossRef]

161. Prabhu, S.; Poulose, E.K. Silver nanoparticles: Mechanism of antimicrobial action, synthesis, medical applications, and toxicity effects. *Int. Nano Lett.* **2012**, *2*, 1–10. [CrossRef]

162. Macwan, D.P.; Dave, P.N.; Chaturvedi, S. A review on nano-TiO_2 sol–gel type syntheses and its applications. *J. Mater. Sci.* **2011**, *46*, 3669–3686. [CrossRef]

163. Suman, T.Y.; Radhika Rajasree, S.R.; Kirubagaran, R. Evaluation of zinc oxide nanoparticles toxicity on marine algae *Chlorella vulgaris* through flow cytometric, cytotoxicity and oxidative stress analysis. *Ecotoxicol. Environ. Saf.* **2015**, *113*, 23–30. [CrossRef]

164. U.S. Environmental Protection Agency. Analysis Phase: Ecological Effects Characterization. In *Technical Overview of Ecological Risk Assessment*; EPA: Washington, DC, USA, 2017. Available online: https://www.epa.gov/pesticide-science-and-assessing-pesticide-risks/technical-overview-ecological-risk-assessment-0 (accessed on 28 November 2019).

165. United Nations. Part 4 Environmental Hazards. In *Globally Harmonized System of Classification and Labelling of Chemicals (GHS)*, 5th ed.; United Nations: New York, NY, USA; Geneva, Switzerland, 2013; pp. 219–243.

166. Sohn, E.K.; Chung, Y.S.; Johari, S.A.; Kim, T.G.; Kim, J.K.; Lee, J.H.; Lee, Y.H.; Kang, S.W.; Yu, I.J. Acute toxicity comparison of single-walled carbon nanotubes in various freshwater organisms. *BioMed Res. Int.* **2015**, *2015*, 323090. [CrossRef]

167. Sanchís, J.À. Occurrence and Toxicity of Nanomaterials and Nanostructures in the Environment. Ph.D. Thesis, Universitat de Barcelona, Barcelona, Spain, July 2015. Available online: http://www.tdx.cat/handle/10803/305366 (accessed on 7 November 2019).

168. Alam, B.; Philippe, A.; Rosenfeldt, R.R.; Seitz, F.; Dey, S.; Bundschuh, M.; Schaumann, G.E.; Brenner, S.A. Synthesis, characterization, and ecotoxicity of CeO_2 nanoparticles with differing properties. *J. Nanopart. Res.* **2016**, *18*, 303. [CrossRef]

169. Brown, D.M.; Johnston, H.J.; Gaiser, B.; Pinna, N.; Caputo, G.; Culha, M.; Kelestemur, S.; Altunbek, M.; Stone, V.; Roy, J.C.; et al. A cross-species and model comparison of the acute toxicity of nanoparticles used in the pigment and ink industries. *NanoImpact* **2018**, *11*, 20–32. [CrossRef]

170. Modlitbova, P.; Novotny, K.; Porizka, P.; Klus, J.; Lubal, P.; Zlamalova-Gargosova, H.; Kaiser, J. Comparative investigation of toxicity and bioaccumulation of Cd-based quantum dots and Cd salt in freshwater plant *Lemna minor* L. *Ecotox Environ. Saf.* **2018**, *147*, 334–341. [CrossRef] [PubMed]

171. Song, L.; Vijver, M.G.; Peijnenburg, W. Comparative toxicity of copper nanoparticles across three Lemnaceae species. *Sci. Total Environ.* **2015**, *518*, 217–224. [CrossRef] [PubMed]

172. Monteiro, C.; Daniel-da-Silva, A.L.; Venancio, C.; Soares, S.F.; Soares, A.; Trindade, T.; Lopes, I. Effects of long-term exposure to colloidal gold nanorods on freshwater microalgae. *Sci. Total Environ.* **2019**, *682*, 70–79. [CrossRef]

173. Zhang, Y.Q.; Dringen, R.; Petters, C.; Rastedt, W.; Koser, J.; Filser, J.; Stolte, S. Toxicity of dimercaptosuccinate-coated and un-functionalized magnetic iron oxide nanoparticles towards aquatic organisms. *Environ. Sci. Nano* **2016**, *3*, 754–767. [CrossRef]

174. Hlavkova, D.; Beklova, M.; Kopel, P.; Havelkova, B. Evaluation of platinum nanoparticles ecotoxicity using representatives of distinct trophic levels of aquatic biocenosis. *Neuroendocr. Lett.* **2018**, *39*, 465–472.

175. Pereira, S.P.P.; Jesus, F.; Aguiar, S.; de Oliveira, R.; Fernandes, M.; Ranville, J.; Nogueira, A.J.A. Phytotoxicity of silver nanoparticles to *Lemna minor*: Surface coating and exposure period-related effects. *Sci. Total Environ.* **2018**, *618*, 1389–1399. [CrossRef]

176. Picado, A.; Paixao, S.M.; Moita, L.; Silva, L.; Diniz, M.; Lourenco, J.; Peres, I.; Castro, L.; Correia, J.B.; Pereira, J.; et al. A multi-integrated approach on toxity effects of engineered TiO_2 nanoparticles. *Front. Environ. Sci. Eng.* **2015**, *9*, 793–803. [CrossRef]

177. Hartmann, N.B.; Gottardo, S.; Sokull-Klüttgen, B. Part II: Toxicity assessment. In *Review of Available Criteria for Non-Aquatic Organisms within PBT/vPvB Frameworks*; Publications Office of the European Union: Luxembourg, 2014.

178. Lahive, E.; Jurkschat, K.; Shaw, B.J.; Handy, R.D.; Spurgeon, D.J.; Svendsen, C. Toxicity of cerium oxide nanoparticles to the earthworm *Eisenia fetida*: Subtle effects. *Environ. Chem.* **2014**, *11*, 268–278. [CrossRef]

179. Ribeiro, M.J.; Amorim, M.J.B.; Scott-Fordsmand, J.J. Cell *in vitro* testing with soil invertebrates-challenges and opportunities toward modeling the effect of nanomaterials: a surface-modified cuo case study. *Nanomaterials* **2019**, *9*, 9. [CrossRef]

180. Yekeen, T.A.; Azeez, M.A.; Akinboro, A.; Lateef, A.; Asafa, T.B.; Oladipo, I.C.; Oladokun, S.O.; Ajibola, A.A. Safety evaluation of green synthesized *Cola nitida* pod, seed and seed shell extract-mediated silver nanoparticles (AgNPs) using an Allium cepa assay. *J. Taibah Univ. Sci.* **2017**, *11*, 895–909. [CrossRef]

181. Kleiven, M.; Rossbach, L.M.; Gallego-Urrea, J.A.; Brede, D.A.; Oughton, D.H.; Coutris, C. Characterizing the behavior, uptake, and toxicity of NM300K silver nanoparticles in *Caenorhabditis elegans*. *Environ. Toxicol. Chem.* **2018**, *37*, 1799–1810. [CrossRef]

182. Mariyadas, J.; Amorim, M.J.B.; Jensen, J.; Scott-Fordsmand, J.J. Earthworm avoidance of silver nanomaterials over time. *Environ. Pollut* **2018**, *239*, 751–756. [CrossRef] [PubMed]

183. Rocheleau, S.; Arbour, M.; Elias, M.; Sunahara, G.I.; Masson, L. Toxicogenomic effects of nano- and bulk-TiO_2 particles in the soil nematode *Caenorhabditis elegans*. *Nanotoxicology* **2015**, *9*, 502–512. [CrossRef] [PubMed]

184. Garcia-Gomez, C.; Babin, M.; Garcia, S.; Almendros, P.; Perez, R.A.; Fernandez, M.D. Joint effects of zinc oxide nanoparticles and chlorpyrifos on the reproduction and cellular stress responses of the earthworm *Eisenia andrei*. *Sci. Total Environ.* **2019**, *688*, 199–207. [CrossRef] [PubMed]

Review

Exposure Route of TiO$_2$ NPs from Industrial Applications to Wastewater Treatment and Their Impacts on the Agro-Environment

Zahra Zahra [1,*], Zunaira Habib [2], Sujin Chung [3] and Mohsin Ali Badshah [4]

[1] Department of Civil & Environmental Engineering, University of California-Irvine, Irvine, CA 92697, USA
[2] Institute of Environmental Sciences and Engineering, School of Civil and Environmental Engineering, National University of Sciences and Technology, Islamabad 44000, Pakistan; zhabib.phdiese@student.nust.edu.pk
[3] Plamica Labs, Batten Hall, 125 Western Ave, Allston, MA 02163, USA; thesujins@gmail.com
[4] Department of Chemical and Biomolecular Engineering, University of California-Irvine, Irvine, CA 92697, USA; badshahm@uci.edu
* Correspondence: nzahra@uci.edu; Tel.: +1-949-627-5126

Received: 18 June 2020; Accepted: 20 July 2020; Published: 27 July 2020

Abstract: The tremendous increase in the production and consumption of titanium dioxide (TiO$_2$) nanoparticles (NPs) in numerous industrial products and applications has augmented the need to understand their role in wastewater treatment technologies. Likewise, the deleterious effects of wastewater on the environment and natural resources have compelled researchers to find out most suitable, economical and environment friendly approaches for its treatment. In this context, the use of TiO$_2$ NPs as the representative of photocatalytic technology for industrial wastewater treatment is coming to the horizon. For centuries, the use of industrial wastewater to feed agriculture land has been a common practice across the globe and the sewage sludge generated from wastewater treatment plants is also used as fertilizer in agricultural soils. Therefore, it is necessary to be aware of possible exposure pathways of these NPs, especially in the perspective of wastewater treatment and their impacts on the agro-environment. This review highlights the potential exposure route of TiO$_2$ NPs from industrial applications to wastewater treatment and its impacts on the agro-environment. Key elements of the review present the recent developments of TiO$_2$ NPs in two main sectors including wastewater treatment and the agro-environment along with their potential exposure pathways. Furthermore, the direct exposure routes of these NPs from production to end-user consumption until their end phase needs to be studied in detail and optimization of their suitable applications and controlled use to ensure environmental safety.

Keywords: TiO$_2$ NPs; applications; wastewater treatment; agro-environment; exposure pathways

---

1. Introduction

Nanotechnology has touched every field by its scientific novelties. Although the use of nanotechnology is at the early stage, it appears to have significant effects in different areas. It offers great potential for the use of nanomaterials (NMs) in various fields related to all public and industrial sectors, including material, energy, agriculture, healthcare, communication, and information technologies. NMs are the materials that have at least one dimension on the nanoscale [1]. Titanium dioxide with formula TiO$_2$ is the most important binary metal oxide material which exists in three naturally occurring solid phases; anatase (3.2 eV), rutile (3.0 eV), and brookite (3.2 eV) [2]. Anatase is mostly used in photocatalysis and recognized as a major phase of commercial TiO$_2$. Degussa P25 is widely used commercial TiO$_2$ nanoparticles (NPs) with an anatase to rutile phase ratio of 4:1. It has always been a hot topic to attain high crystalline TiO$_2$ NPs with tunable functions by manipulating its morphology. In this context, mesocrystal TiO$_2$ has gained much attention with remarkable photocatalytic activity in several applications [3]. Moreover, TiO$_2$ NPs have unique characteristics of a very high refractive index, whiteness, and opacity, efficiency, increased chemical stability, and minimum cost [4]. TiO$_2$ is widely used as a flocculent, disperser, and whitening agent in the paints and coatings industry. TiO$_2$ along with other colored pigments is used in several end-user products such as emulsion paints, automotive coatings, aircraft coatings, etc. In automotive varnishes, the manufactured good is employed as a dispersive agent in conjunction with the highest gloss retention and elevated chalk resistance. These elements contributing to the automotive sector are supposed to have a definite impact on the global market over the projected period. Overall, according to the global market report, TiO$_2$ market is expected to increase from USD 15,405.5 million in 2017 to USD 20,530.1 million by 2024, with a compound annual growth rate (CAGR) of 4.2% [5]. Annually about 4 million tons of TiO$_2$ produced globally, and about 3000 tons of that process in the nano-scale form [6]. Since 2005, the number of products containing NPs available in commercial markets increased from 54 to 2850 in 2016. According to nano-databases, TiO$_2$ NPs are used in about 25% of the products including paints [7], pigments, cosmetics, food processing and packaging (under E code number E171 as a food colorant), nano-fertilizers or nano-pesticides, biomedicine and clean-energy appliances like solar cells and also as part of pollutant removal from wastewater [8].

TiO$_2$ NPs are one of the most extensively used NPs in different sectors [9]. For example, TiO$_2$ NPs are widely used in the agriculture sector for different purposes such as nano-pesticides and nano-fertilizers to introduce sustainable agricultural practices [10]. The availability of these nano-based agrochemicals in the market is expected to rise in near future [11]. Similarly, the use of TiO$_2$ NPs has also gained the utmost importance in other fields, and eventually from different sources, the inevitable release of these NPs into the environment is obvious either through a direct or indirect route. For example, in 2008, the first evidence of TiO$_2$ NPs leaching (3.5×10^7 NPs per L) into the aquatic environment from facade paints was reported [12]. In 2011, TiO$_2$ NPs were first detected in effluents of wastewater treatment plants, which were discharged into freshwater bodies where these NPs can cause unknown ecological risks [13]. TiO$_2$ NPs have also been observed to detach from some textiles and paints due to washing or weathering and to run into wastewater treatment plants [14,15] and especially in sewage sludge reaching the approximate concentration of 2 g·kg^{-1} [16]. Sewage sludge is commonly employed as soil fertilizer in agriculture at the rate of approximately 3 tons per hectare (on a dry weight basis) annually [17–19], and become an ultimate source of TiO$_2$ NPs dissemination in agricultural soils. However, the overall concentration of these NPs in the environment through direct exposure route will be much higher than the indirect release. Interestingly, in both soil and water medium, TiO$_2$ NPs can be used for purification purposes due to their unique characteristics of photocatalysis in the presence of ultraviolet (UV) light [20,21]. Figure 1 below illustrates the brief overview of TiO$_2$ NPs applications, their role in wastewater treatment and their impacts on agro-environment which we have focused on in this review.

Although, TiO$_2$ NPs offer several benefits, their hidden release into the environment poses potential risks to the entire ecosystem. Increased production and consumption of these NPs indicate their uncontrolled release into the environment, which raises serious environmental concerns that need to be studied [22]. There are three main environmental compartments such as air, water, and soil that provide prospective routes for NPs entrance. Plants also offer a potential route for the transfer of NPs to the environment and ultimately pave the way for their bioaccumulation into the food chain. As the environmental exposure of TiO$_2$ NPs is increasing, humans are also more susceptible to these NPs, and they can easily enter the human body via various routes like oral, inhalation, and dermal contact [23,24]. Common people may be exposed to these NPs via drinking water, food ingestion, medications, and dermal contact with consumer products containing NPs. There are three major possible exposure routes of TiO$_2$ NPs such as occupational, consumer, and environmental exposure. The present review summarized the recent developments of TiO$_2$ NPs in the field of wastewater treatment, and the agro-environment, their environmental impacts along with an overview of their possible exposure routes. As nanotechnology is still in its infancy, so it is timely to consider the potential future problems it could cause before large amounts of NMs/products reach the market, and inevitably reach the environment. In this way, we may prevent undesirable large-scale effects through proactive approaches.

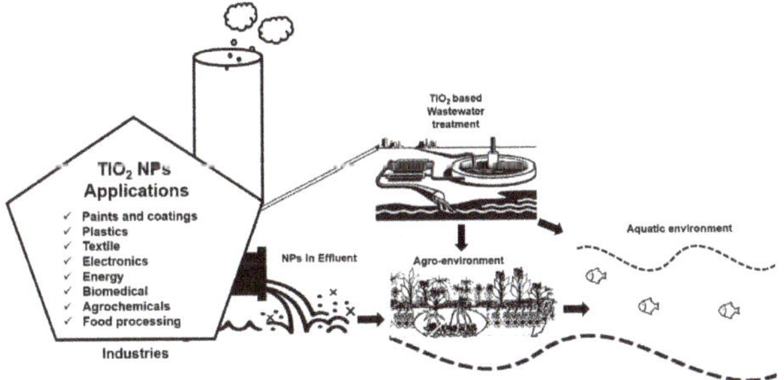

Figure 1. Illustration of the wide range of TiO_2 NPs applications from industries, their release into the wastewater, and their possible exposure routes towards the agro-environment.

2. Wastewater Treatments

With the onset of industrialization, there has been a steady increase in the types and amount of pollutants released in the environment. These environmental problems have garnered much attention on the global scale, especially water scarcity. Global water scarcity is a temporal and graphical mismatch between freshwater resources and the world's water demand. The increasing world population and urban industrialization have made water scarcity more alarming as shown in Figure 2, predicting the gap between supply (4200 billion m^3) and demand (6900 billion m^3) of freshwater in 2030. A major proportion of this water is used for the agriculture sector and then for the industrial sector.

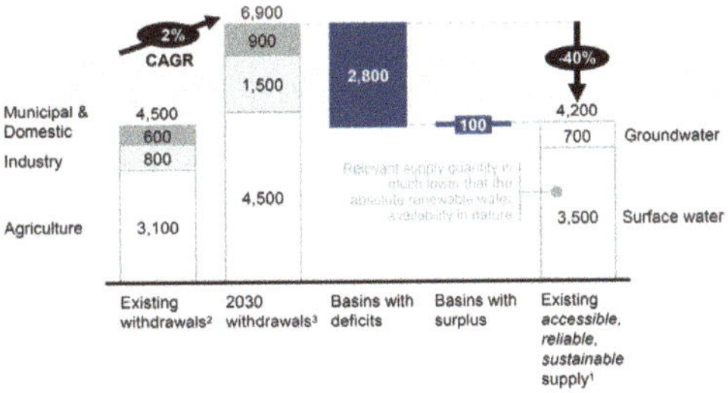

Figure 2. Comparison of current and future water demand, Reproduced with permission from [25], published by McKinsey & Company, New York, NY, USA, 2009.

With a growing world population, an ever-increasing demand for food production and potable water is questionable. The agriculture sector requires a surplus amount of water for irrigation. To avoid water scarcity issues, the reuse of wastewater is tremendously increasing across the planet. Reusing wastewater is a sustainable strategy to manage natural water resources [26]. However, the use of untreated wastewater for irrigation is a usual practice in developing countries causing serious threats to the ecosystem as well as human health. Specifically, carcinogenic pollutants pose a solemn threat to agricultural land, irrigated with industrial effluent without any treatment [27].

The whole world stands as a witness to unintended repercussions caused by rapid industrialization. The wastewater generated from industrial sectors has pronounced effects on humans as well as landmass fertility. Some industrial estates have operational wastewater treatment plants but unfortunately, they cannot handle a large proportion of industrial effluent. To meet the international standards of wastewater discharge, suitable technologies are required for wastewater treatment before discharging to streams. It could help to reduce the burden on freshwater resources by reusing treated water in various industrial processes. Due to the widely used application of nanotechnology, challenges, and opportunities of using engineered nanomaterials (ENMs) in wastewater treatment is a matter of endless concern. Based on the wastewater standards, a technique using TiO_2 NPs for resilient pollutants in the context of wastewater treatment has become popular in recent years. Up to date, TiO_2 NPs have drawn attention over other photocatalysts in every field of life. Over the last few decades, TiO_2 NPs with high photocatalytic efficacy has been tested to reduce the pollution load from various industrial units. The conventional wastewater treatment methods mostly come up with high costs as well as lower efficiencies. However, the advantages of the use of TiO_2 NPs (non-toxic, inexpensive, stable, and reusable NPs) appeared as a promising strategy to save the environment from pollution.

2.1. Slurry-Based Titanium Dioxide (TiO_2) System

To date, a widely used photocatalyst in wastewater treatment is Degussa-P25, a trademark used for commercial TiO_2 NPs. Very fine NPs of P25 TiO_2 have been used in the form of slurry as reference material for comparison of photocatalytic degradation under various conditions. This is because in slurry form, these commercial NPs are always linked with volumetric production of reactive oxygen species (ROS) relative to active surface sites. TiO_2 NPs have also played their role in the treatment of high strength industrial effluent (containing toxic organic and chlorinated compounds) generating from paper and pulp industries. In the pulp industry, the biodegradability index of effluent is 0.02–0.07 during bleaching of pulp which requires further treatment of the biological process for complete removal of these persistent pollutants [28]. Later in a study, where wastewater treatment was carried out using TiO_2 NPs, the biodegradability index increased from 0.16 to 0.35 indicating the use of TiO_2 NPs as an efficient pretreatment process before biological treatment step [29]. Some of the recent applications of TiO_2 NPs in the form of a slurry, for wastewater treatment, are listed in Table 1.

Table 1. TiO_2 NPs applications for photocatalytic degradation of industrial wastewater treatment.

Type of Pollutant	Photo-Catalyst	Experimental Conditions		Light Source	Photocatalytic Activity	Ref.
		Catalyst Dose	Contaminant Conc.			
Dimethyl arsenic acid (DMA)	Mesoporous TiO_2 NPs	0.8 g/L	200 µg/L, 100 mL	300 W Xe-arc lamp	95.12% DMA removal at pH 7.5 and further increase was observed between pH 3–5	[30]
Methylene blue (MB) and Congo red (CR)	TiO_2 NPs	25 mg/mL	15 mL, MB (10 mg/L), CR (20 mg/L)	UV–Vis light (λ = 304–785 nm)	85% MB removal at pH 11.25, 99.7% CR removal at pH 5.40	[31]
Chemical Oxygen Demand (COD) and SO_4^{2-} from oil refinery wastewater	TiO_2 NPs	0.5–1.5 g/L	1 L real refinery effluent	18 W UV lamp (λ = 400 nm)	91.21% of COD and 86% SO_4^{2-} removal after 15 min	[32]
Rhodamine B	Porous TiO_2 NPs	0.100 g/100 mL	400 mg/L	300 W tungsten filament solar lamp	98% degradation within 20 min	[33]
Refinery wastewater	TiO_2 NPs	100 mg/L	150 mL real refinery effluent	6 W low-pressure mercury vapor lamp (λ = 254 nm)	32% Total Organic Carbon (TOC) and 67% Total Nitrogen (TN) after 90 min	[34]
Tannery wastewater	TiO_2 NPs	5 g/L	5 L real tannery effluent	Solar radiations of intensity 985 W/m^2	83% COD and 76% Cr^{+6} after 5 h	[35]
Rhodamine B	TiO_2 NPs	20 g/L	4 mg/L, 400 mL	Visible light (λ ~ 365 nm)	65% Rhodamine B degradation	[36]
1,4-dioxane	Degussa P25-TiO_2	1.24 g/L	(25, 50, 100, 150 and 200 mg L^{-1}), 20 mL	1000 Wm^{-2} Xe lamp (λ = 315–400 nm)	50% COD and 40% TOC after 6 h	[37]
Rhodamine B	Degussa P-25 TiO_2	1.6 g/L	20 mg/L, 25 mL	Blue UV light (λ = 390–410 nm)	96% degradation in 60 min	[38]
Acid Orange 7	Degussa P-25 TiO_2	0.5 g/L	40 mg/L, 800 mL	400 W HP Hg lamp (λ = 253.7 nm)	100% degradation in 120 min	[39]

2.2. TiO$_2$-Based Photocatalytic Reactors

One of the essential aspects after the use of slurry-based TiO$_2$ NPs is regeneration, which is an important concern for the case of economically viable water-treatment technology. The regeneration capability of nanomaterials might be reflected as an additional benefit for their attractiveness in water-treatment technologies. Several techniques have been used to resolve the problem associated with the additional cost of separating NPs from water including immobilization of NMs on adequate substrates and the use of different separation methods. Regeneration of NPs can be achieved efficiently using various photocatalytic reactors. Photocatalytic reactors are classified into two main configurations based on the deployed state of TiO$_2$ NPs: (i) use of NPs in form of suspension and (ii) immobilization of NPs on inert carrier [40]. Downstream separation is required in the first type of configurations as compared to the other one which is a continuous operation. The various types of photoreactors, catalyst employed, and their mode of application are described in Table 2.

Table 2. Overview of various types of reactors, the catalyst employed, and their application in wastewater treatment.

Reactor Type	Reactor Name with Photocatalyst	Target Pollutant Conc.	Findings	Ref.
Suspended	Baffled reactor using Degussa P25-TiO$_2$ NPs	Acid orange 52 (50 mg/L)	Complete mineralization after 30 h at a flow rate of 14.4 L/h	[41]
	Submerged membrane photocatalysis reactor (SMPR) using UV/TiO$_2$	Rhodamine B	95% removal was observed at a catalyst loading of 0.1 g/L under 3 ultraviolet (UV) c lamps at pH 8	[42]
	Slurry photoreactor having mesoporous TiO$_2$ NPs	Dichlorophenol-indophenol (DCPIP) dye (1 to 4×10^{-4} mol/L)	96.4% DCPIP degradation occurred within 3 min at 1×10^{-4} mol/L concentration and pH 3	[43]
	Packed bed photoreactor	Phenazopyridine (10, 20, 30, 40 mg/L)	90% decrease in TOC was observed after 150 min	[44]
	Photocatalytic drum reactor having TiO$_2$ NPs	MB (10 μM) and 4-Chlorophenol (100 μM)	93% MB degradation after 15 min and 94% 4-CP removal after 90 min	[45]
	Rotating drum reactor having Degussa P25-TiO$_2$	MB	98% of MB removal was observed at 30 g/L TiO$_2$ after 60 min	[46]
Immobilized	Thin-film fixed bed reactor having TiO$_2$ NPs	Carmoisine dye (10 mg/L)	97% removal was observed at pH 2 after 45 min at a flow rate of 0.25 L min^{-1}	[47]
	Baffled reactor immobilized with TiO$_2$ NPs	Acid orange 52 (AO52) (50 mg L^{-1})	After 4 h, dye converted into benzene annular compound, intermediates gradually decreased after 10 h and complete mineralization into CO$_2$ and H$_2$O in 30 h	[41]
	Rotating disc photoreactor, TiO$_2$ (P25) immobilized on High Density Polyethylene (HDPE) plate	p-nitrophenol (15 mg L^{-1})	83% removal was observed at pH 5 after 118 min at 800 mL volume	[48]
	Rotating aluminum drum with TiO$_2$-coated corrugated aluminum drum	Tetracycline (0.5, 1, 5, 10, 30, 50, 60, and 80 ppm)	93% Tetracycline was observed after 20 min	[49]
	Spiral photoreactor system sintered with TiO$_2$ thin film	4-tert-octylphenol (4-t-OP) (2, 5, 8 and 10 mg L^{-1})	90% 4-t-OP degradation was observed at 10 mg/L concentration with single layer TiO$_2$ film (13.6% TiO$_2$ precursor)	[50]

2.3. TiO$_2$-Based Electrospun Nanofibers

Nowadays, TiO$_2$ assisted photoreactors have become unfavorable, owing to the proper configuration and artificial light source which is associated with surplus use of electric power as well as treatment costs. The limitations associated with the use of TiO$_2$ NPs in conventional ways paved a path for the synthesis of TiO$_2$-based nanofibers (a one-dimensional form of nanomaterial) by electrospinning. One dimensional nanofibers are superior to NPs owing to intriguing characteristics such as; excellent charge carrier mobility, larger surface area, electrode availability to hole-transporting materials due to pores, improved charge collection as well as transport, and capability to assemble as free stand-alone membrane [51]. They can be synthesized in the form of thin mats and films with a fixed substrate with no need to recover the NPs after treatment.

With the entry of nanotechnology in every field of science, functional NPs can be easily immobilized/impregnated into polymer matrix for avoiding the costly downstream separation step. Furthermore, it also offers an opportunity to fetch priority contaminants that come close to the photocatalytic active sites for efficient utilization of short-lived reactive oxygen species (ROS), commonly known as "bait-hook and destroy strategy" [52]. Polymeric nanofibers can serve as a competent carrier of photocatalytic NPs for efficient industrial wastewater treatment. Photocatalytic degradation of organic contaminants using nanofibers has garnered much attention in recent years.

In the past few decades, several researchers have been devoted to the fabrication and characterization of electrospun TiO_2 nanofibers where the precursor solution (polymeric solution) contains amorphous TiO_2 followed by calcination at 500 °C. After calcination, obtained TiO_2 nanofibers are transformed into crystallized forms (rutile and anatase) for efficient photocatalytic activity. TiO_2 nanofibers were also synthesized using titanium-tetraisopropoxide (TTIP) and tetrabutyl titanate $(Ti(OBu)_4)$ as TiO_2 precursors [53,54]. The commercial-grade TiO_2 NPs Degussa (P25) has been used directly with a polymer blend for the fabrication of TiO_2 nanofibers which did not require a calcination step afterward [55]. TiO_2 NPs can easily be immobilized/supported on polymer nanofibers (either directly in polymer solution or decorated on the surface of nanofibers) with the advantage of efficient recovery after complete mineralization of pollutants [56,57]. Some of the studies on industrial wastewater treatment using TiO_2 nanofibers are summarized in Table 3.

Table 3. Overview of various types of pollutants and TiO_2 based nanofiber photocatalysts employed for wastewater treatment.

Type of Pollutant	Photo-Catalyst	Experimental Conditions		Light Source	Photocatalytic Activity	Ref.
		Catalyst Dose	Contaminant Conc.			
MB	Carbonized TiO_2 nanofibers	2, 4 and 6 mg/40 mL	MB-blue (10 mg/L)	300 W Xenon lamp	At 4 mg dose, 94.98 ± 0.02% degradation was observed after 120 min which decreased up to 83.20 ± 0.01% after 5th cycle	[58]
MB	TiO_2 NPs supported on Polyethylene terephthalate (PET) nanofibers	0.0032 g of TiO_2 adsorbed on 0.011 g of PET nanofibers in 10 mL	MB (10 mg/L)	100 W Xenon lamp	88% degradation after 10 min	[57]
MB, Bisphenol A (BPA) and 17α-ethynylestradiol (EE2)	TiO_2 nanofibers	4×5 cm^2 rectangular coupons/50 mL	MB (6.4 mg/L), BPA and EE2 (C_0 = 5.0 mg/L)	Six UV-A lamps (λ = 365 nm)	97% MB adsorbed in 240 min and degraded completely in less than 90 min, 96% removal for BPA and EE2 within 4 h and 1.5 h, respectively	[56]
MB	Polymethyl methacrylate (PMMA)/TiO_2 nanofibers	3×3 cm^2 rectangular coupons/50 mL	MB (10 mg/L)	8 W UV (λ = 254 nm)	20% degradation after 180 min	[59]
Rhodamine B	TiO_2 nanofibers	0.1 g/100 mL	Rhodamine B (5 mg/L)	500-Watt tungsten halogen lamp (λ ~ 420 nm)	99% of degradation was observed after 2.5 h for nanofibers calcined at 500 °C	[60]
CR	Porous TiO_2 nanofibers after silica leaching	0.5 g/L	CR (20 mg/L)	UV irradiation in a photochemical reactor	76.56 wt% photocatalytic degradation after 1 h	[61]

Heterogeneous photocatalytic degradation using TiO_2 NPs has gained popularity as an effective alternative environment-friendly water treatment approach for a variety of water pollutants including organic and inorganic impurities in industrial effluent. Besides the tremendous use of TiO_2 NPs in wastewater treatment, the inhibitory and biocidal effects of these NPs have been well known. In this context, they have exhibited an excellent broad-spectrum antibacterial activity against various microorganisms especially the pathogenic bacteria [62]. Before discharging industrial effluent into the agro-environment, TiO$_2$ NPs-based treatment technique can be used that could help to completely mineralize water contaminants and eradicate the major concerns of the industrial wastewater treatment. Subsequently, the use of TiO$_2$ NPs is highly anticipated for future studies owing to their effective photocatalytic property, and photo-stability. According to a modeling approach, among the ENMs released from wastewater treatment plants; the concentrations of TiO$_2$ NPs in biosolids constitute about 263–367 mg kg^{-1}, 273–342 mg kg^{-1}, and 70–120 mg kg^{-1} in London, New York City, and Shanghai [63]. The biosolids produced during wastewater treatment are utilized as fertilizers in agriculture [64]. The application of biosolids to agricultural land leads to an increase in the release of TiO$_2$ NPs in the soil [65]. In this context, the potential impacts, fate, and behavior of these NPs need to be investigated in the agro-environment.

3. Impacts of TiO$_2$ Nanoparticles (NPs) in the Agro-Environment

In agro-environment, soil is the main and complex matrix in which analyzing the fate of TiO$_2$ NPs is a challenging task. Furthermore, the impacts of TiO$_2$ NPs are difficult to measure in the soil due to the high geogenic background of Ti ($\approx$0.6% of the terrestrial crust). Up until now, modeling studies had helped to estimate the approximate amount of TiO$_2$ NPs that is accumulating in the environment. According to recent forecasts, TiO$_2$ NPs sludge treated soils (with 45,000 tons) were observed to be the largest sink for NPs release among different environmental compartments [16]. The crop plants served as an entry route for NPs' uptake into the food chain. Presently, there are limited data available about these NPs interactions within the soil matrix. As nanotechnology is emerging in the field of agriculture sector in terms of growing global food production, nutritional contents, quality, food safety, and security [66]. Besides all these aspects, there are several other applications of NPs in agro-environments as shown in Figure 3, such as food processing and production, nano-fertilizer, nano-pesticides, etc. but the important concern arises here is the fate of these NPs.

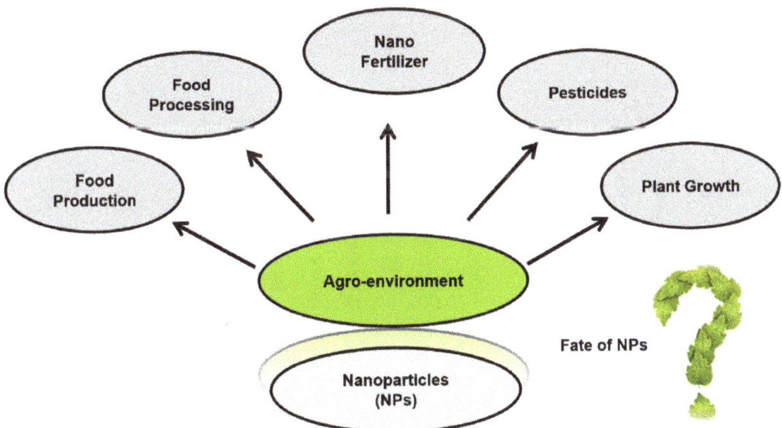

Figure 3. Applications of NPs in agro-environments.

Scientists have investigated the effects of TiO$_2$ NPs on the soil–plant continuum and have observed diverse impacts based on different characteristics of NPs, plant species, experimental conditions, and exposure period. For example, Figure 4, shows the TiO$_2$ NPs effects on plants with respect to different stages, concentration range, and exposure time. In a recent study, experiments were conducted on growth-promoting rhizobacteria (PGPR) inoculation with and without TiO$_2$ NPs in peat soil under the three stress situations. TiO$_2$ NPs were reported to enhance the performance of growth-promoting rhizobacteria which further promotes the solubilization of insoluble phosphates [67]. A grassland soil was treated with TiO_2 NPs at the rate of 0, 500, 1000, and 2000 mg kg^{-1} of soil. These NPs were observed to negatively affect the soil bacterial communities after 60 days of exposure [68]. TiO_2 NPs effects on several bacterial taxa were also studied using incubated soil microcosms having concentrations range of TiO_2 NPs 0, 0.5, 1.0, and 2.0 mg g^{-1} soil. Of the identified taxa that exist in all samples, 9 taxa were found to be positively correlated with TiO_2 NPs, 25 taxa were negatively correlated whereas 135 taxa were not affected by TiO_2 NPs [69]. In another study, TiO_2 NPs effects were investigated at concentrations ranging from 0.05 to 500 mg kg^{-1} dry soil on different bacterial communities. The abundance of ammonia-oxidizing archaea was reported to decrease by 40% in response to TiO_2 NPs whereas *Nitrospira* was not affected at all. Furthermore, the abundance of ammonia-oxidizing bacteria and *Nitrobacter* were also reported to reduce due to TiO_2 NPs treatments [70].

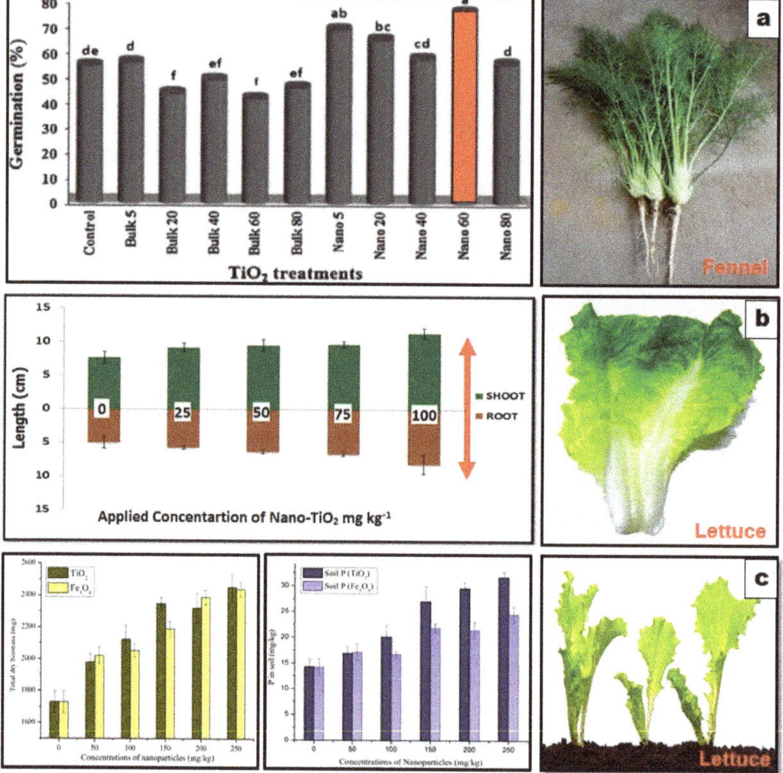

Figure 4. Effects of TiO_2 NPs on plants with respect to different stages, concentration range, and exposure time. (**a**) represents the effects of TiO_2 NPs on germination % of fennel seeds after short term exposure in a petri dish, the lowercase letters show the level of significance such as 'a' represent significant increase in germination percentage at Nano 60 treatment compared to control group. Adapted with permission from [71], published by ELSEVIER, 2013, (**b**) shows the effects of TiO_2 NPs on plant length after short-term exposure in soil Adapted with permission from [72], published by Society for the Advancement of Agricultural Sciences Pakistan, 2015, (**c**) shows the effects of these NPs on lettuce plants after long term exposure of 90 days in soil, Adapted with permission from [73], published by American Chemical Society, 2015.

TiO_2 NPs (0, 5, 20, 40, 60, and 80 mg/kg) were used to study phytotoxicity and stimulatory impacts on fennel after 14 days of exposure. The mean germination percentage was increased by 76% at 60 mg L^{-1}, while the mean germination time was decreased by 31% at 40 mg L^{-1} [71]. Similarly, in another study, plant shoot-root length was increased by 49% and 62%, respectively at 100 mg kg^{-1} of NPs treatment in lettuce after 14 days exposure in soil medium [72]. Another study was performed using TiO$_2$ NPs treatments (0, 50–250 mg kg^{-1}) in soil medium for a period of 90 days. The total dry biomass was observed to increase 1.4-fold and phyto-available phosphorus (P) in soil by 2.2-fold, respectively [73]. Table 4 enlists the recent studies conducted for the investigation of TiO$_2$ NPs effects on different plants.

Table 4. TiO$_2$ NPs applications since 2010 on different plants and their impacts.

Experimental Conditions	Plants	Impacts of TiO$_2$	Ref.
TiO$_2$ NPs Size: 20–30 nm Treatments: 0, 50, 100 and 200 mg L^{-1}) in the growth medium of cocopite and perlite. Period: 60 days	Moldavian balm	Plants cultivated in salt stress conditions were observed to have improved physical traits and increased antioxidant enzyme activity in response to TiO$_2$ NPs treatment compared to control.	[74]
TiO$_2$ NPs Size: 50 and 68 nm Treatments: 100 mg nTiO$_2$/kg on 10 mg kg^{-1} of Cd-spiked soils Period: 14 days	Cowpea	No change in chlorophylls occurred. In leaves and roots, both ascorbate peroxidase and catalase activities were improved by NPs. TiO$_2$ NPs have the potential for soil nano-remediation and could be an environmentally friendly option to tolerate soil Cd toxicity in cowpea plants.	[75]
TiO$_2$ NPs Size: 30 nm Treatments: 0, 30, 50 and 100 mg kg^{-1} Period: 60 days	Wheat	TiO$_2$ NPs without P fertilizer increased Ca (316%), Cu (296%), Al (171%), and Mg (187%) contents in shoots at 50 mg kg^{-1} TiO$_2$ NPs treatment which shows improved grain quality and crop growth.	[76]
TiO$_2$ NPs Treatments: 0, 5, 10, 15, and 20 mg L^{-1} (foliar spray) Medium: Soil Period: 55 days	Rice (*Oryza sativa*)	The foliar spray of TiO$_2$ NPs reduced the soil bioavailable Cd by 10, 14, 28, and 32% in response to 5, 10, 20, and 30 mg/L NPs treatments compared to their control values. These NPs also significantly decreased the Cd concentration in the shoot as well.	[77]
TiO$_2$ NPs Size: <40 nm Treatments: 0, 50, and 100/mg kg^{-1} Medium: Soil Period: 40 days	Wheat (*Triticum aestivum*)	Shoots and root lengths of wheat plants increased by16% and 4%, respectively. Phosphorus in shoots and roots was increased by 23.4% and 17.9% at 50/mg kg^{-1} of soil compare to control.	[78]
TiO$_2$ NPs Size: <40 nm Treatments: 0, 25, 50, 150, 250, 500, 750 and 1000 mg L^{-1} Medium: Soil	Wheat (*Triticum aestivum*)	TiO$_2$ NPs at the highest treatment level of 1000 mg kg^{-1}, plant growth, biomass. Phosphorus content along with other tested parameters did not shown any improvement in the testing soils.	[79]
TiO$_2$ NPs Treatments: 0, 100 and 500 mg kg^{-1} Medium: soil Period: 60 days	Wheat (*Triticum aestivum*)	No effect of phytotoxicity was observed in plant growth, chlorophyll content, and biomass.	[80]
TiO$_2$ NPs Treatments: 0–750 mg kg^{-1} Medium: Soil Period: 90 days	Rice (*Oryza sativa*)	Phosphorus concentration was increased in roots by 2.6-fold, shoots 2.4-fold, and grains 1.3-fold upon 750 mg kg^{-1} of NPs treatment. Metabolomics study revealed that levels of amino acids, glycerol content, and palmitic acid were also improved in grains.	[81]
TiO$_2$ NPs Treatments: 0, 100, 150, 200, 400, 600, and 1000 mg L^{-1} Medium: Hydroponics Period: 7 days	Barley (*Hordeum vulgare* L.)	No adverse effect on shoot growth. Root growth inhibited as the concentration of TiO$_2$ NPs increases. No effect on chlorophyll *a* and *b*. No significant effect on biomass.	[82]
TiO$_2$ NPs Treatments: 0–100 mg kg^{-1} Medium: Soil Period: 60 days	Wheat (*Triticum aestivum*)	NPs treatment at the rate of 20, 40, and 60 mg kg^{-1} increased plant growth and phosphorus uptake. 32.3% of chlorophyll content increased at 60 mg kg^{-1} while 11.1% decrease at 100 mg kg^{-1}.	[83]
TiO$_2$ NPs Size: >20 nm Treatments: 0, 100, 250, 500 and 1000 mg L^{-1} Medium: Soil Period: 5 weeks	Arabidopsis thaliana (*L.*)	Plant biomass and chlorophyll content decreased as the NPs treatment increase. Higher concentrations of NPs improved root growth. NPs treatments from 100 to 1000 μg mL^{-1} affect vitamin E content in plants. Decrease in plant biomass by 3-fold in response to 500 and 1000 mg/mL NPs treatment, whereas, at 100 mg/mL, the biomass decreases to half relative to control.	[84]
TiO$_2$ NPs Treatments: 250 and 500 μg/mL	Cabbage, Cucumber, Onion	The germination of cabbage significantly increased. In cucumber and onion, significant root elongation was observed.	[85]

Table 4. *Cont.*

Experimental Conditions	Plants	Impacts of TiO$_2$	Ref.
TiO$_2$ NPs P25: 29 ± 9 nm, E171: 92 ± 31nm, Non-nanomaterial TiO$_2$: 145 ± 46 nm Treatments: 1, 10, 100, 1000 mg kg^{-1} Period: 12 weeks	Wheat, Red clover	TiO$_2$ NPs showed restricted mobility from soil to leachate. No significant translocation of Ti was observed in both plant species, while average Ti content increased from 4 to 8 mg kg^{-1} at the highest treatments.	[86]
TiO$_2$ NPs Size: 22 and 25 nm Period: 6 weeks	Soya bean	Plant growth significantly decreased which corresponds to the reduced carbon content in leaves.	[87]
TiO$_2$ NPs Treatments: 0, 10, 20, 40 and 80 mg L^{-1} Medium: Petri dish Period: 10 days	*Alyssum homolocarpum, Salvia mirzayanii, Carum copticum, Sinapis alba,* and *Nigella sativa*	TiO$_2$ NPs affected the germination and seedling vigor of 5 medicinal plants. Appropriate concentration levels had improved the germination as well as the vigor index of the subjected plant.	[88]
TiO$_2$ NPs Treatments: 0, 10, 20, 30, and 40 mg mL^{-1}	Parsley	Significant increase in seedlings germination percentage, germination rate index, shoot-root length, fresh biomass, vigor index, and chlorophyll content. 30 mg mL^{-1} was observed to be the optimum concentration of NPs. Increased germination percentage (92.46%) was observed at 40 mg mL^{-1} treatment, relative to the lowest one (44.97%) at control.	[89]
TiO$_2$ NPs Treatments: 0, 0.01%, 0.02%, and 0.03% Medium: Soil Period: 14 days	Wheat (*Triticum aestivum*)	Under the water-stressed conditions, the plant's length, biomass, and seed number along with the other tested traits like gluten and starch content were increased at 0.02% of NPs treatment.	[90]
TiO$_2$ NPs Size: 14–655 nm	Wheat (*Triticum aestivum*)	NPs treatment improved root length. NPs above 140 nm diameter are not accumulated in wheat roots. NPs above 36 nm threshold diameter, can be accumulated (at concentration 109 mg Ti/kg dry weight) in wheat root parenchyma cells but are unable to translocate to the shoot. Enhanced wheat root elongation was observed when exposed to 14 and 22 nm TiO$_2$ NPs.	[91]
TiO$_2$ NPs Size: 5 nm Treatments: 0.25% NPs Medium: Hoagland nutritive fluid Period: 35 days	*Arabidopsis thaliana*	Improved photosynthesis and growth in plants were reported. Generally, the absorption of light in chloroplast and light-harvesting complex II was supposed to be stimulated by TiO$_2$ NPs; thus, enhancing the transformation of light energy to electronic energy, the evolution of oxygen, and water photolysis.	[92]
TiO$_2$ NPs (43%) with sucrose coating Size: >5 nm	*Arabidopsis thaliana*	Results revealed that small NPs entered plant cells and got accumulated in distinct subcellular locations.	[93]
TiO$_2$ NPs Size: <100 nm Treatments: 0, 5, 10 and 20 mg L^{-1} Period: 20 days	*Zea mays* L.	TiO$_2$ NPs treatment significantly reduced the shoot, root biomass, and chlorophyll contents of leaves in a dose-dependent manner. Whereas positive effects were reported on the N, P, K, Zn Mn, and Cu contents except for Fe.	[94]
TiO$_2$ NPs Size: <100 nm Treatments: 15, 30, 60, 120 and 240 mg L^{-1} Period: at different time intervals up to a maximum of 82 days	*Vicia faba*	TiO$_2$ NPs were reported to induce variations in a meiotic activity which results in an increased number of chromosomal abnormalities in the plant's reproductive parts.	[95]
TiO$_2$ NPs Size: <100 nm (tetragonal crystals), <10 nm (spherical shape) Treatments: 50 mg L^{-1} Period: 3 days	*Vicia faba* L.	Based on the characteristics of size and shape, TiO$_2$ NPs can induce different levels of toxicity in terms of seed vigor index, aberration index and oxidative stress in plants.	[96]

Studies have shown the positive effects of TiO$_2$ NPs on the physiology of red bean plants, leaving no negative biochemical impacts in plants [97]. Low concentrations of TiO$_2$ NPs were reported with their positive effects on chickpea cells especially when they were exposed to cold stress. However, TiO$_2$ NPs especially at 5 mg kg^{-1} concentration level was reported to reduce cold-induced damages in sensitive and resistant chickpea genotypes. Such domino effects raised key questions regarding the potential mechanisms. It was supposed that the activation of the defensive mechanisms in chickpea seedlings after the absorption of TiO$_2$ NPs support the plants in cold stress. These results are quite interesting for further practice in cases of environmentally stressed conditions. These new findings could pave the way to increase the use of NPs especially to improve the cold stress tolerance in major crops [98]. Furthermore, in future studies, TiO$_2$ NPs application in combination with fertilizers could be an effective option to search out a way for better application of these agrochemicals in a sustainable way. We further need to explore the potential of nanotechnology by upscaling the present studies by investigating the effects of NPs at different stages in the life cycle of plant species and understand their mechanism of environmental exposure.

4. Understanding the Mechanism of Environmental Exposure of NPs

The increased use of NPs in different fields has raised a worldwide concern regarding their release and impact on human health and the environment. For this reason, in the recent decade, toxicological effects of NPs on human health and the environment also gained attention. The potential for exposure to these NPs begins with the production of these materials until their associated life cycle completion and release into the air, soil, and water [99] as shown in Figure 5.

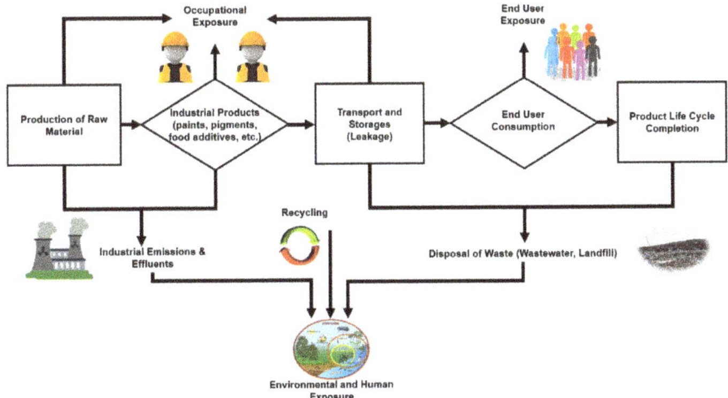

Figure 5. Probable routes of human and environmental exposure.

Among possible exposure routes of TiO$_2$ NPs, there are three major exposure routes including occupational exposure, consumer exposure and environmental exposure.

4.1. Occupational Exposure via Industries

According to a Swiss survey report, the usage of TiO$_2$ NPs increased in amounts of approximately more than 1000 kg per company annually [100]. Occupational exposure to NPs may occur through dermal contact and dust inhalation at workplaces or industries; where these NPs are used or manufactured. The National Institute for Occupational Safety and Health (NIOSH) reported that the workers over the industrial units are at high risk of exposure to NPs due to unintentional hand-to-mouth touch [101]. Usually, the materials at microscale levels are considered to be harmless, however recent studies suggested that frequent inhalation of NPs could be dangerous [102]. The impact of NPs on humans has been investigated using various rodent models through various exposure routes and conditions. For example, inhalation of TiO$_2$ NPs was reported to cause lung damage in mice due to inflammation, pulmonary fibrosis, and initiation of lung tumors [103]. In the human body, the liver is the most susceptible organ targeted by NPs [104]. TiO$_2$ NPs have been reported to induce toxic effects on the liver affecting its functions [105]. In China, a study regarding occupational exposure to TiO$_2$ NPs was conducted in a packaging workshop. Workers were selected on the basis of age (20 years or more) and employment (at least one year). Cardiopulmonary effects through possible biomarkers and physical experiments were conducted to reveal TiO$_2$ NPs exposure. A pattern having time (dose)–response was observed in exposed workers, suggesting that long-term exposure to TiO$_2$ NPs cause serious threats through occupational exposure [106].

TiO$_2$ NPs have the ability to generate reactive oxidative species and oxidative stress even at lower NPs concentration. In a recent study, an acute exposure of TiO$_2$ NPs to human lungs resulted in substantial modifications in gene expression along with long-term effects on progeny cells even after multiple generations via transcriptional changes [107]. Similarly, genotoxicity and cytotoxicity of TiO_2 NPs (having different shapes) in bronchial epithelial cells were studied. Genotoxicity was determined on the basis of cellular-uptake as well as the ability of NPs to aggregate, whereas lesser cytotoxicity of NPs was observed to be significantly influenced by irradiation time and the shape of TiO_2 NPs [108]. Another study reported that in case of acute exposure conditions, TiO_2 NPs did not cause cytotoxicity in human alveolar A549 cells [109], but there is a lack of information related to the magnitude of NPs released and exposed to organisms, which can be better interpreted in future studies [110].

4.2. Consumer Exposure

Consumer products available in markets including i.e., cosmetics, beverages and food, appliances, health and fitness, gardens and homes, etc., contain NPs and are expected to have direct consumer exposure [111]. TiO_2 NPs having different sizes were used to investigate their effects on a human keratinocyte cell line (HaCaT) and reported that all the tested types of TiO_2 NPs increased the superoxide production, and apoptosis in a dose-dependent manner [112]. Another study reported that TiO_2 NPs at the half maximal effective concentration (EC_{50}) ranges from 10^{-4} to 10^{-5} mol L^{-1} induced cytotoxic effects on HaCaT cells [113]. Recently, in another study the authors investigated the TiO_2 NPs in combination with the ingredients from modern lifestyle products like cosmetics, skin-care products, and Henna tattoos. TiO_2 NPs alone were not reported to induce any damage in cell viability upon application of 100 µg mL^{-1} up to 24 h [114]. Recently a social survey was conducted in USA regarding the individual exposure to TiO_2 NPs used in personal care products. From these results, toothpaste and sunscreen were considered as the major source of dermal exposure depending on their usage pattern and amount of TiO_2 NPs in these products. It is estimated that a person can exposed to 2.8 to 21.4 mg TiO_2 per day through dermal exposure. Per day oral exposure is estimated from 0.15 to 3.9 mg TiO_2 via toothpaste [115].

Over the course of history, TiO_2 has been considered to pose low toxicity both for humans and the environment. Since ancient times, it has been the most widely used material as a coloring agent [116]. However, in 2018 the French national assembly revised the guidelines with the amendment to ban the use, import, and sale of nano-scale TiO_2 as a food additive in any kind of food by 2020 [117]. Because of the fact that limited information is available on the safe usage of these NPs in consumer products, their potential hazards for their users need to be assessed. There are several factors involved in the assessment of the consumer exposure to these NPs that constrained due to the limited access of information; (a) list of commercial products containing NPs, (b) amount of NPs used in such products, and (c) behavior of the consumer towards them [118]. Most of the commercial products containing these NPs do not enlist this information on their ingredient lists. Moreover, the number of consumers of such products and industry-derived data are kept hidden from all stakeholders including the governments, public, and private sectors which makes the consumer exposure situation more alarming [119].

4.3. Environmental Exposure

The term environmental exposure is based on the extent of NPs taken up by biota, either in metabolized or degraded form, and their rates of excretion. This is where the least amount of data is available, and particularly data that consider the modifying effects of the environments where organisms live while they are exposed to NPs. In the product life cycle starting from manufacturing until consumer usage, each stage for NPs could result in their release into the environment. The tendency for physicochemical properties varies as these NPs move from different environmental compartments such as water, soil, and air. Understanding the importance, their fate, transport, and transformation need to be emphasized. However, little is known about what governs these processes for NPs in general. So, we tried to summarize what is known about the environmental behavior of these NPs which is as follows.

Upon release into the environment, NPs usually act in one or more of the following ways: (1) stay suspended as an individual particle; (2) form agglomerates (and potentially sorbed onto some surface or experience facilitated transport); (3) dissolve in a liquid; and (4) transform chemically by reacting with organic matter or other natural particles. The extent to which NPs' behavior follows any of the aforementioned patterns depends on their surrounding environment, and several biological, chemical, and physical processes. Nanosize TiO_2 made them extremely mobile in the soil system, but their larger surface areas (compared to their size) enhance their tendency to sorb onto the soil, which restricts their movement or makes them immobile. For example, TiO_2 NPs considered as having low solubility, remained in the soil for long periods which might create potential environmental risks for deeper soil layers. Small-sized TiO_2 NPs (20 nm) were able to penetrate the plant cell wall and have been reported to reduce wheat's biomass [120]. Plants offered a potential route for the transmission of NPs to the environment and ultimately paved the way for their bioaccumulation into the food chain. Different studies have determined the response of NPs to plants growth and their possible mechanism. Plant cell walls do not allow the smooth entrance of any external agent as well as NPs into the plant cells. The screening property of the cell wall depends on the diameter of pores present in the cell wall that mostly ranges from 5 to 20 nm [121]. Therefore, NPs and their accumulates within the stated range could simply cross the cell membrane and transfer to the plant's aerial parts. NPs might generate various morphological changes in the root structures, which increases pore sizes or generates new pores in the cell wall, which further enhances the uptake of NPs and their aggregates [122]. A recent study reported that TiO_2 NPs from the environment undergoes the size selection process during the foliar and root uptake mechanism in *Dittrichia viscosa* wild plants. The study reported that the TiO_2 NPs having a size less than 50 nm were accumulated in plant's leaves (53%), stems (90%), and roots (88.5%) [123]. In another study, TiO_2 NPs of size 4 and 150 nm were reported to be internalized through foliar uptake in lettuce plant leaves via stomata [124]. Another report stated that NPs accumulated on photosynthetic surface-induced foliar heating that can alter the gaseous exchange due to stomatal disturbance. Consequently, altering the different molecular and physiological functions of plants [125]. Therefore, the influence and translocation of different NPs within plants need to be investigated further to understand the whole mechanism and their behavior in plants [126]. As the human food chain instigated with plants, so it is critically important to understand how plants respond differently to these NPs which are frequently concentrating in our ecosystem through various routes.

Aquatic systems are usually considered as the main recipient of NPs. As in the terrestrial environment, transformation in the aquatic system includes several, physical (aggregation/agglomeration and deposition), biological (interaction with macromolecules including polysaccharides, proteins, and surfactants) and chemical processes (dissolution, sorption, and redox reactions). Aside from the intrinsic properties, transformation, and toxicity of TiO_2 NPs also rely on various environmental factors such as temperature, pH, light, and presence of natural organic matter [127]. In natural aquatic systems, many organisms are sensitive to NPs' exposure and exhibit pronounced toxic effects during their transport and transformation. This might be because NPs have surface coatings that help to improve their solubility and suspension and made them more mobile than the other large-sized particles. TiO_2 NPs have been reported to induce a significant decrease in growth parameters of an aquatic plant *Spirodela polyrrhiza*, whereas the increased concentration of TiO_2 NPs was observed to increase the photosynthetic pigmentation and the peroxidase activity [128]. Overall, the negative effects of nanoparticles must not be ignored, especially on human health and the environment, and must be studied in detail to make their use controlled and safe. Table 5 below briefly enlists some studies of TiO_2 NPs on terrestrial and aquatic organisms.

Table 5. TiO$_2$ NPs effects terrestrial and aquatic organisms.

Experimental Conditions	Organisms	Impacts of TiO$_2$	Ref.
Terrestrial Organisms			
TiO$_2$ NPs Size: 25 nm Treatments: 500 and 5000 mg kg^{-1} Period: up to 48 days	Nematodes (*C. elegans*)	Increased generation of intracellular reactive oxygen species. Toxicity reduced and the lifespan of survived nematodes increased in response to TiO$_2$ NPs exposure.	[129]
TiO$_2$ NPs Treatments: 0, 5, 50, and 500 mg kg^{-1} Period: 120 days	Earthworm (*Eisenia fetida*)	Lower glutathione/oxidized glutathione (GSH/GSSG) ratio and significant decrease in superoxide dismutase (SOD) activity was observed for 500 mg/kg TiO$_2$ concentration.	[130]
TiO$_2$ NPs Size: 50–100 nm Treatments: 0, 150 or 300 mg kg^{-1} of dry soil Period: 15, 30, 60 and 90 days	Bacterial community and *Eisenia fetida*	Unamended and earthworm—amended soil increased certain available bacterial groups such as *Firmicutes* and *Acetobacter* whereas decreased *Verrucomicrobia* and *Pedobacter* abundance.	[131]
TiO$_2$ NPs (anatase) Treatments: 10, 50, and 100 nm Period: 2–3 months	Mice	Intestinal inflammation with lower body weight. Mice with removed gut microbiota did not show this phenomenon.	[132]
TiO$_2$ NPs Size: 23 ± 6.8 nm Treatments: 0.5, 2.5, and 10 mg kg^{-1} Period: 2 h and 35 days	Sprague–Dawley rats	Persistent inflammation of lung and liver genotoxicity.	[133]
Aquatic Organisms			
TiO$_2$ NPs Treatments: 0.25, 0.5, and 1.0 mg L^{-1} Period: 21 days	*Daphnia magna*	TiO$_2$ with 20% rutile and 80% anatase had a highest mortality rate as compared to other crystalline forms.	[134]
Biosynthesized TiO$_2$ NPs Size: 43–56 nm Treatments: 0, 2.5, 5, 10, 20, and 40 mg L^{-1}	Zebrafish (*Danio rerio*)	Significant malformations such as tail curvature, egg coagulation, bend the spine and delayed hatching was observed at a concentration of 2.5 mg L^{-1} during 8 to 120 h post fertilized period.	[135]
TiO$_2$ NPs Treatments: 25, 125, and 250/mg L^{-1} Period: 28/days	Red swamp crayfish (*Procambarus clarkia*)	The mortality rate was observed to be 0, 3.3, and 10% in response to 25, 125, and 250/mg L^{-1} of TiO$_2$ NPs, respectively.	[136]
TiO$_2$ NPs Treatments: 1.0 and 5.0 mg L^{-1} Period: 4 and 14 days	Nile tilapia (*Oreochromis niloticus*)	Acute exposure caused oxidative stress with a decrease in catalase (60%), superoxide dismutase (27%), and glutathione peroxidase (37%), while 14 days of exposure elevated the catalase (61%), glutathione-S-transferase (54%), glutathione peroxidase (32%), and glutathione reductase (93%).	[137]
TiO$_2$ NPs Treatments: 500, 1000, 1500, and 2000 mg L^{-1} Period: 24 h	Brine shrimp (*Artemia salina*)	Mortality rate of 5, 20, 20, 53, and 57% was observed in response to 0, 500, 1000, 1500, and 2000 mg L^{-1} TiO$_2$ NPs, respectively.	[138]

5. Conclusions and Future Perspectives

This review briefly discussed the recent developments of TiO$_2$ NPs in wastewater treatment technologies and the agro-environment. The potential exposure pathways of these NPs and their associated environmental risks were also highlighted. In fact, the use of TiO$_2$ NPs will further increase for promising applications in the near future. In wastewater treatment technologies, downstream separation of these NPs after photocatalytic degradation is still a matter of concern which can be minimized by using TiO$_2$ in photocatalytic reactors either in slurry form or immobilized on a solid substrate. Immobilization might result in loss of potential active sites which could be minimized by adding NMs into the polymeric substrate. The polymer can provide firm anchoring to TiO$_2$ NPs, however, there is still a chance of NPs leaching into the treated water and reaching the agricultural soils via irrigation. Since the agriculture sector is the backbone of the economy in most countries, studies based on crop improvement using TiO$_2$ NPs could help to overcome the burden of nutrient deficit in soils providing better crop yield. Apart from the potential benefits of TiO$_2$ NPs there are also some limitations that we could not ignore. At this stage, we could not claim with surety that the use of NPs is fully safe for human health and the environment or if it is harmful. Risks associated with chronic exposure of these NPs, interaction with flora and fauna, and their possible bioaccumulation effects have not been fully considered yet. The other limitations include the lack of information about a safe range of NPs' concentration, scalability of research and development for prototypes, industrial production, and public concern about health and safety issues. Detailed investigations are necessarily required to resolve these concerns and provide conclusive statements. We need to optimize the useful concentration levels of TiO_2 NPs for various applications and limit their usage for environmental safety.

Author Contributions: Z.Z.: Conceptualization, Methodology, Data curation, Supervision, Writing–original draft. Z.H.: Investigation, Data curation, Writing—original draft. S.C.: Investigation, Resources, Data curation. M.A.B.: Data curation, Methodology, Writing, and Proofreading. All authors have read and agreed to the published version of the manuscript.

Funding: This research received no external funding.

Acknowledgments: We thank Maryam Naqvi from The University of California Irvine, for her helpful assistance.

Conflicts of Interest: The authors declare no conflict of interest.

References

1. Holister, P.; Weener, J.-W.; Vas, C.R.; Harper, T. Nanoparticles. *Technol. White Pap.* **2003**, *3*, 1–11.
2. Siroha, P.; Singh, D.; Soni, R.; Gangwar, J. Comparative study on crystallographic representation of transition metal oxides polymorphs nanomaterials using VESTA software: Case study on Fe_2O_3 and TiO_2. In *Proceedings of the AIP Conference*; American Institute of Physics Inc.: College Park, MA, USA, 2018; Volume 2006, p. 030038.
3. Zhang, B.; Cao, S.; Du, M.; Ye, X.; Wang, Y.; Ye, J. Titanium dioxide (TiO_2) mesocrystals: Synthesis, growth mechanisms and photocatalytic properties. *Catalysts* **2019**, *9*, 91. [CrossRef]
4. Rajh, T.; Dimitrijevic, N.M.; Bissonnette, M.; Koritarov, T.; Konda, V. Titanium dioxide in the service of the biomedical revolution. *Chem. Rev.* **2014**, *114*, 10177–10216. [CrossRef] [PubMed]
5. Walia, K. *Global Titanium Dioxide Market Analysis Report 2018–2025*; Value Market Research: Pune, India, 2019.
6. Xu, Y.; Wei, M.T.; Ou-Yang, H.D.; Walker, S.G.; Wang, H.Z.; Gordon, C.R.; Guterman, S.; Zawacki, E.; Applebaum, E.; Brink, P.R.; et al. Exposure to TiO_2 nanoparticles increases Staphylococcus aureus infection of HeLa cells. *J. Nanobiotechnol.* **2016**, *14*, 34. [CrossRef] [PubMed]
7. Consumer Products—The Nanodatabase. Available online: http://nanodb.dk/en/analysis/consumer-products#chartHashsection (accessed on 7 June 2020).
8. Tan, W.; Peralta-Videa, J.R.; Gardea-Torresdey, J.L. Interaction of titanium dioxide nanoparticles with soil components and plants: Current knowledge and future research needs—A critical review. *Environ. Sci. Nano* **2018**, *5*, 257–278. [CrossRef]
9. Hou, J.; Wang, L.; Wang, C.; Zhang, S.; Liu, H.; Li, S.; Wang, X. Toxicity and mechanisms of action of titanium dioxide nanoparticles in living organisms. *J. Environ. Sci.* **2019**, *75*, 40–53. [CrossRef]
10. Kalpana, S.R.; Rashmi, H.B.; Rao, N.H. Nanotechnology Patents as R&D Indicators for Disease Management Strategies in Agriculture. *J. Intellect. Prop. Rights* **2010**, *15*, 197–205.
11. Gogos, A.; Knauer, K.; Bucheli, T.D. Nanomaterials in plant protection and fertilization: Current state, foreseen applications, and research priorities. *J. Agric. Food Chem.* **2012**, *60*, 9781–9792. [CrossRef]
12. Kaegi, R.; Ulrich, A.; Sinnet, B.; Vonbank, R.; Wichser, A.; Zuleeg, S.; Simmler, H.; Brunner, S.; Vonmont, H.; Burkhardt, M.; et al. Synthetic TiO_2 nanoparticle emission from exterior facades into the aquatic environment. *Environ. Pollut.* **2008**, *156*, 233–239. [CrossRef]
13. Westerhoff, P.; Song, G.; Hristovski, K.; Kiser, M.A. Occurrence and removal of titanium at full scale wastewater treatment plants: Implications for TiO_2 nanomaterials. *J. Environ. Monit.* **2011**, *13*, 1195–1203. [CrossRef]
14. Mackevica, A.; Foss Hansen, S. Release of nanomaterials from solid nanocomposites and consumer exposure assessment—A forward-looking review. *Nanotoxicology* **2016**, *10*, 641–653. [CrossRef] [PubMed]
15. Windler, L.; Lorenz, C.; von Goetz, N.; Hungerbühler, K.; Amberg, M.; Heuberger, M.; Nowack, B. Release of Titanium Dioxide from Textiles during Washing. *Environ. Sci. Technol.* **2012**, *46*, 8181–8188. [CrossRef] [PubMed]
16. Sun, T.Y.; Bornhöft, N.A.; Hungerbühler, K.; Nowack, B. Dynamic Probabilistic Modeling of Environmental Emissions of Engineered Nanomaterials. *Environ. Sci. Technol.* **2016**, *50*, 4701–4711. [CrossRef] [PubMed]
17. Gottschalk, F.; Sun, T.; Nowack, B. Environmental concentrations of engineered nanomaterials: Review of modeling and analytical studies. *Environ. Pollut.* **2013**, *181*, 287–300. [CrossRef]

18. Kim, B.; Murayama, M.; Colman, B.P.; Hochella, M.F. Characterization and environmental implications of nano- and larger TiO$_2$ particles in sewage sludge, and soils amended with sewage sludge. *J. Environ. Monit.* **2012**, *14*, 1129. [CrossRef]

19. Sharma, B.; Sarkar, A.; Singh, P.; Singh, R.P. Agricultural utilization of biosolids: A review on potential effects on soil and plant grown. *Waste Manag.* **2017**, *64*, 117–132. [CrossRef]

20. Wu, M.; Deng, J.; Li, J.; Li, Y.; Li, J.; Xu, H. Simultaneous biological-photocatalytic treatment with strain CDS-8 and TiO$_2$ for chlorothalonil removal from liquid and soil. *J. Hazard. Mater.* **2016**, *320*, 612–619. [CrossRef]

21. Zimbone, M.; Cacciato, G.; Boutinguiza, M.; Privitera, V.; Grimaldi, M.G. Laser irradiation in water for the novel, scalable synthesis of black TiOx photocatalyst for environmental remediation. *Beilstein J. Nanotechnol.* **2017**, *8*, 196–202. [CrossRef]

22. Staroń, A.; Dlugosz, O.; Pulit-Prociak, J.; Banach, M. Analysis of the exposure of organisms to the action of nanomaterials. *Materials* **2020**, *13*, 349. [CrossRef]

23. Geraets, L.; Oomen, A.G.; Krystek, P.; Jacobsen, N.R.; Wallin, H.; Laurentie, M.; Verharen, H.W.; Brandon, E.F.A.; de Jong, W.H. Tissue distribution and elimination after oral and intravenous administration of different titanium dioxide nanoparticles in rats. *Part Fibre Toxicol.* **2014**, *11*, 30. [CrossRef]

24. Zhang, R.; Bai, Y.; Zhang, B.; Chen, L.; Yan, B. The potential health risk of titania nanoparticles. *J. Hazard. Mater.* **2012**, *211–212*, 404–413. [CrossRef] [PubMed]

25. Addams, L.; Boccaletti, G.; Kerlin, M.; Stuchtey, M. *Charting Our Water Future: Economic Frameworks to Inform Decision-Making: 2030 Water Resources Group*; McKinsey & Company: New York, NY, USA, 2009.

26. Analouei, R.; Taheriyoun, M.; Safavi, H.R. Risk assessment of an industrial wastewater treatment and reclamation plant using the bow-tie method. *Environ. Monit. Assess.* **2020**, *192*, 1–16. [CrossRef] [PubMed]

27. Cao, H.; Chen, J.; Zhang, J.; Zhang, H.; Qiao, L.; Men, Y. Heavy metals in rice and garden vegetables and their potential health risks to inhabitants in the vicinity of an industrial zone in Jiangsu, China. *J. Environ. Sci.* **2010**, *22*, 1792–1799. [CrossRef]

28. Zhang, T.; Wang, X.; Zhang, X. Recent progress in TiO$_2$-mediated solar photocatalysis for industrial wastewater treatment. *Int. J. Photoenergy* **2014**, *2014*. [CrossRef]

29. Ghaly, M.Y.; Jamil, T.S.; El-Seesy, I.E.; Souaya, E.R.; Nasr, R.A. Treatment of highly polluted paper mill wastewater by solar photocatalytic oxidation with synthesized nano TiO$_2$. *Chem. Eng. J.* **2011**, *168*, 446–454. [CrossRef]

30. Dong, J.; Hu, C.; Qi, W.; An, X.; Liu, H.; Qu, J. Defect-enhanced photocatalytic removal of dimethylarsinic acid over mixed-phase mesoporous TiO$_2$. *J. Environ. Sci.* **2020**, *91*, 35–42. [CrossRef]

31. Sathiyan, K.; Bar-Ziv, R.; Mendelson, O.; Zidki, T. Controllable synthesis of TiO$_2$ nanoparticles and their photocatalytic activity in dye degradation. *Mater. Res. Bull.* **2020**, *126*, 110842. [CrossRef]

32. Tetteh, E.K.; Obotey Ezugbe, E.; Rathilal, S.; Asante-Sackey, D. Removal of COD and SO$_4^{2-}$ from Oil Refinery Wastewater Using a Photo-Catalytic System—Comparing TiO$_2$ and Zeolite Efficiencies. *Water* **2020**, *12*, 214. [CrossRef]

33. Hiremath, V.; Deonikar, V.G.; Kim, H.; Seo, J.G. Hierarchically assembled porous TiO$_2$ nanoparticles with enhanced photocatalytic activity towards Rhodamine-B degradation. *Colloids Surf. A Physicochem. Eng. Asp.* **2020**, *586*, 124199. [CrossRef]

34. Oliveira, C.P.M.D.; Viana, M.M.; Amaral, M.C.S. Coupling photocatalytic degradation using a green TiO$_2$ catalyst to membrane bioreactor for petroleum refinery wastewater reclamation. *J. Water Process Eng.* **2020**, *34*, 101093. [CrossRef]

35. Goutam, S.P.; Saxena, G.; Singh, V.; Yadav, A.K.; Bharagava, R.N.; Thapa, K.B. Green synthesis of TiO$_2$ nanoparticles using leaf extract of *Jatropha curcas* L. for photocatalytic degradation of tannery wastewater. *Chem. Eng. J.* **2018**, *336*, 386–396. [CrossRef]

36. Carneiro, J.O.; Samantilleke, A.P.; Parpot, P.; Fernandes, F.; Pastor, M.; Correia, A.; Luís, E.A.; Chivanga Barros, A.A.; Teixeira, V. Visible Light Induced Enhanced Photocatalytic Degradation of Industrial Effluents (Rhodamine B) in Aqueous Media Using TiO$_2$ Nanoparticles. *J. Nanomater.* **2016**, *2016*, 1–13. [CrossRef]

37. Barndōk, H.; Hermosilla, D.; Han, C.; Dionysiou, D.D.; Negro, C.; Blanco, Á. Degradation of 1,4-dioxane from industrial wastewater by solar photocatalysis using immobilized NF-TiO$_2$ composite with monodisperse TiO$_2$ nanoparticles. *Appl. Catal. B Environ.* **2016**, *180*, 44–52. [CrossRef]

38. Natarajan, T.S.; Thomas, M.; Natarajan, K.; Bajaj, H.C.; Tayade, R.J. Study on UV-LED/TiO$_2$ process for degradation of Rhodamine B dye. *Chem. Eng. J.* **2011**, *169*, 126–134. [CrossRef]

39. Juang, R.S.; Lin, S.H.; Hsueh, P.Y. Removal of binary azo dyes from water by UV-irradiated degradation in TiO$_2$ suspensions. *J. Hazard. Mater.* **2010**, *182*, 820–826. [CrossRef]

40. Gebreyohannes, A.Y.; Mazzei, R.; Poerio, T.; Aimar, P.; Vankelecom, I.F.J.; Giorno, L. Pectinases immobilization on magnetic nanoparticles and their anti-fouling performance in a biocatalytic membrane reactor. *RSC Adv.* **2016**, *6*, 98737–98747. [CrossRef]

41. Ranjbar, P.Z.; Ayati, B.; Ganjidoust, H. Kinetic study on photocatalytic degradation of Acid Orange 52 in a baffled reactor using TiO$_2$ nanoparticles. *J. Environ. Sci.* **2019**, *79*, 213–224. [CrossRef]

42. Vatanpour, V.; Karami, A.; Sheydaei, M. Central composite design optimization of Rhodamine B degradation using TiO$_2$ nanoparticles/UV/PVDF process in continuous submerged membrane photoreactor. *Chem. Eng. Process. Process Intensif.* **2017**, *116*, 68–75. [CrossRef]

43. Hamad, H.A.; Sadik, W.A.; Abd El-latif, M.M.; Kashyout, A.B.; Feteha, M.Y. Photocatalytic parameters and kinetic study for degradation of dichlorophenol-indophenol (DCPIP) dye using highly active mesoporous TiO$_2$ nanoparticles. *J. Environ. Sci.* **2016**, *43*, 26–39. [CrossRef]

44. Shargh, M.; Behnajady, M.A. A high-efficient batch-recirculated photoreactor packed with immobilized TiO$_2$-P25 nanoparticles onto glass beads for photocatalytic degradation of phenazopyridine as a pharmaceutical contaminant: Artificial neural network modeling. *Water Sci. Technol.* **2016**, *73*, 2804–2814. [CrossRef]

45. Salu, O.A.; Adams, M.; Robertson, P.K.J.; Wong, L.S.; McCullagh, C. Remediation of oily wastewater from an interceptor tank using a novel photocatalytic drum reactor. *Desalin. Water Treat.* **2011**, *26*, 87–91. [CrossRef]

46. McCullagh, C.; Robertson, P.K.J.; Adams, M.; Pollard, P.M.; Mohammed, A. Development of a slurry continuous flow reactor for photocatalytic treatment of industrial waste water. *J. Photochem. Photobiol. A Chem.* **2010**, *211*, 42–46. [CrossRef]

47. Elami, D.; Seyyedi, K. Removing of carmoisine dye pollutant from contaminated waters by photocatalytic method using a thin film fixed bed reactor. *J. Environ. Sci. Health Part A Toxic Hazard. Subst. Environ. Eng.* **2020**, *55*, 193–208. [CrossRef] [PubMed]

48. Behnajady, M.A.; Dadkhah, H.; Eskandarloo, H. Horizontally rotating disc recirculated photoreactor with TiO$_2$-P25 nanoparticles immobilized onto a HDPE plate for photocatalytic removal of p-nitrophenol. *Environ. Technol.* **2018**, *39*, 1061–1070. [CrossRef] [PubMed]

49. Bautista, R.; Anderson, W.; Pagsuyoin, S.; Munoz, J. Degradation of tetracycline in synthesized wastewater using immobilized TiO$_2$ on rotating corrugated aluminum drum. In Proceedings of the 2015 Systems and Information Engineering Design Symposium, Charlottesville, VA, USA, 24 April 2015; pp. 115–119.

50. Wu, Y.; Yuan, H.; Jiang, X.; Wei, G.; Li, C.; Dong, W. Photocatalytic degradation of 4-tert-octylphenol in a spiral photoreactor system. *J. Environ. Sci.* **2012**, *24*, 1679–1685. [CrossRef]

51. Liu, J.; McCarthy, D.L.; Cowan, M.J.; Obuya, E.A.; DeCoste, J.B.; Skorenko, K.H.; Tong, L.; Boyer, S.M.; Bernier, W.E.; Jones, W.E., Jr. Photocatalytic activity of TiO$_2$ polycrystalline sub-micron fibers with variable rutile fraction. *Appl. Catal. B Environ.* **2016**, *187*, 154–162. [CrossRef]

52. Brame, J.; Long, M.; Li, Q.; Alvarez, P. Trading oxidation power for efficiency: Differential inhibition of photo-generated hydroxyl radicals versus singlet oxygen. *Water Res.* **2014**, *60*, 259–266.

53. Chen, W.; Chung, C.H.; Hsu, C.H.; Chen, Y.S.; Fu, S.L. Surface morphology and applications of TiO$_2$ nanofiber prepared by electrospinning. In Proceedings of the 2014 International Conference on Electronics Packaging (ICEP), Toyama, Japan, 23–25 April 2014; pp. 787–790.

54. Solcova, O.; Balkan, T.; Guler, Z.; Morozova, M.; Dytrych, P.; Sezai Sarac, A. New preparation route of TiO$_2$ nanofibers by electrospinning: Spectroscopic and thermal characterizations. *Sci. Adv. Mater.* **2014**, *6*, 2618–2624. [CrossRef]

55. Shaham Waldmann, N.; Paz, Y. Photocatalytic Reduction of Cr(VI) by Titanium Dioxide Coupled to Functionalized CNTs: An Example of Counterproductive Charge Separation. *J. Phys. Chem. C* **2010**, *114*, 18946–18952. [CrossRef]

56. Lee, C.G.; Javed, H.; Zhang, D.; Kim, J.H.; Westerhoff, P.; Li, Q.; Alvarez, P.J.J. Porous Electrospun Fibers Embedding TiO$_2$ for Adsorption and Photocatalytic Degradation of Water Pollutants. *Environ. Sci. Technol.* **2018**, *52*, 4285–4293. [CrossRef]

57. Yasin, S.A.; Abbas, J.A.; Ali, M.M.; Saeed, I.A.; Ahmed, I.H. Methylene blue photocatalytic degradation by TiO$_2$ nanoparticles supported on PET nanofibres. *Mater. Today Proc.* **2020**, *20*, 482–487. [CrossRef]

58. Song, M.; Cao, H.; Zhu, Y.; Wang, Y.; Zhao, S.; Huang, C.; Zhang, C.; He, X. Electrochemical and photocatalytic properties of electrospun C/TiO$_2$ nanofibers. *Chem. Phys. Lett.* **2020**, *747*, 137355. [CrossRef]

59. Koysuren, O.; Koysuren, H.N. Photocatalytic activities of poly(methyl methacrylate)/titanium dioxide nanofiber mat. *J. Macromol. Sci. Part A* **2017**, *54*, 80–84. [CrossRef]

60. Li, J.; Qiao, H.; Du, Y.; Chen, C.; Li, X.; Cui, J.; Kumar, D.; Wei, Q. Electrospinning Synthesis and Photocatalytic Activity of Mesoporous TiO$_2$ Nanofibers. *Sci. World J.* **2012**, *2012*, 1–7.

61. Li, Q.; Sun, D.; Kim, H. Fabrication of porous TiO$_2$ nanofiber and its photocatalytic activity. *Mater. Res. Bull.* **2011**, *46*, 2094–2099. [CrossRef]

62. Habib, Z.; Khan, S.J.; Ahmad, N.M.; Shahzad, H.M.A.; Jamal, Y.; Hashmi, I. Antibacterial behaviour of surface modified composite polyamide nanofiltration (NF) membrane by immobilizing Ag-doped TiO$_2$ nanoparticles. *Environ. Technol.* **2019**, 1–13. [CrossRef]

63. Lazareva, A.; Keller, A.A. Estimating potential life cycle releases of engineered nanomaterials from wastewater treatment plants. In *ACS Sustainable Chemistry and Engineering*; American Chemical Society: Washington, DC, USA, 2014; Volume 2, pp. 1656–1665.

64. Simelane, S.; Dlamini, L.N. An investigation of the fate and behaviour of a mixture of WO$_3$ and TiO$_2$ nanoparticles in a wastewater treatment plant. *J. Environ. Sci. China* **2019**, *76*, 37–47. [CrossRef] [PubMed]

65. Nowack, B.; Ranville, J.F.; Diamond, S.; Gallego-Urrea, J.A.; Metcalfe, C.; Rose, J.; Horne, N.; Koelmans, A.A.; Klaine, S.J. Potential scenarios for nanomaterial release and subsequent alteration in the environment. *Environ. Toxicol. Chem.* **2012**, *31*, 50–59. [CrossRef]

66. Singh Sekhon, B. Nanotechnology in agri-food production: An overview. *Nanotechnol. Sci. Appl.* **2014**, *7*, 31–53. [CrossRef]

67. Timmusk, S.; Seisenbaeva, G.; Behers, L. Titania (TiO$_2$) nanoparticles enhance the performance of growth-promoting rhizobacteria. *Sci. Rep.* **2018**, *8*, 1–13. [CrossRef] [PubMed]

68. Ge, Y.; Schimel, J.P.; Holden, P.A. Evidence for negative effects of TiO$_2$ and ZnO nanoparticles on soil bacterial communities. *Environ. Sci. Technol.* **2011**, *45*, 1659–1664. [CrossRef] [PubMed]

69. Ge, Y.; Schimel, J.P.; Holdena, P.A. Identification of soil bacteria susceptible to TiO$_2$ and ZnO nanoparticles. *Appl. Environ. Microbiol.* **2012**, *78*, 6749–6758. [CrossRef] [PubMed]

70. Simonin, M.; Martins, J.M.F.; Le Roux, X.; Uzu, G.; Calas, A.; Richaume, A. Toxicity of TiO$_2$ nanoparticles on soil nitrification at environmentally relevant concentrations: Lack of classical dose–response relationships. *Nanotoxicology* **2017**, *11*, 247–255. [CrossRef] [PubMed]

71. Feizi, H.; Kamali, M.; Jafari, L.; Rezvani Moghaddam, P. Phytotoxity and stimulatory impacts of nanosized and bulk titanium dioxide on fennel (Foeniculum vulgare Mill). *Chemosphere* **2013**, *91*, 506–511. [CrossRef] [PubMed]

72. Hanif, H.U.; Arshad, M.; Ali, M.A.; Ahmed, N.; Qazi, I.A. Phyto-availability of phosphorus to *Lactuca sativa* in response to soil applied TiO$_2$ nanoparticles. *Pak. J. Agric. Sci.* **2015**, *52*, 177–182.

73. Zahra, Z.; Arshad, M.; Rafique, R.; Mahmood, A.; Habib, A.; Qazi, I.A.; Khan, S.A. Metallic Nanoparticle (TiO$_2$ and Fe$_3$O$_4$) Application Modifies Rhizosphere Phosphorus Availability and Uptake by Lactuca sativa. *J. Agric. Food Chem.* **2015**, *63*, 6876–6882. [CrossRef] [PubMed]

74. Gohari, G.; Mohammadi, A.; Akbari, A.; Panahirad, S.; Dadpour, M.R.; Fotopoulos, V.; Kimura, S. Titanium dioxide nanoparticles (TiO$_2$ NPs) promote growth and ameliorate salinity stress effects on essential oil profile and biochemical attributes of Dracocephalum moldavica. *Sci. Rep.* **2020**, *10*, 1–14. [CrossRef] [PubMed]

75. Ogunkunle, C.O.; Gambari, H.; Agbaje, F.; Okoro, H.K.; Asogwa, N.T.; Vishwakarma, V.; Fatoba, P.O. Effect of Low-Dose Nano Titanium Dioxide Intervention on Cd Uptake and Stress Enzymes Activity in Cd-Stressed Cowpea [*Vigna unguiculata* (L.) Walp] Plants. *Bull. Environ. Contam. Toxicol.* **2020**, *104*, 619–626. [CrossRef]

76. Ullah, S.; Adeel, M.; Zain, M.; Rizwan, M.; Irshad, M.K.; Jilani, G.; Hameed, A.; Khan, A.; Arshad, M.; Raza, A.; et al. Physiological and biochemical response of wheat (*Triticum aestivum*) to TiO$_2$ nanoparticles in phosphorous amended soil: A full life cycle study. *J. Environ. Manag.* **2020**, *263*, 110365. [CrossRef]

77. Rizwan, M.; Ali, S.; ur Rehman, M.Z.; Malik, S.; Adrees, M.; Qayyum, M.F.; Alamri, S.A.; Alyemeni, M.N.; Ahmad, P. Effect of foliar applications of silicon and titanium dioxide nanoparticles on growth, oxidative stress, and cadmium accumulation by rice (*Oryza sativa*). *Acta Physiol. Plant.* **2019**, *41*, 1–12.

78. Zahra, Z.; Maqbool, T.; Arshad, M.; Badshah, M.A.; Choi, H.K.; Hur, J. Changes in fluorescent dissolved organic matter and their association with phytoavailable phosphorus in soil amended with TiO$_2$ nanoparticles. *Chemosphere* **2019**, *227*, 17–25. [CrossRef] [PubMed]

79. Zahra, Z.; Ali, M.A.; Parveen, A.; Kim, E.; Khokhar, M.F.; Baig, S.; Hina, K.; Choi, H.-K.; Arshad, M. Exposure–Response of Wheat Cultivars to TiO_2 Nanoparticles in Contrasted Soils. *Soil Sediment Contam. Int. J.* **2019**, *28*, 184–199. [CrossRef]

80. Larue, C.; Baratange, C.; Vantelon, D.; Khodja, H.; Surblé, S.; Elger, A.; Carrière, M. Influence of soil type on TiO_2 nanoparticle fate in an agro-ecosystem. *Sci. Total Environ.* **2018**, *630*, 609–617. [CrossRef] [PubMed]

81. Zahra, Z.; Waseem, N.; Zahra, R.; Lee, H.; Badshah, M.A.; Mehmood, A.; Choi, H.-K.; Arshad, M. Growth and Metabolic Responses of Rice (*Oryza sativa* L.) Cultivated in Phosphorus-Deficient Soil Amended with TiO_2 Nanoparticles. *J. Agric. Food Chem.* **2017**, *65*, 5598–5606. [CrossRef] [PubMed]

82. Kořenková, L.; Šebesta, M.; Urík, M.; Kolenčík, M.; Kratošová, G.; Bujdoš, M.; Vávra, I.; Dobročka, E. Physiological response of culture media-grown barley (*Hordeum vulgare* L.) to titanium oxide nanoparticles. *Acta Agric. Scand. Sect. B Soil Plant Sci.* **2017**, *67*, 285–291. [CrossRef]

83. Rafique, R.; Zahra, Z.; Virk, N.; Shahid, M.; Pinelli, E.; Park, T.J.; Kallerhoff, J.; Arshad, M. Dose-dependent physiological responses of *Triticum aestivum* L. to soil applied TiO_2 nanoparticles: Alterations in chlorophyll content, H_2O_2 production, and genotoxicity. *Agric. Ecosyst. Environ.* **2018**, *255*, 95–101. [CrossRef]

84. Szymańska, R.; Kołodziej, K.; Ślesak, I.; Zimak-Piekarczyk, P.; Orzechowska, A.; Gabruk, M.; Zadło, A.; Habina, I.; Knap, W.; Burda, K.; et al. Titanium dioxide nanoparticles (100–1000 mg/l) can affect vitamin E response in *Arabidopsis thaliana*. *Environ. Pollut.* **2016**, *213*, 957–965. [CrossRef]

85. Andersen, C.P.; King, G.; Plocher, M.; Storm, M.; Pokhrel, L.R.; Johnson, M.G.; Rygiewicz, P.T. Germination and early plant development of ten plant species exposed to titanium dioxide and cerium oxide nanoparticles. *Environ. Toxicol. Chem.* **2016**, *35*, 2223–2229. [CrossRef]

86. Gogos, A.; Moll, J.; Klingenfuss, F.; Heijden, M.; Irin, F.; Green, M.J.; Zenobi, R.; Bucheli, T.D. Vertical transport and plant uptake of nanoparticles in a soil mesocosm experiment. *J. Nanobiotechnol.* **2016**, *14*, 40. [CrossRef]

87. Burke, D.; Pietrasiak, N.; Situ, S.; Abenojar, E.; Porche, M.; Kraj, P.; Lakliang, Y.; Samia, A. Iron Oxide and titanium dioxide nanoparticle effects on plant performance and root associated microbes. *Int. J. Mol. Sci.* **2015**, *16*, 23630–23650. [CrossRef]

88. Hatami, M.; Ghorbanpour, M.; Salehiarjomand, H. Nano-anatase TiO_2 modulates the germination behavior and seedling vigority of some commercially important medicinal and aromatic plants. *J. Biol. Environ. Sci.* **2014**, *8*, 53–59.

89. Dehkourdi, E.H.; Mosavi, M. Effect of anatase nanoparticles (TiO_2) on parsley seed germination (*petroselinum crispum*) in vitro. *Biol. Trace Elem. Res.* **2013**, *155*, 283–286. [CrossRef] [PubMed]

90. Jaberzadeh, A.; Moaveni, P.; Tohidi Moghadam, H.R.; Zahedi, H. Influence of bulk and nanoparticles titanium foliar application on some agronomic traits, seed gluten and starch contents of wheat subjected to water deficit stress. *Not. Bot. Horti Agrobot. Cluj Napoca* **2013**, *41*, 201–207. [CrossRef]

91. Larue, C.; Veronesi, G.; Flank, A.M.; Surble, S.; Herlin-Boime, N.; Carrière, M. Comparative uptake and impact of TiO_2 nanoparticles in wheat and rapeseed. *J. Toxicol. Environ. Health Part A Curr. Issues* **2012**, *75*, 722–734. [CrossRef]

92. Ze, Y.; Liu, C.; Wang, L.; Hong, M.; Hong, F. The Regulation of TiO_2 Nanoparticles on the Expression of Light-Harvesting Complex II and Photosynthesis of Chloroplasts of *Arabidopsis thaliana*. *Biol. Trace Elem. Res.* **2011**, *143*, 1131–1141. [CrossRef]

93. Kurepa, J.; Paunesku, T.; Vogt, S.; Arora, H.; Rabatic, B.M.; Lu, J.; Wanzer, M.B.; Woloschak, G.E.; Smalle, J.A. Uptake and distribution of ultrasmall anatase TiO_2 alizarin reds nanoconjugates in *arabidopsis thaliana*. *Nano Lett.* **2010**, *10*, 2296–2302. [CrossRef]

94. Dağhan, H. Effects of TiO_2 nanoparticles on maize (*Zea mays* L.) growth, chlorophyll content and nutrient uptake. *Appl. Ecol. Environ. Res.* **2018**, *16*, 6873–6883.

95. Kushwah, K.S.; Patel, S. Effect of Titanium Dioxide Nanoparticles (TiO_2 NPs) on Faba bean (*Vicia faba* L.) and Induced Asynaptic Mutation: A Meiotic Study. *J. Plant Growth Regul.* **2019**, 1–12. [CrossRef]

96. Ruffini Castiglione, M.; Giorgetti, L.; Bellani, L.; Muccifora, S.; Bottega, S.; Spanò, C. Root responses to different types of TiO_2 nanoparticles and bulk counterpart in plant model system *Vicia faba* L. *Environ. Exp. Bot.* **2016**, *130*, 11–21. [CrossRef]

97. Jahan, S.; Alias, Y.B.; Bakar, A.F.B.A.; Yusoff, I. Bin Toxicity evaluation of ZnO and TiO_2 nanomaterials in hydroponic red bean (*Vigna angularis*) plant: Physiology, biochemistry and kinetic transport. *J. Environ. Sci.* **2018**, *72*, 140–152. [CrossRef] [PubMed]

98. Mohammadi, R.; Maali-Amiri, R.; Abbasi, A. Effect of TiO$_2$ Nanoparticles on Chickpea Response to Cold Stress. *Biol. Trace Elem. Res.* **2013**, *152*, 403–410. [CrossRef] [PubMed]

99. Robichaud, C.O.; Dicksen, T.; Mark, R.W. Assessing Life-Cycle Risks of Nanomaterials. In *Environmental Nanotechnology: Applications and Impacts of Nanomaterials*; McGraw Hill: New York, NY, USA, 2007; pp. 481–524.

100. Schmid, K.; Riediker, M. Use of Nanoparticles in Swiss Industry: A Targeted Survey. *Environ. Sci. Technol.* **2008**, *42*, 2253–2260. [CrossRef] [PubMed]

101. The National Institute for Occupational Safety and Health (NIOSH). *Safe Nanotechnology in the Workplace, An Introduction for Employers, Managers, and Safety and Health Professionals*; NIOSH: Washington, DC, USA, 2008.

102. Kheiri, S.; Liu, X.; Thompson, M. Nanoparticles at biointerfaces: Antibacterial activity and nanotoxicology. *Colloids Surf. B Biointerf.* **2019**, *184*, 110550. [CrossRef] [PubMed]

103. Bondarenko, O.; Juganson, K.; Ivask, A.; Kasemets, K.; Mortimer, M.; Kahru, A. Toxicity of Ag, CuO and ZnO nanoparticles to selected environmentally relevant test organisms and mammalian cells in vitro: A critical review. *Arch. Toxicol.* **2013**, *87*, 1181–1200. [CrossRef]

104. Vasantharaja, D.; Ramalingam, V.; Reddy, G.A. Oral Toxic Exposure of Titanium Dioxide Nanoparticles on Serum Biochemical Changes in Adult Male Wistar Rats Oral Toxicity of TiO$_2$ Nanoparticles. *Nanomed. J.* **2015**, *2*, 46–53.

105. Iavicoli, I.; Leso, V.; Bergamaschi, A. Toxicological effects of titanium dioxide nanoparticles: A review of in vivo studies. *J. Nanomater.* **2012**, *2012*, 36. [CrossRef]

106. Zhao, L.; Zhu, Y.; Chen, Z.; Xu, H.; Zhou, J.; Tang, S.; Xu, Z.; Kong, F.; Li, X.; Zhang, Y.; et al. Cardiopulmonary effects induced by occupational exposure to titanium dioxide nanoparticles. *Nanotoxicology* **2018**, *12*, 169–184. [CrossRef]

107. Jayaram, D.T.; Kumar, A.; Kippner, L.E.; Ho, P.Y.; Kemp, M.L.; Fan, Y.; Payne, C.K. TiO$_2$ nanoparticles generate superoxide and alter gene expression in human lung cells. *RSC Adv.* **2019**, *9*, 25039–25047. [CrossRef]

108. Gea, M.; Bonetta, S.; Iannarelli, L.; Giovannozzi, A.M.; Maurino, V.; Bonetta, S.; Hodoroaba, V.D.; Armato, C.; Rossi, A.M.; Schilirò, T. Shape-engineered titanium dioxide nanoparticles (TiO$_2$-NPs): Cytotoxicity and genotoxicity in bronchial epithelial cells. *Food Chem. Toxicol.* **2019**, *127*, 89–100. [CrossRef]

109. Moschini, E.; Gualtieri, M.; Colombo, M.; Fascio, U.; Camatini, M.; Mantecca, P. The modality of cell-particle interactions drives the toxicity of nanosized CuO and TiO$_2$ in human alveolar epithelial cells. *Toxicol. Lett.* **2013**, *222*, 102–116. [CrossRef]

110. Alkaladi, A.; El-Deen, N.A.M.N.; Afifi, M.; Zinadah, O.A.A. Hematological and biochemical investigations on the effect of vitamin E and C on Oreochromis niloticus exposed to zinc oxide nanoparticles. *Saudi J. Biol. Sci.* **2015**, *22*, 556–563. [CrossRef] [PubMed]

111. Hansen, S.F.; Michelson, E.S.; Kamper, A.; Borling, P.; Stuer-Lauridsen, F.; Baun, A. Categorization framework to aid exposure assessment of nanomaterials in consumer products. *Ecotoxicology* **2008**, *17*, 438–447. [CrossRef] [PubMed]

112. Wright, C.; Iyer, A.K.V.; Wang, L.; Wu, N.; Yakisich, J.S.; Rojanasakul, Y.; Azad, N. Effects of titanium dioxide nanoparticles on human keratinocytes. *Drug Chem. Toxicol.* **2017**, *40*, 90–100. [CrossRef] [PubMed]

113. Crosera, M.; Prodi, A.; Mauro, M.; Pelin, M.; Florio, C.; Bellomo, F.; Adami, G.; Apostoli, P.; De Palma, G.; Bovenzi, M.; et al. Titanium Dioxide Nanoparticle Penetration into the Skin and Effects on HaCaT Cells. *Int. J. Environ. Res. Public Health* **2015**, *12*, 9282–9297. [CrossRef] [PubMed]

114. Geppert, M.; Schwarz, A.; Stangassinger, L.M.; Wenger, S.; Wienerroither, L.M.; Ess, S.; Duschl, A.; Himly, M. Interactions of TiO$_2$ nanoparticles with ingredients from modern lifestyle products and their effects on human skin cells. *Chem. Res. Toxicol.* **2020**, *33*, 1215–1225. [CrossRef]

115. Wu, F.; Hicks, A.L. Estimating human exposure to titanium dioxide from personal care products through a social survey approach. *Integr. Environ. Assess. Manag.* **2020**, *16*, 10–16. [CrossRef]

116. Hashimoto, K.; Irie, H.; Fujishima, A. TiO$_2$ Photocatalysis: A Historical Overview and Future Prospects. *Jpn. J. Appl. Phys.* **2005**, *44*, 8269–8285. [CrossRef]

117. Hu, J.; Wu, X.; Wu, F.; Chen, W.; Zhang, X.; White, J.C.; Li, J.; Wan, Y.; Liu, J.; Wang, X. TiO$_2$ nanoparticle exposure on lettuce (*Lactuca sativa* L.): Dose-dependent deterioration of nutritional quality. *Environ. Sci. Nano* **2020**, *7*, 501–513. [CrossRef]

118. Hansen, S.F. Regulation and Risk Assessment of Nanomaterials: Too Little, Too Late? Ph.D. Thesis, Technical Unversity of Denmark, Kongens Lyngby, Denmark, 2009.

119. Wijnhoven, S.W.P.; Dekkers, S.; Hagens, W.I.; De Jong, W.H. *Exposure to Nanomaterials in Consumer Products*; National Institute for Public Health and the Environment: Bilthoven, The Netherlands, 2009.

120. Du, W.; Sun, Y.; Ji, R.; Zhu, J.; Wu, J.; Guo, H. TiO_2 and ZnO nanoparticles negatively affect wheat growth and soil enzyme activities in agricultural soil. *J. Environ. Monit.* **2011**, *13*, 822. [CrossRef]

121. Rondeau-Mouro, C.; Defer, D.; Leboeuf, E.; Lahaye, M. Assessment of cell wall porosity in Arabidopsis thaliana by NMR spectroscopy. *Int. J. Biol. Macromol.* **2008**, *42*, 83–92. [CrossRef]

122. Jia, G.; Wang, H.; Yan, L.; Wang, X.; Pei, R.; Yan, T.; Zhao, Y.; Guo, X. Cytotoxicity of Carbon Nanomaterials: Single-Wall Nanotube, Multi-Wall Nanotube, and Fullerene. *Environ. Sci. Technol.* **2005**, *39*, 1378–1383. [CrossRef] [PubMed]

123. Belhaj Abdallah, B.; Andreu, I.; Chatti, A.; Landoulsi, A.; Gates, B.D. Size Fractionation of Titania Nanoparticles in Wild Dittrichia Viscosa Grown in a Native Environment. *Environ. Sci. Technol.* **2020**. [CrossRef] [PubMed]

124. Larue, C.; Castillo-Michel, H.; Sobanska, S.; Trcera, N.; Sorieul, S.; Cécillon, L.; Ouerdane, L.; Legros, S.; Sarret, G. Fate of pristine TiO_2 nanoparticles and aged paint-containing TiO_2 nanoparticles in lettuce crop after foliar exposure. *J. Hazard. Mater.* **2014**, *273*, 17–26. [CrossRef] [PubMed]

125. Silva, L.C.D.; Oliva, M.A.; Azevedo, A.A.; Araújo, J.M. De Responses of restinga plant species to pollution from an iron pelletization factory. *Water. Air. Soil Pollut.* **2006**, *175*, 241–256. [CrossRef]

126. Nair, R.; Varghese, S.H.; Nair, B.G.; Maekawa, T.; Yoshida, Y.; Kumar, D.S. Nanoparticulate material delivery to plants. *Plant Sci.* **2010**, *179*, 154–163. [CrossRef]

127. Amde, M.; Liu, J.F.; Tan, Z.Q.; Bekana, D. Transformation and bioavailability of metal oxide nanoparticles in aquatic and terrestrial environments. A review. *Environ. Pollut.* **2017**, *230*, 250–267. [CrossRef] [PubMed]

128. Movafeghi, A.; Khataee, A.; Abedi, M.; Tarrahi, R.; Dadpour, M.; Vafaei, F. Effects of TiO_2 nanoparticles on the aquatic plant Spirodela polyrrhiza: Evaluation of growth parameters, pigment contents and antioxidant enzyme activities. *J. Environ. Sci.* **2018**, *64*, 130–138. [CrossRef] [PubMed]

129. Hu, C.; Hou, J.; Zhu, Y.; Lin, D. Multigenerational exposure to TiO_2 nanoparticles in soil stimulates stress resistance and longevity of survived C. elegans via activating insulin/IGF-like signaling. *Environ. Pollut.* **2020**, *263*, 114376. [CrossRef] [PubMed]

130. Zhu, Y.; Wu, X.; Liu, Y.; Zhang, J.; Lin, D. Integration of transcriptomics and metabolomics reveals the responses of earthworms to the long-term exposure of TiO_2 nanoparticles in soil. *Sci. Total Environ.* **2020**, *719*, 137492. [CrossRef]

131. Sánchez-López, K.B.; De Los Santos-Ramos, F.J.; Gómez-Acata, E.S.; Luna-Guido, M.; Navarro-Noya, Y.E.; Fernández-Luqueño, F.; Dendooven, L. TiO_2 nanoparticles affect the bacterial community structure and Eisenia fetida (Savigny, 1826) in an arable soil. *PeerJ* **2019**, *2019*, e6939. [CrossRef]

132. Mu, W.; Wang, Y.; Huang, C.; Fu, Y.; Li, J.; Wang, H.; Jia, X.; Ba, Q. Effect of Long-Term Intake of Dietary Titanium Dioxide Nanoparticles on Intestine Inflammation in Mice. *J. Agric. Food Chem.* **2019**, *67*, 9382–9389. [CrossRef] [PubMed]

133. Relier, C.; Dubreuil, M.; Garcìa, O.L.; Cordelli, E.; Mejia, J.; Eleuteri, P.; Robidel, f.; Loret, T.; Pacchierotti, F.; Lucas, S.; et al. Study of TiO_2 P25 nanoparticles genotoxicity on lung, blood, and liver cells in lung overload and non-overload conditions after repeated respiratory exposure in rats. *Toxicol. Sci.* **2017**, *156*, 527–537. [PubMed]

134. Liu, S.; Zeng, P.; Li, X.; Thuyet, D.Q.; Fan, W. Effect of chronic toxicity of the crystalline forms of TiO_2 nanoparticles on the physiological parameters of Daphnia magna with a focus on index correlation analysis. *Ecotoxicol. Environ. Saf.* **2019**, *181*, 292–300. [CrossRef] [PubMed]

135. Srinivasan, M.; Venkatesan, M.; Arumugam, V.; Natesan, G.; Saravanan, N.; Murugesan, S.; Ramachandran, S.; Ayyasamy, R.; Pugazhendhi, A. Green synthesis and characterization of titanium dioxide nanoparticles (TiO_2 NPs) using Sesbania grandiflora and evaluation of toxicity in zebrafish embryos. *Process Biochem.* **2019**, *80*, 197–202. [CrossRef]

136. Abd El-Atti, M.; Desouky, M.M.A.; Mohamadien, A.; Said, R.M. Effects of titanium dioxide nanoparticles on red swamp crayfish, Procambarus clarkii: Bioaccumulation, oxidative stress and histopathological biomarkers. *Egypt. J. Aquat. Res.* **2019**, *45*, 11–18. [CrossRef]

137. Fırat, Ö.; Bozat, R.C. Assessment of biochemical and toxic responses induced by titanium dioxide nanoparticles in Nile tilapia Oreochromis niloticus. *Hum. Ecol. Risk Assess.* **2019**, *25*, 1438–1447. [CrossRef]
138. Kachenton, S.; Jiraungkoorskul, W.; Kangwanrangsan, N.; Tansatit, T. Cytotoxicity and histopathological analysis of titanium nanoparticles via Artemia salina. *Environ. Sci. Pollut. Res.* **2019**, *26*, 14706–14711. [CrossRef]

Article

Iron Oxide Nanoparticles as an Alternative to Antibiotics Additive on Extended Boar Semen

Ioannis A. Tsakmakidis [1],*, Theodoros Samaras [2], Sofia Anastasiadou [1], Athina Basioura [1], Aikaterini Ntemka [1], Ilias Michos [1], Konstantinos Simeonidis [2], Isidoros Karagiannis [1], Georgios Tsousis [1], Mavroeidis Angelakeris [2] and Constantin M. Boscos [1]

[1]   School of Veterinary Medicine, Faculty of Health Sciences, Aristotle University of Thessaloniki, 54627 Thessaloniki, Greece; anastasiadousof@yahoo.gr (S.A.); basioura@vet.auth.gr (A.B.); demkaterina@hotmail.com (A.N.); imichos84@hotmail.com (I.M.); anisivet@yahoo.gr (I.K.); tsousis@vet.auth.gr (G.T.); pboscos@vet.auth.gr (C.M.B.)

[2]   School of Physics, Faculty of Sciences, Aristotle University of Thessaloniki, 54124 Thessaloniki, Greece; theosama@auth.gr (T.S.); ksime@physics.auth.gr (K.S.); agelaker@auth.gr (M.A.)

*   Correspondence: iat@vet.auth.gr; Tel.: +30-2310-994-467

Received: 20 July 2020; Accepted: 7 August 2020; Published: 10 August 2020

Abstract: This study examined the effect of Fe_3O_4 nanoparticles on boar semen. Beltsville thawing solution without antibiotics was used to extend ejaculates from 5 boars (4 ejaculates/boar). Semen samples of control group (C) and group with Fe_3O_4 (Fe; 0.192 mg/mL semen) were incubated under routine boar semen storage temperature (17 °C) for 0.5 h and nanoparticles were removed by a magnetic field. Before and after treatment, aliquots of all groups were cultured using standard microbiological methods. The samples after treatment were stored (17 °C) for 48 h and sperm parameters (computer-assisted sperm analyzer (CASA) variables; morphology; viability; hypo-osmotic swelling test (HOST); DNA integrity) were evaluated at storage times 0, 24, 48 h. Semen data were analyzed by a repeated measures mixed model and microbial data with Student's t-test for paired samples. Regarding CASA parameters, Fe group did not differ from C at any time point. In group C, total motility after 24 h and progressive motility after 48 h of storage decreased significantly compared to 0 h. In group Fe, linearity (LIN) after 48 h and head abnormalities after 24 h of storage increased significantly compared to 0 h. The microbiological results revealed a significant reduction of the bacterial load in group Fe compared to control at both 24 and 48 h. In conclusion, the use of Fe_3O_4 nanoparticles during semen processing provided a slight anti-microbiological effect with no adverse effects on sperm characteristics.

Keywords: bacterial resistance; boar; microbiological analyses; nanoparticles; semen

1. Introduction

The performance of artificial insemination (AI) is the most acknowledged method worldwide to fertilize sows with liquid-extended boar semen. Bacterial contamination affects semen's qualitative characteristics and fertilizing capacity and induces a potential health risk to the females after AI. Previous studies reported that bacteriospermia results in higher estrus returns, early embryo death, endometritis and specific infections in pigs [1–3]. Porcine semen usually contains two to three species of bacteria, most commonly *Staphylococci, Streptococci* and *Pseudomonas* [4]. In this aspect, the antibiotics have been main constituents of semen extenders to control the bacterial growth during the storage time of boar liquid semen. However, Althouse and Lu [5] found bacterial occurrence in one third of the produced boar sperm doses, with bacteria largely resistant to antibiotics such as amoxycillin, gentamycin, lincomycin and tylosin. This fact has led the scientific community to explore alternative strategies to minimize the development of antibiotic resistance. Thus, methods of physically removing bacteria by centrifugation in colloidal solutions [6] and the addition of natural or synthetic peptides with antimicrobial activity to the diluents have been reported in boar semen processing [7,8]. Nanoparticles (NPs) of size 40–60 nm, expressed significant antimicrobial capacity against *Escherichia coli*, *Pseudomonas aeruginosa* and *Staphylococcus aureus* [9]. Li et al. [10], investigating the mechanism of action of Ag NPs, found that 10 µg Ag NPs/mL can completely restrain the growth of 10^7 cfu/mL of *E. coli* cells, by damaging cell membrane structure, limiting the activity of enzymes and inducing bacterial death. The results of the study of Shahverdi et al. [11], where silver NPs significantly increased the antimicrobial activity of vancomycin, amoxycillin and penicillin against *S. aureus* are remarkable. Moreover, iron oxide magnetic NPs provided significant antibacterial properties against *p. aeruginosa* and *S. aureus*, due to the reactive oxygen species (ROS) generation [12,13]. Although the scientific community continuously seeks alternative approaches to the use of antibiotics in semen processing, some studies have reported toxic effects of NPs on live cells. Sahu et al. [14] found that nanotoxicity could be dependent on the type of the cell in terms of a different sensitivity response. For this reason, our research team was the first to investigate the appropriate effective antibacterial concentration and co-incubation time of silver Ag/Fe and Fe_3O_4 NPs in semen [15]. In this study, Ag/Fe NPs demonstrated a detrimental effect on boar spermatozoa. Conversely, Fe_3O_4 NPs at a minimum inhibitory concentration (0.192 mg/mL semen) had no negative effect on computer-assisted sperm analyzer (CASA) motility parameters of boar sperm after 30 min of co-incubation [15]. Therefore, further research on their application for semen handling is necessary. The aim of this research was to extend our knowledge regarding an alternative methodology in order to control the microbial load of boar semen without having detrimental effects on its quality, using iron oxide NPs.

2. Materials and Methods

The semen samples used in the present study were commercially available. No operations on research animals were carried out and no approval by the Ethics Committee on Animal Use of Aristotle University of Thessaloniki (Greece) was necessary.

2.1. Reagents and Media

All the reagents and chemicals used were purchased from Sigma Aldrich, Seelze, Germany, unless otherwise stated. Semen samples were extended with laboratory-produced Beltsville thawing solution (BTS: 205 mM glucose, 20.4 mM 112 sodium citrate, 10.0 mM KCl, 15.0 mM NaHCO$_3$, 3.6 mM ethylenediaminetetraacetic acid (EDTA); pH 7.2–7.4; 290–300 mOsmol/kg) without antibiotics.

2.2. Synthesis and Dispersions of Fe$_3$O$_4$ NPs

2.2.1. Synthesis

Magnetite nanoparticles were synthesized by the oxidative precipitation of FeSO$_4$ in an ethanol/water mixture and NaNO$_3$ and NaOH were added as mild oxidant and acidity controller, respectively [16,17]. A solution of 1 M FeSO$_4$·7H$_2$O was prepared by the reagent's dissolution in 350 mL of 0.01 M H$_2$SO$_4$ solution. Another solution with dissolved NaNO$_3$ (0.25 M) and NaOH (0.52 M) in a 30% ethanol mixture in distilled water was prepared to a total volume of 1400 mL. The two solutions were mixed under intense stirring to form green rust. After agitating the mixture for 15 min, it was transferred to a water bath regulated at 90 °C and allowed for 6 h to ageing of green rust and to the formation of magnetite nanoparticles. When cooled to room temperature, the dispersion was washed/centrifuged several times with distilled water to remove any residuals. The Fe$_3$O$_4$ NPs were sterilized before co-incubation with semen by heating in an autoclave (steam sterilization, 121–124 °C, 3 atm, 20 min).

2.2.2. Characterization

The identification of the structural phases appearing in the nanoparticles was carried out by powder X-ray diffractometry (XRD) using an Ultima+ diffractometer, Rigaku, Sendagaya, Japan, operating with CuKα radiation at 40 kV/30 mA, 0.05° for step size and 3 s as step time. The diffraction diagrams were evaluated after comparison with the Powder Diffraction Files (PDF) database [18]. Scanning electron microscopy (SEM) images were taken by a Quanta 200 ESEM FEG instrument (FEI, Hillsboro, OR, USA) with a field-emission gun adjusted to 30 kV. Magnetic measurements of the nanoparticles were performed in a MPMS XL SQUID magnetometer (Quantum Design, San Diego, CA, USA) at room temperature.

To determine the percentage of bivalent iron in the nanoparticles (Fe^{2+}/Fe^{3+} ratio) as an indicator of Fe_3O_4 formation, the dried sample (0.1 g) was digested in 50 mL 7 M H_2SO_4 under heating and then, titrated by a 0.05 M $KMnO_4$ solution. The appearance of pink color signified the complete reduction of MnO_4^- ions and the end of the titration. The sum of iron in the sample was defined by graphite furnace atomic absorption spectrophotometry (Perkin Elmer AAnalyst 800, Perkin Elmer, Waltham, MA, USA) after dissolving a weighted quantity in HCl.

2.3. Animals, Semen Samples Collection and Dilution

The semen samples were collected from 5 crossbred boars (2–2.5 years of age) from a commercial pig farm with capacity of 700 sows. In total, 20 ejaculates (4 ejaculates/boar) were collected by the gloved hand technique and the gelatinous portion was discarded using a gauze. Two ejaculates were collected on a weekly basis, pooled, and transported in an isothermal vessel (37 °C) to the farm laboratory. The ejaculates were assessed for the basic quality parameters [volume, concentration (SDM1, Minitube®, Tiefenbach, Germany) and motility (subjective microscopic evaluation by a phase contrast microscope, Zeiss, Oberkochen, Germany)], and those with volume >200 mL, concentration >200 × 10^6 sperm/mL, total number of spermatozoa/ejaculate >40 × 10^9, and gross motility >70% were further processed.

Semen samples of good quality were extended in BTS without antibiotics (30 × 10^6 spermatozoa/mL) and re-evaluated microscopically for motility. Extended pooled semen samples with motility >70% were transported (17 °C) within 60 min into a portable semen storage unit (Minitube®, Tiefenbach, Germany) to the Unit of Biotechnology of Reproduction, Clinic of Farm Animals, Faculty of Veterinary Medicine, Aristotle University of Thessaloniki.

2.4. Experimental Design

2.4.1. Semen Processing with Nanoparticles (NPs)

Upon arrival at the Unit of Biotechnology of Reproduction, each semen sample was separated in 2 aliquots and the following two experimental groups were prepared: (1) control group (C): extended semen without any treatment; (2) iron oxide group (Fe): extended semen with Fe_3O_4 NPs (0.192 mg Fe_3O_4/mL semen).

2.4.2. Trial 1: Determination of Non-Detrimental Co-Incubation Time of Semen with NPs

The beneficial/detrimental co-incubation period and the appropriate antibacterial concentration of iron oxide NPs for boar semen handling, was evaluated in a previous trial of our laboratory [15]. Briefly, semen samples were cultured for the detection of bacterial pathogens. Plates containing sheep blood agar and plate count agar (PCA, Oxoid, Thermo Scientific, Lenexa, KS, USA) were inoculated with 100 μL aliquot of extended semen without antibiotics and incubated at 37 °C to estimate the microbial load and select strains for the antibacterial assay. Isolation and identification of picked colonies from blood agar was performed after incubation in 10 mL of tryptone soya broth (TS broth, Oxoid, Thermo Scientific, Lenexa, KS, USA), (37 °C for 24 h) and the conduction of conventional laboratory procedures. *Pseudomonas*, *Staphylococcus* and *Streptococcus* strains were finally selected for further investigation of the antimicrobial activity of nanoparticles. The selected strains were tested against several concentrations of the examined iron oxide NPs using a standard two-fold broth dilution method [19]. Dilutions of the Fe_3O_4 NPs were prepared and dispensed in tubes. Each tube was inoculated with the relevant microbial inoculum (adjusted to 0.5 McFarland scale) of each of the selected strains. Tubes were incubated at 37 °C aerobically and 100 µL were spread on blood agar at time 0, 30 min, 60 min and 24 h of incubation. The results demonstrated that the concentration of iron oxide NPs that prevented the growth of the selected bacteria was different for each strain. The minimum inhibitory concentration (MIC) of NPs for the bacterial growth was defined as the lowest concentration of NPs, which inhibited bacterial growth.

Concentrations of iron oxide NPs were re-assessed and finally selected so as not to be harmful for semen samples. Considering the MIC (0.192 mg/mL semen) of the examined NPs after in vitro antimicrobial activity assessment, they were dissolved in distilled water to prepare a stock solution of 19.2 mg Fe_3O_4 NPs/mL distilled water. A fresh stock solution of NPs was prepared every week. Prior to its use, the NPs solution was sonicated for 20 min to improve its dispersion stability. The C and NPs groups were incubated at 17 °C (appropriate storage temperature of extended boar semen) for 30, 45 and 60 min and afterwards the NPs were removed with a magnetic field (as it is described in Trial 2). Total motility and progressive movement spermatozoa were assessed by the CASA. The incubation period of 45 and 60 min were excluded as the values of the examined CASA parameters were significantly decreased in the NPs groups compared to the control group. The co-incubation period of 30 min had no adverse effects on the evaluated CASA parameters and was selected for further research.

2.4.3. Trial 2: Investigation of the Effect of Iron Oxide (Fe_3O_4) NPs on Boar Semen Quality

The C and Fe groups were incubated (17 °C) for 30 min after iron oxide NPs addition to group Fe. Subsequently NPs were removed with the help of a magnetic field. To achieve that, the tubes were placed in a plexiglass acrylic rack equipped with commercial NdFeB permanent magnets, remained in vertical position for at least 5 min and the post-treated semen was transferred to a new tube, while the NPs were discarded (Figure 1). Three repetitions of this process were applied to completely remove the NPs. Then the control and the post-treatment NPs samples were stored at 17 °C for 48 h. Semen quality and functionality as well as the microbiological tests were performed at 0 (time of NPs removal), 24 and 48 h post treatment.

Figure 1. Plexiglass acrylic rack with 6 Eppendorf tubes' places, equipped with commercial NdFeB permanent magnets.

2.5. Sperm Kinetics/Motility

The Sperm Class Analyser (SCA®, Microptic S.L., Barcelona, Spain) CASA system, a microscope (AXIO Scope A1, Zeiss, Oberkochen, Germany) with a heating stage (37 °C) and a camera (Basler scA780 54fc, Ahrensburg, Germany) were used for the evaluation of sperm motility and kinetics. The analysis by SCA® software (v.6.3.) was performed with the following configurations: 4–6 fields were recorded (×100) for each sample, >500 spermatozoa, 25 frames/s, region of particle control 10–18 microns, progressive movement of >45% of the parameter straightness (STR), circumferential movement <50% linearity (LIN), depth of field 10, and temperature of the microscope plate 37 °C. The debris incorrectly classified as spermatozoa were manually deleted.

A semen sample volume of 10 µL was placed on Makler counting chamber (Makler®, 10 µm, Sefi Medical Instruments, Haifa, Israel), which was preheated at 37 °C, and the following parameters were evaluated: (1) total and progressive motility (%), (2) spermatozoa with slow/medium/rapid movement (10 < slow < 25 < medium < 45 < rapid µm/s; %), (3) straight line velocity (VSL; µm/s), (4) curvilinear velocity (VCL; µm/s), (5) average path velocity (VAP; µm/s), (6) linearity (LIN; VSL/VCL × 100), (7) straightness (STR; VSL/VAP × 100), (8) wobble (WOB; VAP/VCL × 100), (9) amplitude of lateral head displacement (ALH; µm), (10) beat cross-frequency (BCF; Hz), and (11) hyperactivation (LIN < 0.32, VSL > 97 µm/s, ALH > 3.5 µm; %).

2.6. Sperm Viability

Sperm viability was evaluated applying the eosin-nigrosine staining protocol in one step [20]. Two hundred spermatozoa were estimated in of an optical microscope (×1000; Zeiss, Oberkochen, Germany), to calculate the ratio (%) of live-dead cells.

2.7. Sperm Morphology

According to the manufacturer's instructions, the SpermBlue staining method (SpermBlue®, Microptic S.L., Barcelona, Spain) was applied for the assessment of sperm morphology. For each semen sample, 200 spermatozoa were counted microscopically (×400; Zeiss, Oberkochen, Germany) and the results were described as percentage of spermatozoa with normal morphology or with morphological abnormalities including head and integrity of acrosome membrane, midpiece, tail, and cytoplasmic droplets.

2.8. Sperm Membrane Functionality

For the sperm membranes functionality, the hypo-osmotic swelling test (HOST) was applied according to Vazquez et al. [21] after modification. The HOST solution was prepared (75 mmol/L fructose; 32 mmol/L sodium citrate) and the osmolarity was adjusted to 150 mosm/kg (OSMOMAT® 030, Gonotec, Berlin, Germany). Briefly, for each group, a semen sample of 100 µL was mixed with 1 mL of HOST solution and incubated (37 °C) for 1 h. Finally, 200 spermatozoa per sample were evaluated microscopically (×400; Zeiss, Oberkochen, Germany) and the results were indicated as spermatozoa with functional membrane that is with swollen tails (%).

2.9. Sperm DNA Integrity

Sperm DNA integrity was assessed by the acridine orange test [22], which quantifies the metachromatic shift of acridine orange fluorescence [23]. Specifically, spermatozoa with compact chromatin structure fluoresced green, while those with damaged chromatin integrity fluoresced red. For each semen sample, 200 spermatozoa were counted under a fluorescence microscope (×1000; Zeiss, Oberkochen, Germany) and the results were expressed as spermatozoa with damaged DNA (%).

2.10. Microbiological Analysis

Samples from C and Fe experimental groups were subjected to microbiological analysis for bacterial counts and culture using standard protocols. Preparations of all culture media were made according to the manufacturer's recommendations. Samples were diluted up to 10^{-6} in 0.9% normal saline and 100 µL of each dilution were spread into plate count agar (OXOID) and incubated at 37 °C. The plates were read after 24 and 48 h and the number of colonies formed was reported as colony forming units per mL (cfu/mL). For the detection of frequently isolated bacteria in semen such as *Staphylococcus* spp., *Streptococcus* spp., *Enterobacter* spp., *Bacillus* spp., *Proteus* spp., *Escherichia coli*, *Pseudomonas aeruginosa* [4], 100 µL of each dilution was spread on sheep blood agar (OXOID), MacConkey agar (OXOID), Baird Parker medium with egg yolk tellurite emulsion (OXOID), kanamycin aesculin azide agar (OXOID) and incubated at 37 °C. *Pseudomonas* agar base containing CN selective supplement (OXOID) plates were incubated at 25 °C. Growth of bacterial colonies on plates was monitored and recorded after 24 and 48 h of incubation. Bacterial isolates were then identified using standard microbiological procedures, considering production of haemolysin, culture and colonial characteristics, Gram staining, oxidase- and catalase- reaction, coagulase testing and other conventional biochemical tests when needed.

2.11. Statistical Analysis

Statistical Analysis Systems version 9.3 (SAS Institute Inc., 1996, Cary, NC, USA) was used for the performance of the statistical analysis. The Shapiro-Wilk Test (PROC UNIVARIATE) was applied to test the normality of the data. The parameters Head, Midpiece, Tail abnormalities and Cytoplasmic droplets did not follow a normal distribution and were normalized by square root transformation. For reasons of clarity, the means and SEM of the not transformed data are presented. To conduct the statistical analysis a repeated measures mixed model (PROC MIXED) was applied. The model included group, time and their interaction as fixed effects and boar as a random effect. Semen sample was defined as the subject of the repeated observations. Covariance structure was chosen based on the values of the Akaike information criterion (AIC). Six models were run with different structures (variance components, compound symmetry, unstructured, first-order autoregressive, first-order ante dependence and Toeplitz) and the model with the least AIC was chosen. Pairwise comparisons where performed with the PDIFF command incorporating the Tukey adjustment. Regarding microbiological data-paired differences between the control and Fe group were calculated for every variable and time point by extracting the values of group Fe from control. The normality of the differences was tested using the Shapiro–Wilk test and normality was evident in all cases. A paired t-test was applied to examine the null hypothesis that the true mean was zero. Statistically significant difference was defined as $p < 0.05$.

3. Results

3.1. Nanoparticles' Characteristics

The obtained nanoparticles following the described methodology were identified to be iron oxides with inverse spinel structure according to the XRD diagram (Figure 2). However, the saturation magnetization value which approaches 90 emu/g, stands very close to the expected value for magnetite, indicating Fe_3O_4 as the dominant phase (Figure 2b). In addition, chemical analysis and specifically the determination of Fe^{2+}/Fe^{3+} ratio provides further evidence of the Fe_3O_4 presence, which was roughly 43% with the ideal case of Fe_3O_4 stoichiometry being 50%. SEM imaging was used to find the geometrical characteristics of the sample (Figure 2a). Nanoparticles appear to show a narrow size distribution with the average diameter of the observed spheres estimated around 42 nm.

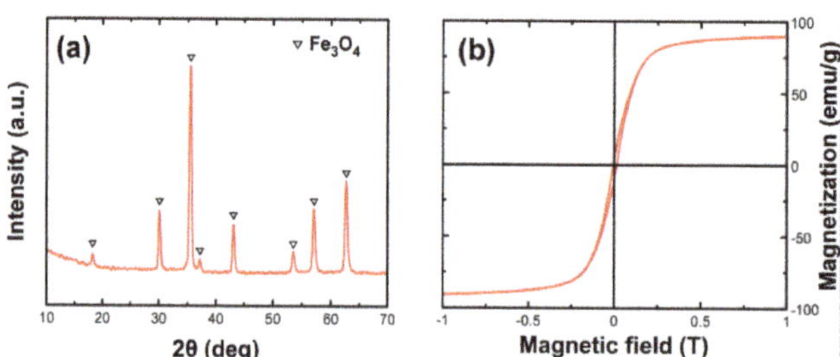

Figure 2. X-ray diffraction (XRD) diagram of the synthesized Fe_3O_4 nanoparticles (**a**) and corresponding magnetic hysteresis loop at room temperature (**b**).

3.2. Efficiency of the Experimental Process

The efficiency of incubation procedure is illustrated by the attachment of magnetic nanoparticles onto the semen. Figure 3b indicates a representative case were nanoparticles aggregates were located in the tail of an isolated spermatozoon.

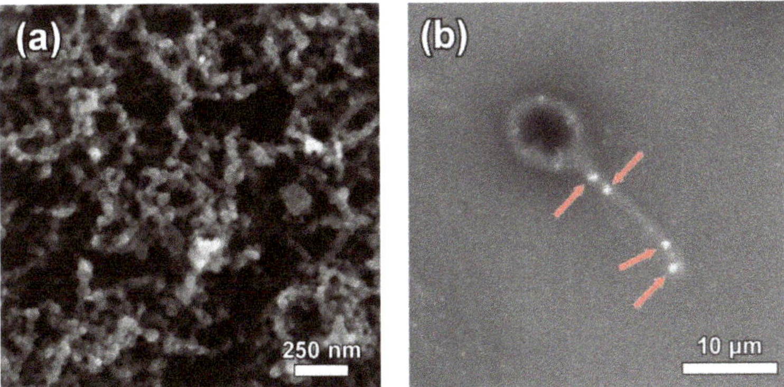

Figure 3. Scanning electron microscope (SEM) image of the synthesized Fe_3O_4 nanoparticles (**a**) and representative picture of nanoparticles attachment on a spermatozoon after incubation (**b**).

3.3. Semen Variables' Assessment

No differences ($p > 0.05$) were observed between the experimental groups in the mean values for all the variables that were assessed in this experiment (Figures 4 and 5, Table 1). However, regarding CASA motility and kinetic parameters, the percentage of total motility ($p = 0.03$) and progressive movement spermatozoa ($p = 0.03$) were less after 24 and 48 h of storage post treatment (0 h) in group C, respectively (Figure 4). For the same group, the percentage of slow movement spermatozoa increased ($p = 0.03$) after 48 h of storage compared to 0 h (Figure 4). In the Fe group, only the LIN decreased ($p = 0.03$) after 48 h of storage compared to 0 h (Figure 4). For the remaining sperm quality and function variables (Table 1 and Figure 5), there were statistical differences only for sperm morphology. Specifically, in the Fe group, the values of spermatozoa with normal morphology decreased ($p = 0.0001$) along the storage period (Figure 5). Regarding the statistical analysis for each category of morphological abnormalities, it was revealed that the deterioration of sperm morphology corresponds only to an increase ($p = 0.0001$) of spermatozoa with head abnormalities in terms of acrosome reacted membrane (Figure 5). Finally, regarding all samples the percentage of DNA fragmentation was 0–1% and no differences were observed between treatments.

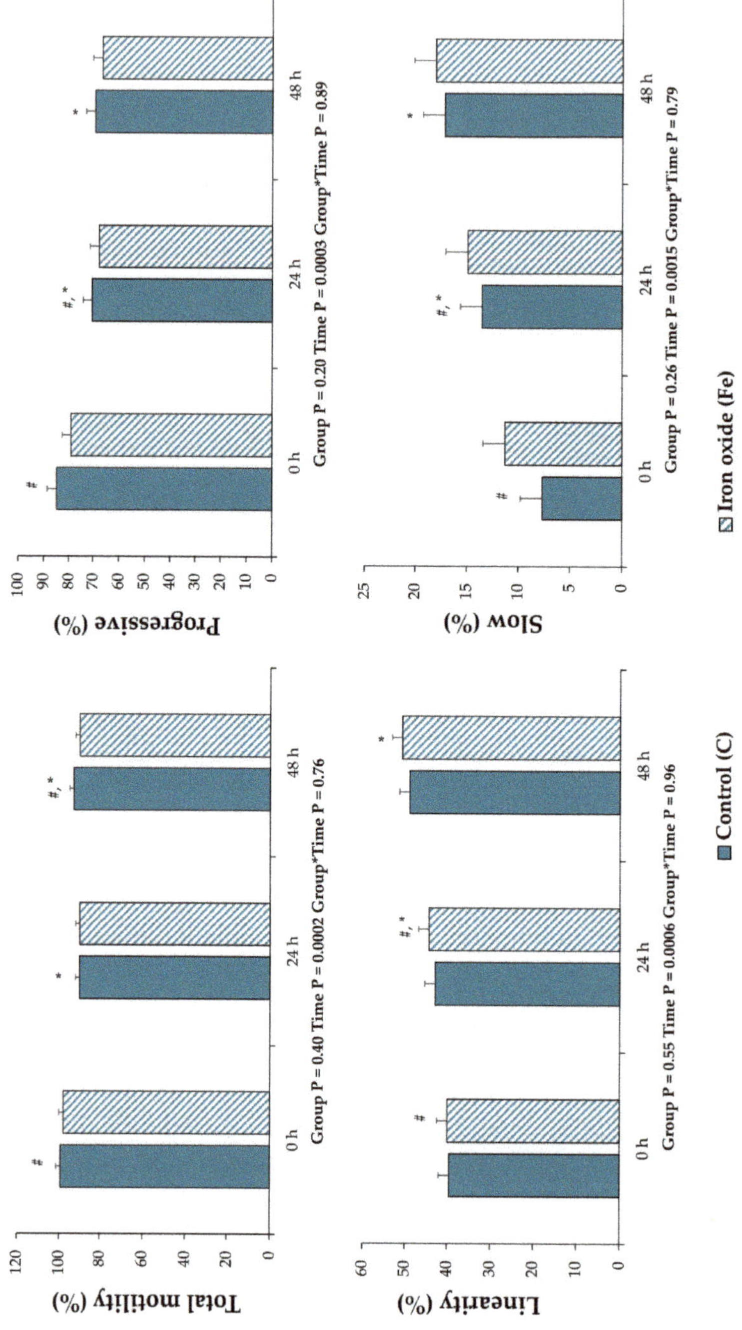

Figure 4. Sperm motility and kinematic parameters of extended boar semen samples at 0, 24 and 48 h of storage post treatment with Fe_3O_4 nanoparticles (NPs). Control group (C): extended boar semen samples without any treatment; Iron oxide group (Fe): extended semen with Fe_3O_4 NPs (0.192 mg Fe_3O_4/mL semen). All the values are expressed as mean ± standard error of the mean (SEM). Different symbols (#, *) denote significant differences between evaluation times within each experimental group.

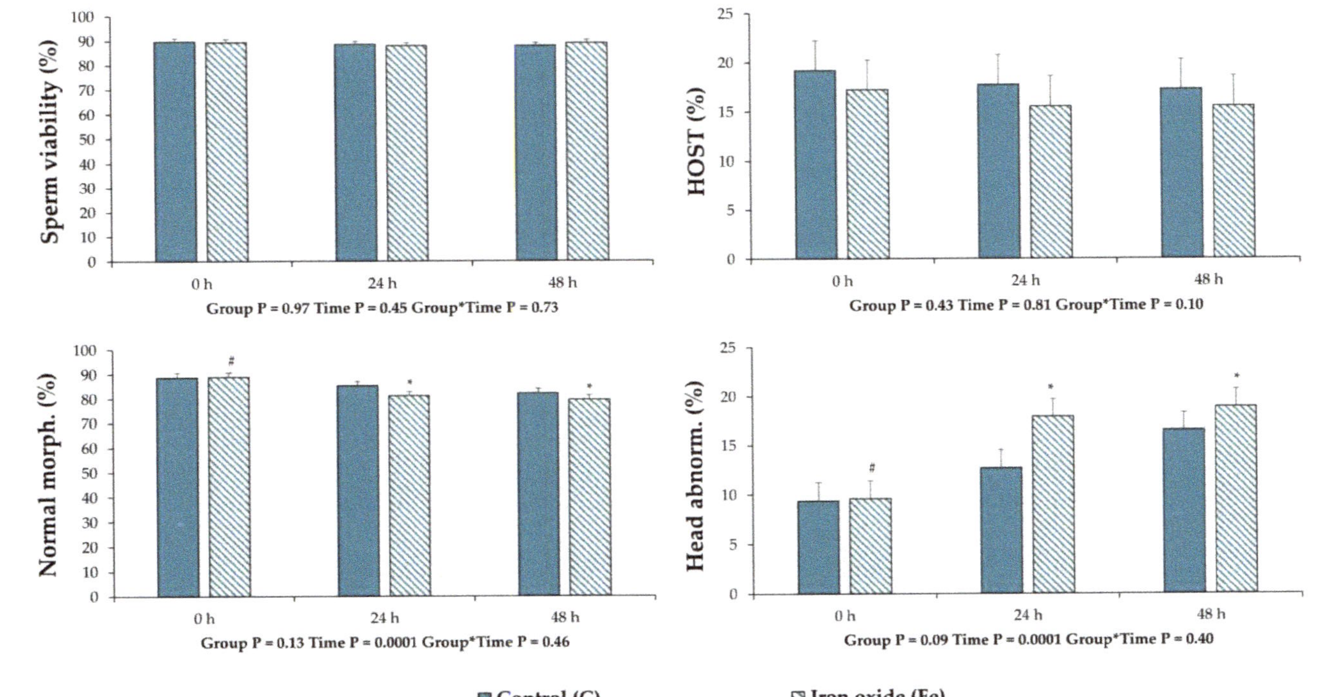

Figure 5. Sperm quality and function variables of extended boar semen samples at 0, 24 and 48 h of storage post treatment with Fe_3O_4 nanoparticles (NPs). Control group (C): extended boar semen samples without any treatment; Iron oxide group (Fe): extended semen with Fe_3O_4 NPs (0.192 mg Fe_3O_4/mL semen). All the values are expressed as mean ± standard error of the mean (SEM). Different symbols (#, *) denote significant differences between evaluation times within each experimental group.

Table 1. Computer-assisted sperm analyzer (CASA) kinematic parameters (mean ± SEM) of extended boar semen samples at 0, 24 h and 48 h of storage post treatment with Fe_3O_4 NPs.

Variable	Group C			Group Fe			p Value		
	0 h	24 h	48 h	0 h	24 h	48 h	Group	Time	G*T
Rapid (%)	68.83 ± 4.68	57.95 ± 4.68	52.17 ± 4.68	64.57 ± 4.68	51.26 ± 4.68	50.19 ± 4.68	0.26	0.0042	0.88
Medium (%)	22.85 ± 2.01	18.60 ± 2.01	23.44 ± 2.01	22.36 ± 2.01	18.72 ± 2.01	21.60 ± 2.01	0.65	0.09	0.88
VCL (μm/sec)	64.65 ± 3.14	62.66 ± 3.14	55.62 ± 3.14	61.57 ± 3.14	58.14 ± 3.14	55.51 ± 3.14	0.32	0.06	0.77
VSL (μm/sec)	25.02 ± 2.52	28.39 ± 2.52	30.39 ± 2.52	24.04 ± 2.52	27.07 ± 2.52	29.66 ± 2.52	0.63	0.10	0.99
VAP (μm/sec)	47.48 ± 3.09	43.32 ± 3.09	36.72 ± 3.09	46.42 ± 3.09	41.82 ± 3.09	37.78 ± 3.09	0.84	0.0107	0.91
ALH (μm)	2.75 ± 0.16	2.52 ± 0.16	2.29 ± 0.16	2.56 ± 0.16	2.30 ± 0.16	2.28 ± 0.16	0.29	0.07	0.80
BCF (Hz)	5.49 ± 2.45	5.57 ± 2.45	5.32 ± 2.45	5.26 ± 2.45	5.52 ± 2.45	5.14 ± 2.45	0.43	0.45	0.93
STR (%)	49.07 ± 2.85	49.76 ± 2.85	53.84 ± 2.85	52.81 ± 2.85	51.12 ± 2.85	55.69 ± 2.85	0.32	0.26	0.91
WOB (%)	72.90 ± 1.77	75.10 ± 1.77	75.77 ± 1.77	74.39 ± 1.77	76.87 ± 1.77	76.27 ± 1.77	0.39	0.31	0.93
Hyper (%)	0.74 ± 0.43	1.08 ± 0.43	0.84 ± 0.43	0.93 ± 0.43	0.99 ± 0.43	1.31 ± 0.43	0.39	0.16	0.53

Group C: untreated extended semen sample, and Group Fe: extended semen sample treated with Fe nanoparticles. Time points 0, 24 and 48 h: storage time post removal of Fe_3O_4 NPs. Rapid Medium: rapid, medium movement spermatozoa (%; 25<medium<45<rapid μm/sec); VCL: curvilinear velocity (μm/sec); VSL: straight line velocity (μm/sec); VAP: average path velocity (μm/sec); STR: straightness (VSL/VAP × 100); WOB: wobble (VAP/VCL × 100); ALH: amplitude of lateral head displacement (μm); BCF: beat/cross-frequency (Hz); Hyper.: hyperactive spermatozoa (%). G*T: Group*Time interaction.

3.4. Microbiological Results

Bacterial growth was present in all semen samples. However, the microbial load varied. The most prevalent bacteria belonged to the Enterobacteriaceae family, *Staphylococcus* spp., *Enterococcus* spp. and *Pseudomonas* spp. The latter was found not to be affected by the examined NPs when detected at the specific concentration (data not shown). Total bacterial count in boar semen was respectively low and the number of cfu/mL demonstrated a wide range of microbial load among samples (from 45 to 1855, min–max, respectively). Treatment with Fe_3O_4 NPs did not eliminate bacterial content (Figure 6). However, a statistically significant reduction of the microbial load of semen was evident ($p = 0.03$) (Table 2). Among the other detected bacteria staphylococci tended to be less on Fe group compared to control, while Enterobacteriaceae (*Enterobacter* spp., *E. coli*, *Proteus* spp) and *Enterococcus* spp. had no significant difference from the control group (Table 2).

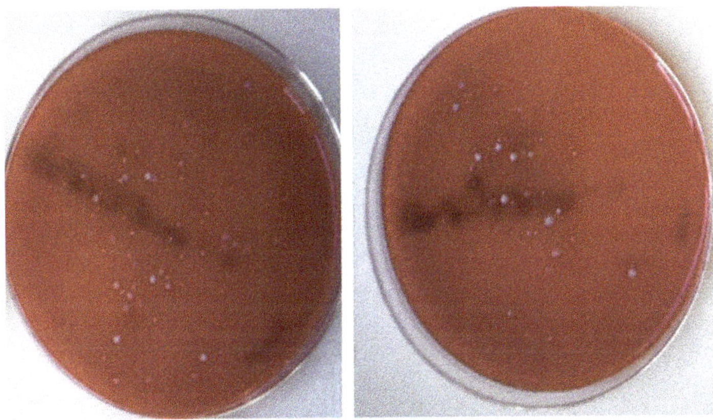

Figure 6. Culture on sheep blood agar [a: control group (C), b: iron oxide group (Fe)]. Reduction of the microbial load after 24 h of incubation (37 °C).

Table 2. Microorganisms (cfu/mL) isolated from boar semen samples 24 h and 48 h after incubation on blood agar and selective culture media.

Variable	Control	Fe	Difference (Control-Fe)	*p*-Value
Blood Agar 24 h	558 ± 455	443 ± 381	115 ± 103	0.03
Blood Agar 48 h	779 ± 651	616 ± 515	164 ± 150	0.03
Staphylococcus spp	314 ± 411	251 ± 380	63 ± 92	0.06
Enterococcus spp	5.4 ± 7.6	4.5 ± 8.7	0.9 ± 7.9	0.72
Enterobacteriaceae	69.9 ± 106.6	60.1 ± 86.5	9.8 ± 27.5	0.29

4. Discussion

Despite the potential biological benefits of NPs, nanotoxicity and its impact on cells' health is a concern [14]. Previous studies reported that Fe_3O_4 NPs affect rainbow trout sperm [24], whilst titanium dioxide TiO_2 NPs negatively affect mouse gene expression of Leydig cells, as well as semen quality parameters [25]. No studies were found to report the effects of Fe_3O_4 NPs on boar semen. Thus, the first aim of this study was to perform a full laboratory assessment of boar semen processing with iron oxide NPs. This study was based on previous findings of our research team [15], that explored the minimum inhibitory concentration of Fe_3O_4 NPs (0.192 mg/mL semen) and the appropriate co-incubation time of semen with NPs (30 min). However, the present study assessed the full profile regarding boar sperm characteristics.

The process of fertilization is a complex of sequencing events, involving the normal movement of the spermatozoa to reach the oviduct, the approach to oocyte, the sperm acrosome reaction, the sperm penetration into the ooplasm, the merging of the gametes, the fusion of the pronuclei and the intermingling of the paternal and maternal chromosomes. Low semen quality, as expressed by variables from semen analysis, is a common cause of subfertility or infertility [24]. This study provided a protocol for co-incubation of sperm with NPs, regarding both time and concentration, which had no detrimental effect on semen parameters. No effects were observed regarding sperm viability, morphology, membrane functionality, DNA integrity and CASA analyzed kinematics. This is an important finding, as it realizes the use of Fe_3O_4 NPs in boar semen handling with no toxicity. The sperm parameters analyzed in this study are of paramount importance for the fertilizing capacity of boar semen. Many researchers highlighted that motility is better correlated to field fertility compared to other kinetic parameters [26,27]. Moreover, Broekhuijse et al. [28,29] showed that CASA parameters, like progressive motility, BCF and VCL, could be related to farrowing rate, while the total number of born piglets could be affected by total motility, ALH, VSL and VAP [26,27]. In accordance with these findings, Holt et al. [30] showed that the VSL could positively affect the litter size. In vitro fertility of boar sperm has been positively correlated with progressive motility, VAP and VSL, and negatively correlated with STR, LIN and ALH [31]. Also, it is well accepted that the more diagnostic tests performed (such as the assessment of sperm morphology, motility and chromatin integrity), the better the prediction of in vitro fertility that can be achieved [32,33]. None of the above-mentioned sperm parameters were affected in our study. It seems that the restricted period of sperm co-incubation with Fe_3O_4 NPs and the gentle removal of them with a magnetic field, protected boar spermatozoa from NPs toxicity. This is in agreement with previous reports, that suggest a hypothesis that the time of interaction between semen and NPs can be crucial regarding their potential toxic activity [34,35]. In accordance with this scientific hypothesis, a significant decrease of VCL, VSL and VAP was observed after a prolonged (24 h) incubation of rainbow trout semen with Fe_3O_4 NPs [24].

In the Fe group, the value of spermatozoa with acrosome-reacted membrane was increased during storage time, which was not the case for the control group. However, even if this effect was statistically significant, the reported numerical values are not indicative of an important biological consequence. This finding could be attributed to the presence of the examined NPs. It is known that NPs penetrate the cells' membranes affecting their physiology [36]. Although it was not within the purposes of the present study, according to the literature, the most prevalent mechanism of action of nanoparticles is related to the induction of oxidative stress [37]. Subsequently, the increased production of ROS can irreversibly affect the membranes of spermatozoa and can be involved in sperm capacitation [38], leading to the perturbation of the acrosome membrane's integrity [39]. Therefore, the reported increase in spermatozoa with reacted acrosome membrane could be an undesirable characteristic for extended liquid semen used in AI programs, but it could be a perspective for IVF protocols, in which the induction of acrosome reaction is a prerequisite.

Semen extenders are cell culture media and thus an ideal environment for bacteria proliferation. There are directives from the European Union [40] and from national governments that specify the antibiotics' category and dose as semen extenders' additives. A variety of antibiotic compounds have been used to control microbial contamination in extenders. In farm animals' AI, including boars, streptomycin and penicillin are the most widely used antibiotics [41]. Additionally, antibiotics, like gentamicin, linco-spectin and clindamycin have been successfully implemented in different semen extenders [42,43]. During the last decades, the bacterial resistance to antibiotics has been a serious problem to humans' as well as animals' health. Moreover, some antibiotics, at certain concentrations, may have a direct detrimental effect on spermatozoa [44]. Contemporary studies have proved that some antimicrobials may have a deleterious effect on bull [45] and equine [46] spermatozoa. In this aspect, NPs have been an interesting alternative for the scientists. In the present study, the examined iron oxide NPs had no toxic effect on boar spermatozoa and showed a slight antibacterial effect, although not for all bacteria species. Boar semen contaminants were present in all samples of the study, but the microbial load varied between them. Total bacterial count for aerobic mesophiles was up to 1.8×10^3, while Gączarzewicz et al. [4] reported findings up to 360×10^6. Despite the different degrees of contamination, a significant reduction regarding the total microbial count was observed in the presence of Fe_3O_4 NPs compared to control after 24 and 48 h of incubation. Given the initial low microbial load, this reduction suggests a promising result for the use of Fe NPs in heavier bacterial contaminations. After culture on selective media, the antimicrobial activity of Fe_3O_4 NPs demonstrated variation depending on the strain. The predominant bacteria on the samples were *Staphylococci* and Enterobacteriaceae, similar to studies previously reported [4,47]. Treatment with Fe_3O_4 NPs reduced the number of viable *Staphylococci*, though had a minimum effect on the Gram negative Enterobacteriaceae. In this field, reports demonstrate the bactericidal properties of iron oxide NPs (50 and 100 μg/mL) against *Shigella dysentery* and *Escherichia coli* [48] as well as their antimicrobial effects on antibiotic-resistant strains of *E. coli* [49]. Others report that iron NPs have only moderate antimicrobial action against *Escherichia coli* and *Bacillus subtilis* but remain potentially useful for application in the pharmaceutical and biomedical industry [50]. Studies have reported NPs to have excellent antimicrobial resistance properties and the ability to inhibit the formation of bacterial biofilms [51,52], which favor their use in drug-delivery systems. Concerning their mechanism of action, under specific conditions iron oxide nanoparticles may provide significant antimicrobial activity in contact with common bacteria. The interaction is stronger when nanoparticles' surface is positively charged and when it involves the presence of both Fe^{2+} and Fe^{3+}. Both properties are met in uncoated Fe_3O_4 nanoparticles distributed in the pH of biological environment. It has been reported that positively-charged Fe_3O_4 nanoparticles indicate stronger interaction with bacteria while the addition of surfactants counterbalances surface charge and limits such tendency [53]. An occurring mechanism involves the attachment of Fe_3O_4 nanoparticles on the bacteria membrane, and the dissolution of Fe^{2+} and Fe^{3+} at their interface which initiates the generation of reactive oxygen species. This triggers hydrogen peroxide release and production of free radicals through a Fenton reaction:

$$Fe^{3+} + H_2O_2 \rightarrow Fe^{2+} + OH^- + OH^\bullet$$

$$Fe^{2+} + H_2O_2 \rightarrow Fe^{3+} + HO_2^\bullet + H^+$$

The presence of toxic free radicals on bacteria membrane causes its electrostatic modification inducing chemical stress that causes the damage of the bacteria unit. The antimicrobial potential of Fe_3O_4 nanoparticles is preserved until their surface gets fully oxidized to γ-Fe_2O_3. Armijo et al. [54] findings suggest that Fe_3O_4 NPs are potential alternatives to silver NPs in several antibacterial applications minimizing the cost and enhancing microbial inactivation and elimination. The results of our research are aligned with these findings, reinforcing the future utilization of the antibacterial activity of Fe_3O_4 NPs as a new perspective to prevent bacteria in semen.

In conclusion, the Fe_3O_4 NPs examined are a potential useful and effective semen supplementation with antibacterial properties. Moreover, the combination of NPs with conventional antibiotics could enhance their antibacterial action and thus reduce the dose demanded. It is suggested that further studies regarding the oxidative status of boar semen treated with Fe_3O_4 NPs should be carried out to investigate the mechanism of action on boar spermatozoa.

Author Contributions: Conceptualization, I.A.T., T.S. and C.M.B.; methodology, I.A.T., T.S., C.M.B. and G.T.; formal analysis, G.T.; investigation, I.A.T., T.S., S.A., A.B., A.N., I.M., K.S., I.K., G.T., M.A. and C.M.B.; data curation, S.A., A.B., A.N., I.M., K.S. and I.K.; writing—original draft preparation, I.A.T., S.A., K.S. and A.B.; writing—review and editing, C.M.B., T.S. and G.T.; visualization, I.A.T., G.T., S.A., A.B. and K.S.; supervision, C.M.B., I.A.T. and T.S.; project administration, C.M.B.; funding acquisition, C.M.B. All authors have read and agreed to the published version of the manuscript.

Funding: This research was co-financed by Greece and the European Union (European Social Fund—ESF) through the Operational Programme Human Resources Development, Education and Lifelong Learning 2014–2020 in the context of the project "Investigation of nanotechnological applications on farm animals" semen processing "(MIS 5004681)". Part of the measurements received funding from the EU H2020 research and innovation program under grant agreement No. 654360, having benefitted from the access provided by ICMAB-CSIC and Universitat Autonoma de Barcelona in Bellaterra-Barcelona within the framework of the NFFA Europe Transnational Access Activity.

Acknowledgments: The authors are grateful to "Karanikas farm" for kindly providing the semen samples and to Theofanis Avramidis for his contribution to semen doses preparation. Also, authors would like to thank Anna Esther Carrillo and Bernat Bozzo for the experimental assistance during SEM observations and SQUID measurements.

Conflicts of Interest: The authors declare no conflict of interest.

References

1. Goldberg, A.M.; Argenti, L.E.; Faccin, J.E.; Linck, L.; Santi, M.; Bernardi, M.L.; Cardoso, M.R.I.; Wentz, I.; Bortolozzo, F.P. Risk factors for bacterial contamination during boar semen collection. *Res. Vet. Sci.* **2013**, *95*, 362–367. [CrossRef] [PubMed]

2. Kuster, C.E.; Althouse, G.C. The impact of bacteriospermia on boar sperm storage and reproductive performance. *Theriogenology* **2016**, *85*, 21–26. [CrossRef] [PubMed]

3. Maes, D.; Nauwynck, H.; Rijsselaere, T.; Mateusen, B.; Vyt, P.; de Kruif, A.; Soom, A.V. Diseases in swine transmitted by artificial insemination: An overview. *Theriogenology* **2008**, *70*, 1337–1345. [CrossRef]

4. Gączarzewicz, D.; Udała, J.; Piasecka, M.; Błaszczyk, B.; Stankiewicz, T. Bacterial contamination of boar semen and its relationship to sperm quality preserved in commercial extender containing gentamicin sulfate. *Pol. J. Vet. Sci.* **2016**, *19*, 451–459. [CrossRef] [PubMed]

5. Althouse, G.C.; Lu, K.G. Bacteriospermia in extended porcine semen. *Theriogenology* **2005**, *63*, 573–584. [CrossRef] [PubMed]

6. Morrell, J.; Wallgren, M. Alternatives to antibiotics in semen extenders: A review. *Pathogens* **2014**, *3*, 934–946. [CrossRef]

7. Schulze, M.; Dathe, M.; Waberski, D.; Müller, K. Liquid storage of boar semen: Current and future perspectives on the use of cationic antimicrobial peptides to replace antibiotics in semen extenders. *Theriogenology* **2016**, *85*, 39–46. [CrossRef]

8. Speck, S.; Courtiol, A.; Junkes, C.; Dathe, M.; Müller, K.; Schulze, M. Cationic synthetic peptides: Assessment of their antimicrobial potency in liquid preserved boar semen. *PLoS ONE* **2014**, *9*, e105949. [CrossRef]

9. Guzman, M.; Dille, J.; Godet, S. Synthesis and antibacterial activity of silver nanoparticles against gram-positive and gram-negative bacteria. *Nanomed. Nanotechnol. Biol. Med.* **2012**, *8*, 37–45. [CrossRef]

10. Li, W.R.; Xie, X.B.; Shi, Q.S.; Zeng, H.Y.; Ouyang, Y.S.; Chen, Y.B. Antibacterial activity and mechanism of silver nanoparticles on Escherichia coli. *Appl. Microbiol. Biotechnol.* **2010**, *85*, 1115–1122. [CrossRef]

11. Shahverdi, A.R.; Fakhimi, A.; Shahverdi, H.R.; Minaian, S. Synthesis and effect of silver nanoparticles on the antibacterial activity of different antibiotics against Staphylococcus aureus and Escherichia coli. *Nanomed. Nanotechnol. Biol. Med.* **2007**, *3*, 168–171. [CrossRef] [PubMed]

12. Irshad, R.; Tahir, K.; Li, B.; Ahmad, A.R.; Siddiqui, A.; Nazir, S. Antibacterial activity of biochemically capped iron oxide nanoparticles: A view towards green chemistry. *J. Photochem. Photobiol. B Biol.* **2017**, *170*, 241–246. [CrossRef] [PubMed]

13. Toledo, L.D.A.S.; Rosseto, H.C.; Bruschi, M.L. Iron oxide magnetic nanoparticles as antimicrobials for therapeutics. *Pharmaceut. Dev. Technol.* **2018**, *23*, 316–323. [CrossRef] [PubMed]

14. Sahu, D.; Kannan, G.M.; Tailang, M.; Vijayaraghavan, R. In vitro cytotoxicity of nanoparticles: A comparison between particle size and cell type. *J. Nanosci.* **2016**, *4023852*. [CrossRef]

15. Basioura, A.; Michos, I.; Ntemka, A.; Karagiannis, I.; Boscos, C.M. Effect of iron oxide and silver nanoparticles on boar semen CASA motility and kinetics. *J. Hellenic Vet. Med. Soc.* **2020**. accepted for publication.

16. Asimakidou, T.; Makridis, A.; Veintemillas-Verdaguer, S.; Morales, M.P.; Kellartzis, I.; Mitrakas, M.; Vourlias, G.; Angelakeris, M.; Simeonidis, K. Continuous production of magnetic iron oxide nanocrystals by oxidative precipitation. *Chem. Eng. J.* **2020**, *393*, 124593. [CrossRef]

17. Luengo, Y.; Morales, M.P.; Gutiérrez, L.; Verdaguer, S.V. Counterion and solvent effects on the size of magnetite nanocrystals obtained by oxidative precipitation. *J. Mater. Chem.* **2016**, *4*, 9482–9488. [CrossRef]

18. *Powder Diffraction File*; International Centre for Diffraction Data: Newtown Square, PA, USA, 2004; Joint center for powder diffraction studies.

19. *M100-S25 Performance Standards for Antimicrobial Susceptibility Testing*; Clinical and Laboratory Standards Institute (CLSI): Wayne, PA, USA, 2015; Twenty-fifth informational supplement.
20. World Health Organization (WHO). *Laboratory Manual for the Examination of Human Sperm and Semen-Cervical Mucus Interaction*, 5th ed.; WHO Press: Geneva, Switzerland, 2010.
21. Vazquez, J.M.; Martinez, E.A.; Martinez, P.; Artiga, C.G.; Roca, J. Hypoosmotic swelling of boar spermatozoa compared to other methods for analysing the sperm membrane. *Theriogenology* **1997**, *47*, 913–922. [CrossRef]
22. Tsakmakidis, I.A.; Lymberopoulos, A.G.; Khalifa, T.A.A.; Boscos, C.M.; Saratsi, A.; Alexopoulos, C. Evaluation of zearalenone and α-zearalenol toxicity on boar sperm DNA integrity. *J. Appl. Toxicol.* **2008**, *28*, 681–688. [CrossRef]
23. Silva, P.F.N.; Gadella, B.M. Detection of damage in mammalian sperm cells. *Theriogenology* **2006**, *65*, 958–978. [CrossRef]
24. Özgür, M.E.; Ulu, A.; Balcıoğlu, S.; Özcan, I.; Köytepe, S.; Ates, B. The toxicity assessment of iron oxide (Fe$_3$O$_4$) nanoparticles on physical and biochemical quality of rainbow trout spermatozoon. *Toxics* **2018**, *6*, 62. [CrossRef] [PubMed]
25. Komatsu, T.; Tabata, M.; Irie, M.K.; Shimizu, T.; Suzuki, K.I.; Nihei, Y.; Takeda, K. The effects of nanoparticles on mouse testis Leydig cells in vitro. *Toxicol. Vitr.* **2008**, *22*, 1825–1831. [CrossRef] [PubMed]
26. Blondin, P.; Beaulieu, M.; Fournier, V.; Morin, N.; Crawford, L.; Madan, P.; King, W.A. Analysis of bovine sexed sperm for IVF from sorting to the embryo. *Theriogenology* **2009**, *71*, 30–38. [CrossRef] [PubMed]
27. Tardif, S.; Laforest, J.P.; Cormier, N.; Bailey, J.L. The importance of porcine sperm parameters on fertility in vivo. *Theriogenology* **1999**, *52*, 447–459. [CrossRef]
28. Broekhuijse, M.L.; Šoštarić, E.; Feitsma, H.; Gadella, B.M. Application of computer-assisted semen analysis to explain variations in pig fertility. *J. Anim. Sci.* **2012**, *90*, 779–789. [CrossRef]
29. Broekhuijse, M.L.; Feitsma, H.; Gadella, B.M. Field data analysis of boar semen quality. *Reprod. Domest. Anim.* **2011**, *46*, 59–63. [CrossRef]
30. Holt, C.; Holt, W.V.; Moore, H.D.M.; Reed, H.C.B.; Curnock, R.M. Objectivily measured boar sperm motility parameters correlate with the outcomes of on-farm inseminations: Results of two fertility trials. *J. Androl.* **1997**, *18*, 312–323. [CrossRef]
31. Kathiravan, P.; Kalatharan, J.; Karthikeya, G.; Rengarajan, K.; Kadirvel, G. Objective sperm motion analysis to assess dairy bull fertility using computer-aided system-a review. *Reprod. Domest. Anim.* **2011**, *46*, 165–172. [CrossRef]
32. Tsakmakidis, I.A.; Lymberopoulos, A.G.; Khalifa, T.A.A. Relationship between sperm quality traits and field-fertility of porcine semen. *J. Vet. Sci.* **2010**, *11*, 151–154. [CrossRef]
33. Gillan, L.; Kroetsch, T.; Chis Maxwell, W.M.; Evans, G. Assessment of in vitro sperm characteristics in relation to fertility in dairy bulls. *Anim. Reprod. Sci.* **2008**, *103*, 201–214. [CrossRef]
34. Feugang, J.M.; Shengfa, F.L.; Crenshaw, M.A.; Clemente, H.; Willard, S.T.; Ryan, P.L. Lectin-functionalized magnetic iron oxide nanoparticles for reproductive improvement. *JFIV Reprod. Med. Genet.* **2014**, *03*, 1–5. [CrossRef]
35. Odhiambo, J.F.; DeJarnette, J.M.; Geary, T.W.; Kennedy, C.E.; Suarez, S.S.; Sutovsky, M.; Sutovsky, P. Increased conception rates in beef cattle inseminated with nanopurified bull semen1. *Biol. Reprod.* **2014**, *91*, 1–10. [CrossRef] [PubMed]
36. Brooking, J.; Davis, S.S.; Illum, L. Transport of nanoparticles across the rat nasal mucosa. *J. Drug Target.* **2001**, *9*, 267–279. [CrossRef] [PubMed]
37. Wang, R.; Song, B.; Wu, J.; Zhang, Y.; Chen, A.; Shao, L. Potential adverse effects of nanoparticles on the reproductive system. *Int. J. Nanomed.* **2018**, *13*, 8487–8506. [CrossRef]
38. Lamirande, E.D.; Tsai, C.; Harakat, A.; Gagnon, C. Involvement of reactive oxygen species in human sperm arcosome reaction induced by A23187, lysophosphatidylcholine, and biological fluid ultrafiltrates. *J. Androl.* **1998**, *19*, 585–594. [CrossRef]
39. Yoshida, Y.; Itoh, N.; Saito, Y.; Hayakawa, M.; Niki, E. Application of water-soluble radical initiator, 2,2′-azobis-[2-(2-imidazolin-2-yl)propane] dihydrochloride, to a study of oxidative stress. *Free Radic. Res.* **2004**, *38*, 375–384. [CrossRef]
40. Council Directive, European Union, 90/429/EEC. Available online: http://faolex.fao.org/docs/pdf/eur110397.pdf (accessed on 28 October 2014).
41. Salamon, S.; Maxwell, W.M. Storage of ram semen. *Anim. Reprod. Sci.* **2000**, *62*, 77–111. [CrossRef]

42. Ahmad, K.; Foote, R.H.; Kaproth, M. Antibiotics for bull semen frozen in milk and egg yolk extenders. *J. Dairy Sci.* **1987**, *70*, 2439–2443. [CrossRef]

43. Gadea, J. Review: Semen extenders used in the artificial inseminarion of swine. *Span. J. Agric. Res.* **2003**, *1*, 17. [CrossRef]

44. Lorton, S.P.; Sullivan, J.J.; Bean, B.; Kaproth, M.; Kellgren, H.; Marshall, C. A new antibiotic combination for frozen bovine semen. 2. Evaluation of seminal quality. *Theriogenology* **1988**, *29*, 593–607. [CrossRef]

45. Azawi, O.I.; Ismaeel, M.A. Influence of addition of different antibiotics in semen diluent on viable bacterial count and spermatozoal viability of awassi ram semen. *Vet. World* **2012**, *5*, 75–79. [CrossRef]

46. Varner, D.D.; Scanlan, C.M.; Thompson, J.A.; Brumbaugh, G.W.; Blanchard, T.L.; Carlton, C.M.; Johnson, L. Bacteriology of preserved stallion semen and antibiotics in semen extenders. *Theriogenology* **1998**, *50*, 559–573. [CrossRef]

47. Oehler, C.; Janett, F.; Schmitt, S.; Malama, E.; Bollwein, H. Development of a flow cytometric assay to assess the bacterial count in boar semen. *Theriogenology* **2019**, *133*, 125–134. [CrossRef] [PubMed]

48. Saqib, S.; Munis, M.F.H.; Zaman, W.; Ullah, F.; Shah, S.N.; Ayaz, A.; Farooq, M.; Bahadur, S. Synthesis, characterization and use of iron oxide nano particles for antibacterial activity. *Microsc. Res. Tech.* **2019**, *82*, 415–420. [CrossRef]

49. Gabrielyan, L.; Hakobyan, L.; Hovhannisyan, A.; Trchounian, A. Effects of iron oxide (Fe_3O_4) nanoparticles on Escherichia coli antibiotic-resistant strains. *J. Appl. Microbiol.* **2019**, *126*, 1108–1116. [CrossRef] [PubMed]

50. Margabandhu, M.; Sendhilnathan, S.; Maragathavalli, S.; Karthikeyan, V.; Annadurai, B. Synthesis characterization and antibacterial activity of iron oxide nanoparticles. *Global. J. Biosci. Biotechnol.* **2015**, *4*, 335–341.

51. Wang, L.; Hu, C.; Shao, L. The antimicrobial activity of nanoparticles: Present situation and prospects for the future. *Int. J. Nanomed.* **2017**, *12*, 1227–1249. [CrossRef]

52. Ramos, M.A.D.S.; Silva, P.B.D.; Spósito, L.; Toledo, L.G.D.; Bonifácio, B.V.; Rodero, C.F.; Santos, K.C.D.; Chorilli, M.; Bauab, T.M. Nanotechnology-based drug delivery systems for control of microbial biofilms: A review. *Int. J. Nanomed.* **2018**, *13*, 1179–1213. [CrossRef]

53. Arakha, M.; Pal, S.; Samantarrai, D.; Panigrahi, T.K.; Mallick, B.C.; Pramanik, K.; Mallick, B.; Jha, S. Antimicrobial activity of iron oxide nanoparticle upon modulation of nanoparticle-bacteria interface. *Sci. Rep.* **2015**, *5*, 14813. [CrossRef]

54. Armijo, L.M.; Wawrzyniec, S.J.; Kopciuch, M.; Brandt, Y.I.; Rivera, A.C.; Withers, N.J.; Cook, N.C.; Huber, D.L.; Monson, T.C.; Smyth, H.D.C.; et al. Antibacterial activity of iron oxide, iron nitride, and tobramycin conjugated nanoparticles against pseudomonas aeruginosa biofilms. *J. Nanobiotechnol.* **2020**, *18*, 35. [CrossRef]

Article

Bio-Mediated Synthesis of Reduced Graphene Oxide Nanoparticles from *Chenopodium album*: Their Antimicrobial and Anticancer Activities

Mohammad Faisal Umar [1], Faizan Ahmad [2], Haris Saeed [3], Saad Ali Usmani [2], Mohammad Owais [3] and Mohd Rafatullah [1,*]

[1]   School of Industrial Technology, Universiti Sains Malaysia, Penang 11800, Malaysia; faisalumar@student.usm.my
[2]   Department of Post Harvest Engineering and Technology, Faculty of Agricultural Sciences, Aligarh Muslim University, Aligarh 202002, India; faizan.phe@amu.ac.in (F.A.); sausmani@myamu.ac.in (S.A.U.)
[3]   Molecular Immunology Lab, Interdisciplinary Biotechnology Unit, Aligarh Muslim University, Aligarh 202002, India; haris6229@gmail.com (H.S.); owais_lakhnawi@yahoo.com (M.O.)
*   Correspondence: mohd_rafatullah@yahoo.co.in or mrafatullah@usm.my; Tel.: +60-465-321-11; Fax: +60-465-6375

Received: 3 May 2020; Accepted: 28 May 2020; Published: 1 June 2020

Abstract: A novel method of preparing reduced graphene oxide (RGOX) from graphene oxide (GOX) was developed employing vegetable extract, *Chenopodium album*, as a reducing and stabilizing agent. *Chenopodium album* is a green leafy vegetable with a low shelf life, fresh leaves of this vegetable are encouraged to be used due to high water content. The previously modified 'Hummers method' has been in practice for the preparation of GOX by using precursor graphite powder. In this study, green synthesis of RGOX was functionally verified by employing FTIR and UV-visible spectroscopy, along with SEM and TEM. Our results demonstrated typical morphology of RGOX stacked in layers that appeared as silky, transparent, and rippled. The antibacterial activity was shown by analyzing minimal inhibitory concentration values, agar diffusion assay, fluorescence techniques. It showed enhanced antibacterial activity against Gram-positive and Gram-negative bacteria in comparison to GOX. It has also been shown that the synthesized compound exhibited enhanced antibiofilm activity as compared to its parent compound. The efficacy of RGOX and GOX has been demonstrated on a human breast cancer cell line, which suggested RGOX as a potential anticancer agent.

Keywords: graphite powder; graphene oxide; reduced graphene oxide; *Chenopodium album*; anticancer; antimicrobial

---

1. Introduction

Ever rising consciousness about human health, the addition of fruits and vegetables are gaining importance in their regular diet [1]. It has been reported that 39% of food waste occurs in the food production industry, and overall waste was approximately 126.2 million tons in 2020 [2]. Some parts of agricultural waste can effectively be utilized for the synthesis of graphene, and the same can also be used to obtain reduced graphene oxide (RGOX) from graphene oxide (GOX) [3].

Chenopodium album is also known as 'bathua', a leafy vegetable having a low shelf life due to its high moisture content [4]. It cannot be preserved for a longer duration, and it goes as a waste product, so liquid extraction of *Chenopodium album* can be stored and used as a reducing agent as well as a stabilizing agent. *Chenopodium album* is an excellent source of high-grade vitamins, proteins, nutrients and antioxidant predominantly retinol and ascorbic acid. The other health benefits of fiber rich *Chenopodium album* leaves include its laxative properties helpful in curing constipation, which in turn are useful in the treatment of piles. Furthermore, it improves the hemoglobin level, appetite, and purifies blood, which is beneficial for the human heart [5].

It was reported that graphene oxide nanosheets with enhanced peroxidize like activity used for electrochemical cancer cell detection and targeted therapeutics [6]. Furthermore, there were many reports which show the efficacy of graphene oxide-graphene quantum dots on cancerous cells [7]. The researchers used ascorbic acid, amino acids, reducing sugars, and protein bovine serum albumin as reducing as well as stabilizing agents [8]. The reducing chemicals are highly toxic, hazardous and may introduce impurities into RGOX, which is harmful to the environment and may impart adverse effects on its biological applications [9,10]. It was also reported that graphene oxide with silver nanocomposites exhibits excellent anticancer properties [11]. Guo and Mei reported that the graphene oxide with silver nanocomposites have bactericidal activity against *E. coli* through disrupting the integrity of the bacterial cell wall and showed a bacteriostatic effect on *S. aureus* cell division [12]. Another report demonstrated that GOX could be used in biomedical applications such as targeted drug delivery to the lung. Additionally, properties such as significant retention, better accumulation, remarkable pathological changes make GOX an efficient drug [13,14]. Contrary to the above facts, GOX is also reported to be involved in oxidative stress formation and induces cell cytotoxicity at high concentrations [15].

The formation of RGOX by different methods such as thermal [16], electrochemical [17], or chemical reductions process [18,19] has been reported in various research. In this study, graphite was oxidized to GOX, while conversion of graphene oxide (GOX) to RGOX was assigned to the reduction method with different strong reducing agents such as hydrazine (N_2H_2) [20] and sodium hydride (NaH) [21]. These reducing agents are highly toxic, hazardous, and pose a serious threat to our environment as well as an effect on biological activities. The assessment consequently attained an existent preparation of RGOX linked by preceding work conveyed designed for a diverse natural reducing agent such as lemon juice [22], tea leaves extract [23], fresh carrots [24], glycylglycine [25], *Euphorbia wallichi* leaf [26], *Abelmoschus esculentus* [27] and L-valine [28]. One of the studies showed that *Olax scandens* leaf extract bio-mediated conversion of silver and copper salts to a nanocomposite structure [29]. Similarly, the novel biological synthesis of RGOX was prepared by graphene oxide with plant extract *Chenopodium album* as a reducing agent. This biological synthesis of RGOX from GOX involving of *Chenopodium album* is cost-effective and less toxic as compared to synthesis by chemical reducing agents. The reduction parameters were investigated by UV-visible spectroscopy, FTIR, SEM and TEM. Moreover, antibacterial and anticancer activities of the synthesized RGOX and GOX activity have been evaluated. RGOX showed promising antibacterial and anticancer activities on breast cancer cell lines, ensuring a promising approach for its future treatment.

2. Materials and Methods

2.1. Chemicals and Reagents

All the chemicals and reagents utilized in the experiment were AR (analytical reagent) grade. Natural graphite fine powder which was used was from the CDH company, concentrated sulfuric acid (H_2SO_4) and potassium permanganate ($KMnO_4$) were from the BDH company, hydrogen peroxide (30% H_2O_2), and $NaNO_3$ (Laboratory reagent) were from qualikems and concentrated HCl was from Loba Chemie Laboratory reagents and fine chemicals, New delhi, India. Phosphate buffer saline (PBS) (pH = 7.4), 3-(4, 5-dimethylthiazol-2-yl)-2, 5-diphenyl-tetrazoliumbromide) MTT dye, Roswell Park Memorial Institute (RPMI) media were purchased from Sigma-Aldrich and Invitrogen (New Delhi, India). Fluorescein isothiocyanate (FITC) dye for fluorescence microscopy was purchased from Sigma Aldrich. Bacterial culture media Agar, Brain Heart Infusion (BHI), and Luria Bertani (LB) were procured from Himedia Laboratories Pvt. Ltd. (New Delhi, India). The bacterial strains used were Gram-negative *Escherichia coli* (ATCC-25922), Gram-positive *Staphylococcus aureus* (ATCC-25923), and fungal strain *Candida albicans* (MTCC-183) was also used. All the chemicals were used as received without any further treatment. Deionized or doubly distilled water (DDW) was used throughout the experiments.

2.2. Preparation of Reduced Graphene Oxide (RGOX)

Initially, graphene oxide was prepared from graphite powder using the Hummers method in which 4 g of graphite were taken with sodium nitrate and concentrated sulfuric acid mixed in a round bottom flask is kept on a magnetic stirrer for 30 min [27,30]. After that, 12 g of potassium permanganate was added slowly while stirring the reaction mixture continuously under ice cooling with fast stirring keeping the temperature at 20 °C. At the end of 30 min, instead of the ice bath, the blends were kept inside an oven at 35 °C for 2 h followed by magnetic stirring. Afterward, 200 mL of deionized water was poured into the flask leading to the appearance of the brownish solution. At the end of 15 min, a mixture of 30% of hydrogen peroxide and toasty DMW was included continuously until the brownish color solution changed into a yellowish-brown color solution. Finally, the product of the solution was filtered by Whatman filter paper and washed with hydrochloric acid and double-distilled water. The obtained brown powdered graphite oxide was kept dry in an oven at 50 °C overnight and then the yielded graphene oxide (100 mg) was suspended in 100 mL of double-distilled water using an ultrasonicator. The supernatant of graphene oxide was obtained by centrifugation and then stored for experimental use. While the synthesis of reduced graphene oxide (RGOX) used a seasonal plant fresh *Chenopodium album*. Firstly, 500 g of vegetable was added into the distilled water followed by boiling for about 15 min until a green color was obtained and vegetable waste was eliminated using Whatman filter paper. Eventually, the suspension of 500 mg graphene oxide in 500 mL DDW was added into the extract of *Chenopodium album*, then the blend was kept on reflux for 12 h at 100 °C for reduction. After filtration, bio-mediated reduction of graphene oxide turned from a brownish solution into a black colored solution, which was further heated to a fine powder of RGOX. It was then washed with DMW thoroughly and thus finally filtered to obtain reduced graphene oxide. Furthermore, the synthesized reduce graphene oxide was further characterized by different techniques.

2.3. Characterization of Graphene Oxide (GOX) and Reduced Graphene Oxide (RGOX)

Four characterization techniques were used for the confirmation of GOX and RGOX. UV-visible spectroscopy (Hitachi-F-2500-, Tokyo, Japan). and Fourier Transforms Infrared Spectroscopy (FTIR) (PerkinElmer, Spectrum-BX, Norwalk, CT, USA). were used for the confirmation of the formation of GOX and RGOX. Superficial framework and magnitude of GOX and RGOX were confirmed by Scanning electron microscopy (SEM) and Transmission Electron Microscopy (TEM), respectively. The absorbance range of the FTIR spectroscopy technique was 400 to 4000 cm^{-1}. SEM images were taken using a (JSM-6510LV, Jeol, Tokyo, Japan) while an advanced JEOL 6510LV model was used to obtain higher resolution images. The SEM was operated with a working energy of 7 keV, a beam size at a value of 3, and a working distance of 10 mm. TEM analysis was performed employing a (JEM 2100, JEOL, Tokyo, Japan) model with a potential of 200 kV FE (Field Emission). Powder X-ray diffraction (XRD) patterns were characterized using XRD on a Bruker D8 Advance with Cu Kα radiation in a scanning range of 5–60 (2 theta) with a scan rate of 12°/min.

2.4. Determination of Minimum Inhibitory Concentration (MIC)

The minimum inhibitory concentration (MIC) value of in situ synthesized RGOX was calculated by the micro-dilution method. MIC is the minimum concentration of synthesized compounds that inhibit the visible growth of microorganisms after a given time. Bacterial and fungal strains such as *Staphylococcus aureus* (ATCC 25923), *Escherichia coli* (ATCC-25922), and *Candida albicans* (MTCC-183) were tested against synthesized compounds GOX and RGOX. The significant values were obtained on the basis of the visibility test conducted in 96-well microdilution plates, as described previously [31]. Positive controls were prepared without compounds and only cells, while negative control was prepared without compound and cells.

2.5. Antimicrobial Activity Determination by Agar Well Diffusion Assay

The antimicrobial activity of GOX and RGOX was determined by agar well diffusion assay against *Staphylococcus aureus, Escherichia coli,* and *Candida albicans.* An aliquot of 100 µL of cells was taken from a stock of 10^5 CFU/mL of different microbial strains that were spread on agar plates and incubated at 37 °C for a stipulated time period. The wells were punched into the agar plates with the help of sterile yellow tips, and 50 µL of synthesized compounds were loaded and incubated overnight at 37 °C. The zone of inhibition of GOX and RGOX and standard drugs (like ampicillin for bacteria and amphotericin B for fungus) was measured after overnight incubation. The readings were taken in triplicates to reduce deviations, and calculation should be done after taking an average of triplicates.

% activity index = zone of inhibition by test compound (diameter)/zone of inhibition on by standard (diameter) × 100%

2.6. Biofilm Inhibition Employing Fluorescence Microscopy

Staphylococcus aureus (ATCC-25923) was cultured overnight in Brain Heart Infusion (BHI) broth. It was then dispended (10^6 CFU/mL) on a sterile coverslip in a six-well culture plate (Eppendorf India Limited). Then the cells were incubated for 24 h at 37 °C for the growth of biofilm. After incubation, washing with PBS was done to remove non-adherent cells and then treated with GOX and RGOX. After a stipulated time period, the plates were washed with PBS, and coverslips were fixed with 4% paraformaldehyde followed by washing and staining with FITC dye. After 30 min, stained coverslips were washed with PBS and analyzed under a fluorescence microscope (100× magnification) [32].

2.7. XTT Biofilm Assay

2,3-bis(2-methoxy-4-nitro-sulfophenyl)-5-[(phenylamino)carbonyl]-2H-tetrazolium hydroxide (XTT) biofilm assay was performed to analyze the behavior of GOX and RGOX according to a previously published protocol [33]. Briefly, mature biofilms were grown on cover slides, and non-adherent cells were washed thoroughly with sterile phosphate saline buffer (PBS). Then mature biofilm was exposed to increasing concentrations of GOX and RGOX incubated for 48 h. After incubation, XTT solution in PBS was added. Previously prepared 2 µL menadione solutions (0.4 mM) were added in each well and incubated at 37 °C in the dark for 4 h. The colorimetric variation was assessed by a micro titer plate reader (BIO-RAD) at 490 nm.

2.8. Cell Cytotoxicity Assay

3-(4,5-dimethylthiazol-2-y1)-2,5-diphenyltetrazolium bromide (MTT) assay was performed on MCF-7 to determine the effect of GOX and RGOX compounds. Cells were cultured in Roswell Park Memorial Institute (RPMI) 1640 medium supplemented with 10% heat-inactivated fetal calf serum, 10 mmol/L glutamine, and 50 µg/mL each of streptomycin and penicillin. The cells were seeded in 96 well plates with a density of 1×10^5 cells and incubated at 37 °C. GOX and RGOX solutions were freshly prepared for the exposure to cells, and increasing concentration of both compounds (0–250 µM) was added to the cells. After overnight incubation, the cells were washed with PBS, and 0.5% MTT solution (5 mg/mL in PBS) was added in each well. After 4 h incubation, formazan crystals were formed, which gives purple color by adding 0.1% DMSO solution. This colored complex was measured by Microplate Reader (Genetix Biotech Asia Pvt. Ltd., New Delhi, India) at 570 nm. Cisplatin was taken as a control in the experiment.

3. Results and Discussion

3.1. Ultraviolet-Visible Spectroscopy (UV-Vis) Analysis

The confirmation of synthesized GOX and then reduction of GOX into RGOX by extract vegetables were investigated through UV visible spectroscopy, as shown in Figure 1a,b. The absorbance peak of GOX due to the $\pi \rightarrow \pi^*$ transition of aromatic C=C at 230 nm, However the hump of GOX alongside

298 nm was observed and assigned as the nonbonding between the π antibonding transition of the carbonyl group [27,34]. Eventually, the reduction of GOX formed RGOX that shows the absorbance peak at 263 nm, while the hump of GOX disappeared. Finally, complete removal of oxygen functional groups in the GOX confirmed the formation of RGOX, and the resultant reduce graphene oxide by the network of π-conjugation of graphite [35]. One more superior distinguishing feature of GOX and RGOX, the appearance of a shoulder peak around 230 nm due to the transition of aromatic compound C-C bonds which resemble sp^3 hybridization, on the other hand after conversion of GOX into RGOX by reduction of vegetable extract which observed a clear peak at 263 nm through sp^2 hybridization. Extract of the *Chenopodium album* consists of many antioxidants such as vitamin A, vitamin C, vitamin B, flavonoids, etc. In this work, the conversion of graphene oxide into reduces graphene oxide was probably achieved due to the reducing nature of vitamin C and other antioxidants present in the extract of the *Chenopodium album*. Some of the studies have also highlighted the green, reducing nature of these antioxidants [36,37].

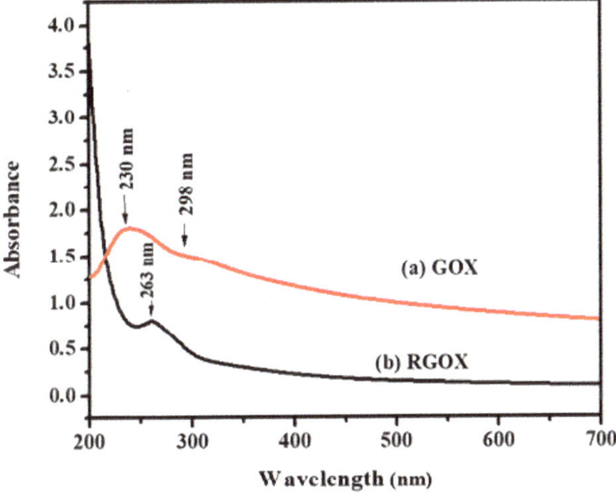

Figure 1. Ultraviolet-visible spectra analysis (**a**) GOX and (**b**) RGOX.

3.2. Fourier Transforms Infrared Spectroscopy (FTIR) Analysis

FTIR spectroscopy was used to verify the synthesis of graphene oxide and material RGOX. FTIR spectrum confirmed of GOX was obtained by complete oxidation of graphite and also confirmed RGOX, as shown in Figure 2. The GOX spectrum displays a broad peak at 3430 cm^{-1} [38], while a number of strong band absorptions at 1010, 1170, 1370, 1580, 1730, 2850 and 2920 cm^{-1}, were observed. The broad absorption peak and 1370 cm^{-1} are attributed to the C-H group, while the peak about 1010 cm^{-1} corresponds to the C–O–C bond of the alkoxy or epoxy group. The IR peak is attributed to 2920, 2850, and 1580 cm^{-1} due to the asymmetric CH$_2$, symmetric CH$_2$ stretching of GOX, and C=C bond stretches without an oxidized graphitic domain respectively [11,39]. The peak around at 1170 cm^{-1} is due to C–OH bonds, and 1730 cm^{-1} is associated with the C=O stretch of carbonyl group [40]. In GOX, the functional group-containing oxygen was probably reduced due to the presence of antioxidants such as vitamin C, ascorbic acid present in *Chenopodium album* as shown in Figure 2. The FTIR peak of RGOX at 1570 cm^{-1} is related to elongating of C=C, besides the broad peak at 1170 cm^{-1}, which is assigned to the C–C stretching. The spectra of RGOX as compared to GOX, the band at 1730 cm^{-1} (associated with C=O in carbonyl group), and 1010 cm^{-1} (associated with C–O–C in the epoxy group) disappeared; this indication confirmed that GOX was reduced to RGOX. Eventually, some strong peaks associated with oxygen-containing functional groups decreased for reduced graphene oxide as compared to graphene oxide.

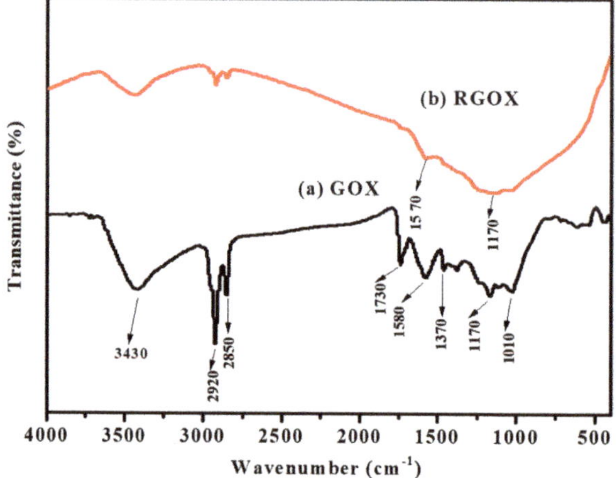

Figure 2. Fourier transforms infrared analysis (**a**) GOX and (**b**) RGOX.

3.3. X-Ray Diffraction (XRD) Analysis.

XRD configuration of GOX and RGOX is depicted in Figure 3a,b. The spectra observed for GOX and RGOX were crystalline and amorphous in nature. However, GOX shows, with respect to its characteristics, a band at $2\theta = 12.50°$ (002) as shown in Figure 3a, after oxidation of graphite that suggests water molecules were inserted into the crystal lattice layer and different functional groups of oxygen were produced between the graphite layers [41]. Thus, after the reduction of GO from vitamin C, most oxygen functional groups (C–OH, C–O and C=O) were removed where a sharp peak at $2\theta = 12.50°$ was extinct. In addition, XRD of RGOX, a diffused peak band appeared at $2\theta = 22.50°$ and corresponded to diffraction (222), as depicted in Figure 3b [42].

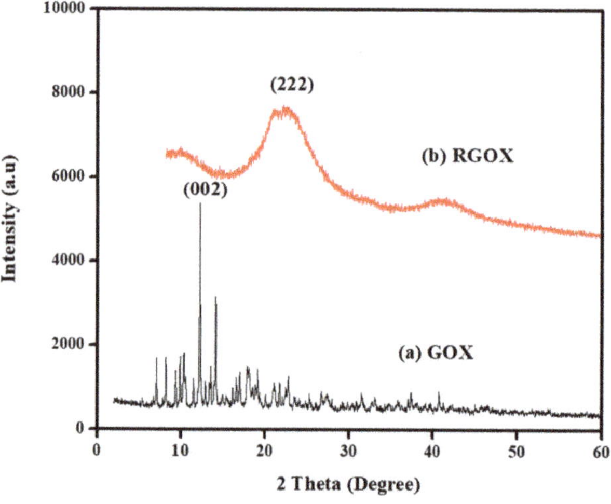

Figure 3. XRD graph of (**a**) GOX and (**b**) RGOX.

3.4. Scanning Electron Microscopy (SEM) and Transmission Electron Microscopy (TEM) Analysis

The morphology of the scanning electron micrograph shows the two types of GOX sheets, independent flat GOX mass with crumples and intact as shown in Figure 4. In Figure 4a at 10,000 times magnification, you can see folding and overlapping and there is a large interspace between the thinner edges of graphene oxide. It is clear from the micrograph that the differential thermal factors amongst graphene sheets and the substrate shrunk throughout the practice formed by annealing [43]. While in Figure 4b at 30,000 times magnification, the GOX sheets overlapped as same, but they were slippery, and there was no interspace between the thinner edges of graphene oxide. Additional SEM examination indicates that the full information on distinct RGOX flecks of layers and edge layers had a distribution of shady graphene flecks. At 3000 times magnification, the entire sheet of RGOX was visible exposed as assigned in Figure 4c. However, at 30,000 times magnification, RGOX overlaid descriptive characteristics of graphene structures and crimped fragments were well scattered and linked with each other as shown in Figure 4d. The number of films can be seen by the distinct contrast obtained in the SEM images. TEM morphology of graphene oxide and reduce graphene oxide are depicted in Figure 5. The surface study of graphene oxide manifested a single layer, without wrinkles, with an irregular pattern meant an amorphous nature and flaw structure. Thus, in the TEM images of RGOX many more wrinkle shapes can be observed with multiple layers stacked on each other. It is signified that the scrolled shape and lateral corrugations investigated by TEM [44].

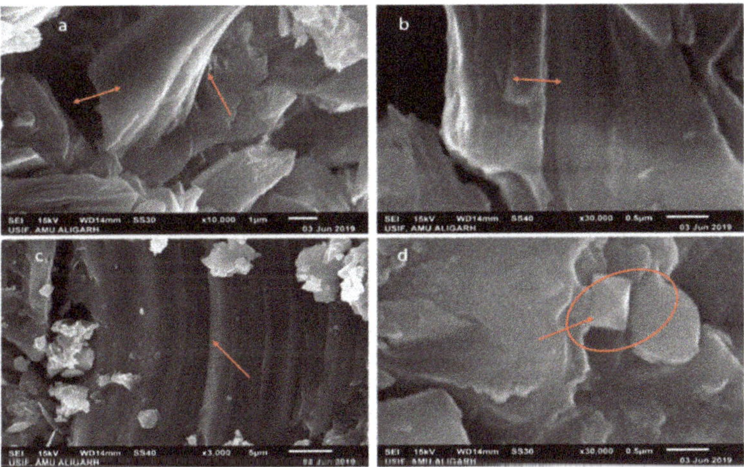

Figure 4. Scanning electron microscopy images at different magnifications (**a**) GOX at 10 K, (**b**) GOX at 30 K, (**c**) RGOX at 3 K and (**d**) RGOX at 30 K.

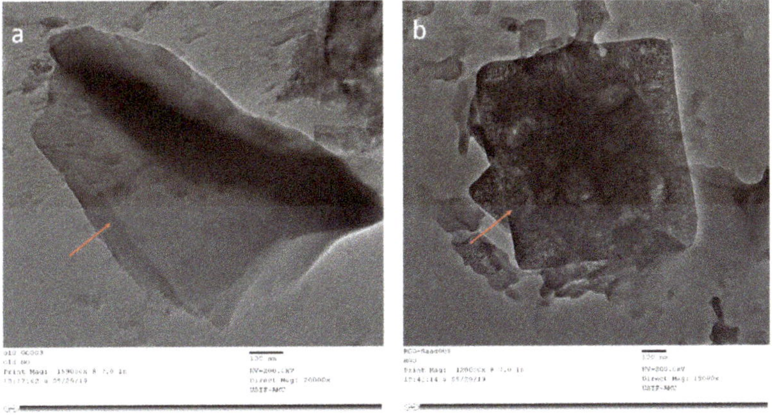

Figure 5. Transmission electron microscopy images of (**a**) GOX and (**b**) RGOX.

3.5. Minimal Inhibitory Concentration of GOX and RGOX against Bacterial and Fungal Isolates

MIC values were determined by the microdilution method in 96-well plates on the basis of a viability test, as previously described [29,45]. Both bacterial and fungal cultures were adjusted to 1×10^5 colony-forming unit (CFU)/mL with their respective media. The stock sample of the formulations was then serially diluted in the 96-well plate. Furthermore, both positive and negative controls were taken without compound, and without compound and organisms, respectively. The plate was then incubated overnight at 37 °C. On the next day, the turbidity of the cells in wells was checked as an indicator of microbial growth. Therefore, the well which showed no growth of microbes was the minimum inhibitory concentration of RGOX and GOX.

The antibacterial potential of GOX and RGOX nanocomposite was determined against Gram-positive and Gram-negative bacteria. Ampicillin has been taken as a control antibiotic. The MIC value of GOX against *Staphylococcus aureus* and *Escherichia coli* was 250 µg/mL, whereas RGOX had a MIC value of 125 µg/mL against both strains. The control ampicillin had MIC values of 4 and 2 µg/mL for *Staphylococcus aureus* and *Escherichia coli*, respectively. Similarly, the MIC values of GOX and RGOX were calculated against *Candida albicans* and found to be 500 µg/mL and 500 µg/mL, respectively, in which amphotericin B was taken as a control.

3.6. Agar Diffusion Assay

The antibacterial efficacy of GOX and RGOX was determined by agar diffusion assay. The formation of a zone of inhibition by these compounds suggested the bactericidal activity of nanocomposites against different bacterial strains. The bactericidal effect was conducted against *E. coli*, *S. aureus*, and *C. albicans* on a nutrient agar plate containing different concentrations of various formulations. It was observed that only the bacterial strains were sensitive to RGOX and GOX while fungal strain was insensitive towards these compounds. The zones of inhibition recorded against these are reported in Table 1. The zones of inhibition of *S. aureus* and *E. coli* increased in RGOX as compared to GOX, while both compounds did not show any effect on *C. albicans*. Furthermore, our results were in concordance with Shen et al., who investigated the antibacterial activity of an Ag-CCG composite against *Colibacillus*, *Staphylococcus aureus* and *Canidia albicans* bacteria [46]. As suggested by the results, *Chenopodium album* synthesized RGOX with considerably better antibacterial activity compared to its parent compound GOX. RGOX and GOX did not show antifungal activity on fungus *Candida albicans*.

Table 1. Zones of inhibition observed (in mm units) against microbial strains.

Bacterial Strains	GOX	RGOX
S. aureus	6.6 ± 2.0	8.6 ± 3.2
E. coli	6.3 ± 2	7.6 ± 2.0
Fungal strain	**GOX**	**RGOX**
C. albicans	1.4 ± 2.4	2.3 ± 2.09

3.7. Antibiofilm Activity Revealed by Fluorescence Microscopy by GOX and RGOX

GOX and RGOX showed inhibition of biofilm formation, Figure 6. As compared to the control, in which no compound was used, the results of GOX and RGOX showed enhanced antibiofilm potential by successfully disrupting *Staphylococcus aureus* biofilm. The FITC labeled fluorescence images (100×) showed the disruptions of *Staphylococcus aureus* biofilm in RGOX and GOX; the panel revealed the antibiofilm activity.

3.8. XTT Assay Employing to Determine Anti-Biofilm Potential of GOX and RGOX

The antibiofilm nature of GOX and synthesized RGOX has been investigated against *S. aureus* with vancomycin antibiotic as a control. The formation of biofilm depends on numerous factors, such as extracellular binding proteins, polysaccharides, etc. The inhibitory action of RGOX on biofilm is caused due to the internal metabolic system, which disrupts the formation of extracellular protein factors responsible for the formation of biofilm. In the present study, *S. aureus* was cultured in a 96-well plate and treated with varying doses of compounds. The dose-dependent effect of GOX and RGOX was found to exhibit antibiofilm activity by employing XTT assay on an *S. aureus* strain. A significant decrease in the bacterial numbers (p-value < 0.05 * and < 0.01 **) was found in both treated groups as compared to vancomycin treatment. This observation is in agreement with the zone of inhibition assay. As the dose of the compound increased, the percentage of biofilm decreased, as shown in Figure 7. Growth inhibition was calculated by comparing the relative metabolic activity (RMA) obtained by taking the untreated control as 100% by XTT.

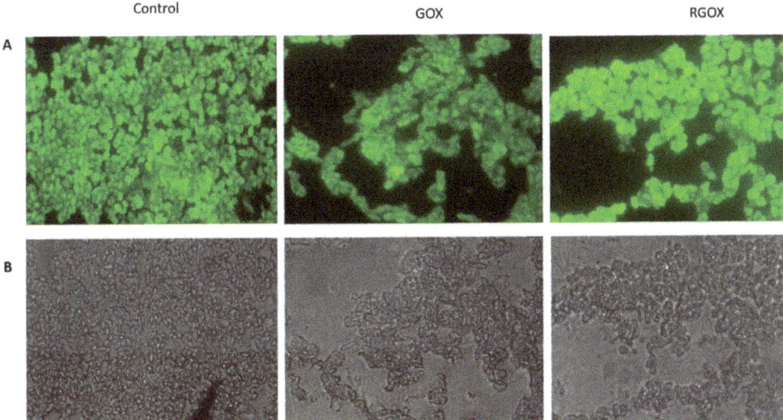

Figure 6. Inhibition of *S. aureus* biofilm by synthesized compounds: (**A**) shows fluorescence microscopic images while (**B**) shows bright-field microscopic images, the three panel shows untreated control image of *S. aureus* biofilm, biofilm treated with GOX, biofilm exposed to RGOX as revealed by fluorescence microscopy (100× magnification).

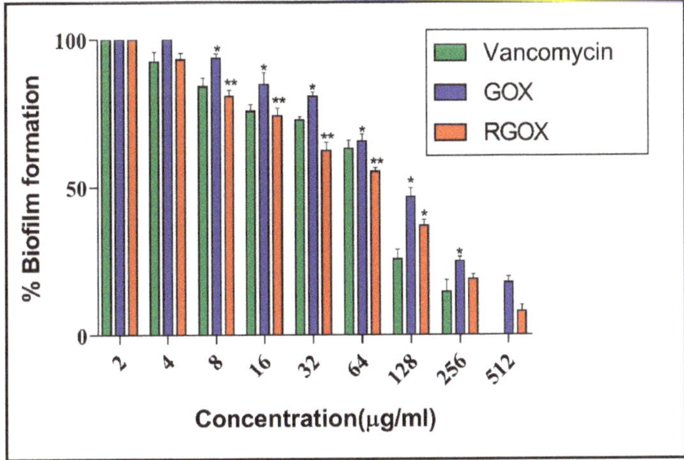

Figure 7. GOX and RGOX effect against *S. aureus* biofilm development. An increasing concentration of compounds decreases the percentage of biofilm formation. Vancomycin served as a control (p-value < 0.05 *, < 0.01 ** and < 0.001 ***).

3.9. Cytotoxicity of GOX and RGOX Towards MCF-7 Cells

In this study, we evaluated the cytotoxicity of RGOX towards MCF-7 (human breast cancer cell line) by employing an MTT assay. The fluorinated graphene oxide showed no cytotoxicity towards breast cancer cells at a concentration of 576 μg/mL [47] Moreover, graphene oxide showed cytotoxicity in HBI.F3 human neuronic cells and BEAS-2B human lung cells at a dose-dependent concentration of 10–100 μg/mL [48]. In most of the previous studies, it was seen that graphene oxide shows a cytotoxic effect on breast carcinoma at a very high concentration above 500 μg/mL.

According to Liao et al., the cytotoxicity of skin fibroblasts enhanced with the increase in the concentration of graphene oxide [49]. In addition to that, an increasing concentration of GOX amplified the cytotoxic effect on HepG2 cells [50]. In one of the studies, it was observed that the synergistic effects of RGOX and ZnO nanorods produced excellent cytotoxic effects in human embryonic kidney cells (HEK293) with the help of enhanced antioxidant properties. This is because zinc ions attached to RGOX sheets contacted with HEK293 cells and cause dthe disruption of cells [51]. In one of the dose-dependent studies, an increasing dose of GO-FA-ZnO from 0 to 100 μg/mL reduced cell viability up to 19% as compared to a control [52]. Previously, it was shown that the GO-Ag toxicity towards breast cancer cells may be synergistic. Breast cancer cells were treated with different concentrations of GO-Ag nanocomposite in a dose-dependent manner (10–100 μg/mL), which decreased cell viability [53]. As compared to earlier studies, the bio-mimetically synthesized RGOX was used to determine its effect on the viability of breast cancer cells. In a dose-dependent study, it was found that the *Chenopodium album* synthesized RGOX showed decreased cell viability from 90% to 35% towards MCF-7 cells at a concentration from 1 to 250 μg/mL. Furthermore, the *Chenopodium album* synthesized RGOX compound was found to be more cytotoxic as compared to the control GOX. Figure 8 shows clearly the loss of cell viability when the concentration increased gradually (in a dose-dependent manner). Therefore, the toxicity of bio-mimetically synthesized RGOX towards breast cancer cells may be synergistic, which disrupted the interaction between cancerous cells.

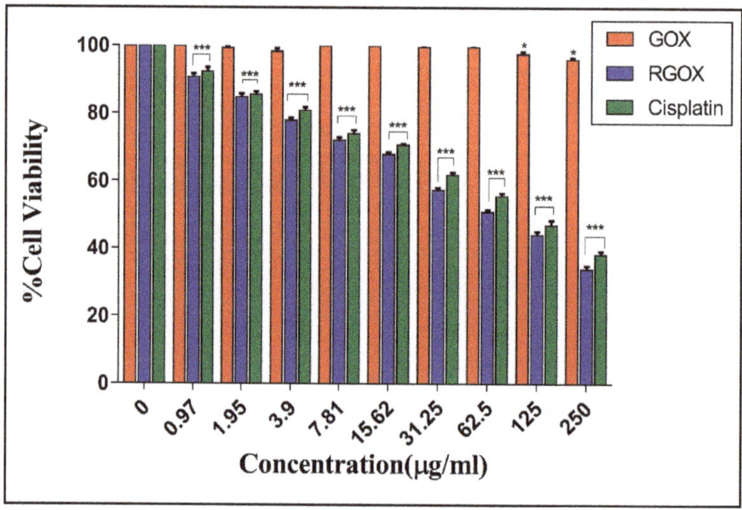

Figure 8. Cytotoxicity assay of MCF-7 cancer cells with increasing concentrations (0–250 μg/mL) of Cisplatin, GOX, and RGOX (p-value < 0.05 *, < 0.01 ** and < 0.001 ***).

4. Conclusions

Graphene oxide (GOX) was successfully prepared using graphite powder by a modified 'Hummers method'. GOX was satisfactorily reduced to RGOX in the presence of *Chenopodium album* leaves extract. The present study is the first to report this green, simple, facile and cost-effective procedure for direct reduction of GOX using *Chenopodium album* leaves extract. The obtained GOX and RGOX were characterized using UV and FTIR techniques, which show the negligible presence of oxygen functional groups in GOX during the reduction process. *Chenopodium album* is a good source of vitamin C, casein, caffeic acid, and polyphenols, which together provide reducing ambiance. UV, FTIR, SEM, and TEM analysis confirmed the formation of RGOX from graphene oxide using *Chenopodium album* leaves extract. The overall results of the experiments prove that *C. album* leaves extract is an important alternative to the traditional chemical reduction method to avoid chemical intervention. The activity of GOX and RGOX against bacteria and fungi has been demonstrated. It was found that RGOX showed an increased antibacterial (Gram-positive and Gram-negative) and antibiofilm activity as compared to GOX. The antifungal activity could not be ascertained using the synthesized compound. Upon evaluation of the anticancer activity of RGOX, it was found that it showed better anti-breast cancer activity than its parent compound GOX. The results of RGOX against MCF-7 cells will pave the way for a new approach for the prevention of breast cancer. RGOX obtained in this work can be further utilized in several potential applications in various fields such as drug delivery, anticancer, antifungal, antibacterial (Gram-positive and Gram-negative), DNA binding interaction, generation of bio-composites and biosensors.

Author Contributions: Conceptualization, M.F.U., S.A.U. and H.S.; Methodology, M.F.U., H.S.; Supervision, F.A. and M.R.; Writing—original Draft preparation, S.A.U., H.S., and M.F.U.; Writing—Review and Editing, M.O., M.R. and F.A. All authors read and approved the final manuscript.

Funding: The authors are grateful to the Universiti Sains Malaysia and financial support through Fundamental Research Grant Scheme (FRGS) grant number (203/PTEKIND/6711822) and Aligarh Muslim University for the support and research facilities for this project.

Conflicts of Interest: The authors declare no conflict of interest.

References

1. Dibsdall, L.; Lambert, N.; Bobbin, R.; Frewer, L. Low-income consumers' attitudes and behaviour towards access, availability and motivation to eat fruit and vegetables. *Public Health Nutr.* **2003**, *6*, 159–168. [CrossRef] [PubMed]

2. Ishangulyyev, R.; Kim, S.; Lee, S.H. Understanding food loss and waste—Why are we losing and wasting food? *Foods* **2019**, *8*, 297. [CrossRef] [PubMed]

3. Somanathan, T.; Prasad, K.; Ostrikov, K.; Saravanan, A.; Krishna, V. Graphene Oxide Synthesis from Agro Waste. *Nanomaterials* **2015**, *5*, 826–834. [CrossRef] [PubMed]

4. Singh, U.; Sagar, V.R. Quality characteristics of dehydrated leafy vegetables influenced by packaging materials and storage temperature. *J. Sci. Ind. Res. (India)* **2010**, *69*, 785–789.

5. Laghari, A.H.; Memon, S.; Nelofar, A.; Khan, K.M.; Yasmin, A. Determination of free phenolic acids and antioxidant activity of methanolic extracts obtained from fruits and leaves of Chenopodium album. *Food Chem.* **2011**, *126*, 1850–1855. [CrossRef]

6. Alizadeh, N.; Salimi, A.; Hallaj, R.; Fathi, F.; Soleimani, F. CuO/WO$_3$ nanoparticles decorated graphene oxide nanosheets with enhanced peroxidase-like activity for electrochemical cancer cell detection and targeted therapeutics. *Mater. Sci. Eng. C* **2019**, *99*, 1374–1383. [CrossRef]

7. Kumawat, M.K.; Thakur, M.; Bahadur, R.; Kaku, T.; Prabhuraj, R.S.; Ninawe, A.; Srivastava, R. Preparation of graphene oxide-graphene quantum dots hybrid and its application in cancer theranostics. *Mater. Sci. Eng. C* **2019**, *103*, 109774. [CrossRef]

8. Bolaños, K.; Kogan, M.J.; Araya, E. Capping gold nanoparticles with albumin to improve their biomedical properties. *Int. J. Nanomed.* **2019**, *14*, 6387–6406. [CrossRef]

9. Wang, Y.; Shi, Z.; Yin, J. Facile synthesis of soluble graphene via a green reduction of graphene oxide in tea solution and its biocomposites. *ACS Appl. Mater. Interfaces* **2011**, *3*, 1127–1133. [CrossRef]

10. Yan, L.; Zhao, F.; Li, S.; Hu, Z.; Zhao, Y. Low-toxic and safe nanomaterials by surface-chemical design, carbon nanotubes, fullerenes, metallofullerenes, and graphenes. *Nanoscale* **2011**, *3*, 362–382. [CrossRef]

11. Gurunathan, S.; Han, J.W.; Park, J.-H.; Kim, E.S.; Choi, Y.-J.; Kwon, D.-N.; Kim, J.-H. Reduced graphene oxide-silver nanoparticle nanocomposite: A potential anticancer nanotherapy. *Int. J. Nanomed.* **2015**, *10*, 6257–6276. [CrossRef] [PubMed]

12. Guo, X.; Mei, N. Assessment of the toxic potential of graphene family nanomaterials. *J. Food Drug Anal.* **2014**, *22*, 105–115. [CrossRef] [PubMed]

13. Zhang, X.; Yin, J.; Peng, C.; Hu, W.; Zhu, Z.; Li, W.; Fan, C.; Huang, Q. Distribution and biocompatibility studies of graphene oxide in mice after intravenous administration. *Carbon N. Y.* **2011**, *49*, 986–995. [CrossRef]

14. Pinto, A.M.; Gonçalves, I.C.; Magalhães, F.D. Graphene-based materials biocompatibility: A review. *Colloids Surf. B Biointerfaces* **2013**, *111*, 188–202. [CrossRef] [PubMed]

15. An, S.S.A.; Nanda, S.S.; Yi, D.K. Oxidative stress and antibacterial properties of a graphene oxide-cystamine nanohybrid. *Int. J. Nanomed.* **2015**, *10*, 549–556. [CrossRef] [PubMed]

16. McAllister, M.J.; Li, J.-L.; Adamson, D.H.; Schniepp, H.C.; Abdala, A.A.; Liu, J.; Herrera-Alonso, M.; Milius, D.L.; Car, R.; Prud'homme, R.K.; et al. Single sheet functionalized graphene by oxidation and thermal expansion of graphite. *Chem. Mater.* **2007**, *19*, 4396–4404. [CrossRef]

17. Ambrosi, A.; Pumera, M. Precise tuning of surface composition and electron-transfer properties of graphene oxide films through electroreduction. *Chem. A Eur. J.* **2013**, *19*, 4748–4753. [CrossRef]

18. Liu, C.; Han, G.; Chang, Y.; Xiao, Y.; Li, Y.; Li, M.; Zhou, H. Capacitive performances of reduced graphene oxide hydrogel prepared by using sodium hypophosphite as reducer. *Chin. J. Chem.* **2016**, *34*, 89–97. [CrossRef]

19. Stankovich, S.; Dikin, D.A.; Piner, R.D.; Kohlhaas, K.A.; Kleinhammes, A.; Jia, Y.; Wu, Y.; Nguyen, S.T.; Ruoff, R.S. Synthesis of graphene-based nanosheets via chemical reduction of exfoliated graphite oxide. *Carbon N. Y.* **2007**, *45*, 1558–1565. [CrossRef]

20. Park, S.; Ruoff, R.S. Chemical methods for the production of graphenes. *Nat. Nanotechnol.* **2009**, *4*, 217–224. [CrossRef]

21. Mohanty, N.; Nagaraja, A.; Armesto, J.; Berry, V. High-throughput, ultrafast synthesis of solution-dispersed graphene via a facile hydride chemistry. *Small* **2010**, *6*, 226–231. [CrossRef] [PubMed]

22. Begum, H.; Ahmed, M.S.; Cho, S.; Jeon, S. Simultaneous reduction and nitrogen functionalization of graphene oxide using lemon for metal-free oxygen reduction reaction. *J. Power Sources* **2017**, *372*, 116–124. [CrossRef]

23. Moosa, A.; Noori Jaafar, J. Green reduction of graphene oxide using tea leaves Extract with applications to lead Ions removal from water. *Nanosci. Nanotechnol.* **2017**, *7*, 38–47.

24. Vusa, C.S.R.; Berchmans, S.; Alwarappan, S. Facile and green synthesis of graphene. *RSC Adv.* **2014**, *4*, 22470–22475. [CrossRef]

25. Zhang, C.; Chen, M.; Xu, X.; Zhang, L.; Zhang, L.; Xia, F.; Li, X.; Liu, Y.; Hu, W.; Gao, J. Graphene oxide reduced and modified by environmentally friendly glycylglycine and its excellent catalytic performance. *Nanotechnology* **2014**, *25*, 135707. [CrossRef] [PubMed]

26. Atarod, M.; Nasrollahzadeh, M.; Sajadi, S.M. Green synthesis of a Cu/reduced graphene oxide/Fe$_3$O$_4$ nanocomposite using Euphorbia wallichii leaf extract and its application as a recyclable and heterogeneous catalyst for the reduction of 4-nitrophenol and rhodamine B. *RSC Adv.* **2015**, *5*, 91532–91543. [CrossRef]

27. Umar, M.F.; Nasar, A. Reduced graphene oxide/polypyrrole/nitrate reductase deposited glassy carbon electrode (GCE/RGO/PPy/NR): Biosensor for the detection of nitrate in wastewater. *Appl. Water Sci.* **2018**, *8*, 211. [CrossRef]

28. Tran, D.N.H.; Kabiri, S.; Losic, D. A green approach for the reduction of graphene oxide nanosheets using non-aromatic amino acids. *Carbon N. Y.* **2014**, *76*, 193–202. [CrossRef]

29. Mujeeb, A.A.; Khan, N.A.; Jamal, F.; Badre Alam, K.F.; Saeed, H.; Kazmi, S.; Alshameri, A.W.F.; Kashif, M.; Ghazi, I.; Owais, M. Olax scandens mediated biogenic synthesis of Ag-Cu nanocomposites: Potential against inhibition of drug-resistant microbes. *Front. Chem.* **2020**, *8*, 103. [CrossRef]

30. Alam, S.N.; Sharma, N.; Kumar, L. Synthesis of graphene oxide (GO) by modified hummers method and its thermal reduction to obtain reduced graphene Oxide (rGO)*. *Graphene* **2017**, *6*, 1–18. [CrossRef]

31. Wiegand, I.; Hilpert, K.; Hancock, R.E.W. Agar and broth dilution methods to determine the minimal inhibitory concentration (MIC) of antimicrobial substances. *Nat. Protoc.* **2008**, *3*, 163–175. [CrossRef] [PubMed]

32. Pratt, L.A.; Kolter, R. Genetic analysis of Escherichia coli biofilm formation: Roles of flagella, motility, chemotaxis and type I pili. *Mol. Microbiol.* **1998**, *30*, 285–293. [CrossRef] [PubMed]

33. Pierce, C.G.; Uppuluri, P.; Tristan, A.R.; Wormley, F.L.; Mowat, E.; Ramage, G.; Lopez-Ribot, J.L. A simple and reproducible 96-well plate-based method for the formation of fungal biofilms and its application to antifungal susceptibility testing. *Nat. Protoc.* **2008**, *3*, 1494–1500. [CrossRef] [PubMed]

34. Zhang, Z.; Chen, H.; Xing, C.; Guo, M.; Xu, F.; Wang, X.; Gruber, H.J.; Zhang, B.; Tang, J. Sodium citrate: A universal reducing agent for reduction / decoration of graphene oxide with au nanoparticles. *Nano Res.* **2011**, *4*, 599–611. [CrossRef]

35. Fernández-Merino, M.J.; Guardia, L.; Paredes, J.I.; Villar-Rodil, S.; Solís-Fernández, P.; Martínez-Alonso, A.; Tascón, J.M.D. Vitamin C is an ideal substitute for hydrazine in the reduction of graphene oxide suspensions. *J. Phys. Chem. C* **2010**, *114*, 6426–6432. [CrossRef]

36. Poonia, A.; Upadhayay, A. Chenopodium album Linn: Review of nutritive value and biological properties. *J. Food Sci. Technol.* **2015**, *52*, 3977–3985. [CrossRef]

37. Aunkor, M.T.H.; Mahbubul, I.M.; Saidur, R.; Metselaar, H.S.C. The green reduction of graphene oxide. *RSC Adv.* **2016**, *6*, 27807–27828. [CrossRef]

38. Mo, Z.; Sun, Y.; Chen, H.; Zhang, P.; Zuo, D.; Liu, Y.; Li, H. Preparation and characterization of a PMMA/Ce(OH)$_3$, Pr$_2$O$_3$/graphite nanosheet composite. *Polymer (Guildf)* **2005**, *46*, 12670–12676. [CrossRef]

39. Guo, H.-L.; Wang, X.-F.; Qian, Q.-Y.; Wang, F.-B.; Xia, X.-H. A green approach to the synthesis of graphene nanosheets. *ACS Nano* **2009**, *3*, 2653–2659. [CrossRef]

40. Shahriary, L.; Athawale, A.A. Graphene oxide synthesized by using modified hummers approach. *Int. J. Renew. Energy Environ. Eng.* **2014**, *2*, 58–63.

41. Paredes, J.I.; Villar-Rodi, S.; Martinez-Alonsa, A.; Tascon, J.M.D. Graphene oxide dispersion in organic solvents. *Langmuirr* **2008**, *24*, 10560–10564. [CrossRef] [PubMed]

42. Khan, M.U.; Siddiqui, S.W.A.; Siddiqui, Z.N. Efficient reduction of graphene oxide to graphene nanosheets using a silica-based ionic liquid: Synthesis, charcterisation and catalytic properties of IMD-Si/FeCl4-@GNS. *New J. Chem.* **2020**, *44*, 4822–4833. [CrossRef]

43. Li, X.; Cai, W.; An, J.; Kim, S.; Nah, J.; Yang, D.; Piner, R.; Velamakanni, A.; Jung, I.; Tutuc, E.; et al. Large-area synthesis of high-quality and uniform graphene films on copper foils. *Science* **2009**, *324*, 1312–1314. [CrossRef] [PubMed]

44. Kuila, B.K.; Zaeem, S.M.; Daripa, S.; Kaushik, K.; Gupta, S.K.; Das, S. Mesoporous Mn$_3$O$_4$ coated reduced graphene oxide for high-performance supercapacitor applications. *Mater. Res. Express* **2018**, *6*, 015037. [CrossRef]

45. Sakoulas, G.; Gold, H.S.; Venkataraman, L.; Moellering, R.C.; Ferraro, J.M.Y.; Eliopoulos, G.M. Introduction of erm (C) into a linezolid- and methicillin-resistant Staphylococcus aureus does not restore linezolid susceptibility. *J. Antimicrob. Chemother.* **2003**, *51*, 1039–1041. [CrossRef] [PubMed]

46. Hu, R.; Cola, B.A.; Haram, N.; Barisci, J.N.; Lee, S.; Stoughon, S.; Wallace, G.; Too, C.; Thomas, M.; Gestos, A.; et al. Harvesting waste thermal energy using a carbon-nanotube-based thermo-electrochemical cell. *Nano Lett.* **2010**, *10*, 838–846. [CrossRef]

47. Romero-Aburto, R.; Narayanan, T.N.; Nagaoka, Y.; Hasumura, T.; Mitcham, T.M.; Fukuda, T.; Cox, P.J.; Bouchard, R.R.; Maekawa, T.; Kumar, D.S.; et al. Fluorintead graphene oxide; a new multimodal material for biological application. *Adv. Mater.* **2013**, *25*, 5632–5637. [CrossRef]

48. Vallabani, N.V.S.; Mittal, S.; Shukla, R.; Pandey, A.; Dhakate, S.; Pasricha, R.; Dhawan, A. Toxicity of graphene in normal human lung cells (BEAS-2B). *J. Biomed. Nanotechnol.* **2011**, *7*, 106–107. [CrossRef]

49. Liao, K.-H.; Lin, Y.-S.; Macosko, C.W.; Haynes, C.L. Cytotoxicity of graphene oxide and graphene in human erythrocytes and skin fibroblastas. *ACS Appl. Mater. Interfaces* **2011**, *3*, 2607–2615. [CrossRef]

50. Lammel, T.; Boisseaux, P.; Fernandez-Cruz, M.L.; Navas, J.M. Internalization and cytotoxicity of graphene oxide and carboxyl graphene nanoplatelets in the human hepatocellular carcinoma cell line help G2. *Part. Fibre Toxicol.* **2013**, *10*, 27. [CrossRef]

51. Rajeshwari, R.; Prabu, H.G. Synthesis characterization, antimicrobial, antioxidant, and cytotoxic activities of ZnO nanorods on reduced graphene oxide. *J. Inorg. Organomet. Polym. Mater.* **2018**, *28*, 679–693. [CrossRef]

52. Hu, Z.; Li, J.; Li, C.; Zhao, S.; Li, N.; Wang, Y.; Wei, F.; Chen, L.; Huang, Y. Folic acid-conjugated graphene–ZnO nanohybrid for targeting photodynamic therapy under visible light irradiation. *J. Mater. Chem. B* **2013**, *1*, 5003. [CrossRef] [PubMed]

53. Shaheen, F.; Hammad Aziz, M.; Fakhar-e-Alam, M.; Atif, M.; Fatima, M.; Ahmad, R.; Hanif, A.; Anwar, S.; Zafar, F.; Abbas, G.; et al. An In Vitro Study of the photodynamic effectiveness of GO-Ag nanocomposites against human breast cancer cells. *Nanomaterials* **2017**, *7*, 401. [CrossRef] [PubMed]

 *nanomaterials*

Article

Glass Ionomer Cement Modified by Resin with Incorporation of Nanohydroxyapatite: In Vitro Evaluation of Physical-Biological Properties

Luis Eduardo Genaro *, Giovana Anovazzi, Josimeri Hebling and Angela Cristina Cilense Zuanon

Department of Morphology, Genetics, Orthodontic and Pediatric Dentistry, Araraquara School of Dentistry, São Paulo State University (UNESP), Araraquara 14801-903, São Paulo, Brazil; giovanaanovazzi@hotmail.com (G.A.); josimeri.hebling@unesp.br (J.H.); cristina.zuanon@unesp.br (A.C.C.Z.)
* Correspondence: luis-genaro@outlook.com; Tel.: +55-(16)-33016325

Received: 6 April 2020; Accepted: 7 May 2020; Published: 19 July 2020

Abstract: Resin-modified glass ionomer cement (RMGIC) has important properties. However, like other restorative materials, it has limitations such as decreased biocompatibility. The incorporation of nanoparticles (NP) in the RMGIC resulted in improvements in some of its properties. The aim of this study was to evaluate the physical-biological properties of RMGIC with the addition of nanohydroxyapatite (HANP). Material and Methods: Vitremer RMGIC was used, incorporating HANP by amalgamator, vortex and manual techniques, totaling ten experimental groups. The distribution and dispersion of the HANP were evaluated qualitatively by field emission scanning electron microscope (SEM-FEG). The evaluation of image porosity (SEM-FEG) with the help of imageJ. Cell viability 3-(4,5-dimethylthiazol-2yl)-2,5-diphenyl tetrazoline bromide (MTT) and cell morphology analyses were performed on MDPC-23 odontoblastoid cells at 24 and 72 h. Results: It was possible to observe good dispersion and distribution of HANP in the samples in all experimental groups. The incorporation of 5% HANP into the vortex stirred RMGIC resulted in fewer pores. The increase in the concentration of HANP was directly proportional to the decrease in cytotoxicity. Conclusions: It is concluded that the use of a vortex with the incorporation of 5% HANP is the most appropriate mixing technique when considering the smallest number of pores inside the material. A higher concentration of HANP resulted in better cell viability, suggesting that this association is promising for future studies of new restorative materials.

Keywords: scanning electron microscopy; porosity; glass ionomer cement; cytotoxicity

---

1. Introduction

Glass ionomer cement (GIC) is a dental material that contains fluoraluminosilicate glass in its powder composition of calcium, basic silicon oxide, aluminum oxide and calcium, magnesium, and sodium fluoride. The liquid is an aqueous solution of polyacrylic acid, tartaric acid, and itaconic acid, which has biological compatibility [1], adherence to the tooth structure [2], and antibacterial properties [3], among other desirable properties. Even so, its u se as restorative material is still limited due to its brittleness and low compressive strength [4] when compared to other restorative materials. In order to overcome these limitations, the resin-modified GIC (RMGIC) was developed. In addition to the acid–base reaction, RMGICs have a polymerization reaction as they have resin monomers in their composition [5,6]. In most cases, this polymerization is photo-initiated.

Recently, the incorporation of hydroxyapatite nanoparticles (HANP) to RMGIC has shown promising results such as the increase in adhesive strength to dentin [7]. In addition, it has been shown to favor increased enamel remineralization [8]. Nanoparticles show good results when added to restorative materials [9].

The incorporation of HANP to the conventional GIC at 12% concentration showed improvement in microhardness, resistance to compression, and diametrical traction [10]. Concentrations of 2% and 8% also provided improvements in fracture resistance [7]. Thus, when using RMGIC, the addition of HANP could improve its biological properties, since in vitro tests demonstrate that the cytotoxic effects of RMGIC are more evident when compared to conventional GIC [11,12].

Del Angel-Mosqueda et al. [13] (2018) carried out a cytotoxicity study analyzing gingival fibroblasts, where the photopolymerized medium was treated with 100 mg/mL of hydroxyapatite of lyophilized bone for 24 h at room temperature in response to Vitrebond RMGIC and found that there was an increase in cell viability, suggesting that HA plays a protective role, decreasing the cytotoxic effect.

Studies on the association of NPs and GIC show promising results [7,14,15], however, few have performed cytotoxicity tests [16], which are fundamental and precede the application of the material to the oral cavity. It is known that the American Dental Association (ADA) and the American National Standards Institute (ANSI-ISO-10993) recommend preclinical tests such as cytotoxicity assays developed in cell culture, laboratory animal tests, and finally used on humans [17,18].

The effect of adding NPs depends on its proper distribution and dispersion in the restorative material and there is no research related to different techniques for incorporating NP into the GIC, a technical step that certainly influences the properties of this material, so it is important to conduct research with this theme. We observed studies analyzing only GIC encapsulated in the powder/liquid mixing method using mechanical manipulators in different potentials [19]. The properties of the material are also strongly related to its microstructure, which in turn depends on the size and homogeneous distribution of its particles [20], which result in the quality of its mechanical resistance [21].

The analysis of different methods of incorporation of NP into the GIC is innovative and important, since it will contribute to adjust the conditions of homogeneity of NP distribution and dispersion within the material.

According to Moshaverinia et al. [7], the concentration of 2% HANP added to the GIC demonstrated improvements in its physical and mechanical properties. Additionally, the 5% and 10% concentrations are new to be tested. Thus, these concentrations were chosen to be used in the tests proposed in this study. The development of research on this subject may, in the near future, improve the properties of the material, facilitate clinical procedures, and make them more effective and long lasting.

In this way, it is essential to carry out studies that, aside from evaluating the physical, chemical, and mechanical properties of the association of NPs to the RMGIC, also evaluate biological properties, making it safe and effective for clinical application.

The objective of this work was to evaluate the effect of different techniques of the incorporation of HANP in different concentrations to the RMGIC regarding the distribution and dispersion, porosity, and cytotoxicity.

The null hypothesis of this study is that there is no improvement in the physical-biological properties of glass ionomer cement after the addition of HANP.

2. Materials and Methods

2.1. Preparation of Test Specimens

In the present work, portions of RMGIC powder from Vitremer (3M-ESPE Dental Products, St. Paul, MN, USA). With the portions of HANP (Sigma-Aldrich, ref. 677418-10g, batch MKBW9108V, St. Louis, MO, USA), both were weighed with the aid of an analytical balance (Gehaka Ltda-model BG 440).

The amount of each material was established so that, upon mixing thereof, the RMGIC associated with 2%, 5%, and 10% by weight HANP could be obtained. This material was stored individually in Eppendorf tubes for the preparation of each test specimen. Afterward, the materials were mixed and the mixing techniques are explained below.

Eight samples were used and the established groups are shown in Table 1.

Table 1. Groups established according to the technique of manipulation and concentration of nanohydroxyapatite and glass ionomer cement modified by resin.

Mixing Technique	HANP (%)	Groups
Control	0	C
Amalgamator (A)	2	A2
	5	A5
	10	A10
Vortex (V)	2	V2
	5	V5
	10	V10
Manual (M)	2	M2
	5	M5
	10	M10

2.2. Mixing Techniques

We mixed the powdered materials according to the techniques below:

Amalgamator. The powder (RMGIC + HANP) was inserted into amalgam capsules. These were submitted to the action of the amalgamator (Ultramat-SDI Brasil Industria e Comercio Ltd., Sao Paulo, Brazil) with vibration for six seconds.

Vortex. The Eppendorf-packed powder (RMGIC + HANP) was placed in vortex (Vortex Phoenix, Ref. 12446, Phoenix Ind. and Com. De Equips Scientists Ltd., Araraquara, Sao Paulo, Brazil) for vigorous mixing for one minute.

Manual. The powder (RMGIC + HANP) was manipulated and mixed in a spatula block using a plastic spatula in all directions for one minute.

After obtaining the HANP-associated GIC (2%, 5%, or 10% by weight), a drop of liquid present in the RMGIC kit was added to the powder portion and spatulation was performed as recommended by the manufacturer. This material was inserted into a silicone matrix that was 3 mm high by 6 mm in diameter, with the aid of a Centrix syringe (DFL and Trade S.A. Rio de Janeiro, RJ, Brazil) to obtain the specimens. A polyester strip with a 1 mm thick glass slide and a weight of 100 g for 30 s.

Then, the photoactivation was performed with the aid of the light curing agent Radii Cal (SDI Brasil Industria e Comercio Ltda., Sao Paulo, SP, Brazil) for one minute with an intensity of 1200 mW/cm^2. These specimens were stored at 100% humidity in an appropriate container.

2.3. Distribution and Dispersion of Nanohydroxyapatite

The distribution assessment after 24 h of sample preparation was performed by scanning electron microscopy (SEM-FEG) (SM-300, Topcon, Tokyo, Japan). The samples were fractured with a surgical hammer and chisel, using a channel made in the center to direct the fracture. The analyzed fragments had a dimension of 2 mm by 2 mm, were coated with a gold–palladium alloy under high vacuum and taken to 500 and 20,000 SEM-FEG magnification for imaging. X-ray spectroscopy was also performed by dispersing energy to identify its chemical elements (EDX) (Shimadzu Corporation, Kyoto, Japan).

2.4. Porosity

The specimens were fractured using a surgical hammer and chisel from a channel made in its center for fracture targeting. The analyzed fragments had a dimension of 2 mm by 2 mm, were coated with a gold–palladium alloy under high vacuum and taken to the 100-fold magnification SEM-FEG for imaging, which were analyzed using the ImageJ program (Rasband WS, ImageJ; U.S. National Institutes of Health, Bethesda, MD, USA). All the pores present in the analyzed sample were selected and the total pore area was computed, then this data were analyzed by statistical analysis. All the images obtained were subdivided into quadrants, the area of the first quadrant being analyzed, to avoid a tendency in the results.

2.5. Cell Viability Analysis

The MDPC-23 odontoblast cell was used for feasibility analysis. MDPC-23 cells were grown in Dulbecco's modified Eagle's medium (DMEM, Gibco, Grand Island, NY, USA) supplemented with 10% heat-inactivated fetal bovine serum (FBS; GIBCO, Grand Island, NY, USA), 100 IU/mL, and 100 mg/mL, respectively, of penicillin and streptomycin and 2 mmol/L of glutamine (GIBCO, Grand Island, NY, USA) in a humidified atmosphere containing 5% CO_2 at 37 °C. The cells were subcultured every three days until the number of cells needed to perform the study were obtained using a density of 3×10^4 cells/cm^2.

The specimens were placed individually in pre-cultured MDPC-23 cells and incubated for 24 h and 72 h. After this period, the extracts were aspirated and the cell viability analysis was performed for the samples. The control group was represented by MDPC-23 cells maintained only in DMEM. The experiment was carried out in duplicate. The samples were stored in an oven at 37 °C with relative humidity.

The cells were incubated for 4 h at 37 °C and 5% CO_2 with the 3-(4,5-dimethylthiazol-2yl)-2,5-diphenyl tetrazoline bromide (MTT) solution (Sigma-Aldrich Corp., St. Louis, MO, USA) diluted in DMEM (1:10). The formazan crystals that formed in viable cells were then dissolved in acidified isopropanol and the absorbance of the resulting solution was read at 570 nm (Synergy H1, BioTek, Winooski, VT, USA). The mean absorbance of the positive control group was considered to be 100% of cell viability, and the percentage values for the experimental groups were calculated based on this parameter [12].

2.6. Analysis of Cell Morphology

In order to perform the cell morphology analysis, the cells were seeded on glass slides placed at the bottom of 24-well plates. In each time interval of analysis ($n = 2$), the cells were fixed in 2.5% glutaraldehyde (Sigma Chemical Co.), and then post-fixed in 1% osmium tetroxide (Sigma Chemical Co.), dehydrated in increasing concentrations of alcohol (30, 50, 70, 90, and 100%) and submitted to chemical drying in HMDS (1,1,1,3,3,3-hexamethyldisilazane; Sigma Chemical Co.). Finally, the slides with the cells were fixed on metal stubs, kept in a desiccator for 72 h, sputter-coated with gold, and finally analyzed by scanning electron microscopy (SEM-FEG) (SM-300, Topcon, Tokyo, Japan).

2.7. Statistical Analysis

The data were analyzed by analysis of variance (ANOVA) with fixed criteria. The Tukey test was applied to the pore area data, and the Kruskal–Wallis and Mann–Whitney tests to the cell viability data. The distribution of values and the homogeneity of the data were previously tested. A descriptive analysis was carried out for cell distribution, dispersion, and morphology. For all tests, the level of significance was 5%.

3. Results

It can be seen that there was a good distribution of HANP within the RMGIC, regardless of the concentration and manipulation technique, as demonstrated in Figure 1.

There was no difference between the distribution of the HANP among the experimental groups. This demonstrates that regardless of the technique for incorporating HANP into the RMGIC, there was good distribution within the material.

X-ray dispersive energy spectroscopy (Figure 2) demonstrates the chemical components of HANP, RMGIC, and the experimental groups HANP + RMGIC, which had the same spectrum. It can be seen that HANP was rich in Ca and P ions, while RMGIC had a greater amount of Al and Si ions.

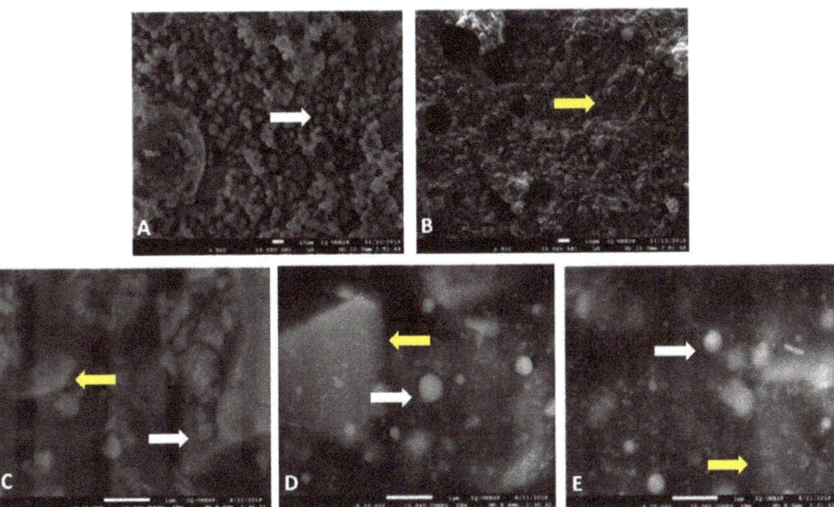

Figure 1. Distribution of nanohydroxyapatite within glass ionomer cement modified by resin. Legend: The white arrows indicate the HANP and the yellow arrows indicate the particles of the RMGIC. (**A**) Microscopic image of spherical HANP structures with magnification × 500. (**B**) Microscopic image in group C with magnification × 500. (**C**) Microscopic image of group A2 with × 20,000 magnification. (**D**) Microscopic image of the V5 group with a magnification of × 20,000. (**E**) Microscopic image of the M10 group with a magnification of × 20,000.

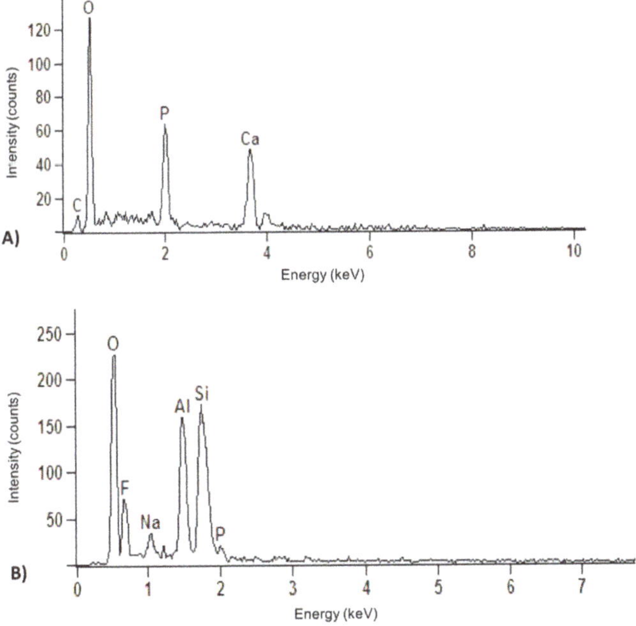

Figure 2. *Cont.*

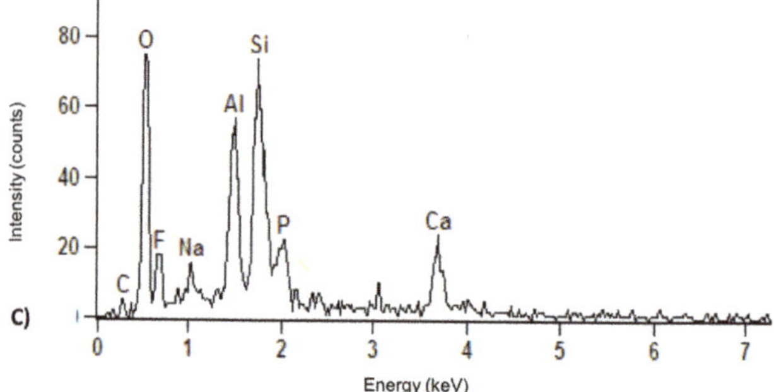

Figure 2. X-ray dispersive energy spectroscopy of the experimental groups. Legend: (**A**) Chemical composition of HANP by means of x-ray dispersive energy spectroscopy. (**B**) Chemical composition of the control group by means of x-ray dispersive energy spectroscopy. (**C**) Chemical composition of groups containing 2%, 5%, and 10% HANP + RMGIC by means of x-ray dispersive energy spectroscopy.

Data related to the number and area occupied by the pores as a function of the HANP concentration incorporated into the RMGIC powder and the mixing technique are shown in Figures 3–5.

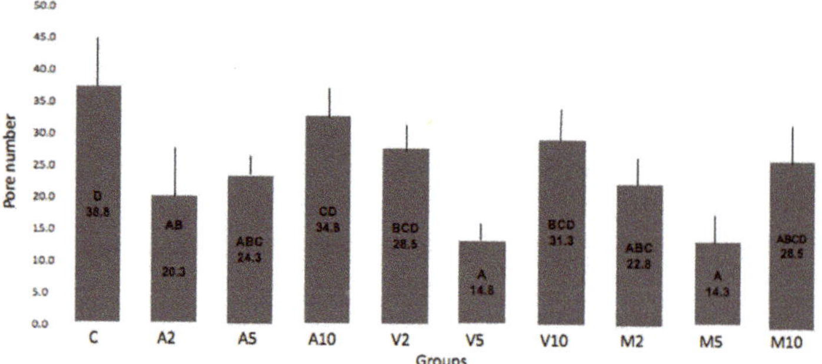

Figure 3. Number of pores of the experimental groups. Legend: Number of pores of the experimental groups. The columns represent averages and the bars represent standard deviations. The columns identified with the same letter do not differ statistically (Tukey, *p* 0.061).

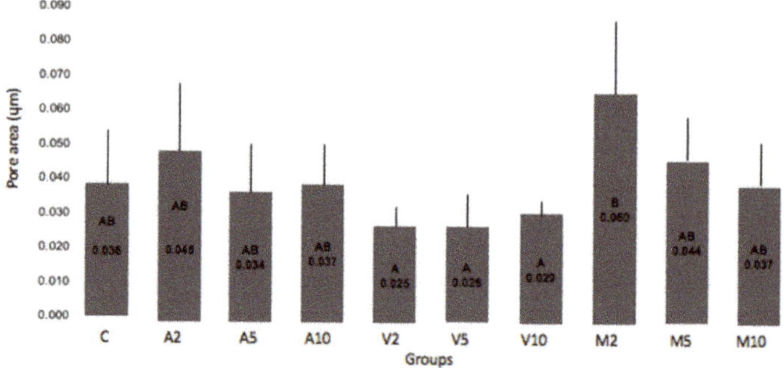

Figure 4. Pore area of experimental groups. Legend: Pore area of experimental groups. The columns represent averages and the bars represent standard deviations. The columns identified with the same letter do not differ statistically (Tukey, *p* 0.068).

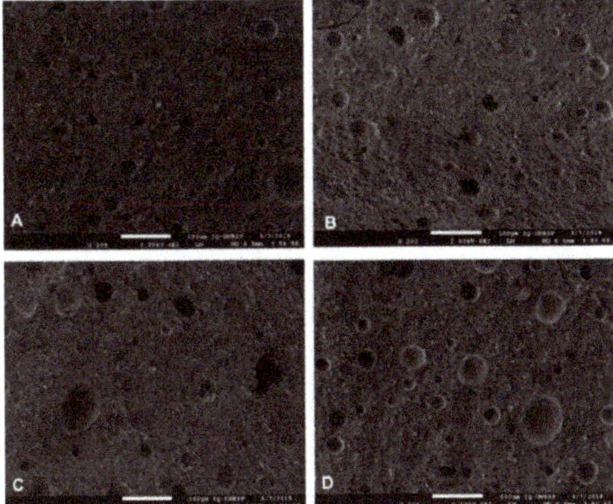

Figure 5. Distribution of the pores of the experimental groups inside the material. Legend: (**A**) Distribution of the pores of the control group. (**B**) Distribution of pores in group V5. (**C**) Distribution of pores in the M5 group. (**D**) Pore distribution in the A10 group.

It can be seen that all groups incorporated with 10% HANP had a number of pores similar to the control. Meanwhile, a significantly smaller number of pores was found for the other groups compared to the control. Groups V5 or M5 resulted in a smaller number of pores than the number observed in group C. None of the groups differed from C in relation to the area occupied by the pores, however, a smaller value was observed among the groups submitted to the vortex.

The evaluation of cytotoxicity was performed in two moments, described in Figures 6 and 7.

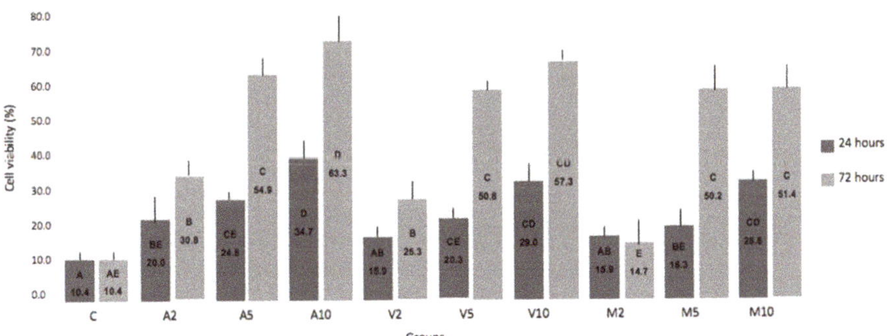

Figure 6. Cellular viability at 24 h and 72 h. Legend: Cell viability after contact of specimens with MDPC-23 cells for 24 and 72 h. No statistical difference between time periods (Kruskal–Wallis and Mann–Whitney tests, *p* 0.67).

There was no statistical difference between the 24 and 72 h periods. However, it can be analyzed that the higher the concentration of HANP incorporated into RMGIC, the greater the cell viability.

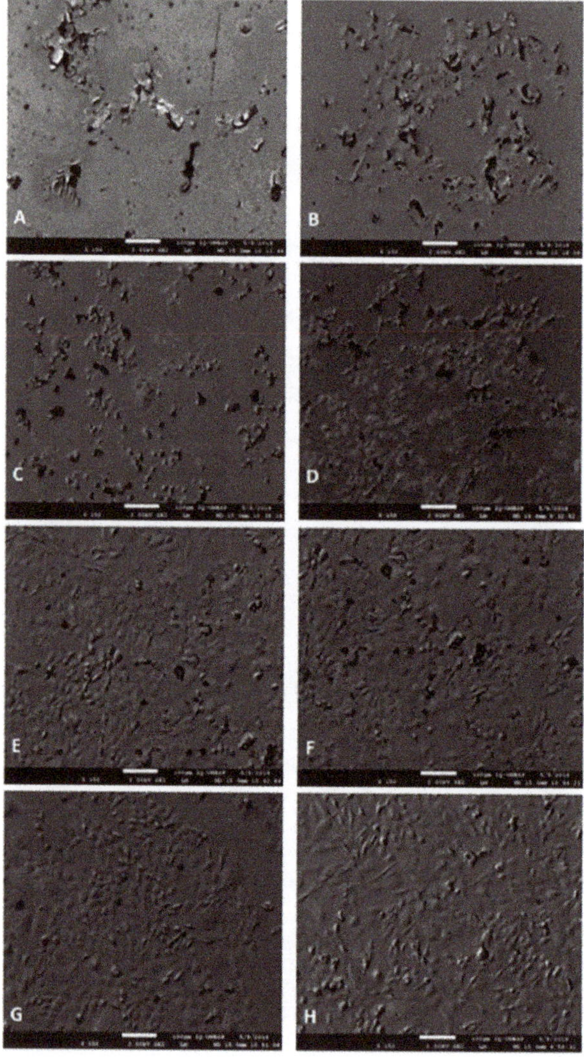

Figure 7. Cell morphology after 24 h and 72 h viability analysis. Legend: (**A**) Group C cell morphology after 24 h. (**B**) Cell morphology of group C after 72 h. (**C**) Cell morphology of the groups with addition of 2% HANP after 24 h. (**D**) Cell morphology of the groups with addition of 2% HANP after 72 h. (**E**) Cell morphology of the groups with addition of 5% HANP after 24 h. (**F**) Cell morphology of the groups with addition of 5% HANP after 72 h. (**G**) Cell morphology of the groups with the addition of 10% HANP after 24 h. (**H**) Cell morphology of the groups with the addition of 10% HANP after 72 h.

4. Discussion

RMGIC has greater mechanical resistance and better physical properties, but it has clinical limitations such as decreased biocompatibility [22–24] due to the presence of the 2-hydroxyethyl methacrylate (HEMA) monomer, which is considered cytotoxic in contact with the pulp tissue, thus limiting its use in deep cavities. This monomer is capable of inducing apoptosis by increasing oxidative stress induced by an excess of reactive oxygen species, damage to DNA, and suppression of cell proliferation [25].

The addition of HANP to the GIC has shown good results [10,11,21,22,25,26]. However, it is extremely important to be careful with the manipulation technique during the incorporation of HANP into the material, since improper mixing can affect its properties such as mechanical resistance [10]. Thus, the microstructural analysis of the GIC associated to NP as well as the observation of the distribution and dispersion of the same, is fundamental. From this form, three forms of manipulation of the HANP + RMGIC were defined (Table 1).

This study used SEM-FEG, which is widely used for the investigation of the surface microstructures of materials, also making possible the chemical analysis of the sample under observation [22]. In addition, this analytical method provides precision, accuracy, sensitivity, and preservation of the sample.

In all experimental groups, regardless of the technique of concentration or manipulation of the HANP, good distribution and dispersion of it was observed in the RMGIC. Due to the size of the nanomaterial, the NP can occupy small spaces, resulting in greater homogeneity of the surface and wide distribution within the material [23]. The control group and the experimental groups with the incorporation of HANP + RMGIC are shown in Figure 1.

The interaction of the chemical components of HANP + RMGIC can also be observed by x-ray dispersive energy spectroscopy, as shown in Figure 2, which shows a large amount of phosphorus (P) and calcium (Ca) in HANP (Figure 2A) compared to RMGIC alone (Figure 2B). After the incorporation of HANP into RMGIC, the chemical components were mixed (Figure 2C).

The good distribution of the NPs allows for an increase in the mechanical resistance as it promotes high density and narrowing among the particles within the ionomeric matrix [3]. Gu et al. [10] (2005) observed good distribution and dispersion of NPs within the ionomeric matrix by adding 4% or 12% HA to an encapsulated RMGIC. The authors reported the low tendency to form agglomerates, ensuring uniform distribution of NP and improvement of the mechanical properties of the material.

The porosity of the material is also an important property to be evaluated as it is related to the degree of dissolution and, consequently, to the resistance of the restorative material, which can alter its clinical durability. The increase in porosity also results in increased material roughness and greater adhesion of microorganisms [24].

In the present study, it can be seen that all groups incorporated with 10% HANP (A10, V10, and M10), regardless of the mixing mode, presented a number of pores similar to the control, so a viable explanation could be that the GIC liquid was unable to bathe all HANP as they have a larger surface area due to their nanometric size, thus making the connections insufficient to reduce the porosity of the material.

A significantly smaller number of pores was found for the other groups compared to the control (Figure 3). Groups V5 or M5 resulted in a smaller number of pores than the number observed in Group C. It is possible that the vortex and the manual technique incorporated less air compared to the technique using the amalgamator; in addition, the 5% concentration of HANP provided better chemical bonds with the GIC, resulting in less pores inside the new material. This comparison can be seen in Figure 5.

None of the groups differed from C in relation to the area occupied by the pores (Figure 4). However, a lower value was observed among the groups submitted to the vortex, indicating that, for this mixing technique, smaller pores were formed. This fact is important due to the longevity of the restorations [10,24].

The large number and area occupied by porosities can result in cracks and fractures of the material [10], and these failures can be minimized by the incorporation of 5% HANP with the help of vortex.

In addition to the physical and mechanical properties of materials, extreme importance represents cytotoxicity for the viability of a restorative material. Regardless of the incorporation technique, there was an improvement in cell viability directly proportional to the increase in the concentration of HANP (Figure 6).

Several in vitro studies have evaluated the cytotoxicity of conventional and resin-modified GICs and found that the leachable components present in the GIC were responsible for adverse effects on cell culture [8,27,28].

The isolated Vitremer showed higher cytotoxic effects compared to the groups containing HANP (Figure 6). Costa et al. (2003) [28], when evaluating five different GICs, also observed higher Vitremer cytotoxicity in MDPC-23 cells, suggesting that HEMA is the main component that contributes significantly to the cytotoxicity of this material.

Thus, the addition of HANP has shown promising results [7,10,11,15]. Pagano et al. (2019) [29] evaluated the addition of 4% HANP to a GIC and observed a significant reduction in the cytotoxicity of the material. The authors reported that this occurrence most likely occurred because HA has excellent biological behavior due to its chemical composition and crystalline structure, similar to apatite in the human skeletal system [15].

GICs are said to have low toxicity of dental pulp in clinical use. These cements exhibit cytotoxicity in the recently established state, but decrease substantially and depend on time [13], as observed in Figures 6 and 7. Additionally, the addition of HANP to RMGIC may have decreased cytotoxicity due to chemical interaction, since HANP has a large amount of calcium and phosphorus (Figure 2). Noorani et al. [30], when analyzing the cytotoxicity of nanohydroxyapatite-silica, found that GIC interacted with HANP through the carboxylate groups in the polyacid [30], which can decrease the cytotoxicity of the material.

5. Conclusions

The HANP added to the RMGIC should be the subject of further studies as it has been shown to have good distribution and dispersion within the material. The use of vortex was the most indicated mixing method and the incorporation of 5% HANP resulted in fewer pores inside the material. Greater cell viability with the addition of higher HANP concentration could also be observed.

Author Contributions: L.E.G. and G.A. conducted the main experiments. A.C.C.Z. and J.H. designed the study. A.C.C.Z. supervised the research. All authors contributed to the analysis of results, conclusions and reviewed the article. All authors have read and agreed to the published version of the manuscript.

Funding: This research was funded by Sao Paulo Research Foundation (FAPESP) (No. 2018/02010-7) and (No. 2020/06919-0).

Conflicts of Interest: The authors declare no conflict of interest.

References

1. Boaventura, J.M.C.; Roberto, A.R.; Becci, A.C.D.O.; Ribeiro, B.C.I.; De Oliveira, M.R.B.; De Andrade, M.F. Importância da biocompatibilidade de novos materiais: Revisão para o cimento de ionômero de vidro. *Rev. Odontol. Univ. Cid. São Paulo* **2017**, *24*, 42–50. [CrossRef]
2. Hotz, P.; McLean, J.W.; Sced, I.; Wilson, A.D. The bonding of glass ionomer cements to metal and tooth substrates. *Br. Dent. J.* **1977**, *142*, 41–47. [CrossRef] [PubMed]
3. Spezzia, S. Cimento de ionômero de vidro: Revisão de literatura. *J. Oral Investig.* **2017**, *6*, 74–88. [CrossRef]
4. Silva, M.R.R.; Sena, N.J.C.; A Cunha, D.; Souza, L.C.; Rodrigues, N.S.; A Saboia, V.P. Materiais Promissores para Melhorar as Propriedades Mecânicas e Físico-Químicas do Cimento de Ionômero de Vidro. *J. Health Sci.* **2018**, *19*. [CrossRef]
5. Matos, A.B.; Da Silva, B.B.; Araújo, E.M.D.S.; Da Silva, B.T.F. Influência do envelhecimento sobre a rugosidade e estabilidade de cor dos cimentos de ionômero de vidro encapsulados. *Clin. Lab. Res. Dent.* **2019**, 1–10. [CrossRef]
6. Popoff, J.M.D.S.; Rodrigues, J.; Aras, W.M.D.F.; Cassoni, A. Influence of Photoactivation Source on Restorative Materials and Enamel Demineralization. *Photomed. Laser Surg.* **2014**, *32*, 274–280. [CrossRef]
7. Moshaverinia, A.; Ansari, S.; Movasaghi, Z.; Billington, R.W.; Darr, J.A.; Rehman, I. Modification of conventional glass-ionomer cements with N-vinylpyrrolidone containing polyacids, nano-hydroxy and fluoroapatite to improve mechanical properties. *Dent. Mater.* **2008**, *24*, 1381–1390. [CrossRef]

8. Huang, S.; Gao, S.; Cheng, L.; Yu, H. Remineralization Potential of Nano-Hydroxyapatite on Initial Enamel Lesions: An in vitro Study. *Caries Res.* **2011**, *45*, 460–468. [CrossRef]

9. Becci, A.C.D.O.; Marti, L.M.; Zuanon, A.C.C.; Brighenti, F.L.; Spolidorio, D.M.P.; Giro, E.M. Influence of the addition of chlorhexidine diacetate on bond strength of a high-viscosity glass ionomer cement to sound and artificial caries-affected dentin. *Rev. Odontol. UNESP* **2014**, *43*, 1–7. [CrossRef]

10. Gu, Y.; Yap, A.; Cheang, P.; Khor, K.A. Effects of incorporation of HA/ZrO$_2$ into glass ionomer cement (GIC). *Biomaterials* **2005**, *26*, 713–720. [CrossRef]

11. Rodriguez, I.A.; A Ferrara, C.; Campos-Sanchez, F.; Alaminos, M.; Echevarría, J.U.; Campos, A. An in vitro biocompatibility study of conventional and resin-modified glass ionomer cements. *J Adhes Dent* **2013**, *15*, 541–546. [PubMed]

12. Mosmann, T. Rapid colorimetric assay for cellular growth and survival: Application to proliferation and cytotoxicity assays. *J. Immunol. Methods* **1983**, *65*, 55–63. [CrossRef]

13. Del Angel-Mosqueda, C.; Hernandez-Delgadillo, R.; Rodriguez-Luis, O.E.; Ramírez-Rodríguez, M.T.; Munguía-Moreno, S.; Zavala-Alonso, N.V.; Solís-Soto, J.M.; Nakagoshi-Cepeda, M.A.A.; Sánchez-Nájera, R.I.; Nakagoshi-Cepeda, S.E.; et al. Hydroxy- apatite decreases cytotoxicity of a glass ionomer cement by calcium fluoride uptake in vitro. *J. Appl. Biomater. Funct. Mater.* **2018**, *16*, 42–46.

14. Cheng, L.; Weir, M.D.; Xu, H.H.K.; Kraigsley, A.M.; Lin, N.J.; Lin-Gibson, S.; Zhou, X. Antibacterial and physical properties of calcium–phosphate and calcium–fluoride nanocomposites with chlorhexidine. *Dent. Mater.* **2012**, *28*, 573–583. [CrossRef]

15. Elsaka, S.E.; Hamouda, I.M.; Swain, M.V. Titanium dioxide nanoparticles addition to a conventiona glass-ionomer restorative: Influence on physical and antibacterial properties. *J. Dent.* **2011**, *39*, 589–598. [CrossRef]

16. Geurtsen, W.; Lehmann, F.; Spahl, W.; Leyhausen, G. Cytotoxicity of 35 dental resin composite monomers/additives in permanent 3T3 and three human primary fibroblast cultures. *J. Biomed. Mater. Res.* **1998**, *41*, 474–480. [CrossRef]

17. Hebling, J.; Giro, E.; Costa, C. Human pulp response after an adhesive system application in deep cavities. *J. Dent.* **1999**, *27*, 557–564. [CrossRef]

18. Krewski, D.; Acosta, D.; Andersen, M.E.; Anderson, H.; Bailar, J.C.; Boekelheide, K.; Brent, R.; Charnley, G.; Cheung, V.G.; Green, S.; et al. Toxicity testing in the 21st century: a vision and a strategy. *J. Toxicol. Environ. Health Part B* **2010**, *13*, 51–138. [CrossRef]

19. Vasconcellos, W.A.; Giovannini, J.F.B.G.; Jansen, W.C. Influência De Diferentes Manipuladores Mecânicos No Preparo De Cimentos Ionoméricos Encapsulados. *Arq. Bras. Odontol.* **2010**, *3*, 3–10.

20. Gu, Y.; Yap, A.; Cheang, P.; Kumar, R. Spheroidization of glass powders for glass ionomer cements. *Biomaterials* **2004**, *25*, 4029–4035. [CrossRef]

21. Alam Moheet, I.; Luddin, N.; Ab Rahman, I.; Kannan, T.P.; Ghani, N.R.N.A.; Masudi, S.M. Modifications of Glass Ionomer Cement Powder by Addition of Recently Fabricated Nano-Fillers and Their Effect on the Properties: A Review. *Eur. J. Dent.* **2019**, *13*, 470–477. [CrossRef]

22. Sanches, J.; Durigan, J.F.; Dos Santos, J.M. Utilização da microscopia eletrônica de varredura como ferramenta de avaliação da estrutura do tecido de abacate 'quintal' após danos mecânicos. *Rev. Bras. Frutic.* **2007**, *29*, 57–60. [CrossRef]

23. Andrade, K.C.A.; Costa, M.S.C.; Coqueiro, R.D.S.; Pithon, M.M.; Simões, F.X.P.C. Avaliação da rugosidade superficial de cimentos ionoméricos modificados por resina. *Rev. Odontol. Univ. Cid. São Paulo* **2017**, *26*, 38–45. [CrossRef]

24. Moberg, M.; Brewster, J.; Nicholson, J.; Roberts, H. Physical property investigation of contemporary glass ionomer and resin-modified glass ionomer restorative materials. *Clin. Oral Investig.* **2018**, *23*, 1295–1308. [CrossRef] [PubMed]

25. Pagano, S.; Coniglio, M.; Valenti, C.; Negri, P.; Lombardo, G.; Costanzi, E.; Cianetti, S.; Montaseri, A.; Marinucci, L. Biological effects of resin monomers on oral cell populations: descriptive analysis of literature. *Eur. J. Paediatr. Dent.* **2019**, *20*, 224–232. [PubMed]

26. Lee, J.-J.; Lee, Y.-K.; Choi, B.-J.; Lee, J.; Choi, H.-J.; Son, H.-K.; Hwang, J.-W.; Kim, S.O. Physical properties of resin-reinforced glass ionomer cement modified with micro and nano-hydroxyapatite. *J. Nanosci. Nanotechnol.* **2010**, *10*, 5270–5276. [CrossRef]

27. Lucas, M.E.; Arita, K.; Nishino, M. Toughness, bonding and fluoride-release properties of hydroxyapatite-added glass ionomer cement. *Biomaterials* **2003**, *24*, 3787–3794. [CrossRef]
28. Costa, C.A.D.S.; Hebling, J.; Garcia-Godoy, F.; Hanks, C.T. In vitro cytotoxicity of five glass-ionomer cements. *Biomaterials* **2003**, *24*, 3853–3858. [CrossRef]
29. Pagano, S.; Chieruzzi, M.; Balloni, S.; Lombardo, G.; Torre, L.; Bodo, M.; Cianetti, S.; Marinucci, L. Biological, thermal and mechanical characterization of modified glass ionomer cements: The role of nanohydroxyapatite, ciprofloxacin and zinc l-carnosine. *Mater. Sci. Eng. C* **2019**, *94*, 76–85. [CrossRef]
30. Noorani, T.; Luddin, N.; Rahman, I.A.; Masudi, S.M. In Vitro Cytotoxicity Evaluation of Novel Nano-Hydroxyapatite-Silica Incorporated Glass Ionomer Cement. *J. Clin. Diagn. Res.* **2017**, *11*, ZC105–ZC109. [CrossRef]

 *nanomaterials*

Article

Silica-Coated Magnetic Nanoparticles Decrease Human Bone Marrow-Derived Mesenchymal Stem Cell Migratory Activity by Reducing Membrane Fluidity and Impairing Focal Adhesion

Tae Hwan Shin [1], Da Yeon Lee [1], Abdurazak Aman Ketebo [2], Seungah Lee [3], Balachandran Manavalan [1], Shaherin Basith [1], Chanyoung Ahn [4], Seong Ho Kang [3], Sungsu Park [2] and Gwang Lee [1,*]

[1] Department of Physiology, Ajou University School of Medicine, Suwon 16499, Korea; catholicon@ajou.ac.kr (T.H.S.); ekdus93@ajou.ac.kr (D.Y.L.); bala.msuman@gmail.com (B.M.); shaherinb@gmail.com (S.B.)

[2] School of Mechanical Engineering, Sungkyunkwan University, Suwon 16419, Korea; anuabman@gmail.com (A.A.K.); nanopark@skku.edu (S.P.)

[3] Department of Applied Chemistry and Institute of Natural Sciences, Kyung Hee University, Yongin-si 17104, Korea; moon11311@naver.com (S.L.); shkang@khu.ac.kr (S.H.K.)

[4] Department of Molecular and Cell Biology, University of California, Berkeley, CA 94720, USA; justinahn99@berkeley.edu

* Correspondence: glee@ajou.ac.kr; Tel.: +82-31-219-4554

Received: 9 September 2019; Accepted: 14 October 2019; Published: 17 October 2019

Abstract: For stem cell-based therapies, the fate and distribution of stem cells should be traced using non-invasive or histological methods and a nanomaterial-based labelling agent. However, evaluation of the biophysical effects and related biological functions of nanomaterials in stem cells remains challenging. Here, we aimed to investigate the biophysical effects of nanomaterials on stem cells, including those on membrane fluidity, using total internal reflection fluorescence microscopy, and traction force, using micropillars of human bone marrow-derived mesenchymal stem cells (hBM-MSCs) labelled with silica-coated magnetic nanoparticles incorporating rhodamine B isothiocyanate (MNPs@SiO$_2$(RITC)). Furthermore, to evaluate the biological functions related to these biophysical changes, we assessed the cell viability, reactive oxygen species (ROS) generation, intracellular cytoskeleton, and the migratory activity of MNPs@SiO$_2$(RITC)-treated hBM-MSCs. Compared to that in the control, cell viability decreased by 10% and intracellular ROS increased by 2-fold due to the induction of 20% higher peroxidized lipid in hBM-MSCs treated with 1.0 µg/µL MNPs@SiO$_2$(RITC). Membrane fluidity was reduced by MNPs@SiO$_2$(RITC)-induced lipid oxidation in a concentration-dependent manner. In addition, cell shrinkage with abnormal formation of focal adhesions and ~30% decreased total traction force were observed in cells treated with 1.0 µg/µL MNPs@SiO$_2$(RITC) without specific interaction between MNPs@SiO$_2$(RITC) and cytoskeletal proteins. Furthermore, the migratory activity of hBM-MSCs, which was highly related to membrane fluidity and cytoskeletal abnormality, decreased significantly after MNPs@SiO$_2$(RITC) treatment. These observations indicated that the migratory activity of hBM-MSCs was impaired by MNPs@SiO$_2$(RITC) treatment due to changes in stem-cell biophysical properties and related biological functions, highlighting the important mechanisms via which nanoparticles impair migration of hBM-MSCs. Our findings indicate that nanoparticles used for stem cell trafficking or clinical applications should be labelled using optimal nanoparticle concentrations to preserve hBM-MSC migratory activity and ensure successful outcomes following stem cell localisation.

Keywords: magnetic nanoparticles; human bone marrow-derived mesenchymal stem cells; membrane fluidity; focal adhesion; cytoskeletal abnormality

1. Introduction

Nanoparticles are being increasingly used for disease diagnosis and therapy and cell tracing [1–3]. Among nanoparticles, magnetic nanoparticles (MNPs) and MNPs coated with biocompatible polymers and silica for safety are used for in vitro cell labelling, fluorescence-based in vivo cell tracking, and magnetic resonance imaging (MRI)-based stem cell-labelled in vivo tracing [4–7]. However, detailed information regarding the biophysical effects of nanoparticles at the cellular level is still limited.

Mesenchymal stem cells (MSCs) are used in biomedical applications (cytotherapy) for multiple sclerosis and for cardiovascular, ischemic, and neurodegenerative disorders [8–12]. In particular, bone marrow-derived MSCs (BM-MSCs) possess useful characteristics, including high degree of plasticity, trophic factor secretion, and immune response suppression capability [13–15]. Thus, human BM-MSCs (hBM-MSCs) are considered promising therapeutic candidates for clinical application [12,16–18]. For successful cytotherapeutic outcomes using stem cells [19,20], nanoparticle-based methods for tracking the localization of transplanted cells in the body are essential to ensure their distribution in the impaired tissue [21]. However, nanoparticle-induced biophysical disturbances caused by reactive oxygen species (ROS) generation, which result in changes in normal physiological redox-regulated functions and cellular alteration, are matters of concern [22]. For example, silica-coated magnetic nanoparticles incorporating rhodamine B isothiocyanate (MNPs@SiO$_2$(RITC)) induce ROS production, leading to endoplasmic reticulum (ER) stress, reduced proteasome activity, and altered cellular metabolism [23–25].

Nanoparticle-induced ROS oxidize proteins to generate protein radicals [22] and induce lipid peroxidation [26], which impairs the functions of the plasma membrane [27,28]. These oxidative cleavage events deplete unsaturated phospholipids and cholesterol in the cell membrane, leading to loss of the fluidity and permeability of the membrane and, thereby, affecting its physiological functions [29–31]. Several studies have reported that lipid peroxidation decreases membrane fluidity, indicating that the membrane fluidity of nanoparticle treated-cells can change due to oxidative stress-induced membrane damage [30,32]. However, the relationship between nanoparticle-induced lipid peroxidation and membrane fluidity remains unclear.

The membrane and the cytoskeleton are tightly linked via phosphoinositides, especially via focal adhesion and actin assembly [33–35]. Therefore, membrane damage, caused by nanoparticle-induced oxidative stress, is highly related to the cytoskeleton. Cell morphology is determined by the balance between adhesion and tension. Nanoparticles disrupt the cytoskeleton to affect focal adhesion proteins and, thus, adhesion [36] initiated by the lamellipodia (branched actin filaments) and filopodia (extended finger-like protrusions) as focal complexes [37]. During cell death; membrane repair; and osmotic-, oxidative-, and heat stress; these structures are abolished, such that cells display a rounded and shrunken morphology [38,39]. However, studies on the effect of nanoparticles on cell adhesion and tension are limited.

The focal adhesion of hBM-MSCs is strongly associated with changes in cellular traction forces [40]. Elastomeric pillar arrays are considered excellent for measuring cellular traction force by calculating the nanometric level of pillar deflection [41,42]. In addition, sub-micron pillar arrays have been shown to mimic continuous substrates of specific rigidity [43]. Thus, biophysical changes in nanoparticle-treated cells have been quantitatively studied using elastomeric submicron pillars [41,43].

In this study, we aimed to investigate the biophysical properties of MNPs@SiO$_2$(RITC)-treated hBM-MSCs, such as membrane fluidity (using total internal reflection fluorescence microscopy (TIRFM)), traction force (using micropillars), cytoskeletal characteristics, and migratory activity.

2. Materials and Methods

2.1. MNPs@SiO$_2$(RITC) and Silica Nanoparticles (NPs)

MNPs@SiO$_2$(RITC) particles, composed of a ~9 nm cobalt ferrite core (CoFe$_2$O$_3$) chemically bonded to rhodamine isothiocyanate dye (RITC) and coated by a silica shell [4], were purchased from BITERIALS (Seoul, Korea). Previously, these nanoparticles have been characterized for confirming their quality [44]. Size distribution and morphology are important factors determining the uniformity of nanoparticles and were analyzed using electron and atomic microscopy [44]. Hydrodynamic size, polydispersity, and surface charge were determined using dynamic light scattering [45]. The purity and contents of nanoparticles are usually analyzed using an X-ray based technique [44]. In this study, X-ray diffraction (XRD) analysis using a high-power X-Ray diffractometer (Ultima III, Rigaku, Japan) confirmed the structure of MNPs@SiO$_2$(RITC) (data not shown). The silica NPs were composed of identical materials and were of a similar size as the MNPs@SiO$_2$(RITC) shell, and their biological effects were similar to those of MNPs@SiO$_2$(RITC) [23,24,46,47]. The diameters of the MNPs@SiO$_2$(RITC) and silica NPs were 50 nm, and the *zeta* potential of MNPs@SiO$_2$(RITC) was between −40 to −30 mV [4,46]. A previous study determined ~10^5 particles of MNPs@SiO$_2$(RITC) per cell in MNPs@SiO$_2$(RITC)-treated MCF-7 cells using inductively coupled plasma atomic emission spectrometry [4]. Furthermore, in previous reports, the dosage was determined by measuring the fluorescence intensity of HEK293 cells treated with MNPs@SiO$_2$(RITC) at concentrations ranging from 0.01 to 2.0 μg/μL for 12 h. The optimal concentration of MNPs@SiO$_2$(RITC) was 0.1 μg/μL for in vitro use, whereas 1.0 μg/μL was the plateau concentration for cellular uptake [24]. Furthermore, MNPs@SiO$_2$(RITC) concentrations ranging from 0 to 1.0 μg/μL have been used for MRI contrasting without toxicological effects on human cord blood-derived MSCs [48], and caused changes in gene expression and metabolic profiles similar to those of the control HEK293 cells at 0.1 μg/μL [24]. In addition, the uptake efficiency of MNPs@SiO$_2$(RITC) almost plateaued at 1.0 μg/μL in HEK293 cells [24,25]. The dose-dependent fluorescence intensity of MNPs@SiO$_2$(RITC)-labelled hBM-MSCs was similar to those of labelled HEK293 cells. In addition, the viability of human cord blood-derived MSCs was determined to assess the cytotoxic effect of MNPs@SiO$_2$(RITC) after 24, 48, and 72 h of treatment with 0–1.0 μg/μL MNPs@SiO$_2$(RITC); compared to the control group, no significant cytotoxic effect was observed [48]. Therefore, in this study, hBM-MSCs were treated with 0.1 μg/μL (low dose) MNPs@SiO$_2$(RITC)or 1.0 μg/μL (high dose), similarly to previous reports [23,24,47].

2.2. Cell Culture

hBM-MSCs were purchased from PromoCell (Heidelberg, Germany) and were cultured as described in previous studies [49,50]. Briefly, the cells were rinsed with phosphate buffered saline (PBS), resuspended, cultured in Dulbecco's low-glucose modified Eagle's medium (DMEM, Gibco, Grand Island, NY, USA) supplemented with 10% fetal bovine serum (Gibco, Grand Island, NY, USA), 100 units/mL penicillin, and 100 μg/mL streptomycin (Gibco, USA), and incubated in a 5% humidified CO$_2$ chamber at 37 °C. The hBM-MSC surface markers, CD73 and CD105, and negative markers of hBM-MSCs, namely, CD34 and CD45, were analyzed and maintained (data not shown).

2.3. Morphological Analysis of hBM-MSCs

To evaluate the MNPs@SiO$_2$(RITC)-induced morphological changes, hBM-MSCs were treated with 0.1 and 1.0 μg/μL of MNPs@SiO$_2$(RITC) for 12 h. Images were acquired with an Axio Vert 200M fluorescence microscope (Zeiss, Jena, Germany). The excitation wavelength for MNPs@SiO$_2$(RITC) was 530 nm.

2.4. Cell Viability Assay

For analysis of cell viability, the CellTiter 96-cell proliferation assay kit (MTS, Promega, Madison, WI, USA) was used, according to the manufacturer's instructions. Briefly, 2×10^4 hBM-MSCs were seeded on 96-well assay plates. After 16 h, the hBM-MSCs were washed with PBS and treated with MNPs@SiO$_2$(RITC) for 12 h. The hBM-MSCs were then washed with PBS to remove excess MNPs@SiO$_2$(RITC), and MTS solution was added to each well (1/10 volume of media). Subsequently, the plate was incubated for 1 h in a 5% CO$_2$ chamber maintained at 37 °C. The absorbance of the soluble formazan was measured using a plate reader (Molecular Devices, San Jose, CA, USA) at 490 nm. Values were normalized relative to the protein absorbance value for each corresponding group.

2.5. Evaluation of Intracellular ROS Levels in hBM-MSCs

Intracellular ROS levels in hBM-MSCs were evaluated using DCFH-DA staining (Cell Biolabs, San Diego, CA, USA) according to the manufacturer's protocol. Briefly, control and MNPs@SiO$_2$(RITC) -treated hBM-MSCs were resuspended in 10 µM DCFH-DA and incubated in a 5% CO$_2$, 37 °C chamber for 1 h. The hBM-MSCs were washed twice with PBS, and DCF fluorescence was measured using a fluorescence microplate reader (Gemini EM, Molecular Devices, Sunnyvale, CA, USA) at 480 nm excitation and 530 nm emission wavelengths.

2.6. Evaluation of Lipid Peroxidation

Peroxidized lipids in control and MNPs@SiO$_2$(RITC)-treated hBM-MSCs were quantified using a lipid peroxidation kit (Cayman Chemical, Ann Arbor, MI, USA) according to the manufacturer's instructions. Briefly, hBM-MSCs were treated with MNPs@SiO$_2$(RITC) for 12 h. Subsequently, hBM-MSCs were trypsinized and washed twice with PBS. The hBM-MSCs were transferred to glass tubes, and the lipids were extracted in crystalline solid-saturated methanol and cold chloroform. The mixture was centrifuged (1500× g, 0 °C, 5 times) to form two layers, and the bottom chloroform layer was collected. The samples were mixed with 2.25 mM ferrous sulphate, 0.1 M hydrochloric acid, and 1.5% ammonium thiocyanate in methanol at a 9:1 ratio. Next, the mixtures were incubated at room temperature for 5 min. Ferric ions were produced in the reaction, and the level of peroxidized unsaturated lipids was evaluated using thiocyanate as a chromogen. Absorbance was measured at 500 nm using a quartz cuvette and microplate reader (Molecular Devices, San Jose, CA, USA).

2.7. Measurement of Membrane Fluidity

Membrane fluidity of hBM-MSCs was measured using a homemade combined differential interference contrast (DIC)-total internal reflection fluorescence microscopy (TIRFM) experimental system [49,51]. The procedure was based on a previously described protocol [52,53]. Briefly, hBM-MSCs were cultured on 0.13–0.16 mm thick cover slips and treated with MNPs@SiO$_2$(RITC) for 12 h. The hBM-MSCs were incubated with media containing 10 µM Laurdan in a 5% CO$_2$, 37 °C chamber for 2 h. The hBM-MSCs were washed twice with PBS and incubated with fixation buffer (Cytofix; BD, San Jose, CA, USA). Subsequently, the hBM-MSC-containing cover slips were mounted onto 0.13–0.16 mm thick cover slips with mounting medium (Prolong Gold antifade; Molecular Probes, Eugene, OR, USA). Laurdan fluorescence was observed with a 100× objective lens (oil-type, Olympus UPLFL 1.3 N.A., W.D. 0.1 mm) and an A CCD camera (QuantEM 512SC, Photometrics, Tucson, AZ, USA). The fluorescence intensity was measured using an excitation wavelength of 405 nm, and emission fluorescence was detected using 420 nm and 473 nm bandpass filter (resolution: ±5 nm). As a parameter of membrane fluidity, the generalized polarizations (GP) = (fluorescence intensity at 420 nm—fluorescence intensity at 473 nm)/(fluorescence intensity at 420 nm + fluorescence intensity at 473 nm) was calculated, and pseudo-colored GP value images were generated using the Image J 1.48v software (NIH, Bethesda, MD, USA) [54]. Gauss distributions were generated using the nonlinear fitting algorithm in Sigma Plot 10.0 (Systat Software Inc., San Jose, CA, USA).

2.8. Immunocytochemistry

hBM-MSCs were seeded on cover slips and treated with 0.1 µg/µL and 1.0 µg/µL MNPs@SiO$_2$(RITC) for 12 h. The cells were, then, fixed in Cytofix buffer (BD, San Jose, CA, USA). To reduce non-specific binding, the cover slips were blocked with PBS containing 2% bovine serum albumin (BSA) and 0.1% Triton-X100 (Sigma-Aldrich, St Louis, MO, USA). For actin labelling, hBM-MSCs were then incubated with Alexa Fluor 488-conjugated phalloidin (Molecular Probe, Carlsbad, CA, USA, 1:200), diluted in blocking buffer, for 1 h at room temperature. For β-tubulin and vinculin labelling, hBM-MSCs were incubated with anti-β-tubulin mouse polyclonal antibody (Molecular Probes, Carlsbad, CA, USA, 1:200) or anti-vinculin rabbit monoclonal antibody (Becton Dickinson, Franklin Lakes, NJ, USA, 1:200) diluted in blocking buffer, for 12 h at 4 °C. Following three times rinsing in PBS containing 0.1% Triton-X100, the cells were incubated with Alexa Fluor 488-conjugated anti-mouse goat polyclonal antibody or Alexa Fluor 647-conjugated anti-rabbit goat polyclonal antibody for 1 h at room temperature. The labelled cells were washed thrice with PBS containing 0.1% Triton-X100 and incubated with PBS containing 10 µg/mL Hoechst 33342 for 10 min at room temperature to label the nuclei. After washing three times with PBS, cover slips were mounted onto slides using Prolong Gold antifade mounting medium (Molecular Probes). Fluorescent images were acquired using confocal laser scanning microscopy (LSM710, Carl Zeiss Microscopy GmbH, Jena, Germany). The excitation wavelengths for Alexa Fluor 488, Hoechst 33342, and MNPs@SiO$_2$(RITC) were 488, 405, and 530 nm, respectively. The attached hBM-MSCs areas were analyzed using the Image J software.

2.9. Western Blotting

hBM-MSCs were seeded at a density of 2×10^5 cells/well in 6-well plates and cultured for 36 h. hBM-MSCs were treated with 0.1 or 1.0 µg/µL MNPs@SiO$_2$(RITC) for 12 h and lysed in radioimmunoprecipitation assay (RIPA) buffer. The lysates were vortexed and incubated at 4 °C for 1 h. Then, the lysates were centrifuged at 14,000× g for 15 min at 4 °C, and the supernatants were collected. Next, 40 µg protein was subjected to sodium dodecyl sulphate-polyacrylamide gel electrophoresis (SDS–PAGE) and transferred onto nitrocellulose membranes. The membranes were blocked with 3% non-fat milk for 1 h at room temperature and incubated with primary antibody overnight at 4 °C. The following primary antibodies were used: t-c-SRC (1:2000, Santa Cruz Technologies, Santa Cruz, CA, USA), p-c-SRC (1:2000, Santa Cruz Technologies, Santa Cruz, CA, USA), t-FAK (1:2000, Santa Cruz Technologies, Santa Cruz, CA, USA), p-FAK (1:2000, Santa Cruz Technologies, Santa Cruz, CA, USA), and β-actin (1:5000, Cell Signaling, La Jolla, CA, USA). Secondary antibodies were used at a dilution of 1:2000 (Santa Cruz Technologies, USA). The blots were developed using enhanced chemiluminescence solution (ECL, Thermo Scientific, Waltham, MA, USA), and luminescence was captured on medical blue X-ray film (AGFA, Mortsel, Belgium) in a dark room.

2.10. Microfabrication of Pillar Arrays

A standard photolithograph was used to fabricate a mold with arrays of holes over a silicon wafer [55]. To fabricate pillar arrays, Polydimethylsiloxane (PDMS) was mixed at 10:1 ratio with its curing agent (Sylgard 184; Dow Corning, Midland, MI, USA) and degassed for 15 min. Next, it was spin-coated over the mold, degassed again for 30 min to remove trapped air bubbles within the mixture, and cured at 80 °C for 4 h and 30 min until the Young modulus of the PDMS reached 2 ± 0.1 MPa. Subsequently, the cured pillar array was carefully removed from the mold. In the case of micron scale pillars, cellular contractions occurred around individual pillars (diameter 2 µm) [43,56]. Thus, we generated linearly arranged pillar arrays of 900 nm diameter, 1 µm height, and a pillar diameter two times the center-to-center distance between pillars. The bending stiffness of the pillar was calculated based on Euler-Bernoulli beam theory [57]:

$$k = \frac{3}{64}\pi E \frac{D^4}{L^3}$$

where D is the diameter, L is the length, and E is the Young modulus of the pillar [43]. The bending stiffness (k) of the pillar arrays used in this study was 28.8 nN/µm.

2.11. Measurement of Traction Force

Images of pillars in hBM-MSCs were captured using a live cell chamber at 1 Hz in a fluorescence microscope (Deltavision, GE Healthcare, Chicago, IL, USA) equipped with a camera (CoolSNAP HQ^2, Photometrics) at 37 °C and 5% humidity. The place of each pillar in each frame was determined using the pillar tracking plugin (PillarTracker 1.1.3 version) of the Image J software. PillarTracker uses the pillar reconstruction algorithm to establish an exact grid of the pillar arrays, thus allowing users to automatically detect and track the locations of the pillars. Throughout this study, pillars with no cell contact were used as reference pillars. To account for stage drift, the average displacement of the reference pillars was deducted from the displacement data of pillars deflected by hBM-MSCs. To avoid unwanted displacement of pillars by MNPs@SiO$_2$(RITC), MNPs@SiO$_2$(RITC)-treated cells were washed five times using Dulbecco's phosphate buffered saline (DPBS) before seeding on a pillar array. The displacement of each pillar was multiplied by its bending stiffness to calculate the traction force.

2.12. Wound Healing Assay

hBM-MSCs were seeded and grown to 100% confluence in 6-well plates, followed by washing in PBS. Cell monolayers were wounded with a 200-µL micropipette tip in two different places in each well, treated with MNPs@SiO$_2$(RITC) in serum-free media, and allowed to migrate for 16 h. Images of the wounded areas were captured under an Axio Vert 200M fluorescence microscope (Zeiss, Jena, Germany) at 0 and 16 h. The excitation wavelength for MNPs@SiO$_2$(RITC) was 530 nm. Migration activity was quantified by analyzing the cell number.

2.13. Invasion Assay

Invasion assays of hBM-MSCs were performed using an 8-µm pore size transwell polycarbonate membrane (Corning, Corning, NY, USA). The upper side of the membrane was coated with Matrigel (1:10 dilution in 0.01 M Tris pH 8.0, 0.7% NaCl) for 2 h at 37 °C. hBM-MSCs were treated with MNPs@SiO$_2$(RITC) for 12 h. Next, 2.5×10^4 hBM-MSCs were transferred to the upper chamber of the transwell in serum-free media, and 10% FBS containing medium was added to the lower chamber as a chemoattractant. The cells were incubated for 12 h at 37 °C. Subsequently, the cells on the upper side of the membrane were removed with a cotton swab, and the invading cells on the lower side of the membrane were fixed in Cytofix buffer (BD, San Jose, CA, USA) and stained with Hoechst 33342. Images were acquired using an Axio Vert 200M fluorescence microscope (Zeiss, Jena, Germany). The excitation wavelengths for MNPs@SiO$_2$(RITC) and Hoechst 33342 were 530 nm and 405 nm, respectively. The number of invading cells was counted using the Image J software.

2.14. Statistical Analysis and Error Correction

The results were analyzed using one-way analysis of variance (ANOVA) with Bonferroni's multiple-comparison test of the IBM-SPSS software (IBM Corp., Armonk, NY, USA). Differences were considered significant for p values < 0.05. In the micropillar experiments, errors of the pillar deflections were corrected by reducing the average deflection of pillars outside the cell.

3. Results

3.1. Decrease in Cell Viability and ROS Generation of MNPs@SiO$_2$(RITC)-Treated hBM-MSCs

To evaluate the viability of and ROS generation in MNPs@SiO$_2$(RITC)-treated hBM-MSCs, hBM-MSCs were treated with MNPs@SiO$_2$(RITC) for 12 h before analysis (Figure 1a). A monolayer of hBM-MSCs was clearly observed for the non-treated control cells, while the monolayer was disintegrated for the MNPs@SiO$_2$(RITC)-treated hBM-MSCs (Figure 1b). Furthermore, compared to that in the non-treated control, the viability of hBM-MSCs decreased by ~10% upon treatment with 0.1 and 1.0 µg/µL MNPs@SiO$_2$(RITC) (Figure 1c). However, the viability of cells treated with 0.1 µg/µL MNPs@SiO$_2$(RITC) was not statistically significantly different from that of cells treated with 1.0 µg/µL MNPs@SiO$_2$(RITC). Intracellular ROS generation was evaluated using 2′,7′-dichlorodihydrofluorescin diacetate (DCFH-DA) staining in MNPs@SiO$_2$(RITC)-treated hBM-MSCs. Treatment with MNPs@SiO$_2$(RITC) and 50 nm-sized silica NPs increased intracellular ROS levels (Figure 1d,e). In particular, compared to that in non-treated control cells and cells treated with 0.1 µg/µL NPs, intracellular ROS levels increased by more than 50% in cells treated with 1.0 µg/µL MNPs@SiO$_2$(RITC) and silica-NPs. Furthermore, the level of ROS-induced peroxidized lipids increased by ~30% in hBM-MSCs treated with 1.0 µg/µL MNPs@SiO$_2$(RITC) (Figure 1f).

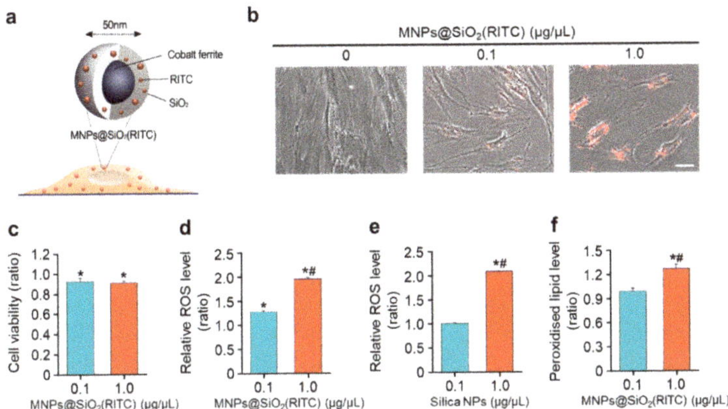

Figure 1. Intracellular reactive oxygen species (ROS) generation and lipid peroxidation in MNPs@SiO$_2$(RITC)-treated hBM-MSCs. (**a**) Schematic showing MNPs@SiO$_2$(RITC) composition. (**b**) Morphological analysis of non-treated control and MNPs@SiO$_2$(RITC)-treated hBM-MSCs. Scale bar = 50 µm (**c**) Cell viability assay with hBM-MSCs treated with MNPs@SiO$_2$(RITC) for 12 h. Evaluation of intracellular ROS generation using DCFH-DA for 12 h in HEK293 cells treated with (**d**) MNPs@SiO$_2$(RITC) and (**e**) silica NPs. The non-oxidized DCFH-DA was used as the blank. (**f**) Evaluation of peroxidized lipids using ferrous thiocyanate. Ferrous thiocyanate was used as the blank. Data represent mean ± SD of three independent experiments. * $p < 0.05$ vs. non-treated control, # $p < 0.05$ for the comparison between 0.1 and 1.0 µg/µL MNPs@SiO$_2$(RITC) or silica NP-treated cells. Data represent mean ± SD of three independent experiments.

3.2. Reduction in Membrane Fluidity of hBM-MSCs after MNPs@SiO$_2$(RITC) Treatment

To analyze changes in membrane fluidity due to peroxidation of lipids, the MNPs@SiO$_2$(RITC)-treated hBM-MSCs were stained with Laurdan and generalized polarization (GP) values were calculated using TIRFM (Figure 2a). The number of high-GP areas on the hBM-MSC surface—Corresponding to rigid domains—Increased upon MNPs@SiO$_2$(RITC) treatment. In particular, the abundantly distributed region of MNPs@SiO$_2$(RITC) majorly co-localized with the high GP-distribution region in a GP scale of −1.0 to 1.0 (Figure 2b). GP frequency distribution values of MNPs@SiO$_2$(RITC) treated-hBM-MSCs were subtracted from the corresponding values of the non-treated control hBM-MSCs to obtain frequency difference curves (Figure 2c) and total mean GP values (Figure 2d).

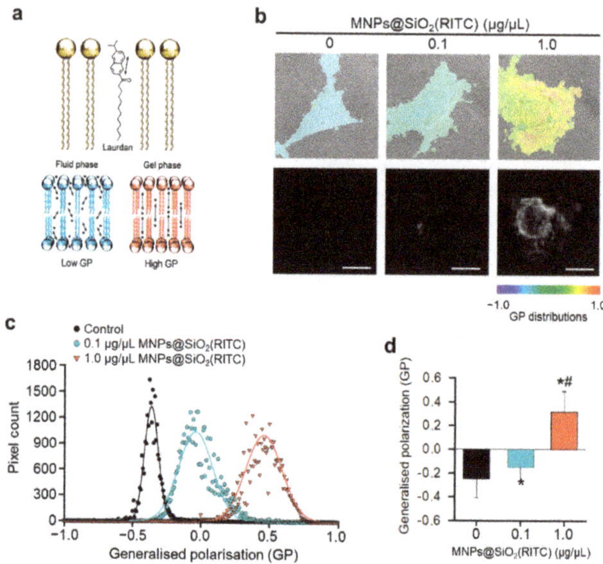

Figure 2. Laurdan generalized polarizations (GP) images and GP frequency distributions of hBM-MSCs treated with MNPs@SiO$_2$(RITC) for 12 h. (**a**) Schematic of Laurdan for measuring membrane GP value. (**b**) Merged differential interference contrast (DIC) and total internal reflection fluorescence microscopy (TIRFM) images of human bone marrow-derived mesenchymal stem cells (hBM-MSCs) in each upper panel. Distributions of magnetic nanoparticles incorporating rhodamine B isothiocyanate (MNPs@SiO$_2$(RITC)) are indicated in each lower panel. GP distributions ranged from −1.0 to 1.0. Scale bar = 2.5 μm. (**c**) GP frequency distributions of cells. GP values of each pixel are represented as dots and were fitted to Gaussian distributions. (**d**) Total GP values. Data represent mean ± SD of three independent experiments ($n = 10$). * $p < 0.05$ vs. non-treated control, $^{\#}$ $p < 0.05$ for the comparison between cells treated with 0.1 and 1.0 μg/μL of MNPs@SiO$_2$(RITC).

3.3. Cytoskeletal Abnormality in MNPs@SiO$_2$(RITC)-Treated hBM-MSCs

We investigated the changes in cell morphology and focal adhesion in MNPs@SiO$_2$(RITC)-treated hBM-MSCs. As shown in Figure 2, cell shrinkage was observed with reduction in the attached area of MNPs@SiO$_2$(RITC)-treated hBM-MSCs, indicating cytoskeletal changes and abnormal adhesion, which are followed by reduced membrane fluidity due to lipid peroxidation. Furthermore, immunocytochemical analysis of β-tubulin, a major cytoskeletal protein, showed shrinkage of the cell and β-tubulin structure, although interaction between MNPs@SiO$_2$(RITC) and β-tubulin was not detected (Figure 3).

Immunocytochemical analysis of F-actin, a major cytoskeletal protein, co-stained with vinculin, a focal adhesion marker, showed that unlike that in non-treated control hBM-MSCs, lamellipodia and filopodia disappeared, and vinculin was abnormally congregated in the central region of hBM-MSCs treated with 1.0 μg/μL MNPs@SiO$_2$(RITC) (Figure 4a). However, specific interaction between MNPs@SiO$_2$(RITC) and F-actin or vinculin was not detected. Furthermore, compared to that in the non-treated control, the attached relative cell area was reduced in 0.1 and 1.0 μg/μL MNPs@SiO$_2$(RITC)-treated hBM-MSCs (Figure 4b). Levels of phosphorylated proto-oncogene tyrosine-protein kinase SRC (c-SRC) and focal adhesion kinase (FAK), which are activated forms of the focal adhesion proteins SRC and FAK, respectively, were reduced in cells treated with 1.0 μg/μL MNPs@SiO$_2$(RITC) (Figure 4c).

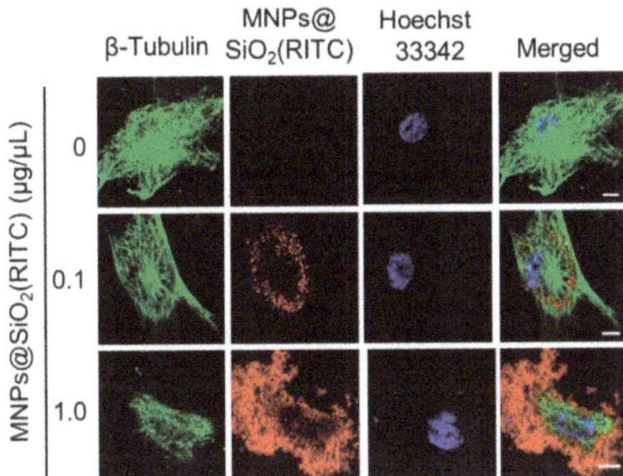

Figure 3. Tubulin-based analysis of the cytoskeleton of hBM-MSCs treated with MNPs@SiO$_2$(RITC) for 12 h. β-Tubulin stained images of hBM-MSCs treated with MNPs@SiO$_2$(RITC). Blue, Hoechst 33342; Green, β-tubulin; red, MNPs@SiO$_2$(RITC). Scale bar = 10 µm.

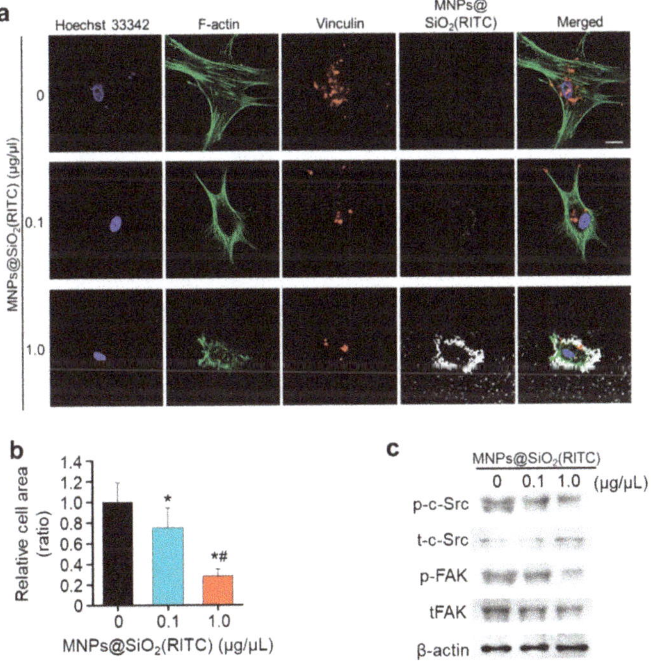

Figure 4. Shrinkage and abnormal focal adhesion of hBM-MSCs treated with MNPs@SiO$_2$(RITC). (**a**) Images of F-actin based cytoskeleton and focal adhesion in MNPs@SiO$_2$(RITC)-treated HEK293 cells. Green, F-actin; red, vinculin; white, MNPs@SiO$_2$(RITC); blue, Hoechst 33342. Scale bar = 20 µm. (**b**) Relative attached area of MNPs@SiO$_2$(RITC)-treated hBM-MSCs compared to non-treated control. Data represent mean ± SD ($n > 30$). * $p < 0.05$ vs. non-treated control. [#] $p < 0.05$ for the comparison between 0.1 and 1.0 µg/µL MNPs@SiO$_2$(RITC)-treated cells. (**c**) Immunoblotting analysis associated with focal adhesion. p-, phosphorylated protein; t, total protein. β-Actin was used as an internal control.

3.4. Reduction in Traction Force of hBM-MSCs after MNPs@SiO$_2$(RITC) Treatment

To analyze, in detail, the parameters of cell adhesion, cell spread areas were measured using images of MNPs@SiO$_2$(RITC)-treated hBM-MSCs and submicron pillars at 12 h after cell seeding (Figure 5a). Compared to that in the non-treated control cells, cell spreading of hBM-MSCs treated with 0.1 and 1.0 µg/µL MNPs@SiO$_2$(RITC) decreased (Figure 5b). Furthermore, the spread areas of hBM-MSCs treated with 0.1 and 1.0 µg/µL MNPs@SiO$_2$(RITC) were significantly smaller than those of the non-treated control cells (Figure 5c).

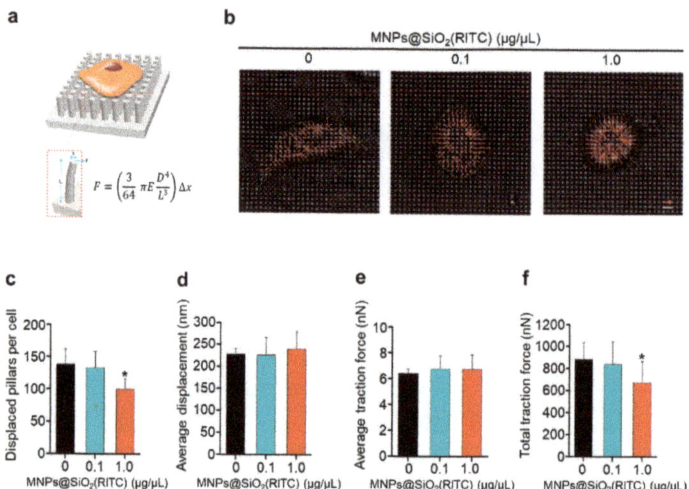

Figure 5. Change in pillar deflection, traction force, aspect ratio, and surface area of hBM-MSCs treated with MNPs@SiO$_2$(RITC). (**a**) Schematic of traction force measurement using a micropillar. F = traction force; E = Young modulus of the pillar; D = diameter of pillar; L = length of pillar; Δx = pillar displacement. (**b**) Representative images showing the concentration of MNPs@SiO$_2$(RITC) inside the cell, pillar deflections, and magnified pillar deflections at the edge of the cell (left to right). The red arrow represents 356 nm of deflection and the white bar represents 8 µm. The yellow line indicates the approximate cell boundary. The direction and length of the red arrow indicate the magnitude and direction of pillar deflection, respectively. (**c**) Displaced pillar number of pillar array (**d**) Average displacement of each pillar under the cell. (**e**) Average traction force of each pillar under the cell and (**f**) total traction force of pillars beneath the cell. Data represent mean ± SD (n = 21). * $p < 0.05$ vs. non-treated control.

The pillar deflection in the magnified images was used to measure pillar displacement (Figure 5d) and calculate traction forces (Figure 5e,f). To calculate the traction force of a pillar, displacement of each pillar was multiplied by the bending stiffness of the pillar [43]. There were no significant changes in pillar displacement and the average traction force of MNPs@SiO$_2$(RITC)-treated hBM-MSCs. However, the total traction force of hBM-MSCs treated with 1.0 µg/µL MNPs@SiO$_2$(RITC) was significantly lower than those of hBM-MSCs treated with 0.1 µg/µL MNPs@SiO$_2$(RITC) and non-treated control hBM-MSCs. The results showed that cell attachment was impaired after treatment with 1.0 µg/µL MNPs@SiO$_2$(RITC). Taken together with the spread area and traction force data, this result implies that the decrease in total traction force of hBM-MSCs results from cell shrinkage upon MNPs@SiO$_2$(RITC) treatment.

3.5. Reduction in Migratory Activity of hBM-MSCs after MNPs@SiO$_2$(RITC) Treatment

To evaluate membrane fluidity and adhesion-related biological functions, we assessed the effect of treatment with 0.1 or 1.0 µg/µL MNPs@SiO$_2$(RITC) on the migratory activity of hBM-MSCs using wound healing and invasion assays. Compared to those of non-treated controls and cells treated with 0.1 μg/μL MNPs@SiO$_2$(RITC), the migratory ability of hBM-MSCs treated with 1.0 μg/μL MNPs@SiO$_2$(RITC) was seen to be significantly impaired through the wound healing assay (Figure 6a). Results of the invasion assay showed that compared to that of the non-treated controls, the invasion ability of hBM-MSCs was significantly impaired by MNPs@SiO$_2$(RITC) treatment in a dose-dependent manner (Figure 6b).

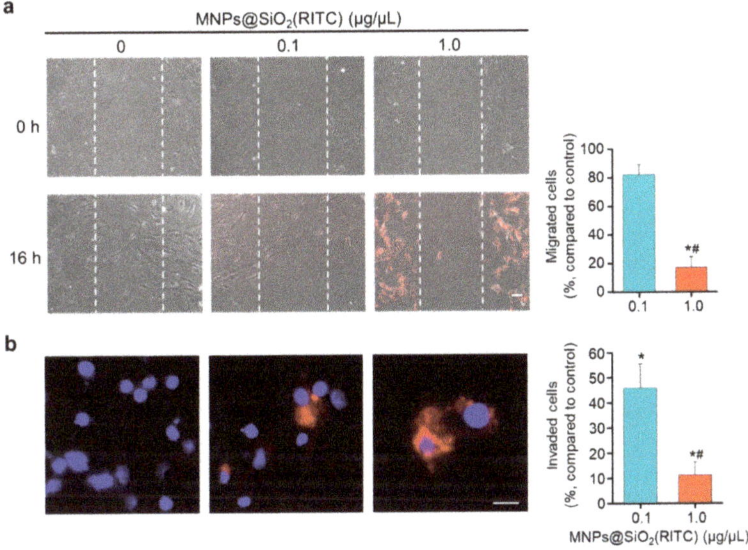

Figure 6. Migratory activity of hBM-MSCs treated with MNPs@SiO$_2$(RITC). (**a**) Representative images of wound healing assay and quantitative image analysis. Images of the initial wounded (0 h) layer are shown the upper panels. Images of cells after MNPs@SiO$_2$(RITC) treatment for 16 h are shown in the lower panels. Scale bar = 100 µm. Quantitative image analysis of migrated cells in MNPs@SiO$_2$(RITC)-treated hBM-MSCs are shown in the bar graph. (**b**) Representative images of hBM-MSCs and quantitative image analysis of the invasion assay results after MNPs@SiO$_2$(RITC) treatment for 12 h. Red, MNPs@SiO$_2$(RITC); blue, Hoechst 33342. Scale bar = 20 µm. Quantitative image analysis of invaded cells in MNPs@SiO$_2$(RITC)-treated hBM-MSCs are shown in the bar graph. Data represent mean ± SD of three independent experiments. Data represent mean ± SD of three independent experiments. * $p < 0.05$ vs. non-treated control, $^{\#}$ $p < 0.05$ for the comparison between cells treated with 0.1 and 1.0 µg/µL of MNPs@SiO$_2$(RITC).

4. Discussion

This study used molecular cellular biology tests, TIRFM measurement of membrane fluidity, micropillar measurement of traction force, and invasion and migration analyses to evaluate the biophysical effects of MNPs@SiO$_2$(RITC) treatment in hBM-MSCs. Our results indicated that MNPs@SiO$_2$(RITC) usage should be minimized in cell labelling to preserve the biophysical effect of hBM-MSCs.

Reduction in cell viability and ROS generation have been reported in MSCs treated with nanoparticles of various types and sizes [58–60]. In this study, cell viability decreased slightly by about 10% in hBM-MSCs treated with both 0.1 and 1.0 µg/µL MNPs@SiO$_2$(RITC) and silica nanoparticles. This result is consistent with that of a previous study on 0–1.0 µg/µL MNPs@SiO$_2$(RITC)-treated human cord blood–derived MSCs [48]. However, the viability of HEK293 cells treated with MNPs@SiO$_2$(RITC) has been shown to decrease by about 1%–3% [24]. These discrepancies may be related to the cell-specific characteristics, as MSCs are more sensitive to excessive ROS induced by nanoparticles than differentiated cell lines [61].

We analyzed the biophysical changes and membrane fluidity in MNPs@SiO$_2$(RITC)-treated hBM-MSCs, the effects of lipid peroxidation by MNPs@SiO$_2$(RITC)-induced ROS, and the physical interaction between the cell membrane and MNPs@SiO$_2$(RITC). Our results suggested that cell membrane damage induced by MNPs@SiO$_2$(RITC) can occur via direct interaction between membrane lipids and nanoparticles [62] and that lipid peroxidation in nanoparticle-treated hBM-MSCs can induce oxidative membrane damage, resulting in biological alterations [63,64]. In addition, the biological effects of MNPs@SiO$_2$(RITC) were caused by their silica shell rather than the cobalt ferrite core compounds, as reported previously [23,24].

We also observed that MNPs@SiO$_2$(RITC)-treated hBM-MSCs showed a rounded and shrunken morphology with disrupted cytoskeletal structure. However, specific interactions between MNPs@SiO$_2$(RITC) and actin or tubulin were not observed. Generally, the cytoskeleton is tightly linked to the membrane via phosphoinositides and linker proteins, such as spectrin, Ezrin/radixin/moesin (ERM), and myosin-I [33,34,65]. Thus, biophysical changes in the membrane are reflected as cytoskeletal changes in hMSCs [65]. Based on the relationship between the cytoskeleton and ROS, high ROS levels have been shown to induce microtubule dysfunction and sever F-actin structure [35]. Furthermore, endocytosis is another potential mechanism underlying cytoskeletal changes. During endocytosis, cells undergo reorganization of the cytoskeleton and the membrane [66]. Thus, the cytoskeletal rearrangement may be caused by ROS generation and bulk endocytosis post-MNPs@SiO$_2$(RITC) treatment.

Previous studies have suggested that the major mechanism underlying nanoparticle-induced cell shrinkage and abnormal formation of focal adhesions involves cytoskeletal depolymerization and an increase in cellular traction force [67–69]. We observed that the lamellipodia and filopodia structures of MNPs@SiO$_2$(RITC)-treated hBM-MSCs were disrupted and the phosphorylation of FAK and c-SRC, which are markers of focal adhesion formation, was reduced. However, there were no changes in the traction forces, although the total traction force was reduced. These results indicated that MNPs@SiO$_2$(RITC)-induced reduction of FAK and c-SRC phosphorylation corresponded with the reduction in the attached area of cells and the number of focal adhesions during cell shrinkage and cytoskeletal structure disruption after MNPs@SiO$_2$(RITC) treatment.

We observed that the changes in biophysical properties, such as reduced membrane fluidity, cell shrinkage with disrupted cytoskeletal structure, and reduced total traction force in MNPs@SiO$_2$(RITC)-treated hBM-MSCs, were tightly linked to each other and contributed to the reduction in the migratory activity of hBM-MSCs. These findings are supported by two previous findings: (i) Actin assembly-based protrusions and generation of traction forces are key processes in cell migration [70]. Furthermore, migratory activity is highly related to cellular biophysical properties, such as membrane fluidity and traction force [71,72]. (ii) Condensation of the cytoskeleton due to MNPs@SiO$_2$(RITC) treatment increases cell stiffness, which impedes cellular movement in the contractile machinery (shrinkage) [71].

Evaluation of the migratory activity is a major requirement for hBM-MSC applications and for successful outcome of stem cell therapy and tracking studies [66]. In addition, appropriate localization of hBM-MSCs in damaged tissues is important for the therapeutic effect of trophic factors and cytokines secreted by hBM-MSCs [73]. Previously, we have showed that PKH-26, a red fluorescent cell labelling dye, labelled hBM-MSCs in a localized, spotted-like pattern, implying secretion of the red dye outside the cell, in the border region of a lesion in a rat model of ischemic stroke [50]. However, MNPs@SiO$_2$(RITC)-labelled human umbilical cord blood–derived MSCs in vivo rarely appeared as spotted particles in a mouse model, suggesting retention of MNPs@SiO$_2$(RITC) in the cell for good tracking efficacy [48]. Hence, we believe that MNPs@SiO$_2$(RITC) is better than the PKH-26 dye for hBM-MSC tracing. Further studies are required (i.e., in vivo experiment) for tracking MNPs@SiO$_2$(RITC)-treated hBM-MSCs.

Based on our analysis of the effect of MNPs@SiO$_2$(RITC) treatment on the biophysical and biological functions of hBM-MSCs, we suggest the mechanism of action of MNPs@SiO$_2$(RITC) as follows: (i) MNPs@SiO$_2$(RITC) are internalized into hBM-MSCs; (ii) intracellular ROS are generated by the internalized MNPs@SiO$_2$(RITC); (iii) the ROS induces peroxidation of lipids, which are major components of the cell membrane; (iv) peroxidation decreases membrane fluidity; (v) cell shrinkage is induced by the reduction in membrane fluidity; (vi) owing to the cell shrinkage, abnormality in adhesion and reduction in total traction force are induced; (vii) owing to the reduction in membrane fluidity and abnormality in adhesion, fundamentally induced by intracellular ROS, the migratory activity of MNPs@SiO$_2$(RITC)-treated hBM-MSCs decreases. The effect of this treatment on other cell functions will be addressed in future investigations.

In conclusion, our findings suggest that high-dose MNPs@SiO$_2$(RITC) can alter biophysical properties and reduce the migratory activity of MNPs@SiO$_2$(RITC)-treated hBM-MSCs. Thus, nanoparticles used for stem cell trafficking or clinical applications should be labelled using optimal nanoparticle concentrations to preserve hBM-MSC migratory activity and ensure successful outcomes following stem cell localization.

Author Contributions: Conceptualization, T.H.S. and G.L.; methodology, B.M., S.B. and C.A.; software, T.H.S.; validation, S.B., T.H.S. and A.A.K.; formal analysis, D.Y.L. and S.L.; investigation, S.B. and C.A.; resources, S.P. and S.H.K.; data curation, T.H.S. and A.A.K.; Writing—Original draft preparation, T.H.S. and G.L.; Writing—Review and editing, T.H.S., S.P. and G.L.; visualization, T.H.S.; supervision, G.L.; project administration, G.L.; funding acquisition, G.L.

Funding: This work was supported by grants from the National Research Foundation (NRF) funded by the Ministry of Science and ICT (MSIT) in Korea [2018R1D1A1B07049494] and [2016M3C7A1904392], a grant from the Korea Basic Science Institute (KBSI) National Research Facilities & Equipment Center (NFEC) funded by the Korea government (Ministry of Education) [2019R1A6C1010003], and the BioNano Health-Guard Research Center as Global Frontier Project (NRF-2018M3A6B2057299) funded by the Ministry of Science and ICT (MSIT) in Korea.

Acknowledgments: The authors thank the Ajou Core-Facility Center for technical support regarding cell image analysis.

Conflicts of Interest: The authors declare no conflict of interest. The funders had no role in the design of the study; in the collection, analyses, or interpretation of data; in the writing of the manuscript, or in the decision to publish the results.

References

1. Kim, D.; Kim, J.; Park, Y.I.; Lee, N.; Hyeon, T. Recent Development of Inorganic Nanoparticles for Biomedical Imaging. *ACS Cent. Sci.* **2018**, *4*, 324–336. [CrossRef] [PubMed]
2. Parveen, S.; Misra, R.; Sahoo, S.K. Nanoparticles: A boon to drug delivery, therapeutics, diagnostics and imaging. *Nanomedicine* **2012**, *8*, 147–166. [CrossRef] [PubMed]
3. Kim, J.; Chhour, P.; Hsu, J.; Litt, H.I.; Ferrari, V.A.; Popovtzer, R.; Cormode, D.P. Use of Nanoparticle Contrast Agents for Cell Tracking with Computed Tomography. *Bioconjug. Chem.* **2017**, *28*, 1581–1597. [CrossRef] [PubMed]
4. Yoon, T.J.; Kim, J.S.; Kim, B.G.; Yu, K.N.; Cho, M.H.; Lee, J.K. Multifunctional nanoparticles possessing a "magnetic motor effect" for drug or gene delivery. *Angew. Chem. Int. Ed.* **2005**, *44*, 1068–1071. [CrossRef]
5. Ding, Y.F.; Li, S.; Liang, L.; Huang, Q.; Yuwen, L.; Yang, W.; Wang, R.; Wang, L.H. Highly Biocompatible Chlorin e6-Loaded Chitosan Nanoparticles for Improved Photodynamic Cancer Therapy. *ACS Appl. Mater. Interfaces* **2018**, *10*, 9980–9987. [CrossRef]
6. Santoso, M.R.; Yang, P.C. Magnetic Nanoparticles for Targeting and Imaging of Stem Cells in Myocardial Infarction. *Stem Cells Int.* **2016**, *2016*, 4198790. [CrossRef]
7. Guldris, N.; Argibay, B.; Gallo, J.; Iglesias-Rey, R.; Carbo-Argibay, E.; Kolen'ko, Y.V.; Campos, F.; Sobrino, T.; Salonen, L.M.; Banobre-Lopez, M.; et al. Magnetite Nanoparticles for Stem Cell Labeling with High Efficiency and Long-Term in Vivo Tracking. *Bioconjug. Chem.* **2017**, *28*, 362–370. [CrossRef]
8. Lindvall, O.; Kokaia, Z. Stem cells in human neurodegenerative disorders–time for clinical translation? *J. Clin. Investig.* **2010**, *120*, 29–40. [CrossRef]
9. Li, S.; Wang, X.; Li, J.; Zhang, J.; Zhang, F.; Hu, J.; Qi, Y.; Yan, B.; Li, Q. Advances in the Treatment of Ischemic Diseases by Mesenchymal Stem Cells. *Stem Cells Int.* **2016**, *2016*, 5896061. [CrossRef]

10. Rivera, F.J.; de la Fuente, A.G.; Zhao, C.; Silva, M.E.; Gonzalez, G.A.; Wodnar, R.; Feichtner, M.; Lange, S.; Errea, O.; Priglinger, E.; et al. Aging restricts the ability of mesenchymal stem cells to promote the generation of oligodendrocytes during remyelination. *Glia* **2019**, *67*, 1510–1525. [CrossRef]

11. Shim, W.S.; Jiang, S.; Wong, P.; Tan, J.; Chua, Y.L.; Tan, Y.S.; Sin, Y.K.; Lim, C.H.; Chua, T.; Teh, M.; et al. Ex vivo differentiation of human adult bone marrow stem cells into cardiomyocyte-like cells. *Biochem. Biophys. Res. Commun.* **2004**, *324*, 481–488. [CrossRef] [PubMed]

12. Bang, O.Y.; Lee, J.S.; Lee, P.H.; Lee, G. Autologous mesenchymal stem cell transplantation in stroke patients. *Ann. Neurol.* **2005**, *57*, 874–882. [CrossRef] [PubMed]

13. Veronesi, F.; Giavaresi, G.; Tschon, M.; Borsari, V.; Nicoli Aldini, N.; Fini, M. Clinical use of bone marrow, bone marrow concentrate, and expanded bone marrow mesenchymal stem cells in cartilage disease. *Stem Cells Dev.* **2013**, *22*, 181–192. [CrossRef] [PubMed]

14. Zhou, Y.; Tsai, T.L.; Li, W.J. Strategies to retain properties of bone marrow-derived mesenchymal stem cells ex vivo. *Ann. N. Y. Acad. Sci.* **2017**, *1409*, 3–17. [CrossRef]

15. Kuci, S.; Kuci, Z.; Schafer, R.; Spohn, G.; Winter, S.; Schwab, M.; Salzmann-Manrique, E.; Klingebiel, T.; Bader, P. Molecular signature of human bone marrow-derived mesenchymal stromal cell subsets. *Sci. Rep.* **2019**, *9*, 1774. [CrossRef]

16. Lee, P.H.; Kim, J.W.; Bang, O.Y.; Ahn, Y.H.; Joo, I.S.; Huh, K. Autologous Mesenchymal Stem Cell Therapy Delays the Progression of Neurological Deficits in Patients with Multiple System Atrophy. *Clin. Pharmacol. Ther.* **2007**, *83*, 723–730. [CrossRef]

17. Suk, K.T.; Yoon, J.H.; Kim, M.Y.; Kim, C.W.; Kim, J.K.; Park, H.; Hwang, S.G.; Kim, D.J.; Lee, B.S.; Lee, S.H.; et al. Transplantation with Autologous Bone Marrow-Derived Mesenchymal Stem Cells for Alcoholic Cirrhosis: Phase 2 Trial. *Hepatology* **2016**, *64*, 2185–2197. [CrossRef]

18. Tse, H.F.; Kwong, Y.L.; Chan, J.K.; Lo, G.; Ho, C.L.; Lau, C.P. Angiogenesis in ischaemic myocardium by intramyocardial autologous bone marrow mononuclear cell implantation. *Lancet* **2003**, *361*, 47–49. [CrossRef]

19. de Mel, A.; Murad, F.; Seifalian, A.M. Nitric oxide: A guardian for vascular grafts? *Chem. Rev.* **2011**, *111*, 5742–5767. [CrossRef]

20. Lee, S.; Yoon, H.I.; Na, J.H.; Jeon, S.; Lim, S.; Koo, H.; Han, S.S.; Kang, S.W.; Park, S.J.; Moon, S.H.; et al. In vivo stem cell tracking with imageable nanoparticles that bind bioorthogonal chemical receptors on the stem cell surface. *Biomaterials* **2017**, *139*, 12–29. [CrossRef]

21. Pinho, S.; Macedo, M.H.; Rebelo, C.; Sarmento, B.; Ferreira, L. Stem cells as vehicles and targets of nanoparticles. *Drug Discov. Today* **2018**, *23*, 1071–1078. [CrossRef] [PubMed]

22. Fu, P.P.; Xia, Q.; Hwang, H.M.; Ray, P.C.; Yu, H. Mechanisms of nanotoxicity: Generation of reactive oxygen species. *J. Food Drug Anal.* **2014**, *22*, 64–75. [CrossRef] [PubMed]

23. Phukan, G.; Shin, T.H.; Shim, J.S.; Paik, M.J.; Lee, J.K.; Choi, S.; Kim, Y.M.; Kang, S.H.; Kim, H.S.; Kang, Y.; et al. Silica-coated magnetic nanoparticles impair proteasome activity and increase the formation of cytoplasmic inclusion bodies in vitro. *Sci. Rep.* **2016**, *6*, 29095. [CrossRef] [PubMed]

24. Shim, W.; Paik, M.J.; Nguyen, D.T.; Lee, J.K.; Lee, Y.; Kim, J.H.; Shin, E.H.; Kang, J.S.; Jung, H.S.; Choi, S.; et al. Analysis of changes in gene expression and metabolic profiles induced by silica-coated magnetic nanoparticles. *ACS Nano* **2012**, *6*, 7665–7680. [CrossRef] [PubMed]

25. Shin, T.H.; Lee, D.Y.; Lee, H.S.; Park, H.J.; Jin, M.S.; Paik, M.J.; Manavalan, B.; Mo, J.S.; Lee, G. Integration of metabolomics and transcriptomics in nanotoxicity studies. *BMB Rep.* **2018**, *51*, 14–20. [CrossRef]

26. Potter, T.M.; Neun, B.W.; Stern, S.T. Assay to detect lipid peroxidation upon exposure to nanoparticles. *Methods Mol. Biol.* **2011**, *697*, 181–189. [CrossRef]

27. Girotti, A.W. Lipid hydroperoxide generation, turnover, and effector action in biological systems. *J. Lipid Res.* **1998**, *39*, 1529–1542.

28. Lehman, S.E.; Morris, A.S.; Mueller, P.S.; Salem, A.K.; Grassian, V.H.; Larsen, S.C. Silica Nanoparticle-Generated ROS as a Predictor of Cellular Toxicity: Mechanistic Insights and Safety by Design. *Environ. Sci. Nano* **2016**, *3*, 56–66. [CrossRef]

29. Negre-Salvayre, A.; Coatrieux, C.; Ingueneau, C.; Salvayre, R. Advanced lipid peroxidation end products in oxidative damage to proteins. Potential role in diseases and therapeutic prospects for the inhibitors. *Br. J. Pharmacol.* **2008**, *153*, 6–20. [CrossRef]

30. de la Haba, C.; Palacio, J.R.; Martinez, P.; Morros, A. Effect of oxidative stress on plasma membrane fluidity of THP-1 induced macrophages. *Biochimica et Biophysica Acta* **2013**, *1828*, 357–364. [CrossRef]

31. Blokhina, O.; Virolainen, E.; Fagerstedt, K.V. Antioxidants, oxidative damage and oxygen deprivation stress: A review. *Ann. Bot.* **2003**, *91*, 179–194. [CrossRef] [PubMed]

32. Cazzola, R.; Rondanelli, M.; Russo-Volpe, S.; Ferrari, E.; Cestaro, B. Decreased membrane fluidity and altered susceptibility to peroxidation and lipid composition in overweight and obese female erythrocytes. *J. Lipid Res.* **2004**, *45*, 1846–1851. [CrossRef] [PubMed]

33. Saarikangas, J.; Zhao, H.; Lappalainen, P. Regulation of the actin cytoskeleton-plasma membrane interplay by phosphoinositides. *Physiol. Rev.* **2010**, *90*, 259–289. [CrossRef]

34. Geissler, K.J.; Jung, M.J.; Riecken, L.B.; Sperka, T.; Cui, Y.; Schacke, S.; Merkel, U.; Markwart, R.; Rubio, I.; Than, M.E.; et al. Regulation of Son of sevenless by the membrane-actin linker protein ezrin. *Proc. Natl. Acad. Sci. USA* **2013**, *110*, 20587–20592. [CrossRef] [PubMed]

35. Wilson, C.; Gonzalez-Billault, C. Regulation of cytoskeletal dynamics by redox signaling and oxidative stress: Implications for neuronal development and trafficking. *Front. Cell. Neurosci.* **2015**, *9*, 381. [CrossRef]

36. Ibrahim, M.; Schoelermann, J.; Mustafa, K.; Cimpan, M.R. TiO$_2$ nanoparticles disrupt cell adhesion and the architecture of cytoskeletal networks of human osteoblast-like cells in a size dependent manner. *J. Biomed. Mater. Res. A* **2018**, *106*, 2582–2593. [CrossRef]

37. Hall, A. Rho GTPases and the actin cytoskeleton. *Science* **1998**, *279*, 509–514. [CrossRef]

38. Fulda, S.; Gorman, A.M.; Hori, O.; Samali, A. Cellular stress responses: Cell survival and cell death. *Int. J. Cell Biol.* **2010**, *2010*, 214074. [CrossRef]

39. Block, E.R.; Matela, A.R.; SundarRaj, N.; Iszkula, E.R.; Klarlund, J.K. Wounding induces motility in sheets of corneal epithelial cells through loss of spatial constraints: Role of heparin-binding epidermal growth factor-like growth factor signaling. *J. Biol. Chem.* **2004**, *279*, 24307–24312. [CrossRef]

40. Notbohm, J.; Banerjee, S.; Utuje, K.J.C.; Gweon, B.; Jang, H.; Park, Y.; Shin, J.; Butler, J.P.; Fredberg, J.J.; Marchetti, M.C. Cellular Contraction and Polarization Drive Collective Cellular Motion. *Biophys. J.* **2016**, *110*, 2729–2738. [CrossRef]

41. Tan, J.L.; Tien, J.; Pirone, D.M.; Gray, D.S.; Bhadriraju, K.; Chen, C.S. Cells lying on a bed of microneedles: An approach to isolate mechanical force. *Proc. Natl. Acad. Sci. USA* **2003**, *100*, 1484–1489. [CrossRef] [PubMed]

42. du Roure, O.; Saez, A.; Buguin, A.; Austin, R.H.; Chavrier, P.; Silberzan, P.; Ladoux, B. Force mapping in epithelial cell migration. *Proc. Natl. Acad. Sci. USA* **2005**, *102*, 2390–2395. [CrossRef] [PubMed]

43. Ghassemi, S.; Meacci, G.; Liu, S.; Gondarenko, A.A.; Mathur, A.; Roca-Cusachs, P.; Sheetz, M.P.; Hone, J. Cells test substrate rigidity by local contractions on submicrometer pillars. *Proc. Natl. Acad. Sci. USA* **2012**, *109*, 5328–5333. [CrossRef] [PubMed]

44. Mourdikoudis, S.; Pallares, R.M.; Thanh, N.T.K. Characterization techniques for nanoparticles: Comparison and complementarity upon studying nanoparticle properties. *Nanoscale* **2018**, *10*, 12871–12934. [CrossRef] [PubMed]

45. Fornaguera, C.; Lazaro, M.A.; Brugada-Vila, P.; Porcar, I.; Morera, I.; Guerra-Rebollo, M.; Garrido, C.; Rubio, N.; Blanco, J.; Cascante, A.; et al. Application of an assay Cascade methodology for a deep preclinical characterization of polymeric nanoparticles as a treatment for gliomas. *Drug Deliv.* **2018**, *25*, 472–483. [CrossRef]

46. Beck, G.R., Jr.; Ha, S.W.; Camalier, C.E.; Yamaguchi, M.; Li, Y.; Lee, J.K.; Weitzmann, M.N. Bioactive silica-based nanoparticles stimulate bone-forming osteoblasts, suppress bone-resorbing osteoclasts, and enhance bone mineral density in vivo. *Nanomedicine* **2012**, *8*, 793–803. [CrossRef]

47. Shin, T.H.; Seo, C.; Lee, D.Y.; Ji, M.; Manavalan, B.; Basith, S.; Chakkarapani, S.K.; Kang, S.H.; Lee, G.; Paik, M.J.; et al. Silica-coated magnetic nanoparticles induce glucose metabolic dysfunction in vitro via the generation of reactive oxygen species. *Arch. Toxicol.* **2019**, *93*, 1201–1212. [CrossRef]

48. Park, K.S.; Tae, J.; Choi, B.; Kim, Y.S.; Moon, C.; Kim, S.H.; Lee, H.S.; Kim, J.; Kim, J.; Park, J.; et al. Characterization, in vitro cytotoxicity assessment, and in vivo visualization of multimodal, RITC-labeled, silica-coated magnetic nanoparticles for labeling human cord blood-derived mesenchymal stem cells. *Nanomedicine* **2010**, *6*, 263–276. [CrossRef]

49. Shin, T.H.; Lee, S.; Choi, K.R.; Lee, D.Y.; Kim, Y.; Paik, M.J.; Seo, C.; Kang, S.; Jin, M.S.; Yoo, T.H.; et al. Quality and freshness of human bone marrow-derived mesenchymal stem cells decrease over time after trypsinization and storage in phosphate-buffered saline. *Sci. Rep.* **2017**, *7*, 1106. [CrossRef]

50. Shin, T.H.; Phukan, G.; Shim, J.S.; Nguyen, D.T.; Kim, Y.; Oh-Lee, J.D.; Lee, H.S.; Paik, M.J.; Lee, G. Restoration of Polyamine Metabolic Patterns in In Vivo and In Vitro Model of Ischemic Stroke following Human Mesenchymal Stem Cell Treatment. *Stem Cells Int.* **2016**, *2016*, 4612531. [CrossRef]

51. Lee, S.; Cho, N.P.; Kim, J.D.; Jung, H.; Kang, S.H. An ultra-sensitive nanoarray chip based on single-molecule sandwich immunoassay and TIRFM for protein detection in biologic fluids. *Analyst* **2009**, *134*, 933–938. [CrossRef] [PubMed]

52. Owen, D.M.; Rentero, C.; Magenau, A.; Abu-Siniyeh, A.; Gaus, K. Quantitative imaging of membrane lipid order in cells and organisms. *Nat. Protoc.* **2012**, *7*, 24–35. [CrossRef] [PubMed]

53. Schneckenburger, H.; Stock, K.; Strauss, W.S.; Eickholz, J.; Sailer, R. Time-gated total internal reflection fluorescence spectroscopy (TG-TIRFS): Application to the membrane marker laurdan. *J. Microsc.* **2003**, *211*, 30–36. [CrossRef] [PubMed]

54. Hansen, J.S.; Helix-Nielsen, C. An epifluorescence microscopy method for generalized polarization imaging. *Biochem. Biophys. Res. Commun.* **2011**, *415*, 686–690. [CrossRef] [PubMed]

55. Tan, J.L.; Liu, W.; Nelson, C.M.; Raghavan, S.; Chen, C.S. Simple approach to micropattern cells on common culture substrates by tuning substrate wettability. *Tissue Eng.* **2004**, *10*, 865–872. [CrossRef] [PubMed]

56. Cui, Y.; Hameed, F.M.; Yang, B.; Lee, K.; Pan, C.Q.; Park, S.; Sheetz, M. Cyclic stretching of soft substrates induces spreading and growth. *Nat. Commun.* **2015**, *6*, 6333. [CrossRef] [PubMed]

57. Landau, L.D.; Lifshitz, E.M.; Kosevich, A.M.; Pitaevskii, L.P. *Theory of Elasticity*, 3rd English ed.; Butterworth-Heinemann: Oxford, UK, 1986. [CrossRef]

58. Rajanahalli, P.; Stucke, C.J.; Hong, Y. The effects of silver nanoparticles on mouse embryonic stem cell self-renewal and proliferation. *Toxicol. Rep.* **2015**, *2*, 758–764. [CrossRef]

59. Ickrath, P.; Wagner, M.; Scherzad, A.; Gehrke, T.; Burghartz, M.; Hagen, R.; Radeloff, K.; Kleinsasser, N.; Hackenberg, S. Time-Dependent Toxic and Genotoxic Effects of Zinc Oxide Nanoparticles after Long-Term and Repetitive Exposure to Human Mesenchymal Stem Cells. *Int. J. Environ. Res. Public Health* **2017**, *14*, 1590. [CrossRef]

60. Novotna, B.; Turnovcova, K.; Veverka, P.; Rossner, P., Jr.; Bagryantseva, Y.; Herynek, V.; Zvatora, P.; Vosmanska, M.; Klementova, M.; Sykova, E.; et al. The impact of silica encapsulated cobalt zinc ferrite nanoparticles on DNA, lipids and proteins of rat bone marrow mesenchymal stem cells. *Nanotoxicology* **2016**, *10*, 662–670. [CrossRef]

61. Zhou, D.; Shao, L.; Spitz, D.R. Reactive oxygen species in normal and tumor stem cells. *Adv. Cancer Res.* **2014**, *122*, 1–67. [CrossRef]

62. Van Lehn, R.C.; Ricci, M.; Silva, P.H.; Andreozzi, P.; Reguera, J.; Voitchovsky, K.; Stellacci, F.; Alexander-Katz, A. Lipid tail protrusions mediate the insertion of nanoparticles into model cell membranes. *Nat. Commun.* **2014**, *5*, 4482. [CrossRef] [PubMed]

63. Chen, F.; Liu, Y.; Wong, N.K.; Xiao, J.; So, K.F. Oxidative Stress in Stem Cell Aging. *Cell Transplant.* **2017**, *26*, 1483–1495. [CrossRef] [PubMed]

64. Shaban, S.; El-Husseny, M.W.A.; Abushouk, A.I.; Salem, A.M.A.; Mamdouh, M.; Abdel-Daim, M.M. Effects of Antioxidant Supplements on the Survival and Differentiation of Stem Cells. *Oxid. Med. Cell. Longev.* **2017**, *2017*, 5032102. [CrossRef] [PubMed]

65. Titushkin, I.; Cho, M. Regulation of cell cytoskeleton and membrane mechanics by electric field: Role of linker proteins. *Biophys. J.* **2009**, *96*, 717–728. [CrossRef] [PubMed]

66. Accomasso, L.; Gallina, C.; Turinetto, V.; Giachino, C. Stem Cell Tracking with Nanoparticles for Regenerative Medicine Purposes: An Overview. *Stem Cells Int.* **2016**, *2016*, 7920358. [CrossRef] [PubMed]

67. Bhattacharya, S.; Ahir, M.; Patra, P.; Mukherjee, S.; Ghosh, S.; Mazumdar, M.; Chattopadhyay, S.; Das, T.; Chattopadhyay, D.; Adhikary, A. PEGylated-thymoquinone-nanoparticle mediated retardation of breast cancer cell migration by deregulation of cytoskeletal actin polymerization through miR-34a. *Biomaterials* **2015**, *51*, 91–107. [CrossRef]

68. Rape, A.; Guo, W.H.; Wang, Y.L. Microtubule depolymerization induces traction force increase through two distinct pathways. *J. Cell Sci.* **2011**, *124*, 4233–4240. [CrossRef]

69. Tay, C.Y.; Cai, P.; Setyawati, M.I.; Fang, W.; Tan, L.P.; Hong, C.H.; Chen, X.; Leong, D.T. Nanoparticles strengthen intracellular tension and retard cellular migration. *Nano Lett.* **2014**, *14*, 83–88. [CrossRef]

70. Ridley, A.J.; Schwartz, M.A.; Burridge, K.; Firtel, R.A.; Ginsberg, M.H.; Borisy, G.; Parsons, J.T.; Horwitz, A.R. Cell migration: Integrating signals from front to back. *Science* **2003**, *302*, 1704–1709. [CrossRef]

71. Lange, J.R.; Fabry, B. Cell and tissue mechanics in cell migration. *Exp. Cell Res.* **2013**, *319*, 2418–2423. [CrossRef]

72. Matsuzaki, T.; Matsumoto, S.; Kasai, T.; Yoshizawa, E.; Okamoto, S.; Yoshikawa, H.Y.; Taniguchi, H.; Takebe, T. Defining Lineage-Specific Membrane Fluidity Signatures that Regulate Adhesion Kinetics. *Stem Cell Rep.* **2018**, *11*, 852–860. [CrossRef] [PubMed]

73. de Lucas, B.; Perez, L.M.; Galvez, B.G. Importance and regulation of adult stem cell migration. *J. Cell. Mol. Med.* **2018**, *22*, 746–754. [CrossRef] [PubMed]

 *nanomaterials*

Article

Combination of Roll Grinding and High-Pressure Homogenization Can Prepare Stable Bicelles for Drug Delivery

Seira Matsuo [1], Kenjirou Higashi [2], Kunikazu Moribe [2], Shin-ichiro Kimura [1], Shigeru Itai [1], Hiromu Kondo [1] and Yasunori Iwao [1,*]

[1] Department of Pharmaceutical Engineering, School of Pharmaceutical Sciences, University of Shizuoka, 52-1 Yada, Suruga-ku, Shizuoka 422-8526, Japan; s17119@u-shizuoka-ken.ac.jp (S.M.); s-kimura@u-shizuoka-ken.ac.jp (S.-i.K.); s-itai@u-shizuoka-ken.ac.jp (S.I.); hkondo@u-shizuoka-ken.ac.jp (H.K.)

[2] Graduate School of Pharmaceutical Sciences, Chiba University, 1-8-1, Inohana, Chuo-ku, Chiba 260-8675, Japan; ken-h@faculty.chiba-u.jp (K.H.); moribe@faculty.chiba-u.jp (K.M.)

* Correspondence: yasuiwao@u-shizuoka-ken.ac.jp; Tel.: +81-54-264-5612; Fax: +81-54-264-5615

Received: 26 October 2018; Accepted: 30 November 2018; Published: 3 December 2018

Abstract: To improve the solubility of the drug nifedipine (NI), NI-encapsulated lipid-based nanoparticles (NI-LNs) have been prepared from neutral hydrogenated soybean phosphatidylcholine and negatively charged dipalmitoylphosphatidylglycerol at a molar ratio of 5/1 using by roll grinding and high-pressure homogenization. The NI-LNs exhibited high entrapment efficiency, long-term stability, and enhanced NI bioavailability. To better understand their structures, cryo transmission electron microscopy and atomic force microscopy were performed in the present study. Imaging from both instruments revealed that the NI-LNs were bicelles. Structures prepared with a different drug (phenytoin) or with phospholipids (dimyristoylphosphatidylcholine, dipalmitoylphosphatidylcholine, and distearoylphosphatidylcholine) were also bicelles. Long-term storage, freeze-drying, and high-pressure homogenization did not affect the structures; however, different lipid ratios, or the presence of cholesterol, did result in liposomes (5/0) or micelles (0/5) with different physicochemical properties and stabilities. Considering the result of long-term stability, standard NI-LN bicelles (5/1) showed the most long-term stabilities, providing a useful preparation method for stable bicelles for drug delivery.

Keywords: lipid nanoparticles; cryo transmission electron microscopy; atomic force microscopy; bicelle; micelle; liposome

---

1. Introduction

About 40% of marketed drugs and 70% or more of drug candidates have exhibited poor water solubility [1]. These drugs also have problems such as non-constant absorption in vivo and low bioavailability (BA) [2]. Approaches to improve solubility and absorption include drug miniaturization [3–5], amorphization [6–8], and solid dispersion [9–11]. In particular, nanoparticle formulations have been useful in drug delivery [12]. As the particle size decreases, the dissolution rate increases because of the increased surface area, according to the Noyes–Whitney equation [13,14]; and solubility increases according to the Ostwald–Freundlich equation [15]. In addition, nanoparticle formulations can pass through mucosal layers that can be a barrier to drug absorption, resulting in enhanced oral BA [16,17]. Ideal drug carriers are composed of biocompatible and biodegradable materials with a small size, a high drug-loading capacity, high stability, and easily modified surfaces. They include emulsions [18], liposomes [19], micelles [20], lipid nanoparticles [21], polymer nanoparticles [22], hydrogels [23], and carbon nanotubes [24]. Lipid nanoparticles for drug delivery systems (DDS) have been used for various dosage preparations such as oral, parenteral, dermal, pulmonary, rectal, and ocular [25].

Previously, we reported that the water solubility of nifedipine (NI, class II in the biopharmaceutics classification system [26]) was increased by encapsulation in lipid-based nanoparticle (NI-LN) suspensions. These were fabricated with a neutral phospholipid (hydrogenated soybean phosphatidylcholine, HSPC), and a negatively charged phospholipid (dipalmitoylphosphatidylglycerol, DPPG), using co-grinding by roll milling and subsequent high-pressure homogenization [27]. The mean particle size was 50 nm, with a narrow particle distribution (polydispersity index: PDI) of less than 0.3. In addition, they were stable for four months in cool, dark storage [28]. To test long-term stability, freeze-drying with sugar was performed; the mean particle size was maintained after reconstitution [28]. Furthermore, when NI-LN suspensions were administrated orally to rats, the area under the curve increased fourfold relative to that for NI suspensions, and the oral BA of NI (59%) was higher than that of NI bulk (50%) [29]. This LN preparation could increase the solubility of other water-insoluble drugs as well, regardless of the drug structure. Compared with other reported lipid nanoparticles, they had superior long-term stability without water-dispersible stabilizers, and high entrapment efficiency (EE) for many drugs [30–34].

Here, cryo transmission electron microscopy (cryo-TEM) and atomic force microscopy (AFM) imaging revealed that the NI-LNs were found to be bicelles, which are flat, disk-like lipid bilayers. Generally, bicelles are usually prepared from two phospholipids having different alkyl chain lengths (bilayers from long-chain lipids and edges from short-chain lipids) that are dissolved in organic solvent, evaporated, and hydrated [35,36]. Here, the NI-LNs used HSPC and DPPG that have similar acyl chain lengths, and were prepared without organic solvents via a combination of co-grinding with a roll mill and high-pressure homogenization. According to the structural analysis of lipid nanoparticles prepared under various conditions, the lipid ratios strongly affected the structures. The simple preparation for stable bicelles will be potentially useful for future DDS.

2. Materials and Methods

2.1. Materials

HSPC (COATSOME® NC-21), and 1,2-distearoyl-sn-glycero-3-phosphocholine (DSPC, COATSOME® MC-8080) were provided by Nippon Fine Chemical Co., Ltd. (Osaka, Japan). DPPG, sodium salt (COATSOME® MG-6060LS), 1,2-dimyristoyl-sn-glycero-3-phosphocholine (DMPC, COATSOME® MC-4040), and 1,2-dipalmitoyl-sn-glycero-3-phosphocholine (DPPC, COATSOME® MC-6060) were purchased from Nippon Oil and Fats Co., Ltd. (Tokyo, Japan). NI (Japanese Pharmacopeia XVII) and cholesterol (Chol) were purchased by Sigma-Aldrich, Co. (St. Louis, MO, USA). Methanol (HPLC grade), formic acid (HPLC grade), and sucrose were purchased from Wako Pure Chemical Industries, Ltd (Osaka, Japan). The membrane filters (pore size: 0.20-μm) were purchased from Toyo Roshi Kaisha Ltd. (Tokyo, Japan). Phenytoin (PHT, Japanese Pharmacopeia XVII) was purchased from Sumitomo Dainippon Pharma Co., Ltd. (Osaka, Japan). All of the reagents were of the highest grade commercially available, and all of the solutions were prepared using deionized distilled water.

2.2. Preparation of Drug–Lipid Nanoparticle Suspensions

The drug (NI or PHT)-lipid nanoparticle suspensions were prepared as described previously with a slight modification [27]. Briefly, 40 mg of NI or PHT and 1000 mg of lipid (HSPC/DPPG (5/1) for standard NI-LNs, or (5/0), (5/0.5), and (0/5) molar ratios), were added to a mortar and physically mixed with a pestle. The mixture was then co-ground for five minutes with a roll mill (Model: R3-1R, Kodaira Seisakusho Co., Ltd., Tokyo, Japan) having three grinding rollers rotating at velocity ratios of 1/2.5/5.8. The sample mostly adhered to the rollers, but the mill was stopped every

30 seconds to collect fallen samples. The co-grinding cycle was repeated 10 times, and the mixture was then dispersed in 200 mL of distilled water and premixed with a Speed Stabilizer® (Kinematica Co., Luzern, Switzerland) at 9000 rpm for 10 min. The suspension was then subjected to high-pressure homogenization (Microfluidizer®, M110-E/H; Microfluidics, Co., Newton, MA, USA) with a pass cycle of 100 at 175 MPa. The NI-LN suspension was filtered by suction aspirator with a 0.20-µm polytetrafluoroethylene or cellulose acetate membrane filter (Toyo Roshi Kaisha Ltd., Tokyo, Japan) to remove the particles whose size was over 200 nm. LNs with no drug loading were prepared in the same way.

2.3. Freeze-Drying and Reconstitution of NI–Lipid Nanoparticles

Freeze-drying was performed with two mL of the NI-LN suspensions placed in a vial containing 100 mg of sucrose (5%, *w/v*) as a cryoprotectant. Sucrose was selected based on previous reports [28,37]. The vial was frozen at −70 °C overnight and then freeze-dried in a glass chamber for 48 h with a vacuum pump equipped with a vapor condenser (−20 °C, 0.0225 Torr). Reconstitution was performed immediately after the freeze-drying; two mL of deionized distilled water was added to the vial, and the sample was rehydrated via vortex agitation.

2.4. Measurement of Mean Particle Size and Zeta Potential

The mean particle size and the zeta potential of LN suspensions were measured with dynamic light scattering (DLS) (Zetasizer nano ZS, Malvern Instruments Ltd., Worcestershire, UK) at a scattering angle of 90° at room temperature (25 °C). Mean particle sizes were based on the scattering intensity, while the zeta potential was based on electrophoretic mobility. Since the values of mean count rate of the samples were 200–500 kcps, which are suitable for this measurement, intact samples without dilution were used in this measurement. We repeated measurement three times per sample.

2.5. Measurement of the Concentration and the Entrapment Efficiency of Drugs in Drug–Lipid Nanoparticle Suspensions

The filtered lipid nanoparticles were defined as the soluble state, and the NI concentration was determined as follows. Aliquots of 200-µL NI-LN suspensions were dissolved and diluted with methanol and then analyzed with high-pressure liquid chromatography (HPLC, LC-20AT; Shimadzu, Kyoto, Japan). The analytical column was a three-µm Cadenza CD C18 (1.6 mm × 150 mm, Imtakt Corp., Kyoto, Japan). The detector used a 236-nm ultraviolet (UV) source, the column temperature was 40 °C, the mobile phase was methanol/water, 0.2% formic acid (60/40, *v/v*), 60–90% methanol/10 min, the flow rate was 0.45 mL/min, and the injection volume was two µL. For PHT concentrations, the UV source was 213 nm, the column temperature was 40 °C, the mobile phase was methanol/water, 0.2% formic acid (70/30, *v/v*), 70–95% methanol/10 min, the flow rate was 0.40 mL/min, and the injection volume was five µL.

The drug entrapment efficiency (EE) of the LNs was determined by the amount of free drug after ultrafiltration. Drug–LN suspensions (500-µL) were placed on an ultrafilter in a centrifuge tube (Amicon® Ultra-0.5 Centrifugal Filter Devices, 10 K device 10,000 MNWL; Merck Millipore Ltd., Billerica, MA, USA) and centrifuged at 10,000 rpm at 4 °C for 10 min. This was repeated after 500 µL of water was added. The ultrafiltrate containing the free drug was analyzed by HPLC. The entrapment efficiency was thus:

$$\text{Entrapment efficiency (\%)} = (\text{total drug content} - \text{free drug content})/\text{total drug content} \times 100$$

2.6. Structural Analysis of Lipid Nanoparticles

Cryo-TEM images were acquired on a JEM-2100F microscope (JEOL Co., Ltd., Tokyo, Japan). Hydrophilic treatment of a 200-mesh copper grid covered with a perforated polymer film (Nisshin EM Co. Ltd., Tokyo, Japan) was performed for 60 s using an HDT-400 device (JEOL Co., Ltd. Tokyo, Japan).

A two-µL aliquot of each lipid nanoparticle suspension was then applied to the hydrophilic grid. The grid was then blotted with filter paper for three seconds and immediately vitrified in liquid ethane cooled with liquid nitrogen, using a Leica EM CPC cryofixation system (Leica Microsystems GmbH, Wetzlar, Germany). Frozen samples were maintained at $-170\,^{\circ}$C, using a Gatan 626 cryo-holder (Gatan, Inc., Pleasanton, CA, USA). The cryo-TEM was operated at 120 kV and a provided a magnification of $\times 50{,}000$.

AFM imaging in alternating tapping mode was outperformed with an MFP-3D microscope (Asylum Research, Santa Barbara, CA, USA), equipped with a silicon nitride cantilever OMCL-TR400PSA (Olympus Co., Ltd., Tokyo, Japan). Mica modified with 3-(aminopropyl)triethoxysilane (APTES) was prepared by exposing freshly cleaved mica to an APTES atmosphere for one hour at room temperature. This strengthened the weak binding of the lipid nanoparticles to the surface in an aqueous environment. The lipid nanoparticle suspensions were diluted 300 times with deionized distilled water, and then deposited onto the APTES-modified mica surfaces. After incubation for one minute at room temperature, excess nanoparticles were removed by flushing with deionized distilled water. All of the 1024 × 128-pixel AFM images with areas of $2 \times 2\ \mu m^2$ were recorded at a scan speed of 0.50 Hz at $20 \pm 2\,^{\circ}$C. The height and diameter of each lipid nanoparticle was analyzed by the MFP-3D program, written in IGOR-PRO software (Wavemetrics, Portland, OR, USA).

3. Results

3.1. Stractural Feature of Standard NI-LNs

3.1.1. Cryo-TEM Images of Standard NI-LNs

Cryo-TEM provided images of the suspensions directly without negative staining or drying, and the sample stage could be tilted for imaging at various angles [38]. Figure 1 shows cryo-TEM images of NI-LNs prepared with HSPC/DPPG (5/1) before and after tilting the sample stage by 20 degrees. The shape of most particles changed after the tilting. For example, the rod-like particle before tilting (circled in Figure 1A) became circular (Figure 1B); whereas, the opposite occurred for the particles indicated by arrows. These changes indicated flat particles, as previously reported [39]. There were also small micelles about 10 nm in diameter, as indicated by arrowheads. Thus, flat particles and micelles coexisted in standard NI-LN suspensions.

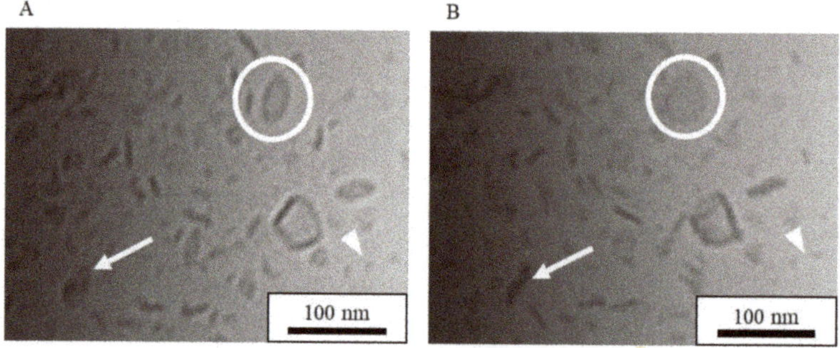

Figure 1. Cryo transmission electron microscopy (cryo-TEM) images of standard nifedipine-encapsulated lipid-based nanoparticles (NI-LNs) prepared with hydrogenated soybean phosphatidylcholine (HSPC) and dipalmitoylphosphatidylglycerol (DPPG) (5/1). (**A**) Before sample stage tilting, and (**B**) after tilting 20 degrees.

3.1.2. AFM Images of Standard NI-LNs

AFM provides three-dimensional images. Figure 2A shows AFM images of standard NI-LNs, and Figure 2B exhibits the cross-sectional profile taken along the red line in Figure 2A, where the blue point corresponds to zero in the profile. The NI-LNs were five-nm high and about 50 nm in diameter (Figure 2B), which was consistent with the cryo-TEM images. The thickness of a lipid bilayer is four to six nm [40,41]; thus, the NI-LNs were a single lipid bilayer. Spherical vesicles such as liposomes sometimes strongly adsorbed onto the substrate, resulting in heights over 10 nm that indicated double lipid bilayers [42]. The phospholipids used here were saturated and relatively rigid, and would be difficult to deform during adsorption. Therefore, the images suggested that the standard NI-LNs seem to be disk-like bicelles (Figure 3).

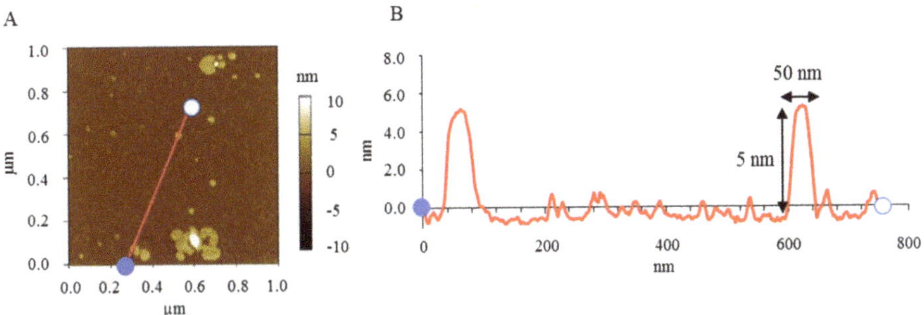

Figure 2. Atomic force microscopy (AFM) imaging of standard NI-LNs. (**A**) AFM image and (**B**) cross-sectional profile.

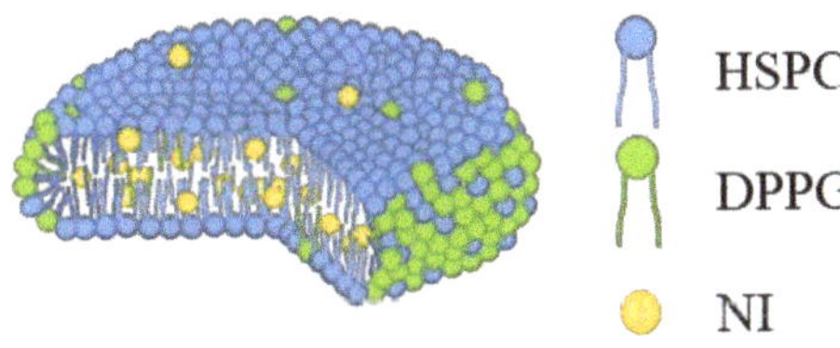

Figure 3. Schematic of an NI-LN bicelle.

3.2. Effect of Long-Term Storage and Stabilization on LN Structure

The mean particle size and drug concentration without stabilizers did not changed for four months when NI-LNs were stored in a cool, dark place [28]. There was no aggregation of particles or drug leakage, indicating that the NI-LNs had long-term stability. However, it was unclear whether the bicelle structure was maintained, considering that the lipids in the plane and in the edges area were miscible and could coalescence into unilamellar vesicles [43]. However, NI-LNs stored for four months maintained their mean particle size and structure (Figure 4A), and remained dispersed because of the electrostatic repulsion of the negatively-charged DPPG. Hence, the bicelle structure may have increased the LN stability. In addition, cryo-TEM images of NI-LNs that were freeze-dried with sucrose (5 *w/v*%) for 48 h and then re-hydrated in distilled water are shown in Figure 4B. The particle size looked slightly bigger after freeze-drying, but the bicelle structure was preserved. Thus, freeze-drying and the addition of a lyophilization stabilizer did not affect the NI-LN structure.

As for determination of the mean particle size, the following two ways were used; one is by cryo-TEM images (Table 1), and the other is by zetasizer using DLS (Table 2). Especially, because bicelles and micelles coexisted in NI-LNs, analysis of the mean particle size for each particle would be useful by cryo-TEM images, and 100 particles in cryo-TEM images printed on paper were measured with a ruler (Table 1). The ratio of bicelle/micelle in NI-LNs just after preparation was about 8/2, and the mean particle size of bicelles was 26.8 nm. On the other hand, the data by DLS showed the value of total particles, including bicelles and micelles, as 47.5 nm. These results suggested that the mean particle size obtained from the cryo-TEM images (Table 1) was smaller than that obtained from DLS (Table 2). Cryo-TEM and DLS data were based on the number reference and on scattering intensity from Brownian motion, respectively. Previously, it was reported that DLS tends to overestimate the hydrodynamic diameter because large particles strongly influences scattering intensity [44]. In addition, DLS furnished the hydrodynamic diameter depending on the assumption that the particles are spherical [45], but the bicelles in this study were not spherical. Therefore, the difference of measurement principle would be involved in the differences of particle size.

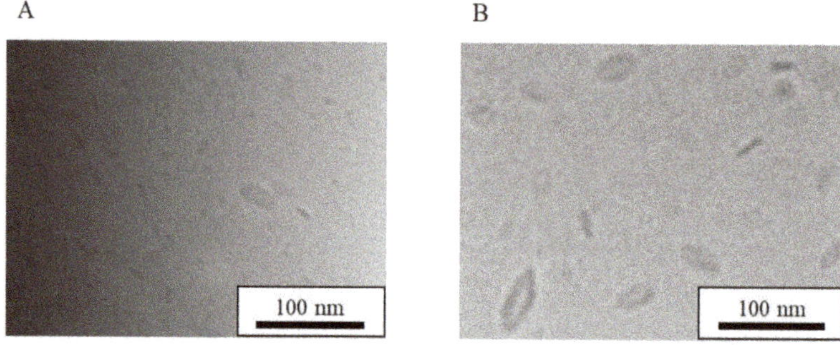

Figure 4. Cryo-TEM images of NI-LNs (**A**) stored for four months in a cold, dark place, and (**B**) re-hydrated after freeze-drying with sucrose.

Table 1. Effect of long-term storage and freeze-drying on mean particle size measured by cryo-TEM images.

Formulation	Bicelle (nm)	Micelle (nm)	Mean Particle Size (nm)
Just after preparation	26.8 ± 8.4 ($n = 81$)	8.0 ± 1.7 ($n = 19$)	23.2 ± 10.6 ($n = 100$)
Stored for four months at a cold dark place	23.4 ± 10.9 ($n = 59$)	7.6 ± 1.8 ($n = 41$)	17.0 ± 11.5 ($n = 100$)
Re-hydrated after freeze-drying with sucrose	39.7 ± 12.1 ($n = 91$)	11.1 ± 1.0 ($n = 9$)	37.1 ± 14.2 ($n = 100$)

Data obtained by cryo-TEM images are average values of particle length (mean $\pm$ S.D.).

Table 2. Effect of long-term storage and freeze-drying on physiochemical properties of NI-LNs by dynamic light scattering (DLS) and high-pressure liquid chromatography (HPLC) measurements. EE: entrapment efficiency, PDI: polydispersity index.

Formulation	Mean Particle Size (nm)	PDI	Zeta Potential (mV)	Drug Concentration (µg/mL)	EE (%)
Just after preparation	47.5 ± 2.9	0.307 ± 0.027	-46.8 ± 9.6	67.3 ± 11.3	97.1 ± 1.6
Stored for four months at a cold dark place	46.3 ± 1.0	0.310 ± 0.033	-52.1 ± 2.8	67.4 ± 6.2	98.1 ± 0.3
Re-hydrated after freeze-drying with sucrose	63.7 ± 1.2	0.351 ± 0.007	-47.0 ± 4.1	30.7 ± 7.6	97.8 ± 0.5

After re-hydration, PDI increased as well as mean particle size (Table 2), so particle size distribution was broader to the bigger value. In addition, drug concentration was decreased to half (30.7 µg/mL) after re-hydration (Table 2). Freeze-drying with cryoprotectant has been reported to cause leakage of the encapsulated drug [46], suggesting that the insertion of sucrose molecules in the phospholipid molecules of the lipid bilayer might be involved in the drug leakage without any significant structural change. Furthermore, the mean particle size was 70.1 nm one month after re-hydration (data not shown), and it exhibited good redispersibility. Therefore, freeze-drying was useful to give long-term stability to LNs.

3.3. Effect of Encapsulated Drugs on LN Structure

Next, to determine the effect of encapsulated drugs on the LN structure, LNs without any drugs or with other water-insoluble drugs was prepared. A preliminary study demonstrated that eight water-insoluble drugs encapsulated in LNs (e.g., ibuprofen and indomethacin) improved their solubility (100 µg/mL drugs), while PHT-encapsulated LNs (PHT-LNs) had only a 35 µg/mL drug concentration, suggesting that different structures might be observed, and PHT was chosen as the other drug.

Figure 5 shows cryo-TEM images of LNs (5A) and PHT-LNs (5B) where bicelles and micelles coexisted in both suspensions, with no difference in each structure or mean particle size compared with that of standard NI-LNs (Table 3). Furthermore, AFM images of LNs revealed low heights with respect their size (Figure 5C,D), indicating bicelle structures. The increase in PDI of LNs might result from the increase in the ratio of micelle (Tables 3 and 4). The PHT-LNs had physicochemical properties that differed most from those of the standard NI-LNs, but they still had bicelle structures. Therefore, the other drug-encapsulated LNs were probably bicelles as well. The PHT-LN preparation replicated the physicochemical properties, indicating lower drug concentrations relative to NI. A lower *EE* meant that the PHT was difficult to encapsulate, and there was much free PHT (Table 4); however, the results here suggested that the structure of the particles was unchanged with or without encapsulated drugs, irrespective of the drugs.

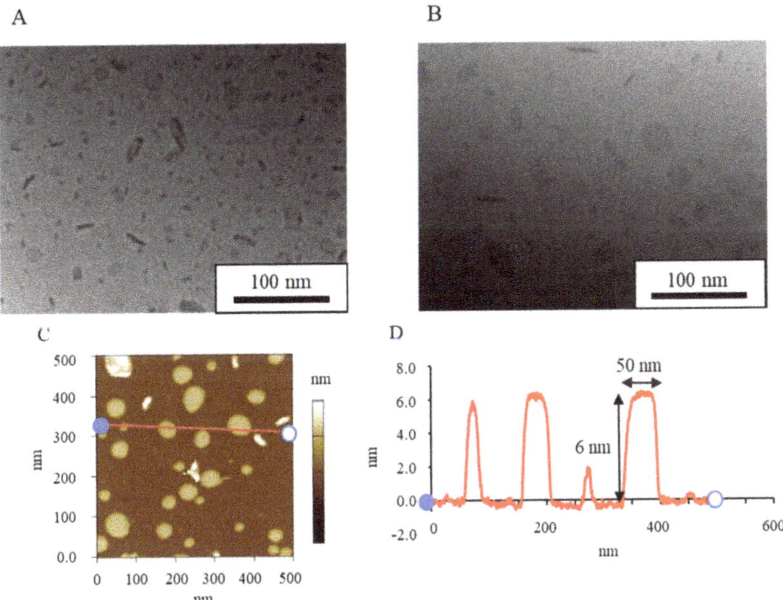

Figure 5. Structural analysis of LNs and phenytoin (PHT)-LNs. Cryo-TEM image of (**A**) LNs and (**B**) PHT-LNs. (**C**) AFM image of LNs and (**D**) the cross-sectional profile.

Table 3. Effect of encapsulated drug on mean particle size measured by cryo-TEM images.

Model Drug	Bicelle (nm)	Micelle (nm)	Mean Particle Size (nm)
NI	26.8 ± 8.4 ($n = 81$)	8.0 ± 1.7 ($n = 19$)	23.2 ± 10.6 ($n = 100$)
None	21.1 ± 5.6 ($n = 68$)	6.1 ± 2.1 ($n = 32$)	16.3 ± 8.7 ($n = 100$)
PHT	26.5 ± 9.3 ($n = 81$)	8.1 ± 1.6 ($n = 19$)	23.0 ± 11.1 ($n = 100$)

Data obtained by cryo-TEM images are average values of particle length (mean ± S.D.).

Table 4. Effect of encapsulated drug on physiochemical properties of LNs by DLS and HPLC measurements.

Model Drug	Mean Particle Size (nm)	PDI	Zeta Potential (mV)	Drug Concentration (µg/mL)	EE(%)
NI	47.5 ± 2.9	0.307 ± 0.027	−46.8 ± 9.6	67.3 ± 11.3	97.1 ± 1.6
None	43.3 ± 0.7	0.357 ± 0.086	−55.6 ± 4.2	-	-
PHT	46.0 ± 4.4	0.267 ± 0.003	−30.7 ± 6.4	30.5 ± 2.8	59.4 ± 7.6

3.4. Effect of Lipid Composition Ratio on LN Structure

NI-LNs were prepared with HSPC/DPPG ratios of (5/0), (5/0.5), (5/1), and (0/5), and LNs (no drug loading) was with HSPC/DPPG (5/0), (5/1), and (0/5). Figure 6 shows cryo-TEM images, and Table 5 lists the physicochemical properties of each lipid nanoparticle. The shapes of NI-LNs prepared with HSPC/DPPG (5/0) did not change markedly when the sample stage was rotated by 20 degrees (Figure 6A,B). Thus, the particles were not flat, and could have been polygonal liposomes [47]. Additionally, a week after preparation, slight aggregation occurred because there was little electrostatic repulsion for lack of DPPG. The DLS mean particle size was about 350 nm two weeks after preparation (data not shown), which was more than three times that just after preparation. The NI concentration was low (2.9 µg/mL) (Table 6), because NI might exist in only the lipid bilayer area of liposomes. Furthermore, as for HSPC/DPPG (5/0), the appearance of LNs suspensions just after 100-pass high-pressure homogenization was still cloudy, indicating that it was difficult to make a particle made from only HSPC smaller, so big particles were trapped by a 200-nm filter membrane. In fact, lipid concentration pass through filter membrane was low (data not shown). So, NI concentration would be low. There was no difference in structure for the no-drug LNs prepared with HSPC/DPPG (5/0) (Figure 6C), relative to that for the NI-LNs. They could not be prepared reproducibly, which indicated large S.D. values of their mean particle size, and their zeta potentials were almost neutral because of lacking DPPG (Table 6). NI-LNs prepared with (5/0.5) (Figure 6D) exhibited morphologies that were between liposomes (5/0) and bicelles (5/1) that corresponded to the DPPG ratio, and the ratio of bicelle/micelle/liposome was about 7/2/1 (Table 5), suggesting that because the amount of DPPG was not enough to form bicelles, the liposomes derived from HSPC still existed. NI-LNs prepared with (0/5) (Figure 6E) were only micelles with a few nanometers in size; there were no bicelles or liposomes, indicating that there were no micelles in the lipid nanoparticle suspensions without DPPG. For HSPC/DPPG (0/5) (Figure 6F), there was no difference in the micellar structure between NI-LNs and LNs. Therefore, the encapsulated drugs did not affect the particle structures. Although big differences in the mean particle size of HSPC/DPPG (0/5) between Tables 5 and 6 were seen, large particles were not observed in cryo-TEM images, and the value in Table 5 might be more accurate, because DLS is commonly influenced by fewer large particles. Moreover, the highest NI concentration (98.0 µg/mL) (Table 6) was reduced almost by half three weeks after preparation (data not shown), and crystalline NI was observed, suggesting an instability of micelles. Therefore, standard NI-LNs had high NI concentrations and good stability. These results suggested that the lipid composition ratio dramatically affected the structure, the physicochemical properties, and the stability of lipid nanoparticles.

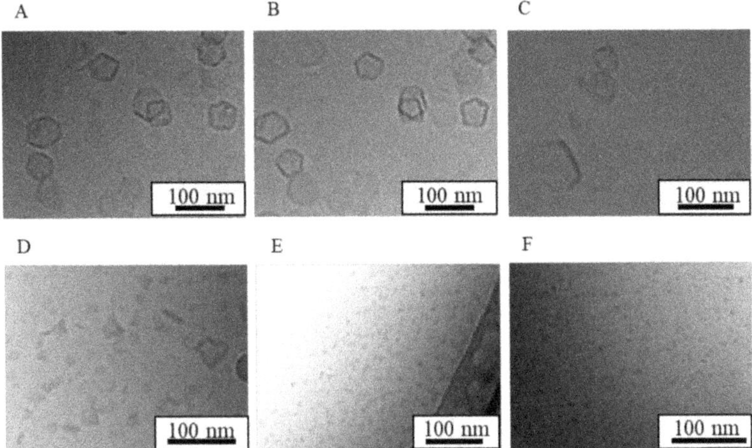

Figure 6. Cryo-TEM images of NI-LNs and LNs structures made with different HSPC/DPPG ratios. HSPC/DPPG (5/0) NI-LNs with the sample stage tilted 20 degrees (**A**) before and (**B**) after. (**C**) HSPC/DPPG (5/0) LNs. NI-LNs made from (**D**) HSPC/DPPG (5/0.5) and (**E**) HSPC/DPPG (0/5). (**F**) HSPC/DPPG (0/5) LNs.

Table 5. Effect of HSPC/DPPG molar ratio on mean particle size measured by cryo-TEM images.

HSPC/DPPG (Molar Ratio)		Bicelle (nm)	Micelle (nm)	Liposome (nm)	Mean Particle Size (nm)
NI-LNs	(5/0)	-	-	77.4 ± 31.5 ($n = 100$)	77.4 ± 31.5 ($n = 100$)
	(5/0.5)	24.1 ± 9.4 ($n = 71$)	9.0 ± 2.1 ($n = 22$)	43.9 ± 10.1 ($n = 7$)	22.2 ± 12.0 ($n = 100$)
	(5/1)	26.8 ± 8.4 ($n = 81$)	8.0 ± 1.7 ($n = 19$)	-	23.2 ± 10.6 ($n = 100$)
	(0/5)	-	8.6 ± 2.4 ($n = 100$)	-	8.6 ± 2.4 ($n = 100$)
LNs	(5/0)	-	-	80.5 ± 40.6 ($n = 20$)	80.5 ± 40.6 ($n = 20$)
	(5/1)	21.1 ± 5.6 ($n = 68$)	6.1 ± 2.1 ($n = 32$)	-	16.3 ± 8.7 ($n = 100$)
	(0/5)	-	10.9 ± 1.7 ($n = 100$)	-	10.9 ± 1.7 ($n = 100$)

Data obtained by cryo-TEM images are average values of particle length (mean $\pm$ S.D.).

Table 6. Effect of HSPC/DPPG molar ratio on physiochemical properties on LNs by DLS and HPLC measurements. PDI: polydispersity index.

HSPC/DPPG (Molar Ratio)		Mean Particle Size (nm)	PDI	Zeta Potential (mV)	Drug Concentration (μg/mL)	EE (%)
NI-LNs	(5/0)	93.1 ± 102.4	0.443 ± 0.049	-5.5 ± 9.1	2.9 ± 2.5	63.4 ± 55.1
	(5/0.5)	50.7 ± 3.1	0.294 ± 0.025	-43.6 ± 7.9	58.0 ± 25.8	97.0 ± 1.8
	(5/1)	47.5 ± 2.9	0.307 ± 0.027	-46.8 ± 9.6	67.3 ± 11.3	97.1 ± 1.6
	(0/5)	54.5 ± 28.4	0.793 ± 0.210	-49.9 ± 3.7	98.0 ± 24.2	99.3 ± 0.1
LNs	(5/0)	217.7 ± 135.1	0.238 ± 0.026	-13.5 ± 0.4	-	-
	(5/1)	48.7 ± 9.4	0.357 ± 0.086	-55.6 ± 4.2	-	-
	(0/5)	92.4 ± 10.3	0.540 ± 0.029	-55.4 ± 5.4	-	-

3.5. Effect of Phosphatidylcholine Acyl Chain Length on LN Structure

HSPC is a mixture of phosphatidylcholines with different acyl chain lengths (C12–20). Thus, to assess chain-length effects, DPPG (C16) was used as a phosphatidylglycerol and the phosphatidylcholine (PC) chain length was varied. NI-LNs using DMPC (C14), DPPC (C16), and DSPC (C18) were prepared with the ratio PC/DPPG (5/1). In Figure 7, cryo-TEM images of each NI-LN showed that the structure of all the particles made from each PC were bicelles, although the particle sizes of the bicelles were different, while that of the micelles was same between all of the PCs (Table 7). Hence, the PC acyl chain length did not affect the bicelle structure through just changing the particle sizes.

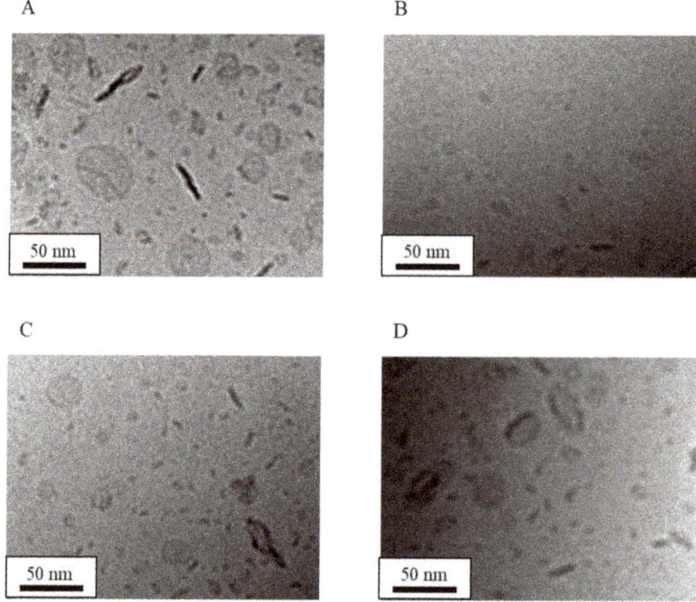

Figure 7. Cryo-TEM images of NI-LN structures having various alkyl chain lengths. (**A**) HSPC/DPPG (5/1), (**B**) 1,2-dimyristoyl-sn-glycero-3-phosphocholine (DMPC)/DPPG (5/1), (**C**) 1,2-dipalmitoyl-sn-glycero-3-phosphocholine (DPPC)/DPPG (5/1) and (**D**) 1,2-distearoyl-sn-glycero-3-phosphocholine (DSPC)/DPPG (5/1).

Table 7. Effect of different alkyl chain lengths on mean particle size measured by cryo-TEM images.

Formulation	Bicelle (nm)	Micelle (nm)	Mean Particle Size (nm)
HSPC/DPPG (5/1)	26.8 ± 8.4 ($n = 81$)	8.0 ± 1.7 ($n = 19$)	23.2 ± 10.6 ($n = 100$)
DMPC/DPPG (5/1)	15.6 ± 3.3 ($n = 40$)	6.7 ± 1.0 ($n = 15$)	13.2 ± 4.9 ($n = 55$)
DPPC/DPPG (5/1)	25.7 ± 11.3 ($n = 82$)	7.2 ± 1.4 ($n = 18$)	22.4 ± 12.5 ($n = 100$)
DSPC/DPPG (5/1)	26.4 ± 8.8 ($n = 77$)	6.1 ± 0.6 ($n = 23$)	21.7 ± 11.5 ($n = 100$)

Data obtained by TEM images are average values of particle length (mean $\pm$ S.D.).

NI-LNs made with DMPC had substantially smaller mean particle size and bigger PDI relative to other the particles (Table 8). Previously, it was reported that the size distribution had bimodal peaks, and that NI crystal precipitated one month after preparation [48]; thus, NI-LNs using DMPC had different properties from the other particles. Longer alkyl chain lengths result in higher phase transition temperatures and stronger van der Waals interactions. Therefore, the pressure-induced unstable lipid bilayers (electrostatic repulsion from polar head groups) made it difficult to form stable particles with DMPC.

Table 8. Effect of different alkyl chain lengths on physiochemical properties of NI-LNs by DLS and HPLC measurements.

Formulation	Mean Particle Size (nm)	PDI	Zeta Potential (mV)	Drug Concentration (µg/mL)	EE (%)
HSPC/DPPG (5/1)	47.5 ± 2.9	0.307 ± 0.027	−46.8 ± 9.6	67.3 ± 11.3	97.1 ± 1.6
DMPC/DPPG (5/1)	34.4 ± 17.2	0.668 ± 0.194	−35.1 ± 10.0	61.0 ± 19.0	98.9 ± 0.2
DPPC/DPPG (5/1)	50.4 ± 11.1	0.361 ± 0.086	−53.1 ± 1.8	61.5 ± 19.6	97.9 ± 1.9
DSPC/DPPG (5/1)	62.5 ± 5.9	0.285 ± 0.041	−47.6 ± 3.7	59.1 ± 17.9	97.7 ± 2.3

3.6. Effect of High-Pressure Homogenization on LN Structure

Cryo-TEM images in Figure 8 show NI-LNs prepared under different high-pressure homogenization conditions. For a 100-MPa/100 pass under reduced pressure, there were bicelles and micelles similar to standard NI-LNs, but some particles remained large because of the weaker applied shearing force (Figure 8A, Table 9). There were variations that suggested Ostwald ripening. In this case, the stability was inferior to that of NI-LNs prepared under standard conditions. Low NI concentration (Table 10) resulted from the particles having a large size, which were removed by the 200-nm filter membrane. For a 175 MPa/150 pass under increased the pass number, the size became smaller than that of NI-LNs prepared under standard conditions (175 MPa/100 pass). It was possible to observe bicelles that had different widths (Figure 8B). When the pass number increased, the particles had smaller absolute value of zeta potentials and low NI concentration (Table 10). This might be due to the adhesion of the sample (especially DPPG) to the flow path of Microfluidizer®. Therefore, the standard condition was the most efficient method. These results suggested that the high-pressure homogenization conditions did not affect the structure, but did affect the size and entrapped drug concentration of the lipid nanoparticles.

A

B

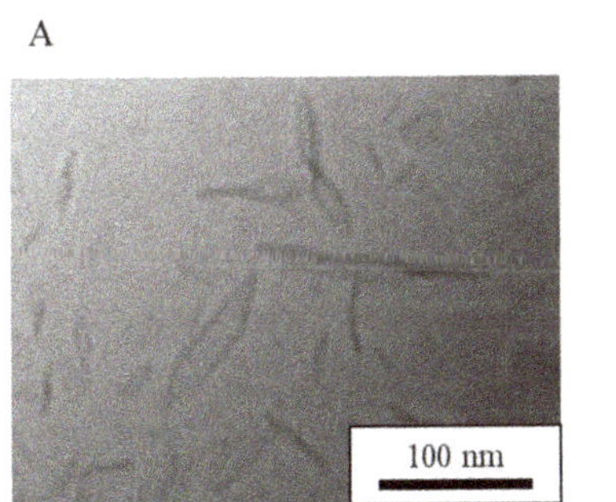
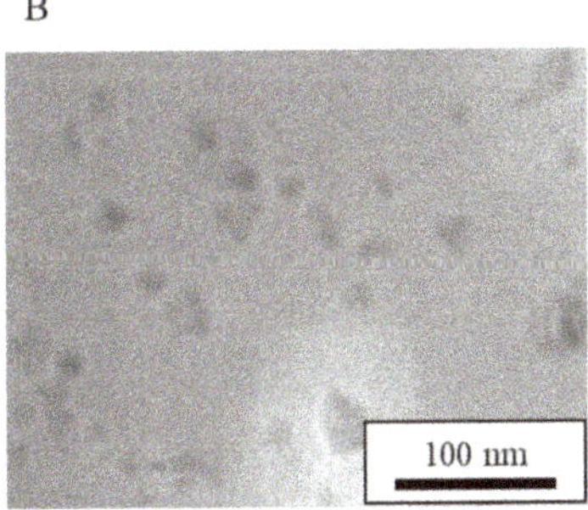

Figure 8. Cryo-TEM images of NI-LN structures made from HSPC/DPPG (5/1) under different high-pressure homogenization conditions. (**A**) 100 MPa/100 pass and (**B**) 175 MPa/150 pass.

Table 9. Effect of different conditions of high-pressure homogenization on mean particle size measured by cryo-TEM images.

Conditions	Bicelle (nm)	Micelle (nm)	Mean Particle Size (nm)
175 MPa/100 pass	26.8 ± 8.4 (n = 81)	8.0 ± 1.7 (n = 19)	23.2 ± 10.6 (n = 100)
100 MPa/100 pass	47.0 ± 21.5 (n = 84)	9.4 ± 1.3 (n = 16)	41.0 ± 24.0 (n = 100)
175 MPa/150 pass	23.1 ± 6.0 (n = 49)	10.4 ± 1.9 (n = 11)	20.7 ± 7.3 (n = 60)

Data obtained by cryo-TEM images are average values of particle length (mean ± S.D.).

Table 10. Effect of different conditions of high-pressure homogenization on physiochemical properties of NI-LNs by DLS and HPLC measurements.

Conditions	Mean Particle Size (nm)	PDI	Zeta Potential (mV)	Drug Concentration (µg/mL)	EE (%)
175 MPa/100 pass	47.5 ± 2.9	0.307 ± 0.027	−46.8 ± 9.6	67.3 ± 11.3	97.1 ± 1.6
100 MPa/100 pass	67.9 ± 8.5	0.251 ± 0.004	−41.2 ± 19.7	24.6 ± 6.7	98.7 ± 0.6
175 MPa/150 pass	39.1 ± 14.4	0.264 ± 0.004	−28.1 ± 31.1	37.0 ± 3.3	99.2 ± 0.4

3.7. Effect of Cholesterol on LN Structure

As discussed above, lipid nanoparticles prepared with HSPC/DPPG (5/0) were polygonal and thought to be liposomes (Figure 5A,E). Cholesterol (Chol) inserts into PC liposomes and increases the packing between the lipids, which decreases membrane permeability [49]. Some liposomes with Chol had smoother and rounder surfaces relative to those without Chol [47]. Here, Chol was added to NI-LNs prepared with HSPC/DPPG (5/0) to form liposomes, and it was added to standard NI-LNs to compare structural and physicochemical properties.

Cryo-TEM images in Figure 9 revealed NI-LNs prepared from HSPC/DPPG/Chol (5/0/0) and (5/1/0), as well as from HSPC/DPPG/Chol (5/0/2) and (5/1/2). For HSPC/DPPG/Chol (5/0/2), the surface asperities disappeared, and the particles became smooth (Figure 9B). Since there was no difference in the widths of identical particles before and after sample stage rotation, the structures were spherical. Thus, lipid nanoparticles prepared from HSPC/DPPG/Chol (5/0/0) were liposomes. NI-LNs prepared with HSPC/DPPG/Chol (5/1/2) were bicelles and micelles, similar to standard NI-LNs, but liposomes were also present (Figure 9D). It was reported that when the Chol content was 20 mol% or less in liposomes, there were two phases: a miscible phase composed of Chol and lipids, and a lipid phase composed only of lipids [50]. For a Chol concentration of more than 20 mol%, there were spherical liposomes formed by the miscible phase, and micelles and bicelles formed by the lipid phase.

As for physicochemical properties (Tables 11 and 12), NI-LNs prepared with (5/0/0) had almost neutral potential, but the zeta potential increased when prepared with (5/0/2) (Table 12). Furthermore, NI-LNs prepared with (5/1/2) had negative charge, so DPPG had a larger contribution to zeta potential than Chol. NI-LNs prepared with (5/1/2) had bigger PDI values because of the mixture of three structures (Tables 11 and 12). Moreover, NI-LNs with Chol had increased mean particle sizes and slightly higher NI concentrations (Table 12). Standard NI-LNs stayed dispersed for four months, while aggregation was observed in NI-LNs with Chol two months after preparation (data not shown). This suggested that fluidity was enhanced and that particle-to-particle adsorptions were easier with Chol in the membrane. Robust and stable lipid membranes could be formed without Chol, and NI-LNs had highly stable *EE* values. This was because the difference in curvature in inside and outside small unilamellar vesicles disordered the lipid molecules and affected their permeability. However, the edge curvature in bicelles was large, while the plane curvature was almost zero. The lipids in the latter were packed by high pressure and shearing forces during preparation. Therefore, NI-LNs had good properties without Chol.

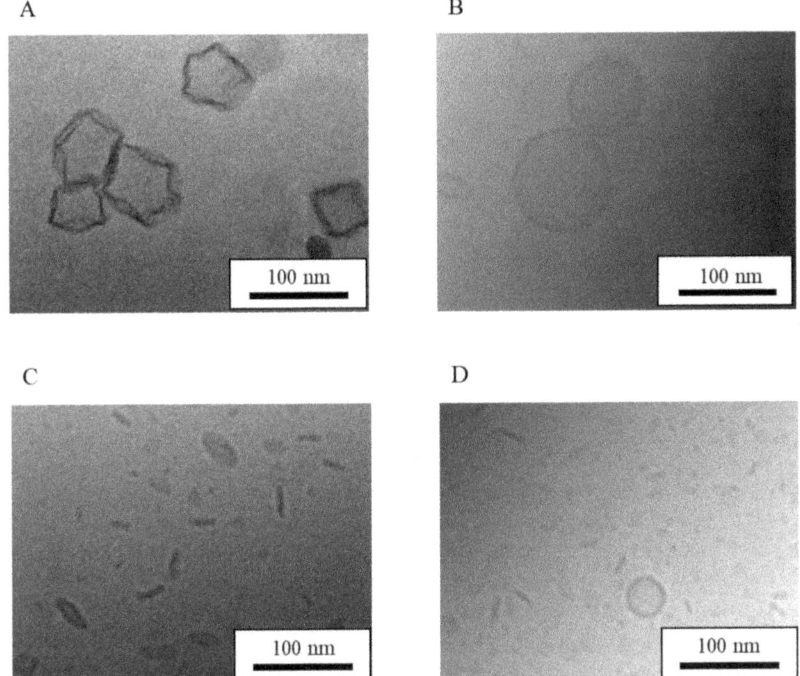

Figure 9. Cryo-TEM images of NI-LNs structures with/without cholesterol (Chol). (**A**) HSPC/DPPG/Chol (5/0/0), (**B**) HSPC/DPPG/Chol (5/0/2), (**C**) HSPC/DPPG/Chol (5/1/0) and (**D**) HSPC/DPPG/Chol (5/1/2).

Table 11. Effect of cholesterol on mean particle size measured by cryo-TEM images.

Formulation	Bicelle (nm)	Micelle (nm)	Liposome (nm)	Mean Particle Size (nm)
HSPC/DPPG/Chol (5/0/0)	-	-	77.4 ± 31.5 ($n = 100$)	77.4 ± 31.5 ($n = 100$)
HSPC/DPPG/Chol (5/0/2)	-	-	46.3 ± 9.6 ($n = 100$)	46.3 ± 9.6 ($n = 100$)
HSPC/DPPG/Chol (5/1/0)	26.8 ± 8.4 ($n = 81$)	8.0 ± 1.7 ($n = 19$)	-	23.2 ± 10.6 ($n = 100$)
HSPC/DPPG/Chol (5/1/2)	22.8 ± 8.2 ($n = 60$)	7.6 ± 1.0 ($n = 32$)	34.7 ± 8.9 ($n = 8$)	25.8 ± 8.8 ($n = 100$)

Data obtained by cryo-TEM images are average values of particle length (mean $\pm$ S.D.).

Table 12. Effect of cholesterol on physiochemical properties of LNs by DLS and HPLC measurements.

HSPC/DPPG/Chol	Mean Particle Size (nm)	PDI	Zeta Potential (mV)	Drug Concentration (μg/mL)	EE (%)
(5/0/0)	93.1 ± 102.4	0.443 ± 0.049	-5.5 ± 9.1	2.9 ± 2.5	63.4 ± 55.1
(5/0/2)	69.2 ± 6.2	0.368 ± 0.004	44.0 ± 1.6	12.6 ± 1.0	98.8 ± 0.6
(5/1/0)	47.5 ± 2.9	0.307 ± 0.027	-46.8 ± 9.6	67.3 ± 11.3	97.1 ± 1.6
(5/1/2)	80.0 ± 21.3	0.468 ± 0.004	-55.2 ± 1.1	109.2 ± 17.1	98.2 ± 1.5

4. Discussion

NI-LNs had shown improved solubility, high *EE*, long-term stability, and good redispersibility. To the best of our knowledge, other lipid nanoparticles having these all properties have not been reported yet, suggesting the possibility that NI-LNs might have unique structural features from other lipid nanoparticles. In this study, the structure of the particles was focused on, and cryo-TEM and

AFM analyses were performed. From a series of experiments, the structure of NI-LNs was found to be bicelles.

In bicelles, long-chain lipids (C16–20) generally form the planar area, and short-chain lipids (C6–8) form the edge area [51,52]. Diheptanoylphosphatidylcholine (DHPC, C6) is usually used as short-chain lipids forming the edge area, but there are some examples except for short-chain lipids; e.g., dimyristoylphosphoethanolamine (C14) introduced by lanthanide to the head group [53], n-dodecylphosphocholine (C12) that had one alkyl chain [54], and ceramide (C16) bound to polyethylene glycol (PEG)-5000 [55]. As the feature of these materials, they had large head areas. It was reported that hydrophilic surfaces with high curvature can be formed by anionic lipids because the head volume is expanded by strong electrostatic repulsion in low ionic strength media [56]. In the present study, the bicelles formed here used phospholipids with similar alkyl chain lengths: HSPC (C12–20) and DPPG (C16). However, anionic DPPG might have a large head area, because there were micelles when only DPPG was contained in the formulation in cryo-TEM images and PG with a negatively charged head group could be self-assembled at high curvatures. Therefore, we speculated that DPPG was localized in the edge area, and HSPC was in the flat area to form bicelles (Figure 3).

It was reported that spherical liposomes were prepared using PC and PG by other methods [57,58]. The general preparation method of bicelles [54,59] is quite different from our method using the roll mill and microfluidizer. Nevertheless, the disk structure was formed in this work, and this may be explained by the following two reasons: one is a high shearing force, and the other is heating by microfluidization. First, the drugs, PC and PG, were miscible by roll milling, and liposome-like particles assembled in distilled water during pre-dispersion were collapsed by the high shearing force during microfluidizer miniaturization. The PG was subsequently assembled along the edge as a cap for the flat lipid bilayer, forming the disk-like particle. In addition, it was reported that thermosensitive liposomes containing micelle-forming components such as distearoylphosphoethanolamine-PEG2000 or lysolipids changed to bilayer discs after a repeated cycling through the phase transition temperature (T_c) [60]. Therefore, the spherical particles may have changed to be bicelles, because of the heating caused by the miniaturization process.

In addition to temperature, the determining factors for whether lipid assembly forms bicelle are lipid concentration and the molar ratio of two phospholipids. Lipid concentration (20% or more) is necessary to form bicelles. It has been reported that when the DMPC/DHPC bicellar system was diluted to 0.07%, the disk form changed to be vesicle [61]. This transformation mechanism was that DHPC, having a relatively high solubility, was separated from the edge of the disk; then, some bilayers integrated, and the size became bigger, resulting in the growing and unstable bilayers transform vesicle. To protect this transformation of bicelles under dilute conditions, bicelle-encapsulated liposomes named "bicosomes" were developed [62]. However, our bicelles can form disk structures at low lipid concentration (0.5% lipid concentration) and be stable for four months without any significant change of the structure (Figure 4A). We speculated that due to the strong hydrophobic interaction between HSPC and the longer acyl chain of DPPG in this work, DPPG might strongly cap the lipid bilayer of HSPC and prevent the leakage of encapsulated drug, giving the relatively long-term stability, contrary to previously reported DMPC/DHPC bicelles. In addition, the concentrated bicelles by centrifugation can also maintain their disk structure, suggesting that our methodology can prepare bicelles with arbitrary concentrations. Furthermore, it is suggested that our bicelles can maintain their disk structure even when intravitally administrated in vivo, where a high amount of liquid exists.

As for the molar ratio of two phospholipids, when the ratio of long-chain to short-chain (the ratio q) fatty acids is 2.5 or less, the bicelles are isotropic; while, when the ratio of q is 3.0 or more, the bicelles are anisotropic in the magnetic field [63], indicating that the particle size of bicelles is small and big, respectively. In our case, when assumed with HSPC as the long chain and DPPG as the short chain, the ratio of q made with HSPC/DPPG (5/1, for standard) and (5/0.5) is 5.0 and 10, respectively, and no significant differences in mean particle size between them are seen (Table 6). Therefore, when the ratio of q is changed to be smaller, bicelles with smaller mean particle size would be obtained. There are some reports about bicelles for the DDS field. For example, Barbosa-Barros reported that bicelles have been used for drug penetration carrier through skin [64]. In addition, organic–inorganic hybrid bicelles named "cerasomes" were also developed [65]. These hybrid bicelles were reported to show a unique drug release profile under an acidic pH environment. However, the amount of published literatures is small. For our bicelles, although the drug release study, membrane permeation study using caco-2, in vivo biocompatibility, and pharmacokinetic study should be determined, further applications using our stable bicelles are also expected for the DDS field. In addition, the differences between our bicelles and other bicelles prepared with general materials using general methods would be necessary to compare in the future.

5. Conclusions

Roll-mill grinding and high-pressure homogenization were used to prepare lipid nanoparticles from a neutral PC lipid and a negatively charged PG lipid under various conditions. Standard NI-LNs with HSPC/DPPG (5/1) were disk-like bicelles. No changes were observed after long-term storage, and NI-LNs freeze-dried with sucrose and re-hydration. Liposomes were formed for HSPC/DPPG (5/0), and micelles were formed for HSPC/DPPG (0/5). The addition of Chol created smooth particle surfaces for HSPC/DPPG (5/0/2). A serious of experiments and results demonstrated that the standard NI-LN bicelles have the best physicochemical properties, and further DDS study would be desired using these stable bicelles.

Author Contributions: Conceptualization, S.M., and Y.I.; Data Curation, S.M.; Investigation, S.M., K.H., S.i.K., and Y.I.; Methodology, S.M., and S.i.K.; Project Administration, Y.I.; Validation, S.I., H.K., and Y.I.; Visualization, S.M., and K.H.; Supervision, K.M., S.I., H.K., and Y.I.; Writing-Original Draft Preparation, S.M., and Y.I.; Writing-Review & Editing, S.M., H.K., and Y.I.

Funding: This work was supported by JSPS Grant-in-Aid for Scientific Research (C) Grant Number JP 17K08455.

Conflicts of Interest: The authors declare no conflicts of interest.

Abbreviations

AFM	Atomic force microscopy
APTES	3-Aminopropyltriethoxysilane
Chol	Cholesterol
CPP	Critical packing parameter
Cryo-TEM	Cryo transmission electron microscopy
DDS	Drug delivery system
DLS	Dynamic light scattering
DHPC	Diheptanoylphosphatidylcholine
DMPC	Dimyristoylphosphatidylcholine
DPPC	Dipalmitoylphosphatidylcholine
DPPG	Dipalmitoylphosphatidylglycerol
DSPC	Distearoylphosphatidylcholine
EE	Entrapment efficiency
HSPC	Hydrogenated soybean phosphatidylcholine
LNs	Lipid-based nanoparticles
NI	Nifedipine
NI-LNs	Nifedipine-encapsulated lipid nanoparticles
PC	Phosphatidylcholine
PDI	Polydispersity index
PEG	Polyethylene glycol
PG	Phosphatidylglycerol
PTFE	Polytetrafluoroethylene
PHT	Phenytoin
PHT-LNs	Phenytoin-encapsulated lipid nanoparticles
Tc	Phase transition temperature

References

1. Di Maio, S.; Carrier, R.L. Gastrointestinal contents in fasted state and post-lipid ingestion: In vivo measurements and in vitro models for studying oral drug delivery. *J. Control. Release* **2011**, *151*, 110–122. [CrossRef] [PubMed]
2. Shegokar, R.; Müller, R.H. Nanocrystals: Industrially feasible multifunctional formulation technology for poorly soluble actives. *Int. J. Pharm.* **2010**, *399*, 129–139. [CrossRef] [PubMed]
3. Yamada, T.; Saito, N.; Imai, T.; Otagiri, M. Effect of grinding with hydroxypropyl cellulose on the dissolution and particle size of a poorly water-soluble drug. *Chem. Pharm. Bull.* **1999**, *47*, 1311–1313. [CrossRef] [PubMed]
4. Sugimoto, M.; Okagaki, T.; Narisawa, S.; Koida, Y.; Nakajima, K. Improvement of dissolution characteristics and bioavailability of poorly water-soluble drugs by novel cogrinding method using water-soluble polymer. *Int. J. Pharm.* **1998**, *160*, 11–19. [CrossRef]
5. Gao, L.; Zhang, D.; Chen, M. Drug nanocrystals for the formulation of poorly soluble drugs and its application as a potential drug delivery system. *J. Nanoparticle Res.* **2008**, *10*, 845–862. [CrossRef]
6. Zhang, P.; Forsgren, J.; Strømme, M. Stabilisation of amorphous ibuprofen in Upsalite, a mesoporous magnesium carbonate, as an approach to increasing the aqueous solubility of poorly soluble drugs. *Int. J. Pharm.* **2014**, *472*, 185–191. [CrossRef]
7. Baghel, S.; Cathcart, H.; Redington, W.; O'Reilly, N.J. An investigation into the crystallization tendency/kinetics of amorphous active pharmaceutical ingredients: A case study with dipyridamole and cinnarizine. *Eur. J. Pharm. Biopharm.* **2016**, *104*, 59–71. [CrossRef]
8. Skrdla, P.J.; Floyd, P.D.; Dell'orco, P.C. Practical Estimation of Amorphous Solubility Enhancement Using Thermoanalytical Data: Determination of the Amorphous/Crystalline Solubility Ratio for Pure Indomethacin and Felodipine. *J. Pharm. Sci.* **2016**, *105*, 2625–2630. [CrossRef]
9. Chiou, W.L.; Riegelman, S. Preparation and dissolution characteristics of several fast-release solid dispersions of griseofulvin. *J. Pharm. Sci.* **1969**, *58*, 1505–1510. [CrossRef]
10. Leuner, C.; Dressman, J. Improving drug solubility for oral delivery using solid dispersions. *Eur. J. Pharm. Biopharm.* **2000**, *50*, 47–60. [CrossRef]
11. Wang, B.; Wang, D.; Zhao, S.; Huang, X.; Zhang, J.; Lv, Y.; Liu, X.; Lv, G.; Ma, X. Evaluate the ability of PVP to inhibit crystallization of amorphous solid dispersions by density functional theory and experimental verify. *Eur. J. Pharm. Sci.* **2017**, *96*, 45–52. [CrossRef] [PubMed]
12. Dunn, S.S.; Beckford Vera, D.R.; Benhabbour, S.R.; Parrott, M.C. Rapid microwave-assisted synthesis of sub-30 nm lipid nanoparticles. *J. Colloid Interface Sci.* **2017**, *488*, 240–245. [CrossRef] [PubMed]
13. Dokoumetzidis, A.; Macheras, P. A century of dissolution research: From Noyes and Whitney to the Biopharmaceutics Classification System. *Int. J. Pharm.* **2006**, *321*, 1–11. [CrossRef] [PubMed]
14. Azad, M.; Moreno, J.; Bilgili, E.; Davé, R. Fast dissolution of poorly water soluble drugs from fluidized bed coated nanocomposites: Impact of carrier size. *Int. J. Pharm.* **2016**, *513*, 319–331. [CrossRef] [PubMed]
15. Müller, R.H.; Peters, K. Nanosuspensions for the formulation of poorly soluble drugs. *Int. J. Pharm.* **1998**, *160*, 229–237. [CrossRef]
16. Fabiano, A.; Zambito, Y.; Bernkop-Schnürch, A. About the impact of water movement on the permeation behaviour of nanoparticles in mucus. *Int. J. Pharm.* **2017**, *517*, 279–285. [CrossRef] [PubMed]
17. Sharma, A.; Pandey, R.; Sharma, S.; Khuller, G.K. Chemotherapeutic efficacy of poly (DL-lactide-co-glycolide) nanoparticle encapsulated antitubercular drugs at sub-therapeutic dose against experimental tuberculosis. *Int. J. Antimicrob. Agents* **2004**, *24*, 599–604. [CrossRef] [PubMed]
18. Hörmann, K.; Zimmer, A. Drug delivery and drug targeting with parenteral lipid nanoemulsions—A review. *J. Control. Release* **2016**, *223*, 85–98. [CrossRef]
19. Barenholz, Y. Doxil®—The first FDA-approved nano-drug: Lessons learned. *J. Control. Release* **2012**, *160*, 117–134. [CrossRef]
20. Sakai-Kato, K.; Nishiyama, N.; Kozaki, M.; Nakanishi, T.; Matsuda, Y.; Hirano, M.; Hanada, H.; Hisada, S.; Onodera, H.; Harashima, H.; et al. General considerations regarding the in vitro and in vivo properties of block copolymer micelle products and their evaluation. *J. Control. Release* **2015**, *210*, 76–83. [CrossRef]

21. Beloqui, A.; Solinís, M.Á.; Rodríguez-Gascón, A.; Almeida, A.J.; Préat, V. Nanostructured lipid carriers: Promising drug delivery systems for future clinics. *Nanomed. Nanotechnol. Biol. Med.* **2016**, *12*, 143–161. [CrossRef] [PubMed]

22. Kowalczuk, A.; Trzcinska, R.; Trzebicka, B.; Müller, A.H.E.; Dworak, A.; Tsvetanov, C.B. Progress in Polymer Science Loading of polymer nanocarriers: Factors, mechanisms and applications. *Prog. Polym. Sci.* **2014**, *39*, 43–86. [CrossRef]

23. Concheiro, A.; Alvarez-Lorenzo, C. Chemically cross-linked and grafted cyclodextrin hydrogels: From nanostructures to drug-eluting medical devices. *Adv. Drug Deliv. Rev.* **2013**, *65*, 1188–1203. [CrossRef] [PubMed]

24. Wong, B.S.; Yoong, S.L.; Jagusiak, A.; Panczyk, T.; Ho, H.K.; Ang, W.H.; Pastorin, G. Carbon nanotubes for delivery of small molecule drugs. *Adv. Drug Deliv. Rev.* **2013**, *65*, 1964–2015. [CrossRef]

25. Sánchez-López, E.; Espina, M.; Doktorovova, S.; Souto, E.B.; García, M.L. Lipid nanoparticles (SLN, NLC): Overcoming the anatomical and physiological barriers of the eye—Part II—Ocular drug-loaded lipid nanoparticles. *Eur. J. Pharm. Biopharm.* **2017**, *110*, 58–69. [CrossRef]

26. Amidon, G.L.; Lennernas, H.; Shah, V.P.; Crison, J.R. A theoretical basis for a biopharmaceutic drug classification: The correlation of in vitro drug product dissolution and in vivo bioavailability. *Pharm. Res.* **1995**, *12*, 413–420. [CrossRef]

27. Kamiya, S.; Yamada, M.; Kurita, T.; Miyagishima, A.; Arakawa, M.; Sonobe, T. Preparation and stabilization of nifedipine lipid nanoparticles. *Int. J. Pharm.* **2008**, *354*, 242–247. [CrossRef]

28. Ohshima, H.; Miyagishima, A.; Kurita, T.; Makino, Y.; Iwao, Y.; Sonobe, T.; Itai, S. Freeze-dried nifedipine-lipid nanoparticles with long-term nano-dispersion stability after reconstitution. *Int. J. Pharm.* **2009**, *377*, 180–184. [CrossRef]

29. Funakoshi, Y.; Iwao, Y.; Noguchi, S.; Itai, S. Lipid nanoparticles with no surfactant improve oral absorption rate of poorly water-soluble drug. *Int. J. Pharm.* **2013**, *451*, 92–94. [CrossRef]

30. Shi, L.L.; Cao, Y.; Zhu, X.Y.; Cui, J.H.; Cao, Q.R. Optimization of process variables of zanamivir-loaded solid lipid nanoparticles and the prediction of their cellular transport in Caco-2 cell model. *Int. J. Pharm.* **2015**, *478*, 60–69. [CrossRef]

31. Gonçalves, L.M.D.; Maestrelli, F.; Mannelli, L.C.; Ghelardini, C.; Almeida, A.J.; Mura, P. Development of solid lipid nanoparticles as carriers for improving oral bioavailability of glibenclamide. *Eur. J. Pharm. Biopharm.* **2016**, *102*, 41–50. [CrossRef] [PubMed]

32. Shah, B.; Khunt, D.; Bhatt, H.; Misra, M.; Padh, H. Application of quality by design approach for intranasal delivery of rivastigmine loaded solid lipid nanoparticles: Effect on formulation and characterization parameters. *Eur. J. Pharm. Sci.* **2015**, *78*, 54–66. [CrossRef] [PubMed]

33. Doktorovova, S.; Shegokar, R.; Rakovsky, E.; Gonzalez-Mira, E.; Lopes, C.M.; Silva, A.M.; Martins-Lopes, P.; Muller, R.H.; Souto, E.B. Cationic solid lipid nanoparticles (cSLN): Structure, stability and DNA binding capacity correlation studies. *Int. J. Pharm.* **2011**, *420*, 341–349. [CrossRef] [PubMed]

34. Kim, J.K.; Park, J.S.; Kim, C.K. Development of a binary lipid nanoparticles formulation of itraconazole for parenteral administration and controlled release. *Int. J. Pharm.* **2010**, *383*, 209–215. [CrossRef] [PubMed]

35. Crowell, K.J.; Macdonald, P.M. Surface charge response of the phosphatidylcholine head group in bilayered micelles from phosphorus and deuterium nuclear magnetic resonance. *Biochim. Biophys. Acta—Biomembr.* **1999**, *1416*, 21–30. [CrossRef]

36. Diller, A.; Loudet, C.; Aussenac, F.; Raffard, G.; Fournier, S.; Laguerre, M.; Grélard, A.; Opella, S.J.; Marassi, F.M.; Dufourc, E.J. Bicelles: A natural "molecular goniometer" for structural, dynamical and topological studies of molecules in membranes. *Biochimie* **2009**, *91*, 744–751. [CrossRef] [PubMed]

37. Barman, R.K.; Iwao, Y.; Funakoshi, Y.; Ranneh, A.-H.; Noguchi, S.; Wahed, M.I.I.; Itai, S. Development of Highly Stable Nifedipine Solid–Lipid Nanoparticles. *Chem. Pharm. Bull.* **2014**, *62*, 399–406. [CrossRef] [PubMed]

38. Kuntsche, J.; Horst, J.C.; Bunjes, H. Cryogenic transmission electron microscopy (cryo-TEM) for studying the morphology of colloidal drug delivery systems. *Int. J. Pharm.* **2011**, *417*, 120–137. [CrossRef]

39. Yasuhara, K.; Miki, S.; Nakazono, H.; Ohta, A.; Kikuchi, J. Synthesis of organic–inorganic hybrid bicelles–lipid bilayer nanodiscs encompassed by siloxane surfaces. *Chem. Commun.* **2011**, *47*, 4691–4693. [CrossRef]

40. Glazier, R.; Salaita, K. Supported lipid bilayer platforms to probe cell mechanobiology. *Biochim. Biophys. Acta—Biomembr.* **2017**, *1859*, 1465–1482. [CrossRef]

41. Emami, S.; Azadmard-Damirchi, S.; Peighambardoust, S.H.; Hesari, J.; Valizadeh, H.; Faller, R. Molecular dynamics simulations of ternary lipid bilayers containing plant sterol and glucosylceramide. *Chem. Phys. Lipids* **2017**, *203*, 24–32. [CrossRef] [PubMed]

42. Tian, B.; Al-Jamal, W.T.; Al-Jamal, K.T.; Kostarelos, K. Doxorubicin-loaded lipid-quantum dot hybrids: Surface topography and release properties. *Int. J. Pharm.* **2011**, *416*, 443–447. [CrossRef] [PubMed]

43. Mahabir, S.; Small, D.; Li, M.; Wan, W.; Kučerka, N.; Littrell, K.; Katsaras, J.; Nieh, M.P. Growth kinetics of lipid-based nanodiscs to unilamellar vesicles—A time-resolved small angle neutron scattering (SANS) study. *Biochim. Biophys. Acta—Biomembr.* **2013**, *1828*, 1025–1035. [CrossRef] [PubMed]

44. Mariusz, K. Interactions of serum with polyelectrolyte-stabilized liposomes: Cryo-TEM studies. Interactions of serum with polyelectrolyte-stabilized liposomes: Cryo-TEM studies. *Colloids Surf. B Biointerfaces* **2014**, *120*, 152–159.

45. Doroty, C.; Jonathan, C.; Jenna, B.; William, J.M.; Maria, A.O.; Sheng, Q. Disc-shaped polyoxyethylene glycol glycerides gel nanoparticles as novel protein delivery vehicles. *Int. J. Pharm.* **2015**, *496*, 1015–1025.

46. Fonte, P.; Reis, S.; Sarmento, B. Facts and evidences on the lyophilization of polymeric nanoparticles for drug delivery. *J. Control. Release* **2016**, *225*, 75–86. [CrossRef] [PubMed]

47. Dos Santos, N.; Mayer, L.D.; Abraham, S.A.; Gallagher, R.C.; Cox, K.A.K.; Tardi, P.G.; Bally, M.B. Improved retention of idarubicin after intravenous injection obtained for cholesterol-free liposomes. *Biochim. Biophys. Acta—Biomembr.* **2002**, *1561*, 188–201. [CrossRef]

48. Funakoshi, Y.; Iwao, Y.; Noguchi, S.; Itai, S. Effect of Alkyl Chain Length and Unsaturation of the Phospholipid on the Physicochemical Properties of Lipid Nanoparticles. *Chem. Pharm. Bull.* **2015**, *63*, 731–736. [CrossRef] [PubMed]

49. McElhaney, R.N.; Gier, J.De; Van Der Neut-Kok, E.C.M. The effect of alterations in fatty acid composition and cholesterol content on the permeability of Mycoplasrno laidlawii B cells and derived liposomes. *Biochim. Biophys. Acta—Biomembr.* **1973**, *298*, 500–512. [CrossRef]

50. Mabrey, S.; Mateo, P.L.; Sturtevant, J.M.; Mabrey, S.; Mateo, P.L.; Sturtevant, J.M. High-Sensitivity Scanning Calorimetric Study of Mixtures of Cholesterol with Dimyristoyl- and Dipalmitoylphosphatidylcholines. *Biochemistry* **1978**, *17*, 2464–2468. [CrossRef]

51. Prosser, R.S.; Evanics, F.; Kitevski, J.L.; Al-Abdul-Wahid, M.S. Current Applications of Bicelles in NMR Studies of Membrane- Associated Amphiphiles and Proteins. *Biochemistry* **2006**, *45*, 8453–8465. [CrossRef] [PubMed]

52. Yamamoto, K.; Pearcy, P.; Ramamoorthy, A. Bicelles exhibiting magnetic alignment for a broader range of temperatures: A solid-state NMR study. *Langmuir* **2014**, *30*, 1622–1629. [CrossRef]

53. Isabettini, S.; Liebi, M.; Kohlbrecher, J.; Ishikawa, T.; Windhab, E.J.; Fischer, P.; Walde, P.; Kuster, S. Tailoring bicelle morphology and thermal stability with lanthanide-chelating cholesterol conjugates. *Langmuir* **2016**, *32*, 9005–9014. [CrossRef] [PubMed]

54. Nolandt, O.V.; Walther, T.H.; Grage, S.L.; Ulrich, A.S. Magnetically oriented dodecylphosphocholine bicelles for solid-state NMR structure analysis. *Biochim. Biophys. Acta—Biomembr.* **2012**, *1818*, 1142–1147. [CrossRef]

55. Zetterberg, M.M.; Reijmar, K.; Pränting, M.; Engström, Å.; Andersson, D.I.; Edwards, K. PEG-stabilized lipid disks as carriers for amphiphilic antimicrobial peptides. *J. Control. Release* **2011**, *156*, 323–328. [CrossRef]

56. Meyer, H.W.; Richter, W.; Rettig, W.; Stumpf, M. Bilayer fragments and bilayered micelles (bicelles) of dimyristoylphosphatidylglycerol (DMPG) are induced by storage in distilled water at 4 °C. *Colloids Surfaces A Physicochem. Eng. Asp.* **2001**, *183–185*, 495–504. [CrossRef]

57. Dichello, G.A.; Fukuda, T.; Maekawa, T.; Whitby, R.L.D.; Mikhalovsky, S.V.; Alavijeh, M.; Pannala, A.S.; Sarker, D.K. Preparation of liposomes containing small gold nanoparticles using electrostatic interactions. *Eur. J. Pharm. Sci.* **2017**, *105*, 55–63. [CrossRef] [PubMed]

58. Von White, G.; Chen, Y.; Roder-Hanna, J.; Bothun, G.D.; Kitchens, C.L. Structural and thermal analysis of lipid vesicles encapsulating hydrophobic gold nanoparticles. *ACS Nano* **2012**, *6*, 4678–4685. [CrossRef]

59. Yamaguchi, T.; Suzuki, T.; Yasuda, T.; Oishi, T.; Matsumori, N. Bioorganic & Medicinal Chemistry NMR-based conformational analysis of sphingomyelin in bicelles. *Bioorg. Med. Chem.* **2012**, *20*, 270–278.

60. Ickenstein, L.M.; Arfvidsson, M.C.; Needham, D.; Mayer, L.D.; Edwards, K. Disc formation in cholesterol-free liposomes during phase transition. *Biochim. Biophys. Acta—Biomembr.* **2003**, *1614*, 135–138. [CrossRef]

61. Barbosa-Barros, L.; Rodríguez, G.; Cócera, M.; Rubio, L.; López-Iglesias, C.; de la Maza, A.; López, O. Structural Versatility of Bicellar Systems and Their Possibilities as Colloidal Carriers. *Pharmaceutics* **2011**, *3*, 636–664.

62. Rodríguez, G.; Soria, G.; Coll, E.; Rubio, L.; Barbosa-Barros, L.; López-Iglesias, C.; Planas, A.M.; Estelrich, J.; De La Maza, A.; López, O. Bicosomes: Bicelles in dilute systems. *Biophys. J.* **2010**, *99*, 480–488. [CrossRef]

63. Vold, R.R.; Prosser, R.S. Magnetically oriented phospholipid bilayered micelles for structural studies of polypeptides. Does the ideal bicelle exist? *J. Magn. Reson. Ser. B* **1996**, *113*, 267–271. [CrossRef]

64. Barbosa-Barros, L.; Rodríguez, G.; Barba, C.; Cócera, M.; Rubio, L.; Estelrich, J.; López-Iglesias, C.; De La Maza, A.; López, O. Bicelles: Lipid nanostructured platforms with potential dermal applications. *Small* **2012**, *8*, 807–818. [CrossRef]

65. Hameed, S.; Bhattarai, P.; Dai, Z. Cerasomes and Bicelles: Hybrid Bilayered Nanostructures With Silica-Like Surface in Cancer Theranostics. *Front. Chem.* **2018**, *6*, 1–17. [CrossRef] [PubMed]

 *nanomaterials*

Review

Current Applications of Nanoemulsions in Cancer Therapeutics

Elena Sánchez-López [1,2,3,*], Mariana Guerra [4], João Dias-Ferreira [4], Ana Lopez-Machado [1,2,3], Miren Ettcheto [4,5], Amanda Cano [1,2,3], Marta Espina [1,2], Antoni Camins [4,5], Maria Luisa Garcia [1,2,3] and Eliana B. Souto [1,6]

[1] Department of Pharmacy and Pharmaceutical Technology and Physical Chemistry, Faculty of Pharmacy, University of Barcelona, 08028 Barcelona, Spain; lora_ana@hotmail.com (A.L.-M.); acanofernandez@ub.edu (A.C.); m.espina@ub.edu (M.E.); marisagarcia@ub.edu (M.L.G.); souto.eliana@gmail.com (E.B.S.)
[2] Institute of Nanoscience and Nanotechnology (IN2UB), University of Barcelona, 08028 Barcelona, Spain
[3] Centro de Investigación Biomédica en Red de Enfermedades Neurodegenerativas (CIBERNED), University of Barcelona, 08028 Barcelona, Spain
[4] Department of Pharmaceutical Technology, Faculty of Pharmacy, University of Coimbra (FFUC), Polo das Ciências da Saúde, Azinhaga de Santa Comba, 3000-548 Coimbra, Portugal; marianags1994@gmail.com (M.G.); j.dias.ferreira@outlook.pt (J.D.-F.); e_miren60@hotmail.com (M.E.); camins@ub.edu (A.C.)
[5] Department of Pharmacology, Toxicology and Therapeutic Chemistry, Faculty of Pharmacy and Food Sciences, University of Barcelona, 08028 Barcelona, Spain
[6] CEB—Centre of Biological Engineering, University of Minho, Campus de Gualtar, 4710-057 Braga, Portugal
* Correspondence: esanchezlopez@ub.edu; Tel.: +34-93-93-402-45-52

Received: 3 May 2019; Accepted: 28 May 2019; Published: 31 May 2019

Abstract: Nanoemulsions are pharmaceutical formulations composed of particles within a nanometer range. They possess the capacity to encapsulate drugs that are poorly water soluble due to their hydrophobic core nature. Additionally, they are also composed of safe gradient excipients, which makes them a stable and safe option to deliver drugs. Cancer therapy has been an issue for several decades. Drugs developed to treat this disease are not always successful or end up failing, mainly due to low solubility, multidrug resistance (MDR), and unspecific toxicity. Nanoemulsions might be the solution to achieve efficient and safe tumor treatment. These formulations not only solve water-solubility problems but also provide specific targeting to cancer cells and might even be designed to overcome MDR. Nanoemulsions can be modified using ligands of different natures to target components present in tumor cells surface or to escape MDR mechanisms. Multifunctional nanoemulsions are being studied by a wide variety of researchers in different research areas mainly for the treatment of different types of cancer. All of these studies demonstrate that nanoemulsions are efficiently taken by the tumoral cells, reduce tumor growth, eliminate toxicity to healthy cells, and decrease migration of cancer cells to other organs.

Keywords: multifunctional nanoemulsions; targeted delivery; cancer

---

1. Introduction

Nanoemulsions are colloidal dispersions that can be used as drug vehicles mainly used for molecules with low water solubility constituted of safe grade excipients [1,2]. This dosage form is composed of a heterogeneous dispersion of a nanometer droplet in another liquid, which leads to high stability and solubility [3]. The encapsulation protects the drug from degradation and increases its half-life in the plasma [4].

To disperse the droplets in an aqueous phase, emulsifying agents are used to stabilize the system. Emulsifying agents are compounds with an amphiphilic profile that reduce the interfacial tension between two immiscible phases as they are constituted of hydrophobic bicarbonate tails that tend to place themselves in non-polar liquids and a polar head that usually places itself in polar liquids (Figure 1) [5].

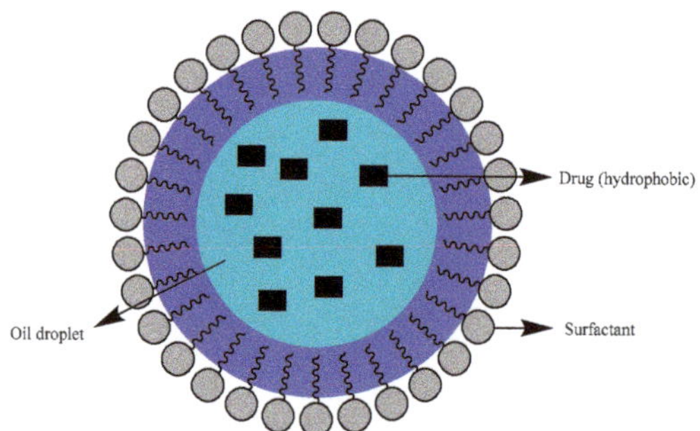

Figure 1. Nanoemulsions structure (based on [5]).

The research in cancer therapy has become more focused on nanoemulsions since they hold characteristics essential to achieve an efficient therapeutic effect: large surface area, superficial charge, elevated half-life of circulation, specific targeting, and the imaging capacity of the formulation. Since cancer cells are surrounded by vascularized tissues, nanoemulsions can easily accumulate in these tissues with their size as an advantage for passing through barriers. Above all this, they can also be designed to define their function, encapsulate distinct types of drugs, and select specific targets [6].

The tumor microenvironment is composed of extracellular matrix (ECM), fibroblasts, epithelial cells, immune cells, pericytes, adipocytes, glial cells (present only in the nervous system), proteins, vascular cells, and lymphatic cells [7]. ECM is essential in processes of growth, structure, migration, invasion, and metastasis of the tumor cells which have specific surface markers that can be targeted by drugs. The delivery of oxygen and nutrients to the tumor is achieved by simple diffusion, but when the tumor becomes larger than 2.0 mm^3, the oxygen levels decrease, leading to hypoxia conditions and angiogenic development of new blood vessels [8]. Therefore, by inhibiting the angiogenic process, cell growth can also decrease. In recent years many anti-angiogenic drugs have been developed: bevacizumab (vascular endothelial growth factor, VEGF-neutralizing antibody), sorafenib (VEGF signaling pathway blockers), sunitinib, and pazopanib. However, these inhibitors of angiogenesis are characterized by marked toxicity, enhanced resistance, and barriers to delivery of compounds [9]. Nanoemulsions can be used to encapsulate the drug inside its core, reducing toxicity and enhancing payload delivery.

Cancer cells get their energy balance using glycolysis. Nonetheless, as a consequence of hypoxia conditions, the final metabolite, pyruvate, is transformed into lactate which is eliminated by a monocarboxilate transporter using H$^+$ and generating tumor acidification. The hypoxia environment also increases the expression of carbonic anhydrase IX, resulting in the production of bicarbonate from carbon dioxide (CO$_2$) which ends up in the uptake process of the weakly-basic tumor cells, leading to a gradient between the extracellular and intracellular milieu of the tumor. This way, pH-responsive lipids might play an interesting role since they are stable at a pH of 7.4 but can change their chemical behavior when settled in an acidic pH, with further release of the therapeutic load [1].

The disordered and heterogeneous profile of tumors stroma leads to fluctuations in the presence of oxygen, drugs, and essential molecules in the tumor microenvironment [10]. This results in the previously referenced hypoxia and neovascularization leading to metastasis. However, the instability of lymphatic vessels might enhance the retention time of drugs since their clearance rate decreases. This set of effects result in high vascular permeability and low lymphatic drainage, named Enhanced

Permeability and Retention (EPR) [11]. Macromolecular and hydrophobic drugs can take advantage of EPR. This therapeutic method is called Passive Targeting [4]. Nanoemulsions with sizes between 20 and 100 nm can be encapsulated and accumulated in tumor tissues, being small enough to pass through blood vessels but big enough to avoid fast renal clearance. Nevertheless, with this range of sizes, the probability of opsonization by the Mononuclear Phagocytic System (MPS) increases [12]. Coating the nanoemulsions with hydrophilic polymers can avoid this problem [13]. Positively charged particles are more likely to be retained through longer periods of time by cancer cells, due to a negatively charged molecule in the tumor cell surface, phosphatidyl-serine [14]. Passive targeting is, however, unable to differentiate healthy tissues from cancerous ones [15]. Active targeting is known as the process through which ligands are associated with the surface of nanoemulsions becoming able to recognize a certain molecule on the tumor tissue. It also takes advantage of the environment surrounding the tumor. What makes it more efficient than passive targeting is the fact that it also generates a new strategy to deliver the drug specifically to cancer cells and, within those, to specific types of cancer cells [16]. The established bond can be of distinct types such as ligand–receptor and antigen–antibody [17]. Active targeting moieties connect to over-expressed receptors in cancer cells like folate [18], transferrin [19], epidermal growth factor (EFGR) [20], or prostate-specific membrane antigen (PSMA) [21]. The targeted delivery causes a specific toxicity in tumor cells and diminished side effects, and is also capable of resorting to surface changes in order to enhance sensitivity to stimuli [22].

The chief mechanisms of multidrug resistance (MDR) are a consequence of the overexpression of multidrug transporters and modifications in the course of apoptosis [23]. Transporter-dependent MDR originates from an overexpression of drug-efflux pumps of the ATP-binding cassette (ABC) family that exports drugs from the cell, removing several anti-cancer drugs. P-glycoprotein (P-gp) encoded by the ABC1 gene was the first ABC transporter identified. It can pump vinblastine, colchicine, etoposide, and paclitaxel (PCX) from the cell [24,25]. Moreover, MDR related with the apoptotic pathway is responsible for enhancing expression of anti-apoptotic genes, such as Bcl-2 and nuclear factor kappa B (NF-kB) [26]. Among other strategies, MDR can be reduced with P-gp inhibitors, with diminishment of Bcl-2 and NF-kB expression, and with nanocarriers (passive or active targeting) [2]. Moreover, more ABC transporters have been described for the resistance of several drugs. In humans, it was estimated that there are 49 ABCs which are ubiquitously distributed in the central nervous system, lung, liver, pancreas, stomach, intestine, and kidney and several anatomical cellular barriers [27].

Small interfering RNA (siRNA) molecules were developed as auxiliary chemotherapy by reducing MDR proteins expression or downregulating anti-apoptotic genes. The problem is that there are few suitable vectors to co-deliver siRNA and drugs. Nanoemulsions might be the solution to this issue, since their combined use with P-gp modulators or Bcl-2 inhibitors can surpass MDR [28]. Ceramides are a family of apoptotic molecules produced in environmental stress situations and they play the role of programmed cell death messenger. There are MDR cells able to avoid apoptosis by over expressing glucosylceramide, responsible for transforming ceramide into its inactive glycosylated form. Ceramide can be delivered by a nanoemulsion with a targeted purpose, increasing apoptotic effects in the tumor tissue [2].

Nanoemulsions can be conjugated with antibodies (Abs) or their fragments for targeting purposes, which is most valuable since antigen–Ab binding is specific and selective. Several studies indicate that this conjugation leads to internalization by cancer cells and successful delivery of drug loaded nanoemulsions. The nanocarrier–Ab complex can be stimuli responsive to make it even more specific to cancer tissues [29]. A process called SELEX provides a library of ssDNA and ssRNA that can be selected to form DNA or RNA oligonucleotides, resulting in aptamers [30]. In comparison to Abs, aptamers have smaller size, no immunogenicity, easy production, and fast penetration. They efficiently bind to the compound of interest and fold into secondary and tertiary DNA/RNA structures. SELEX also allows the selection of aptamers selective for tumor cells based on receptor and biomarker recognition [31]. Folic acid and folate receptors have a large affinity for one another, making folic acid an ideal targeting moiety [32]. Cancers in the brain, lung, pancreas, breast, ovary, cervix, endometrium, prostate, and colon have high folate receptor expression [33], while in normal tissues it is only located on polarized epithelia in the apical surface [34]. Studies indicate that, as cancer progresses, so does the number of folate receptors [35]. Folic acid is not expensive, not toxic, not immunogenic, easy to pair with nanocarriers, and possesses a high binding affinity, being stable both in circulation and storage [36]. Nanoemulsions can also be conjugated with oligonucleotides. However, oligonucleotides are unstable, have a very short half-life in biological fluids, and weak intracellular penetration, making them a rare choice for conjugation with nanoemulsions. Changing phosphodiester to phosphorothionate can raise defenses against enzymatic degradation [37]. Conjugation with polyethylene glycol (PEG) [38], cationic liposomes [39], micelle polyelectrolyte complex (PEC) [40], lipidic and plasmidic DNA complexes [41], and pH-sensitive nanocarriers [42] may help with the remaining problems, enhancing the success of cancer therapy.

The monitoring procedures in cancer treatment with none or very little invasion and tissue damage are possible due to the improvement of imaging techniques. These techniques are often based on the conjugation of nanoemulsions with fluorophores and, seldom, radioisotopes since they can be toxic to humans [2].

2. Nanoemulsions—A Brief Overview

2.1. Composition of the Nanoemulsions

Nanoemulsions are colloidal dispersions consisting of oil, surfactant, and an aqueous phase. The nanoemulsion core will have an impact on the therapeutic payload of the drug, physico–chemical properties, particle size, and stability [43,44]. The formulation can be formed by long chain triglycerides (LCT) which create larger sized particles, with a diameter of 120 nm, or short chain triglycerides (SCT) leading to smaller particles, around 40 nm. Regarding LCT, soybean oil is very often used due to its high content of essential C18 fatty acids like linoleic acid. Medium chain triglycerides, usually from coconut oil, could be also used alone or in a mixture with LCT to overcome their possible immunosuppressive actions and inhibition of lymphocytes [44,45].

The size of the particles will impact the final aspect of the formulation. In general, the smaller the particle, the higher the stability of the formulation. This particular issue is useful against flocculation, gravitational force, and Brownian motion. However, SCT oils are very soluble in water, facilitating Ostwald's ripening [46].

The ideal emulsifying agent should reduce interfacial tension, be rapidly adsorbed at the interface, and stabilize the surface by electrostatic or stearic interactions. An emulsifier is an amphiphilic molecule such as surfactants (Tween® 80), amphiphilic proteins (caseinate), phospholipids (soy lecithin), polysaccharides (modified starch), or polymers (PEG) [46]. PEG-modified nanoemulsions are used to enable specific targeting and longer circulation time [18]. Texture modifiers, weighting agents or ripening retarders can also be used [47,48]. Sorbitan fatty acid esters such as Spans® could also be used as non-ionic surfactants [49].

Nanoemulsions responsive to external stimuli can also be developed using temperature and pH-sensible materials aiming the induction of a conformational change in the formulation which furthers the release of the payload [50,51].

2.2. Physical and Chemical Characterization of Nanoemulsions

Several physical and chemical properties can influence the behavior of nanoemulsions. Average size is one of the crucial parameters regarding this system [52]. Also, size distribution measured as polydispersity index (PDI), is of extreme relevance since the range of nanoemulsion size should be known to evaluate further biological responses [52]. Both parameters could be measured by dynamic light scattering (DLS) which makes use of the Brownian motion of colloidal particles—since they scatter the light to find the diffusion coefficient of the particle [1].

Moreover, modifications of the cellular response due to surface charge is also a known phenomena [52]. Therefore, surface charge of the nanoemulsions should be characterized. The charge of the system's surface influences stability and electrostatic interactions and its measurement demands the presence of a magnetic field. The particles' electrophoretic movement follows Henry's equation (Equation (1)).

$$\mu = \frac{2\varepsilon\zeta f\,(ka)}{3\eta} \tag{1}$$

where μ is the electrophoretic mobility, ε the dielectric constant, ζ the zeta potential (ZP), f(ka) the Henry's function and η the viscosity.

The morphology of the oil droplet is essential in the future definition of the stability of the formulation. The lipophilic–hydrophilic behavior of the nanoemulsion also has significant impact on drug loading, directly influencing the success of drug encapsulation [1]. Optical microscopy, even using differential interference contrast or other phase contrast methods, is generally not a viable method for examining nanoemulsions [33]. However, microscopic techniques are essential in order to obtain reliable data about the actual morphology of the system. In this sense, transmission electron microscopy (TEM) and scanning electron microscopy (SEM) had proven to be useful in order to observe the structure of the nanoemulsions [53]. In addition, SEM gives a three-dimensional image of the droplets [33].

The amount of drug loaded into nanoemulsions constitutes a critical parameter of the formulations aimed to deliver the drug to a target tumoral tissue. In this sense, different approaches in order to carry out the measurement have been performed. It could be measured by separating the free drug from the encapsulated drug and, in this sense, either the free drug could be measured (indirect entrapment efficiency) or the drug encapsulated into the nanoemulsions could be measured by previously dissolving the nanoemulsions core using suitable organic solvents [54]. The separation of the non encapsulated drug from the nanoemulsion could be carried out using a filtration procedure, filtration–centrifugation device, an ultracentrifuge, or dialysis membrane [55,56]. The amount of loaded drug can be measured as the percentage (entrapment efficiency, EE) or as the concentration of drug per nanoemulsion droplet.

2.3. Stability Studies

Stability studies are mandatory to characterize the nanoemulsions. An accelerated stability study can be performed by centrifugation of nanoemulsions in such way that the creaming process is accelerated. The traditional approach used to measure the stability of the formulations by is their storage at 4.0 and 25.0 °C for a period of three to six months. Sample characteristics such as average size, polydispersity index, surface charge, and efficiency of encapsulation are usually analyzed once a month. No significant modifications of the nanoemulsion parameters should be measured in order to confirm that the nanometric emulsion is stable under the storing conditions. Moreover, the Food and Drug Administration (FDA) states that stability should be evaluated over long term, intermediate, and accelerated times. The peroxide value (PV), anisidine value (AV) and Total Oxidation Value (TOTOX) are used to assess degradation products and, therefore, the stability of each formulation. The pH value can be modified by oil oxidation processes quantified by the parameters mentioned [1].

2.4. Nanoemulsion Drug Release

Drug release process is responsible for drugs bioavailability, absorption, and kinetics. Its evaluation is usually carried out using either Franz diffusion cells or using the dialysis bag method. The latter consists of placing the sample in a dialysis bag and the sample is dialyzed using a buffer in the sink receiver compartment, under stirring conditions at 37 °C [57]. In Franz diffusion cells the nanoemulsion is placed on the donor compartment and it is separated from a receptor chamber (usually filled with phosphate buffer) by a dialysis membrane. The amount of drug that is released from the nanoemulsions to the receptor chamber is analyzed at different time points, obtaining the drug release profile [34].

2.5. Nanoemulsion Production

The two phases of nanoemulsion production are the heated and mixed phases (with controlled temperature and agitation) so that the dispersion becomes as homogeneous as possible. Then, the emulsion goes through a process of shear force homogenization to achieve minimal particle size. In the end, particles will have a layer of emulsifiers separating the lipophilic interior from the aqueous phase. This specific layer is a barrier and displays repulsive forces (electrostatic, steric, or electrosteric, depending on the emulsifier) to stabilize the formulation [1].

High shear methods use high pressure homogenizers, microfluidizers, and ultrasonicators [58]. The size of the particle is associated with the instruments and variables like energy, time, temperature, and formulation composition. High energy operations have the plus side of potentially being scaled-up, but they might be inappropriate for certain heat sensitive drugs. In this case, low energy and temperature methods must be used like self-emulsification phase transition and phase inversion [59].

2.5.1. High Pressure Homogenization

This technique makes use of a high-pressure homogenizer/piston homogenizer to produce nanoemulsions (Figure 2). In the High Pressure Homogenization (HPH) method, an aqueous phase containing the emulsifier is added to an organic phase and ultraturrax is used to form an emulsion. Afterwards, the emulsion is added to the HPH in order to reduce droplet size [60]. Several homogenization cycles and pressures can be applied to obtain the desired nanoemulsion parameters. Therefore, this technique is able to produce small sized particles. Over the course of HPH, a variety of forces contribute to get to a small size of particle: hydraulic, turbulence, and cavitation [1].

The main advantage of this method is that it can be applied several times in order to obtain a suitable droplet size [32,33,59].

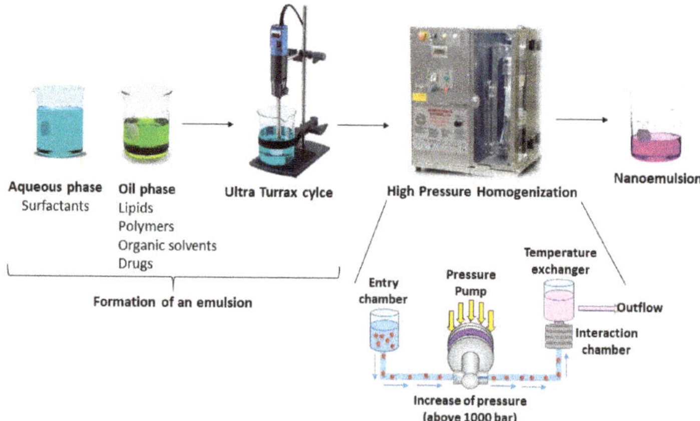

Figure 2. High pressure homogenization technique.

2.5.2. Microfluidization

The microfluidization process requires a microfluidizer instrument [61]. This instrument is patented and consists of a high-pressure positive displacement pump (500–20,000 psi) that makes the product go through the interaction chamber, consisting of small channels (Figure 3). The product flows through the micro-channels on to an impingement area resulting in very fine particles of a submicron range. The two solutions (aqueous phase and oily phase) are combined together [33]. The coarse emulsion is introduced into a microfluidizer where it is further processed to obtain a stable nanoemulsion. The coarse emulsion is passed through the interaction chamber of the microfluidizer repeatedly until the desired particle size is obtained. The bulk emulsion is then filtered through a filter under nitrogen to remove large droplets resulting in a uniform nanoemulsion. This technique can be used in order to produce nanoemulsions at the industrial scale [33].

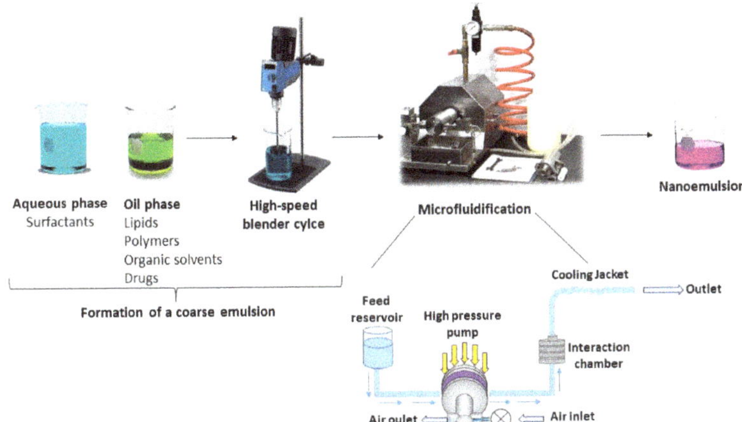

Figure 3. Microfluidification technique.

2.5.3. Phase Inversion Temperature Technique

The phase inversion temperature (PIT) method has highlighted a relationship between minimum droplet size and complete solubilization of the oil in a microemulsion bicontinuous phase independently of whether the initial phase equilibrium is single or multiphase.

Due to their small droplet size, nanoemulsions possess stability against sedimentation or creaming with Ostwald ripening forming the main mechanism of nanoemulsion breakdown [7]. Phase inversion in emulsions can be one of two types: transitional inversion induced by changing factors which affect the hydrophile-lipophile balance (HLB) of the system, e.g., temperature and/or electrolyte concentration, and catastrophic inversion, which can also be induced by changing the HLB number of the surfactant at a constant temperature using surfactant mixtures.

The PIT method employs temperature dependent solubility of nonionic surfactants, such as polyethoxylated surfactants, to modify their affinities for water and oil as a function of the temperature (Figure 4) [62]. It has been observed that polyethoxylated surfactants tend to become lipophilic on heating owing to the dehydration of polyoxyethylene groups. This phenomenon forms the basis of nanoemulsion fabrication using the PIT method. In the PIT method, oil, water, and nonionic surfactants are mixed together at room temperature. This mixture typically comprises o/w microemulsions coexisting with excess oil, and the surfactant monolayer exhibits a positive curvature. When this macroemulsion is heated gradually, the polyethoxylated surfactant becomes lipophilic and at higher temperatures, the surfactant gets completely solubilized in the oily phase and the initial o/w emulsion undergoes phase inversion to w/o emulsion. The surfactant monolayer has a negative curvature at this stage. This method involves heating of the components and it may be difficult to incorporate thermolabile drugs, such as tretinoin and peptides, without affecting their stability. It may be possible to reduce the PIT of the dispersion using a mixture of components (surfactants) with suitable characteristics, in order to minimize degradation of thermolabile drugs [33].

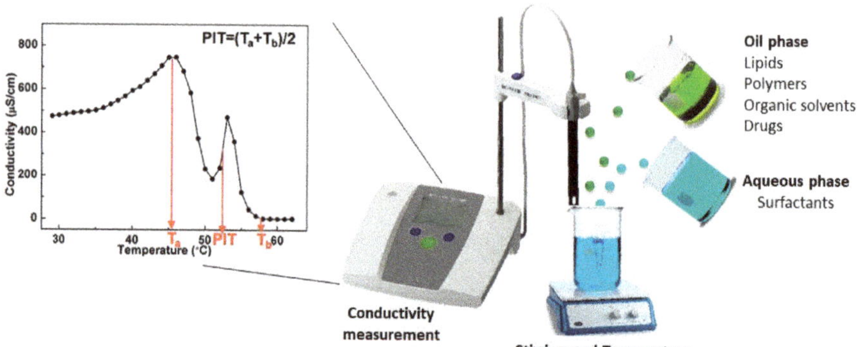

Figure 4. Phase inversion temperature technique (based on [62]).

2.5.4. Solvent Displacement Method

The solvent displacement method for spontaneous fabrication of nanoemulsions has been adopted from the nanoprecipitation method used for polymeric nanoparticles (Figure 5A).

In this method, the oily phase is dissolved in water-miscible organic solvents, such as acetone, ethanol, and ethyl methyl ketone. The organic phase is poured into an aqueous phase containing surfactant to yield spontaneous nanoemulsions by rapid diffusion of organic solvent [63]. The organic solvent is removed from the nanoemulsions by a suitable means, such as vacuum evaporation.

Solvent displacement methods can yield nanoemulsions at room temperature and require simple stirring for the fabrication. Hence, researchers in pharmaceutical sciences are employing this technique for fabricating nanoemulsions mainly for parenteral use. However, the major drawback of this method is the use of organic solvents, such as acetone, which require additional inputs for their removal from the nanoemulsion. Furthermore, a high ratio of solvent to oil is required to obtain a nanoemulsion with a desirable droplet size. This may be a limiting factor in certain cases. In addition, the process of solvent removal may appear simple at the laboratory scale but can pose several difficulties during scale-up [33]. Therefore, reproducibility and scale-up are the major drawbacks of this method.

2.5.5. Phase Inversion Composition Method (Self-Nanoemulsification Method)

This method generates nanoemulsions at room temperature without the use of organic solvents and without increasing temperature.

Kinetically stable nanoemulsions with small droplet sizes (~50 nm) can be generated by the addition of a water phase into a solution of surfactant in oil, with gentle stirring and at a constant temperature (Figure 5B) [64]. The spontaneous nanoemulsification has been related to the phase transitions during the emulsification process and involves lamellar liquid crystalline phases or D-type bicontinuous microemulsion during the process. Nanoemulsions obtained from the spontaneous nanoemulsification process are not thermodynamically stable, although they might have high kinetic energy and long-term colloidal stability [33].

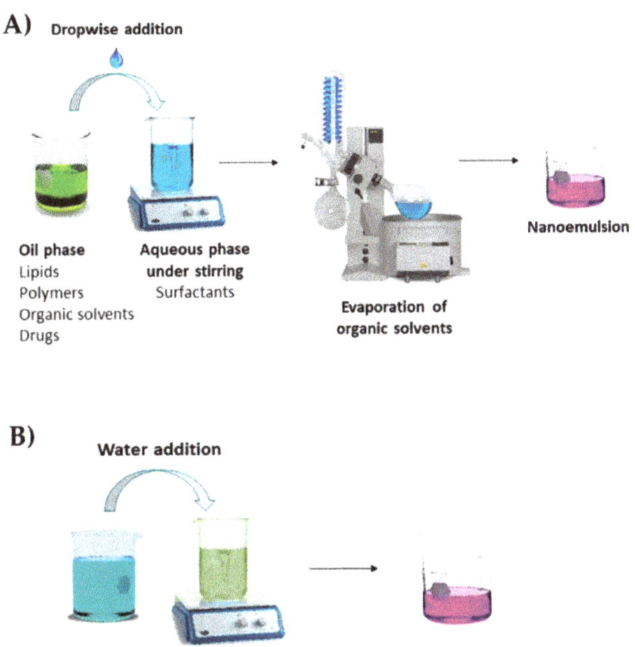

Figure 5. (**A**) The solvent displacement method, and (**B**) the phase inversion composition method.

2.6. Metabolism

Upon systemic administration, the nanoemulsions have to escape from the mononuclear phagocitiyc system (MPS) and the renal clearance pathway in order to reach the tumor tissue [65]. The MPS constitutes a biological barrier made by phagocytic cells that could capture nanoemulsions [65].

The nanoemulsion in the bloodstream interacts with erythrocytes, opsonins, monocytes, platelets, leukocytes, dendritic cells, tissue macrophages, Kupffer cells of the liver, lymph nodes, and B cells of the spleen. The formulation is very likely to interact with erythrocytes since they represent the largest fraction of blood cells, resulting in possible hemolysis and removal by macrophages [66]. The extended half-life of this type of formulation increases the chance of interaction with blood cells and, therefore, events of thrombogenicity can occur resulting in blood vessel occlusion [67]. Opsonins can be adsorbed to the surface of nanoemulsions, facilitating the uptake by macrophages, which reduces the drug delivery to the desired place. This activation of immune cells can also result in anaphylactic, allergic, and hypersensitivity reactions [66]. Size, charge, and surface properties of the nanoformulation also influence erythrocytes. Large cationic or anionic particles have a higher tendency to go through phagocytosis. In addition, cationic surfaces are more likely to damage erythrocytes and cause hemolysis [67,68]. Opsonization can be reduced by adding PEG [68], poloxamer [69], or poloxamine [70] to the surface of the nanoemulsion as they create a "steric shield" around the formulation. Cell uptake of nanoparticles is attained through phagocytosis, macropinocytosis, or endocytosis, accumulating in lysosomes, vacuoles, or the cytoplasm [1]. For the formulations aimed to treat brain tumors, the brain–blood barrier (BBB) stands as a great obstacle to drug delivery that can be surpassed with the use of nanoemulsions targeted to reach receptors expressed in the place of action [71].

The main problem in the metabolism of nanoemulsions is the hepatic clearance [72]. Those nanoparticles taken by hepatocytes are eliminated by the biliary system and the ones taken by Kupffer cells are subjected to phagocytosis, degradation, and further elimination. Renal clearance represents an eminent portion of metabolism of nanoemulsions. Glomerural filtration depends on the particle size; particles smaller than 6 nm are filtered in the kidney and the larger ones return to systemic circulation [73].

3. Nanoemulsions Applied to Cancer Therapy

3.1. Nanoemulsions as a Strategy to Overcome MDR

MDR tumors are a major barrier to effective cancer therapy and, along with metastasis, are estimated to be major contributors to death by cancer. The expression of multifunctional efflux transporters from the ABC gene family have been known to play a crucial role in MDR of tumor cells [25].

ABCs are responsible for the efflux of various endogenous ligands such as proteins, lipids, metabolic products, and drugs such as cytotoxic antibiotics by using energy produced from the hydrolysis of ATP [27].

There are several transporters expressed by the ABC family causing MDR of different antitumoral drugs. P-gp also referred as MDR-1 or ABCB-1, encoded by the ABC1 gene was the first ABC transporter identified and it can pump vinblastine, colchicine, etoposide, and paclitaxel [25,27]. In addition, there are other relevant MDR transporters such as ABCA2 encoded by the *ABCA2* gene, which is for estramustine resistance. MRP1 encoded by *ABCC1* is responsible for doxorubicin, vincristine, etoposide, colchicine, campothethin, and methotrexate resistance and it is one of the most widely expressed transporters in tumoral cells. MRP2 is another transporter of this family and it is responsible for vinblastine, cisplatin, doxorubicin, and methotrexate resistance. It is located on the cell membrane of polarized cells such as kidney, liver, and intestinal epithelium [25,27]. MRP3 encoded by *ABCC3* is responsible for the transportation of organic anions and pumps antitumoral drugs such as methotrexate and etoposide [25,27]. MRP4 encoded by *ABCC4* pumps methotrexate, 6-mercaptopurine, 6-TG 6-thioguanine. In some cases one drug can be the substrate of more than one ABC transporter such as 6-mercaptopurine and 6-TG 6-thioguanine which are also pumped by MRP5 and etoposide which is pumped by MRP6 in addition to MRP1 and MRP3. MRP8 is encoded by *ARCC11* genes and pumps 5-fluorouracil whereas MXR/BCRP (multixenobiotic resistance and breast cancer resistance proteins, respectively) encoded by *ABCG2* genes causes mitoxantrone, topotecan, doxorubicin, daunorubicin, CPT-11, imatinib, and methotrexate resistance [25].

As these data highlight, in oncology, the search for new compounds for the inhibition of these hyperactive ABC pumps is a growing interest in order to increase chemotherapeutic effects. In this sense, several ABC pump inhibitor/modulators functionalizing nanoemulsions have been explored to address the cancer associated MDR [27].

Ganta and colleagues developed nanoemulsions functionalized with folate in order to efficiently deliver docetaxel to ovarian cancer cells overcoming docetaxel MDR [74]. The nanoemulsions were targeted with folate because the folate receptor is poorly expressed the majority of normal tissues. However, it is overexpressed in many cancers especially in epithelial ovarian cancer. In addition, the author's evaluated P-gp expression and demonstrate that folate receptor mediated endocytosis is capable of bypassing MDR present in ovarian cancer cell lines. Therefore, the nanoemulsions targeted with folate were able to successfully deliver docetaxel by receptor mediated endocytosis that showed enhanced cytotoxicity capable of overcoming ABC transporter mediated taxane resistance [74]. In recent years the taxanes, such as paclitaxel and docetaxel, have emerged as fundamental drugs in the treatment of breast cancer although overcoming MDR is a major issue [75]. In order to overcome this obstacle, Meng and colleagues used baicalein in order to inhibit P-gp and also to increase oxidative stress [76]. Increasing oxidative stress is claimed as a suitable strategy to improve cell sensitivity to paclitaxel due to the fact that cellular reactive oxygen species (ROS) and gluthatione are extremely important for cellular redox reactions. Using this strategy, the author's co-encapsulate paclitaxel and baicalein in nanoemulsions in order to treat breast cancer. The developed nanoemulsions were able to increase ROS, decrease cellular GSH and enhance caspase-3 activity in MCF-7/Tax cells. More importantly, an in vivo antitumor study demonstrated that baicalein–paclitaxel nanoemulsions exhibited a much higher antitumor efficacy than other paclitaxel formulations. These findings suggest that co-delivery of paclitaxel and baicalein in nanoemulsions might be a potential combined therapeutic strategy for overcoming MDR [76]. A different strategy to overcome MDR of paclitaxel was used by Zheng and colleagues [77]. They aimed to prepare nanoemulsions able to alter the levels of Bax and Bcl-2 expression and also inhibit the P-gp transport function [77]. In order to achieve this goal, they used a vitamin E derivative. Vitamin E is an antioxidant and its mechanisms consist of reducing peroxyl radicals and eliminating the chain reaction of fatty acid radical propagation [77]. It has also been demonstrated that the vitamin E derivative used in this study, TPGS, is one of the most potent and commercially available surfactants that serves as a P-gp inhibitor, and it can reverse MDR in cancer [78]. Vitamin E can disrupt Bcl–xL–Bax interactions, activates Bax, and thus mediates mitochondrial-centered apoptotic cell death. Therefore, vitamin E based nanoemulsions containing paclitaxel were suitable for study of paclitaxel-resistant human ovarian carcinoma cell lines [77].

3.2. Nanoemulsions for Different Types of Cancer

3.2.1. Nanoemulsions for Cancer Treatment

Nanotechnology has been shown to be a suitable strategy for cancer treatment and, therefore, researchers focused their efforts on the treatment of several types of cancer. The following section summarizes the most recent advances regarding the most common forms of cancer which can also be observed in Table 1.

3.2.2. Nanoemulsion for Colon Cancer Therapy

Colon cancer represents a large portion of cancer related deaths in the world [79]. Within this category, the subclassification includes familial adenomatous polyposis, hereditary nonpolyposis colorectal cancer, sporadic colon cancer, and colitis-associated cancer [80]. Operation combined with herbs, immunotherapy, radiotherapy, and/or chemotherapy tend to be the choices in colon cancer treatment. Still, survival rate decreases about five years after surgery due to metastasis and recurrence, which means the main cause of death is not the tumor itself [81]. Cancer invasion and migration is possible due to epithelial mesenchymal transition (EMT), a mechanism where epithelial cells transform into mesenchymal cells, changing cell structure and increasing adhesion and migration [82].

Lycopene (LP), present in tomatoes, has several functional features—protection against chronic diseases, anti-proliferation activity against leukemia, and colon cancer cells and further triggering of cell cycle arrest on some tumor cells [83]—and its mechanism could be of use in cancer therapy, if it was not for its low stability and bioavailability [84]. The described study focused on developing a nanoemulsion with LP, in order to find a solution for the presented problem. The nanoemulsion formulation also encapsulates gold nanoparticles (AN). AN act only as a drug carrier but is also available to be incorporated with cell receptor ligands contributed for specific cell targeting [85]. However, AN can become toxic in high doses by promotion of human fibroblast cell migration [86]. This effect can be reduced by incorporating liposomes, polymeric substances, or other lipid-based assemblies, such as LP-derived compounds [87].

In this specific formulation, the oil phase is oil with LP, the water phase is an aqueous AN solution, and the emulsifier is Tween 80®. It can be used in a human colon cancer line, HT-29. The evaluation of AN alone, LP alone, and the combination of AN and LP effects on the cell line proves the formulation efficiency [79].

The size of AN impacts toxicity; the lower the size, the stronger the effect on HT-29 cells. The enhancement of the volume of LP incorporated can increase the presence of early apoptotic cells. Both the combined treatment and nanoemulsion increase the levels of early apoptotic, late apoptotic, and necrotic cells. Nevertheless, treatment with nanoemulsions induces apoptotic and necrotic cells. The emulsifier does not impact cells in a significant way, only contributing to nanoemulsion stability. Treatment with nanoemulsions with low AN and LP doses results in low expression of procaspases 3 and 8, and Bcl-2 (tumoral markers), while enhancing Bax and PARP-1 expression, with apoptotic effects on cells [79].

3.2.3. Nanoemulsions for Ovarian Cancer Therapy

Platinum (Pt) chemotherapeutics are used in a vast array of cancer treatments. Carboplatin and cisplatin—molecules containing Pt in their composition—increase survival better than any other ovarian cancer treatment [88]. Pt compounds form intra-strand and inter-strand cross links, shattering DNA structure [89]. The problem with Pt action is related to its effects on cells other than cancer cells, as it also ends up killing healthy cells. Furthermore, cancer cells can develop resistance mechanisms to Pt, like promoting membrane pump presence, or inducing enzymes or DNA repair pathways. Therefore, therapeutics in ovarian cancer always consider whether the tumor is Pt-sensitive or Pt-resistant [90].

Nanomedicinal evolution allowed the design of formulations to overcome toxicity and resistance problems. However, the process is not easy due to Pt properties, particularly due to its lipophilicity [91]. Nanoemulsions can improve delivery and efficiency of Pt-related drugs, since they are able to incorporate enormous amounts of hydrophobic drugs and add specific ligands to their surface in order to achieve a targeted delivery [92]. A nanoemulsion encapsulating myrisplastin (novel platinum Pt-based drug) and C6-ceramide (pro-apoptotic substance) with a surface ligand EGFR-binding peptide and gadolinium (imaging agent) was developed in this study to understand its effect in the following ovarian cancer cells: SKOV3, A2780, and A2780CP [90].

A cytotoxicity screening revealed that SKOV3 cells, expressing epidermal growth factor receptor (EGFR), were resistant to cisplastin presenting an inhibitory concentration, IC_{50}, of 18 µM. By encapsulating myrisplatin instead, cytotoxicity increased in a very significant way, both in targeted and non-targeted nanoemulsions. The targeted nanoemulsions possess 2-fold more toxicity in comparison to non-targeted formulations. The biggest change in cytotoxicity occurs when ceramide is also encapsulated, with the combination confirming its synergistic behavior. The targeted nanoemulsion containing ceramide and myrisplatin is 50.5-fold more effective than cisplastin. A2780 and A2780CP (not expressing EGFR) presented more toxic effects with myrisplatin than with cisplatin [90].

Zeng and colleagues developed vitamin E nanoemulsions containing paclitaxel able to modulate the levels of Bac and BCL-2 expression (related to tumor drug resistance) and inhibit the P-gp transport. They assessed their effect in paclitaxel-resistant human ovarian carcinoma cell line A2780. Taxol enhanced the antiproliferation effect and decreased the mitochondrial potential. The authors claim that the association of anticancer drugs with vitamin E derivative multifunctional nanoemulsions could be a suitable solution for cancer multidrug resistance [93].

3.2.4. Nanoemulsion for Prostate Cancer Therapy

The number of deaths related to prostate cancer (PrC) has grown in the last decade with 70% of treated patients facing recurrence and transition to an untreatable state [94]. Cancer stem cells (CSCs) or tumor initiating cells (TICs) are the root of cancer development, metastasis, and resistance to therapies [77]. Studies show that cancer cells expressing CSC markers, specially CD133 and CD4, are not only associated with drug resistance, but also proliferate after therapy [95]. Drug resistance in CSCs might be due to up-regulation of drug efflux transporters, activation of anti-apoptotic pathways, inactivation of apoptotic mechanisms, and more efficient response to DNA damage and repair processes [96].

The problem with prostate cancer therapy is that it targets populations of fast-growing cancer cells but not subpopulations like CSCs. Also, anti-prostate cancer drug development resorts to cell lines with high passage numbers for preclinical studies to evaluate anti-cancer agents. These cell lines end up acquiring genomic and epigenomic properties with low or no match with the original tumor [97]. The research team responsible for this study uses a cell line (PPT2), derived from a prostate cancer patient, with a very low passage number and, by this way, immaturity and stem-like properties are kept. PPT2 cells have genes associated with anti-apoptotic signaling and resistance to drugs, being a perfect model for CSC-targeted therapy studies [98].

One drug that is often used in prostate cancer treatment is Abraxane®. Abraxane® is a paclitaxel pro-drug, developed to increase its solubility with human serum albumin-bound nanoparticle formulation. However, paclitaxel has shown problems with MDR cancer cells [96]. A new generation taxoid, SBT-1214, is efficient against drug-resistance. This agent can be conjugated with docosahexaenoic acid (DHA), a natural polyunsaturated fatty acid (PUFA) with high affinity for its main bloodstream transporter (human serum albumin) that helps direct toxicity to cancer. Combination of DHA with paclitaxel resulted in a weak decrease in P-gp and ABC transporters [99]. The DHA–SBT-1214 nanoemulsion formulation developed in this study includes phospholipids and fish oil. The affinity of the drug to fish oil will improve drug encapsulation. It is theorized that the nanoemulsion will act on the CSC-initiated PPT2 cell line, taking advantage of EPR effect and resulting in cancer cell apoptosis [96].

The use of patient-derived CSC enriched PPT2 cells can help in the development of drugs that target cells specific for tumor initiation. Combining DHA with SBT-1214 allows the formulation more time in blood circulation. Encapsulating the conjugated hydrophobic drug in a nanoemulsion formulation results in effective delivery. Thus, surface modification with PEG also enhances the time of drug circulation, which increases accumulation due to the EPR effect. The successful cellular uptake means that the nanoemulsion formulation can deliver its payload more efficiently than the drug solution [96].

3.2.5. Nanoemulsions for Leukemia

Cancer is among the leading causes of death worldwide and leukemia is the most the leading cause of cancer-related death in children. In this context, nanoemulsions have been used as biocompatible systems in order to encapsulate drugs and increase the therapeutic effects by decreasing toxic adverse effects.

Lipid nanoemulsions have been used by several authors as a suitable strategy for drug encapsulation for cancer treatment. Moura and colleagues developed lipid nanoemulsions, able to bind to LDL receptors, with the aim of concentrating the chemotherapeutic agents in tissues with low-density lipoprotein receptor overexpression such as tumoral tissues. The authors encapsulate methothraxete for leukemia treatment and assess them in vitro, as the uptake of the nanoemulsions is significantly higher than the free drug increasing the toxicity against tumoral cells [100].

Winter and colleagues develop nanoemulsions encapsulating chalcones for leukemia and assess them in vitro and in vivo. The authors demonstrate that the developed nanoemulsions cause apoptosis of the cancer cells in vitro showing similar anti-leukemic effects both for the free chalcones and nanoemulsion. However, free chalcones induced higher toxicity in VERO cells than chalcones-loaded nanoemulsions. Similar results were observed in vivo. Free chalcones induced a reduction in weight gain and liver injuries, evidenced by oxidative stress, as well as an inflammatory response [101].

3.2.6. Nanoemulsions for Breast Cancer

Breast cancer accounts for 23% of all newly occurring cancers in women worldwide and represents 13.7% of all cancer deaths. Available chemotherapeutic agents are limited mainly due to the low accumulation of chemotherapeutics at the tumors relative to their accumulation at other organs thus leading to increased toxicities. Several strategies have been developed in order to improve the treatment of this in patients.

Nanoemulsions based on natural compounds could constitute a suitable strategy for breast cancer such as the nanoemulsion developed by Periasamy and colleagues using the essential oil of *Nigella sativa* L [102]. This nanoemulsion shows anti-cancer properties in vitro in MCF-7 breast cancer cells by inducing their apoptosis. This nanoemulsion could be useful for the entrapment of active drugs in order to treat breast cancer [102].

Local administration in addition to C6 ceramide nanoemulsion development has also been used as strategy for breast cancer treatment. The authors target cancers and pre-tumor lesions locally by reducing systemic adverse effects using both nanoemulsion drug delivery and local administration. They developed bioadhesive ceramide loaded nanoemulsions and modified their surface with chitosan. The C6 ceramide concentration necessary to reduce MCF-7 cell viability to 50% (EC50) decreased by 4.5-fold with its nanoencapsulation compared to it in solution; a further decrease (2.6-fold) was observed when tributyrin (a pro-drug of butyric acid) was part of the oil phase of the nanoemulsion. Intraductal administration of the nanoemulsion prolonged drug localization for more than 120 h in the mammary tissue compared to its solution [103]. Natesan and colleagues also used chitosan in order to develop nanoemulsions [104]. They encapsulate camptothecin in nanoemulsions and assess them in vitro and in vivo showing the efficacy of the formulations compared to the free drug [104].

3.2.7. Nanoemulsions for Melanoma

Melanoma is the most serious form of skin cancer causing more than 80% of skin cancer-related deaths. The main problem associated with the treatment of melanoma is low response rate to the existing treatment modalities, which in turn is due to the incomplete response by chemotherapeutic agents and inherent resistance of melanoma cells. Standard treatments for late-stage melanoma and metastatic melanoma usually present poor results, leading to life-threatening side effects and low overall survival. For this reason, newer combinations of anti-melanoma drugs and newer strategies utilizing nanotechnology are being studied, such as the use of nanoemulsions.

Kretxer and colleagues developed lipid nanoemulsions encapsulating paclitaxel that are able to bind to bind to low-density lipoprotein (LDL) receptors, decreasing drug associated toxicity and increasing antitumoral action. Moreover, simvastatin association was also assessed in melanoma bearing mice, demonstrating that this drug associated with paclitaxel nanoemulsions increased antitumoral activity, but not with free paclitaxel. This might due to the fact that statins increase LDL receptor expression and these receptors are responsible for the lipid nanoemulsion internalization [105]. Other authors encapsulated cholesterol derivatives, such as 7-ketocholesterol, into lipid core nanoemulsions and assess them in vivo in a murine melanoma cell line where it was demonstrated that the nanoemulsion decreases the tumor size more than 50%, enlarged the necrotic area, and reduced intratumoral vasculature. The in vitro uptake into tumor cells was LDL-receptor-mediated cell internalization and demonstrated that a single dose of the cholesterol nanoemulsions killed 10% of melanoma cells [106].

A different approach was used by Monge-Fuentes and colleagues, such as the photodynamic therapy using acai oil in nanoemulsion. They used this nanoemulsion as a photosensitizer in vitro and in vivo. NIH/3T3 normal cells and B16F10 melanoma cell lines were treated and presented 85% cell death for melanoma cells, while maintaining high viability in normal cells. Tumor bearing C57BL/6 mice treated with acai oil nanoemulsion showed tumor volume reduction of 82% [107].

3.2.8. Nanoemulsion for Lung Cancer Therapy

Paclitaxel (PTX) is an anticancer drug often used to treat lung, breast, pancreatic, and ovarian cancer. It has the ability to interfere with the breakdown of microtubules during cellular division, leading to apoptosis, mitotic arrest, and inhibition of cell functions [108]. PTX has very low water solubility, which is why many formulations, like Taxol® containing Cremophor-EL® and ethanol, have been developed. However, Cremophor-EL® is known for its toxicity, demanding the investigation of targeting molecules for this formulation [109].

Hyaluronic acid (HA) has been investigated for its use in the active delivery of PTX to cancer cells. It is negatively charged, binding specifically to the cluster of differentiation 44 (CD44), a highly expressed tumor cell marker [110]. The development of a nanoemulsion carrier for PTX and HA—HA-complexed PTX nanoemulsion (HPNs)—has the goal of testing the efficiency of the formulation against tumors expressing CD44 in a non-small lung carcinoma cell line (NCI-H460) [111].

HPN demonstrated great physical–chemical particle properties, with a size that allows a long half-life and zeta potential that is suited to stabilize the formulation, a low polydispersity index confirming homogeneity of the formulation, and the desired spherical morphology. Evaluation of tumor weight showed that both PTX nanoemulsions and HPN reduced tumor growth, however the targeted moiety of HPN, HA, enhances the therapeutic efficiency. Results on body weight reveal no significant changes in groups treated with PTX nanoemulsions and HPN, confirming these therapies are less toxic for healthy tissues [111].

Chang and colleagues study the anticancer activity of the curcuminoid extracts of *Curcuma longa* Linnaeus. They prepared nanoemulsions and explored the inhibition mechanism implicated for the anticancer activity against lung cancer cells. The cell cycle of lung cancer cells was retarded at G2/M for both the curcuminoid extract and nanoemulsion treatments, though the cellular pathway may differ. Among different cancer cell lines, H460 cells show an increased susceptibility to apoptosis compared to A549 cells for both curcuminoid extract and nanoemulsion treatments [112].

3.3. Nanoemulsions for Nanotheragnostics

Nanotheragnostics in a new strategy consisting of using the power of nanotechnology for imaging and diagnosis purposes. The aim of this strategy is to produce nanoscale agents affording both therapeutic and diagnostic functions [113].

This strategy is associated with different nanotechnological approaches such as nanoemulsions and researchers have focused with special emphasis on cancer treatment/diagnosis [113]. The recent advancements in this field have enabled the characterization of individual tumors, prediction of nanoemulsions–tumor interactions, and the creation of nanomedicines for individualized treatment [113].

In this sense, Fernandes and colleagues developed perfluorohexane nanoemulsions as novel drug-delivery vehicles and contrast agents for ultrasound and photoacoustic imaging of cancer in vivo, offering higher spatial resolution and deeper penetration of tissue when compared to conventional optical techniques. This NE provides a non-invasive cancer imaging and therapy alternative for patients [56].

Wu and colleagues used a similar strategy, developing magnetic nanoemulsions hydrogels inducing magnetic tumor regression based on a ferrofluid-based magnetic hyperthermia of cancers [114]. Moreover, Niravkumar et al. developed nanoemulsions encapsulating three difatty acid platins, dimyrisplatin, dipalmiplatin, and distearyplatin. They developed fatty acid nanoemulsions that selectively bind the folate receptor α (FR-α) and utilize receptor mediated endocytosis to deliver Pt past cell surface resistance mechanisms (FR-α is overexpressed in a number of oncological conditions including ovarian cancer) [115]. Roberts and colleagues used sonophore molecules for multi-spectral optoacoustic tomography (MSOT) of tumors. They combined near-infrared and highly absorbing dyes loaded into nanoemulsions, enabling the non-invasive in vivo MSOT detection of tumors.

Table 1. Summary of some recent nanoemulsions developed and applications for cancer.

Nanoemulsion Constituents	Active Compound	Production Technique and Physicochemical Parameters	Type of Cancer	Therapeutic Efficacy and Other Observations	Ref.
Nanoemulsions carrying gold nanoparticles Tween 80®	Lychopene	Production using ultrasonication method; Average size: 25.0 nm; Zeta potential: −32.2 mV	Colon cancer	Nanoemulsions reduced the expressions of procaspases 8, 3, and 9 and PARP-1 and Bcl-2; Nanoemulsions enhanced Bax expression; Nanoemulsions increase HT-29 cell apoptosis and reduced their migration capability; Upregulation of epithelial marker E-cadherin and downregulation of Akt, nuclear factor kappa B, pro-matrix metalloproteinase (MMP)-2, and active MMP-9 expressions.	[79]
EGFR-targeted nanoemulsion; EGFR binding peptide; Lipidated gadolinium (Gd) chelate; Lipidated EGFRbp; Egg lecithin; PEG2000DSPE; Glycerol	Myrisplatin: novel platinum pro-drug; C6-ceramide: pro-apoptotic agent	High shear microfluidization process; Average size: <150 nm; Stable in plasma for 24 h	Ovarian cancer	Efficacy was 50-fold drop in the IC_{50} in SKOV3 cells as compared to cisplatin alone; Improved efficacy over cisplatin nanoemulsions.	[90,116]
Vitamin E nanoemulsions: composed of α-TOS, and vitamin E Brij 78 and TPGS	Paclitaxel	Preparation using emulsification–evaporation method; Average size: 236.7 nm; Polydispersity index: 0.29; Zeta potential: −23.9 mV; Drug loading: ≈1.04%	Multidrug resistance cancers	30% of paclitaxel is release in vitro for the first 24 h; Nanoemulsions increase Bax cell levels of and decrease Bcl-2 expression and inhibit the transport function of P-gp and decrease mitochondrial potential in paclitaxel-resistant human ovarian carcinoma cell line A2780/Taxol.	[93]
Taxoid pro-drug nanoemulsions: Lipoid E80®; Polysorbate 80; DSPE-PEG2000	DHA-SBT-1214 (omega-3 fatty acid conjugated taxoid pro-drug)	Production using HPH technique; Average size: 228 ± 7 nm; Zeta potential: −27 mV; Entrapment efficiency: 97%	Prostate cancer	Nanoemulsion surface was modified with PEG; Weekly intravenous administration of nanoemulsions in mice bearing subcutaneous PPT2 tumor xenografts suppressed tumor growth compared to Abraxane®; Nanoemulsions show significant activity against prostate CD133high/CD44+/high tumor-initiating cells in vitro and in vivo.	[96]
Lipid nanoemulsion: mixture of phosphatidylcholine, triolein, and cholesteryloleate	Didodecyl methotrexate (ddMTX, esterification reaction between methotrexate and dodecyl bromide)	Production using ultrasonication method; Average size: 60 nm; Entrapment efficiency: 98%; Nanoemulsions were stable at 4 °C for 45 days	Leukemia	After 48 h of incubation with plasma, approximately 28% ddMTX was released; Nanoemulsion uptake by neoplastic cells was higher than free methotrexate which resulted in markedly greater cytotoxicity; Nanoemulsions cytotoxicity against neoplastic cells was higher than free methotrexate.	[100]

Table 1. *Cont.*

Nanoemulsion Constituents	Active Compound	Production Technique and Physicochemical Parameters	Type of Cancer	Therapeutic Efficacy and Other Observations	Ref.
Lipid nanoemulsions: Miglyol 812; Lipoid S75; Polysorbate 80	Chalcone	Production using ultrasonication method; Average size: 110 nm; PI: 0.17; ZP: −19 mV, 93% EE	Leukemia	Nanoemulsions maintained the antileukemic effect of chalcones; Nanoemulsions decreased chalcone toxic effects in non-tumoral cells and in animals.	[101]
Hyaluronic acid complexed nanoemulsions: DL-α-tocopheryl acetate; Soybean oil; Polysorbate 80; Ferric chloride	Paclitaxel	Production using HPH technique; Average size: 85.2 nm; ZP: −35.7 mV; EE: ≈100%	Lung cancer	Hyaluronic acid nanoemulsions inhibited tumor growth, probably because of the specific tumor-targeting affinity of HA for CD44-overexpressed cancer cells.	[111]
Lipid nanoemulsion (7KCLDE): Egg phosphatidylcholine; Triolein; Cholesteryl oleate; Cholesterol	7-ketocholesterol	Average size: 20–50 nm	Melanoma	Single 7KCLDE injection killed ≈10% of melanoma cells; 7KCLDE was injected into B16 melanoma tumor-bearing mice, was accumulated in the liver and tumor. In melanoma tumor in mice 7KCLDE promoted a >50% tumoral size reduction, enlarged the necrotic area, and reduced intratumoral vasculature. 7KCLDE increased the survival rates of animals, without hematologic or liver toxicity.	[106]
Folic acid targeted albumin nanoemulsions; Albumin; Folic acid; Poloxamer 407	Carbon monoxide releasing molecule-2 (CORM-2)	Production using HPH technique; Average size: <100 nm	Lymphoma	Nanoemulsions increased survival of BALB/c mice bearing subcutaneous A20 lymphoma tumors.	[117]
Perfluorohexane nanoemulsions	Perfluorocarbon (contrast agent)	Production using ultrasonication method; Average size: <100 nm; Suitable long-term stability	Ultrasound and photoacoustic imaging of cancer in vivo	Higher spatial resolution and deeper tissue (compared to conventional optical techniques); Non-invasive cancer imaging and therapy alternative for patients.	[115]
Carotenoid; Nanoemulsions; CapryolTM 90; Transcutol®HP; Tween 80	Carotenoid extract from *Lycium barbarum* L.	Production using ultrasonication method; Average size: 15.1 nm	Colon cancer	Nanoemulsions release carotenoids in the acidic environment (characteristic of tumors) but not at physiological pH; Nanoemulsions IC50 of 4.5 µg/mL; Nanoemulsions upregulate p53 and p21 expression and down-regulate CDK2, CDK1, cyclin A, and cyclin B expression and arrest the cell cycle at G2/M in HT-29 colon cancer cells.	[118]
Perfluorocarbon nanoemulsions; Perfluorodecalin; Fluorinated poly(ethylenimine)	siRNA to silence the expression of Bcl2 gene	Production using ultrasonication method (for nanoemulsions); Formation of polyplexes using nanoemulsions and siRNA; Average size: ≈150 nm; ZP: +50 mV; One week stability	Melanoma	Nanoemulsions-based polyplexes induced apoptosis and inhibited tumor growth in a melanoma mouse model; Nanoemulsions-based polyplexes showed potential for in vivo ultrasound imaging.	[119]
Curcumin nanoemulsions; Medium chain tryglicerides; Cremophor RH40; Glycerol	Curcumin	Self-microemulsifying method; Average size: 34.5 nm; Polidispersity index: 0.129; ZP: −8.54 mV	Prostate cancer	Curcumin nanoemulsions enhance the cellular cytotoxicity, cellular uptake, cell cycle arrest, and apoptosis against prostate cancer cells.	[120]

3.4. Clinical Trials

Nanoemulsions could constitute a suitable alternative to ensure a better treatment for cancer patients. However, as can be observed in Table 2, just a few clinical trials have been reported using these colloidal carriers [121].

Nanoemulsions have been used for superficial basal cancer cell photodynamic therapy in an on-going clinical trial [122]. It compares three photosensitizers using randomized prospective double blinded design (phase 2). The photodynamic therapy is combined with methylaminolevulinate (MAL/Metvix®) or with hexylaminolevulinate (HAL/Hexvix®) and aminolevulinic acid nanoemulsion (BF-200 ALA/Ameluz®) in superficially growing basal cell carcinomas. BF-200 ALA nanoemulsion contains 7.8% of 5-aminolevulinic acid, the first compound in the porphyrin synthesis pathway. It has been formulated with soy phosphatidylcholine and propylene glycol to increase its affinity with epidermal tissue [123]. This formulation has been employed in several clinical trials for the treatment of lentigo maligna (ClinicalTrials.gov ID: NCT02685592), multiple actinic keratosis (ClinicalTrials.gov ID: NCT01893203) and actinic keratosis (ClinicalTrials.gov ID: NCT01966120 and NCT02799069).

As reported earlier, curcumin has been extensively studied for cancer. Curcumin is a natural polyphenolic compound extracted from the rhizomes of *Curcuma longa* and shows different biological activities in an antioxidant and anti-inflammatory capacity, among others [124]. To date, no clinical trial aimed directly to cancer therapy has been undertaken. Instead, curcumin loaded nanoemulsions are being assessed in a randomized, double blinded phase 1 controlled study (ClinicalTrial.gov ID: NCT03865992) [125]. Curcumin nanoemulsions aim to reduce the joint pain in breast cancer survivors with aromatase inhibitor-induced joint disease. Curcumin nanoemulsions will be administered orally to women who have primary invasive adenocarcinoma of the breast taking a third-generation aromatase inhibitor. This anticancer drug has a wide incidence of skeletal adverse events such as bone loss and arthralgia [126].

A second clinical trial using curcumin nanoemulsions has been described for the treatment of women with obesity and high risk for breast cancer (ClinicalTrial.gov ID: NCT01975363). This constitutes a randomized pilot study aimed to determine the tolerability, adherence, and safety of different doses (50 or 100 mg) of nanoemulsion curcumin in obese women at high risk for developing breast cancer. Due to the anti-inflammatory activity of curcumin in breast tissue and fat, the risk of developing breast cancer may be reduced [127].

Table 2. Clinical trials using nanoemulsions for cancer drug delivery.

ClinicalTrial.gov ID	Active Compound	Nanoemulsion Constituents	Sponsor and Collaborators	Description	Status	Ref.
NCT02367547	5-Aminolevulinic acid	Soy phosphatidyl-choline; Propylene glycol	Joint Authority for Päijät-Häme Social and Health Care; Tampere University; University of Jyvaskyla	Photodynamic therapy against superficial basal cell cancer.	Active, not recruiting	[122]
NCT03865992	Curcumin	Data not available	City of Hope Medical Center; National Cancer Institute (NCI)	Oral curcumin nanoemulsion for joint pain reduction in breast cancer survivors caused by treatment with aromatase inhibitors.	Recruiting	[125]
NCT01975363	Curcumin	Data not available	Ohio State University Comprehensive Cancer Center	Oral curcumin nanoemulsion to modulate pro-inflammatory biomarkers in plasma and breast adipose tissue.	Active, not recruiting	[127]

4. Nanoemulsions in the Drug Delivery Field

Nanotechnology has notably improved safety and effectiveness of cancer therapy by developing drug delivery systems such as nanocarriers. Due to their nanometric size, they are suitable for chemotherapeutic passive targeting via the enhanced permeability retention (EPR) effects. Moreover, it can achieve active targeting by receptor-mediated uptake to specific cell types and host tissues [128]. Besides this, it provides a controlled release, an increase of drug stability, and solves water solubility problems related to hydrophobic drugs [129]. Among these nanostructured systems, polymeric nanoparticles, nanostructured lipid carriers, liposomes, and nanoemulsions have shown to be a great approach to achieve drug delivery for cancer treatment [128]. Despite the decades of research about the medical use of inorganic nanoparticles, they have demonstrated a lack of safety and biocompatibility [130]. There are some recently investigations in which overcome this drawback by modifying the particle surface with biocompatible molecules; nevertheless, it needs more research for developing more effective coatings and drug delivery strategies [131]. In contrast, the main advantages of nanoemulsion are its composition. It is formulated using biocompatible components and generally recognized as safe (GRAS) and its easy to scale-up and manufacture [132]. In addition to the advantages mentioned above, common with most nanosystems, nanoemulsions possess a high encapsulation capacity for hydrophobic drugs, great physicochemical stability, potentially improved bioavailability, and the drug pharmacokinetics show lower inter- and intra-individual variability [104,133–135]. This system is available for many administration routes. By oral administration, nanoemulsions could protect drug molecules from gastric and gut wall degradation and avoid first-pass metabolism. Nanoemulsions possess a stability similar to liposomes, ethosomes, or microspheres but they possess the advantage of enhanced solubility and absorption of poorly bioavailable molecules [136]. An in vivo study which compared the accumulation in the brain of lipid nanoparticles and nanoemulsions showed that nanoemulsions enhanced the retention time in a significant manner compared to lipid nanoparticles [137]. Moreover, comparative studies about the effect on skin permeation between liposomes, solid lipid nanoparticles, and nanoemulsions have demonstrated that solid lipid nanoparticles tend to release the drug in superficial skin layers, while liposomes and nanoemulsions are able to permeate to deeper skin layers. However, the encapsulation into nanostructure lipid carriers offers increased protection for photosensitive drug than nanoemulsions [138]. In the same way, a nebulized lipid-based nanoemulsion for lung cancer treatment was carried out to explore the possibility of dissolving a large amount of hydrophobic drugs and to increase the resistance towards hydrolysis and enzymatic degradation [139].

5. Limitations of Nanoemulsions

Nanoemulsions can be of great utility in the delivery of drugs to cancer cells due to the fact that these systems have been demonstrated to be safe and able to deliver the drug to the target tissue, increasing drug effects and avoiding toxicity. To the present, no formulation of this type has been approved by the FDA. A variety of concepts are decisive since they can limit the success of a system [2].

The production of this formulation might involve high temperature and pressure conditions, depending on the drug and excipients. Therefore, not all types of starting materials are suitable for some particular manufacturing processes. When this happens, the design of an appropriate production method, or even optimization of an already existing one, might be necessary and take a great deal of time. It is crucial to guarantee that labile drugs are viable and can be produced at a larger scale. Conceiving multifunctional nanoemulsions in large scale production might be particularly hard as there is a fair number of variables to consider [2]. To investigate a suitable method for a particular nanoformulation, the material safety, scale-up, and all parameters of quality control need to be taken into account [1].

The system of nanoemulsion and its moieties' behavior and even its in vivo metabolism need to be carefully evaluated [2]. Every substance behaves in a unique way and so does the metabolism. Absorption, distribution, and excretion can affect parameters like efficiency and safety of drugs and need to be in continuous evaluation [2]. In this sense, targeting nanoemulsions is a major issue for cancer drug delivery since it has been reported that only 0.7% of the drug dose using nanotechnological based strategies is found to be delivered to the solid tumor [65]. In this sense, metabolism is a crucial factor since only the nanoemulsions able to escape from the MPS and renal clearance biological barriers have the opportunity to interact with the tumor tissue [65].

Every time a new material is considered in a formulation, its long-term stability and safety have to be studied. The problem is related to the fact that models used to evaluate toxicity are frequently questionable since results might not be valid due to the absence of real dynamic interactions, normally acting in real human tissues, in a real human body. Research often begins with cellular models or animal species with characteristics and metabolisms that differ from the human body in very important aspects. Also, the way in which an organism behaves varies from person to person and within the person's state, depending on sex, race, age, environmental features, and a lot of other conditions. All of this complicates the translation from in vitro/in vivo to real-life treatment [1].

Due to these fact, the main limitation of nanoemulsions for cancer drug delivery relies on the low clinical translation of the formulations [65].

6. Future Perspectives

Nanoemulsions are drug delivery systems able to encapsulate hydrophilic and hydrophobic molecules designed in order to satisfy a variety of requests [2].

The main challenge of nanoemulsions in future developments is to keep finding mechanisms to improve nanoemulsion efficiency, differentiating them from other formulations. This continuous process of research has to keep in mind the interactions of the drug with the other components in the system, the impacts of the manufacturing mechanism, and drug stability. Along with production, the interaction of nanoemulsions with target cells is also a main study point in future drug development, exploring different ways to induce drug release and uptake. Different routes of administration for nanoemulsions carrying cancer drugs can also be investigated. The key thinking is to come up with new insights on nanoformulation, creating new opportunities for anticancer drug delivery.

The use of nanoemulsions as imaging agents is emerging, as it provides real time monitoring of cancer with minimum destruction and invasion. Traditional imaging techniques involve X-ray tomography, magnetic resonance, and ultrasound and they are all based on marking a targeting nanocmulsion with a radioactive isotope or a fluorophore [1].

Also, in late development are vaccine carriers in nanoemulsion formulations to target tumors. Nanoemulsions, as previously described here, are able to deliver macromolecules, such as antigens which can lead to a useful antigen specific response from the immune system. Nanoemulsions allow a long circulation time and uptake by cells with the specific antibody on their surface for that antigen, or vice-versa, resulting in a highly specific interaction [2].

7. Conclusions

Nanoemulsions represent a new and promising strategy in cancer therapy. The employment of a hydrophobic core allows the encapsulation of lipophilic drugs, coming up with a solution for one of the main problems related to cancer treatment drugs. The presence of an emulsifying agent and GRAS excipients allow the engineering of a stable and safe alternative. They are composed by small sized particles, allowing them to be retained for a long time in circulation.

The main advantage of nanoemulsions, in relation to other drug carriers, is that they can be designed to target tumor cells and avoid MDR. This is a significant development in cancer therapy, since its major problem remains in the fact that most anti-cancer drugs fail due to their marked toxicity in healthy cells/tissues or even because cancer cells end up developing mechanisms to resist treatment.

Passive targeting delivery takes advantage of the ERP effect, typical in tumor tissues. However, active targeting might bring even more positive features to the formulation, as it uses not only the

EPR effect but also specific targeting moieties for cancer cells. Multifunctional nanoemulsions can co-encapsulate, or bind to their surface, compounds that fight MDR mechanisms.

The examples described in this paper demonstrate diverse methodologies through which nanoemulsions can be designed to achieve successful therapeutic outcomes in several types of cancer.

All of these positive features are useless if the manufacturing process and the metabolism of the drug and excipients are not carefully evaluated. These two parameters represent the chief barriers in nanoemulsion development. To achieve this type of formulation, go through every phase of clinical trials, and be approved, all variables must be considered and innovative solutions have to be studied in order to create anticancer drugs that are safe and efficient. The development of this type of formulation is critical in cancer therapy as this multifactorial illness results in a sizable portion of deaths and no completely viable therapy has been found to this day.

Author Contributions: E.S.-L., M.G. and J.D.-F. have written and formatted the review. A.C. (Amanda Cano) and A.L.-M. have significantly contributed to the writing and preparation of tables and figures. M.E. (Miren Ettcheto), M.E. (Marta Espina) and A.C. (Antoni Camins) have structured, edited and supervised the data included on the review. E.B.S. and M.L.G. have contextualized, structured, reviewed and supervised the present manuscript.

Funding: This work was supported by the Spanish Ministry of Science and Innovation (SAF-2016-33307). M.L.G., M.E. (Marta Espina), A.C. (Amanda Cano) and E.S.L. belong to 2014SGR-1023. The first author, E.S.L., acknowledges the support of Institute of Nanoscience and Nanotechnology (ART2018 project). The authors want to acknowledge the Portuguese Science and Technology Foundation (FCT/MCT) and European Funds (PRODER/COMPETE) under the project UID/AGR/04033/2013, M-ERA-NET-0004/2015-PAIRED, co-funded by FEDER, under the partnership Agreement PT2020.

Conflicts of Interest: The authors declare no conflict of interest.

References

1. Ganta, S.; Talekar, M.; Singh, A.; Coleman, T.P.; Amiji, M.M. Nanoemulsions in Translational Research—Opportunities and Challenges in Targeted Cancer Therapy. *AAPS PharmSciTech* **2014**, *15*, 694–708. [CrossRef]

2. McClements, D.J. Nanoemulsions versus microemulsions: Terminology, differences, and similarities. *Soft Matter* **2012**, *6*, 1719–1729. [CrossRef]

3. Gi, H.J.; Chen, S.N.; Hwang, J.S.; Tien, C.; Kuo, M.T. Studies of Formation and Interface of Oil-Water Microemulsion. *Chin. J. Phys.* **1992**, *30*, 665–678.

4. Maeda, H.; Wu, J.; Sawa, T.; Matsumura, Y.; Hori, K. Tumor vascular permeability and the EPR effect in macromolecular therapeutics: A review. *J. Control. Release* **2000**, *65*, 271–284. [CrossRef]

5. Tadros, T.; Izquierdo, P.; Esquena, J.; Solans, C. Formation and stability of nano-emulsions. *Adv. Colloid Interface Sci.* **2004**, *108–109*, 303–318. [CrossRef]

6. Tiwari, S.; Tan, Y.-M.; Amiji, M. Preparation and In Vitro Characterization of Multifunctional Nanoemulsions for Simultaneous MR Imaging and Targeted Drug Delivery. *J. Biomed. Nanotechnol.* **2006**, *2*, 217–224. [CrossRef]

7. Arruebo, M.; Valladares, M.; González-Fernández, Á. Antibody-conjugated nanoparticles for biomedical applications. *J. Nanomater.* **2009**, *2009*, 1–24. [CrossRef]

8. Keefe, A.D.; Pai, S.; Ellington, A. Aptamers as therapeutics. *Nat. Rev. Drug Discov.* **2010**, *9*, 537–550. [CrossRef]

9. Tan, W.; Wang, H.; Chen, Y.; Zhang, X.; Zhu, H.; Yang, C.; Yang, R.; Liu, C. Molecular aptamers for drug delivery. *Trends Biotechnol.* **2011**, *29*, 634–640. [CrossRef]

10. Gu, F.X.; Karnik, R.; Wang, A.Z.; Alexis, F.; Levy-Nissenbaum, E.; Hong, S.; Langer, R.S.; Farokhzad, O.C. Targeted nanoparticles for cancer therapy. *Nano Today* **2007**, *2*, 14–21. [CrossRef]

11. Parker, N.; Turk, M.J.; Westrick, E.; Lewis, J.D.; Low, P.S.; Leamon, C.P. Folate receptor expression in carcinomas and normal tissues determined by a quantitative radioligand binding assay. *Anal. Biochem.* **2005**, *338*, 284–293. [CrossRef]

12. Kumar, G.P.; Divya, A. Nanoemulsion Based Targeting in Cancer Therapeutics. *Med. Chem.* **2015**, *5*, 272–284.

13. Elnakat, H. Distribution, functionality and gene regulation of folate receptor isoforms: Implications in targeted therapy. *Adv. Drug Deliv. Rev.* **2004**, *56*, 1067–1084. [CrossRef]

14. Low, P.S.; Antony, A.C. Folate receptor-targeted drugs for cancer and inflammatory diseases. *Adv. Drug Deliv. Rev.* **2004**, *56*, 1055–1058. [CrossRef]

15. Toub, N.; Malvy, C.; Fattal, E.; Couvreur, P. Innovative nanotechnologies for the delivery of oligonucleotides and siRNA. *Biomed. Pharmacother.* **2006**, *60*, 607–620. [CrossRef]

16. Bertrand, N.; Wu, J.; Xu, X.; Kamaly, N.; Farokhzad, O.C. Cancer nanotechnology: The impact of passive and active targeting in the era of modern cancer biology. *Adv. Drug Deliv. Rev.* **2014**, *66*, 2–25. [CrossRef]

17. Zhang, C.; Tang, N.; Liu, X.; Liang, W.; Xu, W.; Torchilin, V.P. siRNA-containing liposomes modified with polyarginine effectively silence the targeted gene. *J. Control. Release* **2006**, *112*, 229–239. [CrossRef]

18. Jain, R.K. Normalization of Tumor Vasculature: An Emerging Concept in Antiangiogenic Therapy. *Science* **2005**, *307*, 58–62. [CrossRef]

19. Al-Abd, A.M.; Lee, S.H.; Kim, S.H.; Cha, J.-H.; Park, T.G.; Lee, S.J.; Kuh, H.-J. Penetration and efficacy of VEGF siRNA using polyelectrolyte complex micelles in a human solid tumor model in-vitro. *J. Control. Release* **2009**, *137*, 130–135. [CrossRef]

20. Lin, A.; Slack, N.; Ahmad, A.; Koltover, I.; George, C.; Samuel, C.; Safinta, C. Structure and Structure—Function Studies of Lipid/Plasmid DNA Complexes. *J. Drug Target.* **2000**, *8*, 13–27. [CrossRef]

21. Fattal, E.; Couvreur, P.; Dubernet, C. "Smart" delivery of antisense oligonucleotides by anionic pH-sensitive liposomes. *Adv. Drug Deliv. Rev.* **2004**, *56*, 931–946. [CrossRef]

22. Maranhao, R.; De Almeida, C.P.; Vital, C.G.; Contente, T.; Maria, D.A. Modification of composition of a nanoemulsion with different cholesteryl ester molecular species: Effects on stability, peroxidation, and cell uptake. *Int. J. Nanomed.* **2010**, *5*, 679. [CrossRef]

23. Wooster, T.J.; Golding, M.; Sanguansri, P. Impact of Oil Type on Nanoemulsion Formation and Ostwald Ripening Stability. *Langmuir* **2008**, *24*, 12758–12765. [CrossRef]

24. McClements, D.J. Edible nanoemulsions: Fabrication, properties, and functional performance. *Soft Matter* **2011**, *7*, 2297–2316. [CrossRef]

25. Dean, M. ABC Transporters, Drug Resistance, and Cancer Stem Cells. *J. Mammary Gland Boil. Neoplasia* **2009**, *14*, 3–9. [CrossRef]

26. Talekar, M.; Ganta, S.; Singh, A.; Amiji, M.; Kendall, J.; Denny, W.A.; Garg, S. Phosphatidylinositol 3-kinase Inhibitor (PIK75) Containing Surface Functionalized Nanoemulsion for Enhanced Drug Delivery, Cytotoxicity and Pro-apoptotic Activity in Ovarian Cancer Cells. *Pharm. Res.* **2012**, *29*, 2874–2886. [CrossRef]

27. Mohammad, I.S.; He, W.; Yin, L. Understanding of human ATP binding cassette superfamily and novel multidrug resistance modulators to overcome MDR. *Biomed. Pharmacother.* **2018**, *100*, 335–348. [CrossRef]

28. Chanamai, R.; McClements, D.J. Impact of Weighting Agents and Sucrose on Gravitational Separation of Beverage Emulsions. *J. Agric. Food Chem.* **2000**, *18*, 5561–5565. [CrossRef]

29. McClements, D.J. Emulsion Design to Improve the Delivery of Functional Lipophilic Components. *Annu. Rev. Food Sci. Technol.* **2010**, *1*, 241–269. [CrossRef]

30. Ganta, S.; Devalapally, H.; Shahiwala, A.; Amiji, M. A review of stimuli-responsive nanocarriers for drug and gene delivery. *J. Control. Release* **2008**, *126*, 187–204. [CrossRef]

31. Wang, C.Y.; Huang, L. pH-sensitive immunoliposomes mediate target-cell-specific delivery and controlled expression of a foreign gene in mouse. *Proc. Natl. Acad. Sci. USA* **1987**, *84*, 7851–7855. [CrossRef]

32. Modi, S.; Anderson, B.D. Determination of Drug Release Kinetics from Nanoparticles: Overcoming Pitfalls of the Dynamic Dialysis Method. *Mol. Pharm.* **2013**, *10*, 3076–3089. [CrossRef] [PubMed]

33. Qian, C.; McClements, D.J. Formation of nanoemulsions stabilized by model food-grade emulsifiers using high-pressure homogenization: Factors affecting particle size. *Food Hydrocoll.* **2011**, *25*, 1000–1008. [CrossRef]

34. Lovelyn, C.; Attama, A.A. Current State of Nanoemulsions in Drug Delivery. *J. Biomater. Nanobiotechnol.* **2011**, *2*, 626–639. [CrossRef]

35. Douglas, S.J.; Davis, S.S.; Illum, L. Nanoparticles in drug delivery. *Crit. Rev. Ther. Drug Carrier Syst.* **1987**, *3*, 233–261. [PubMed]

36. Wehrung, D.; Geldenhuys, W.J.; Oyewumi, M.O. Effects of gelucire content on stability, macrophage interaction and blood circulation of nanoparticles engineered from nanoemulsions. *Colloids Surf. B Biointerfaces* **2012**, *94*, 259–265. [CrossRef] [PubMed]

37. Zahr, A.S.; Davis, C.A.; Pishko, M.V. Macrophage Uptake of Core–Shell Nanoparticles Surface Modified with Poly(ethylene glycol). *Langmuir* **2006**, *22*, 8178–8185. [CrossRef]

38. Stolnik, S.S.; Daudali, B.; Arien, A.; Whetstone, J.; Heald, C.; Garnett, M.; Davis, S.; Illum, L.; Garnett, M. The effect of surface coverage and conformation of poly(ethylene oxide) (PEO) chains of poloxamer 407 on the biological fate of model colloidal drug carriers. *Biochim. et Biophys. Acta (BBA)—Biomembr.* **2001**, *1514*, 261–279. [CrossRef]

39. Alvarez-Lorenzo, C.; Rey-Rico, A.; Sosnik, A.; Taboada, P.; Concheiro, A. Poloxamine-based nanomaterials for drug delivery. *Front. Biosci.* **2010**, *2*, 424–440. [CrossRef]

40. Ganta, S.; Deshpande, D.; Korde, A.; Amiji, M. A review of multifunctional nanoemulsion systems to overcome oral and CNS drug delivery barriers. *Mol. Membr. Boil.* **2010**, *27*, 260–273. [CrossRef]

41. Longmire, M.; Choyke, P.L.; Kobayashi, H. Clearance properties of nano-sized particles and molecules as imaging agents: Considerations and caveats. *Nanomedicine* **2008**, *3*, 703–717. [CrossRef] [PubMed]

42. Choi, H.S.; Liu, W.; Misra, P.; Tanaka, E.; Zimmer, J.P.; Itty Ipe, B.; Bawendi, M.G.; Frangioni, J.V. Renal clearance of quantum dots. *Nat. Biotechnol.* **2007**, *25*, 1165–1170. [CrossRef]

43. Mason, T.G.; Wilking, J.N.; Meleson, K.; Chang, C.B.; Graves, S.M. Nanoemulsions: Formation, structure, and physical properties. *Phys. Condens. Matter* **2006**, *18*, R635–R666. [CrossRef]

44. Qi, K.; Al-haideri, M.; Seo, T.; Carpentier, Y.A. Effects of particle size on blood clearance and tissue uptake of lipid emulsions with different triglyceride compositions. *J. Parenter. Enter. Nutr.* **2003**, *27*, 58–64. [CrossRef] [PubMed]

45. Lawler, J. Introduction to the tumour microenvironment Review Series. *J. Cell. Mol. Med.* **2009**, *13*, 1403–1404. [CrossRef] [PubMed]

46. Danhier, F.; Feron, O.; Préat, V. To exploit the tumor microenvironment: Passive and active tumor targeting of nanocarriers for anti-cancer drug delivery. *J. Control. Release* **2010**, *148*, 135–146. [CrossRef] [PubMed]

47. Carmeliet, P.; Jain, R.K. Molecular mechanisms and clinical applications of angiogenesis. *Nature* **2011**, *473*, 298–307. [CrossRef]

48. Padera, T.P.; Stoll, B.R.; Tooredman, J.B.; Capen, D.; di Tomaso, E.; Jain, R.K. Pathology: Cancer cells compress intratumour vessels. *Nature* **2004**, *427*, 695. [CrossRef]

49. Qadir, A.; Faiyazuddin, M.; Hussain, M.T.; Alshammari, T.M.; Shakeel, F. Critical steps and energetics involved in a successful development of a stable nanoemulsion. *J. Mol. Liq.* **2016**, *214*, 7–18. [CrossRef]

50. Caron, W.P.; Lay, J.C.; Fong, A.M.; La-Beck, N.M.; Kumar, P.; Newman, S.E.; Zhou, H.; Monaco, J.H.; Clarke-Pearson, D.L.; Brewster, W.R.; et al. Translational Studies of Phenotypic Probes for the Mononuclear Phagocyte System and Liposomal Pharmacology. *J. Pharmacol. Exp. Ther.* **2013**, *347*, 599–606. [CrossRef]

51. Allen, T.M.; Martin, F.J. Advantages of liposomal delivery systems for anthracyclines. *Semin. Oncol.* **2004**, *31*, 5–15. [CrossRef] [PubMed]

52. Faria, M.; Björnmalm, M.; Thurecht, K.J.; Kent, S.J.; Parton, R.G.; Kavallaris, M.; Johnston, A.P.R.; Gooding, J.J.; Corrie, S.R.; Boyd, B.J.; et al. Europe PMC Funders Group Minimum Information Reporting in Bio—Nano Experimental Literature. *Nat. Nanotechnol.* **2019**, *13*, 777–785. [CrossRef]

53. Yang, R.; Han, X.; Shi, K.; Cheng, G.; Shin, C.; Cui, F. Cationic formulation of paclitaxel-loaded poly D,L-lactic-co-glycolic acid (PLGA) nanoparticles using an emulsion-solvent diffusion method. *J. Pharm. Sci.* **2009**, *4*, 89–95.

54. Kim, B.; Pena, C.D.; Auguste, D.T. Targeted Lipid Nanoemulsions Encapsulating Epigenetic Drugs Exhibit Selective Cytotoxicity on CDH1[−]/FOXM1[+] Triple Negative Breast Cancer Cells. *Mol. Pharm.* **2019**, *16*, 1813–1826. [CrossRef] [PubMed]

55. Najlah, M.; Kadam, A.; Wan, K.; Ahmed, W.; Taylor, K.M.G.; Elhissi, A.M.A. Novel paclitaxel formulations solubilized by parenteral nutrition nanoemulsions for application against glioma cell lines. *Int. J. Pharm.* **2016**, *506*, 102–109. [CrossRef] [PubMed]

56. Chang, H.-B.; Chen, B.-H. Inhibition of lung cancer cells A549 and H460 by curcuminoid extracts and nanoemulsions prepared from Curcuma longa Linnaeus. *Int. J. Nanomed.* **2015**, *10*, 5059–5080.

57. Klang, V.; Matsko, N.B.; Valenta, C.; Hofer, F. Electron microscopy of nanoemulsions: An essential tool for characterisation and stability assessment. *Micron* **2012**, *43*, 85–103. [CrossRef] [PubMed]

58. Bae, Y.H. Drug targeting and tumor heterogeneity. *J. Control Release* **2009**, *133*, 2–3. [CrossRef]

59. Peer, D.; Karp, J.M.; Hong, S.; Farokhzad, O.C.; Margalit, R.; Langer, R. Nanocarriers as an emerging platform for cancer therapy. *Nat. Nanotechnol.* **2007**, *2*, 751–760. [CrossRef]

60. Cappellani, M.R.; Perinelli, D.R.; Pescosolido, L.; Schoubben, A.; Cespi, M.; Cossi, R.; Blasi, P. Injectable nanoemulsions prepared by high pressure homogenization: Processing, sterilization, and size evolution. *Appl. Nanosci.* **2018**, *8*, 1483–1491. [CrossRef]

61. Mahd, S.; Yinghe, J.; Bhesh, H. Optimization of nano-emulsions production by microfluidization. *Eur. Food Res. Technol.* **2007**, *225*, 733–741.

62. Ren, G.; Sun, Z.; Wang, Z.; Zheng, X.; Xu, Z.; Sun, D. Nanoemulsion formation by the phase inversion temperature method using polyoxypropylene surfactants. *J. Colloid Interface Sci.* **2019**, *540*, 177–184. [CrossRef] [PubMed]

63. Trimaille, T.; Chaix, C.; Delair, T.; Pichot, C.; Teixeira, H.; Dubernet, C.; Couvreur, P. Interfacial deposition of functionalized copolymers onto nanoemulsions produced by the solvent displacement method. *Colloid Polym. Sci.* **2001**, *279*, 784–792. [CrossRef]

64. Lefebvre, G.; Riou, J.; Bastiat, G.; Roger, E.; Frombach, K.; Gimel, J.-C.; Saulnier, P.; Calvignac, B. Spontaneous nano-emulsification: Process optimization and modeling for the prediction of the nanoemulsion's size and polydispersity. *Int. J. Pharm.* **2017**, *534*, 220–228. [CrossRef]

65. Wilhelm, S.; Tavares, A.J.; Dai, Q.; Ohta, S.; Audet, J.; Dvorak, H.F.; Chan, W.C.W. Analysis of nanoparticle delivery to tumours. *Nat. Rev. Mater.* **2016**, *1*, 16014. [CrossRef]

66. Haley, B.; Frenkel, E. Nanoparticles for drug delivery in cancer treatment. *Urol. Oncol. Semin. Orig. Investig.* **2008**, *26*, 57–64. [CrossRef]

67. Yuan, Y.; Gao, Y.; Zhao, J.; Mao, L. Characterization and stability evaluation of β-carotene nanoemulsions prepared by high pressure homogenization under various emulsifying conditions. *Food Res. Int.* **2008**, *41*, 61–68. [CrossRef]

68. Mulik, R.S.; Mönkkönen, J.; Juvonen, R.O.; Mahadik, K.R.; Paradkar, A.R. Transferrin mediated solid lipid nanoparticles containing curcumin: Enhanced in vitro anticancer activity by induction of apoptosis. *Int. J. Pharm.* **2010**, *398*, 190–203. [CrossRef]

69. Milane, L.; Duan, Z.; Amiji, M. Development of EGFR-targeted polymer blend nanocarriers for combination paclitaxel/lonidamine delivery to treat multi-drug resistance in human breast and ovarian tumor cells. *Mol. Pharm.* **2011**, *8*, 185–203. [CrossRef]

70. Xu, W.; Siddiqui, I.A.; Nihal, M.; Pilla, S.; Rosenthal, K.; Mukhtar, H.; Gong, S. Aptamer-conjugated and doxorubicin-loaded unimolecular micelles for targeted therapy of prostate cancer. *Biomaterials* **2013**, *34*, 5244–5253. [CrossRef]

71. Jiang, G.; Tang, S.; Chen, X.; Ding, F. Enhancing the receptor-mediated cell uptake of PLGA nanoparticle for targeted drug delivery by incorporation chitosan onto the particle surface. *J. Nanopart.* **2014**, *16*, 2453. [CrossRef]

72. Gottesman, M.M. Mechanisms of Cancer Drug Resistance. *Annu. Med.* **2002**, *53*, 615–627. [CrossRef]

73. Hennessy, M.; Spiers, J. A primer on the mechanics of P-glycoprotein the multidrug transporter. *Pharmacol. Res.* **2007**, *55*, 1–15. [CrossRef]

74. Ganta, S.; Singh, A.; Rawal, Y.; Cacaccio, J.; Patel, N.R.; Kulkarni, P.; Ferris, C.F.; Amiji, M.M.; Coleman, T.P. Formulation development of a novel targeted theranostic nanoemulsion of docetaxel to overcome multidrug resistance in ovarian cancer Formulation development of a novel targeted theranostic nanoemulsion of docetaxel to overcome multidrug resistance in ovarian cancer. *Drug Deliv.* **2016**, *23*, 958–970.

75. Crown, J.; O'Leary, M.; Ooi, W.-S. Docetaxel and Paclitaxel in the treatment of breast cancer: A review of clinical experience. *Oncologist* **2004**, *9*, 24–32. [CrossRef]

76. Meng, L.; Xia, X.; Yang, Y.; Ye, J.; Dong, W.; Ma, P.; Jin, Y.; Liu, Y. Co-encapsulation of paclitaxel and baicalein in nanoemulsions to overcome multidrug resistance via oxidative stress augmentation and P-glycoprotein inhibition. *Int. J. Pharm.* **2016**, *513*, 8–16. [CrossRef]

77. Zheng, N.; Gao, Y.; Ji, H.; Wu, L.; Qi, X.; Liu, X.; Tang, J. Vitamin E derivative based multifunctional nanoemulsions for overcoming multidrug resistance in cancer. *J. Drug Target* **2016**, *24*, 1–27. [CrossRef]

78. Tang, J.; Fu, Q.; Wang, Y.; Racette, K.; Wang, D.; Liu, F. Vitamin E Reverses Multidrug Resistance In Vitro and In Vivo. *Cancer Lett.* **2013**, *336*, 149–157. [CrossRef] [PubMed]

79. Hu, C.-M.J.; Zhang, L. Nanoparticle-based combination therapy toward overcoming drug resistance in cancer. *Biochem. Pharmacol.* **2012**, *83*, 1104–1111. [CrossRef]

80.	Yu, H.; Xu, Z.; Chen, X.; Xu, L.; Yin, Q.; Zhang, Z.; Li, Y. Reversal of lung cancer multidrug resistance by pH-responsive micelleplexes mediating co-delivery of siRNA and paclitaxel. *Macromol. Biosci.* **2014**, *14*, 100–109. [CrossRef]

81.	Huang, R.-F.S.; Wei, Y.-J.; Inbaraj, B.S.; Chen, B.-H. Inhibition of colon cancer cell growth by nanoemulsion carrying gold nanoparticles and lycopene. *Int. J. Nanomed.* **2015**, *10*, 2823–2846.

82.	Cunningham, D.; Atkin, W.; Lenz, H.J.; Lynch, H.T.; Minsky, B.; Nordlinger, B.; Starling, N. Colorectal cancer. *Lancet* **2010**, *375*, 1030–1047. [CrossRef]

83.	Gottlieb, L.S.; Sternberg, S.S.; Bond, J.H.; Schapiro, M.; Panish, J.F.; Kurtz, R.C.; Shike, M.; Ackroyd, F.W.; Stewart, E.T.; Skolnick, M.; et al. Risk of Colorectal Cancer in the Families of Patients with Adenomatous Polyps. *N. Engl. J. Med.* **1996**, *334*, 82–87.

84.	Thiery, J.P.; Sleeman, J.P. Complex networks orchestrate epithelial–mesenchymal transitions. *Nat. Rev. Mol. Cell Boil.* **2006**, *7*, 131–142. [CrossRef]

85.	Mein, J.R.; Lian, F.; Wang, X.-D. Biological activity of lycopene metabolites: Implications for cancer prevention. *Nutr. Rev.* **2008**, *66*, 667–683. [CrossRef]

86.	Chen, Y.J.; Inbaraj, B.S.; Pu, Y.S.; Chen, B.H. Development of lycopene micelle and lycopene chylomicron and a comparison of bioavailability. *Nanotechnology* **2014**, *25*, 155102. [CrossRef] [PubMed]

87.	Lu, W.; Zhang, G.; Zhang, R.; Flores, L.G.; Huang, Q.; Gelovani, J.G.; Li, C. Tumor site-specific silencing of NF-kappaB p65 by targeted hollow gold nanosphere-mediated photothermal transfection. *Cancer Res.* **2010**, *70*, 3177–3188. [CrossRef] [PubMed]

88.	Leu, J.G.; Chen, S.A.; Chen, H.M.; Wu, W.M.; Hung, C.F.; Yao, Y.D.; Tu, C.S.; Liang, Y.J. The effects of gold nanoparticles in wound healing with antioxidant epigallocatechin gallate and alpha-lipoic acid. *Nanomedicine* **2012**, *8*, 767–775. [CrossRef]

89.	Zhao, C.; Feng, Q.; Dou, Z.; Yuan, W.; Sui, C.; Zhang, X.; Xia, G.; Sun, H.; Ma, J. Local Targeted Therapy of Liver Metastasis from Colon Cancer by Galactosylated Liposome Encapsulated with Doxorubicin. *PLoS ONE* **2013**, *8*, e73860. [CrossRef]

90.	Neijt, J.P.; ten Bokkel Huinink, W.W.; van der Burg, M.E.L.; van Oosterom, A.T.; Willemse, P.H.B.; Vermorken, J.B.; van Lindert, A.C.M.; Heintz, A.P.M.; Aartsen, E.; van Lent, M.; et al. Long-term survival in ovarian cancer: Mature data from The Netherlands joint study group for ovarian cancer. *Eur. J. Cancer Clin. Oncol.* **1991**, *27*, 1367–1372. [CrossRef]

91.	Jamieson, E.R.; Lippard, S.J. Structure, Recognition, and Processing of Cisplatin–DNA Adducts. *Chem. Rev.* **1999**, *99*, 2467–2498. [CrossRef]

92.	Ganta, S.; Singh, A.; Patel, N.R.; Cacaccio, J.; Rawal, Y.H.; Davis, B.J.; Amiji, M.M.; Coleman, T.P. Development of EGFR Targeted Nanoemulsion for Imaging and Novel Platinum Therapy of Ovarian Cancer. *Pharm. Res.* **2014**, *31*, 2490–2502. [CrossRef]

93.	Cronin, M.; Dearden, J.; Duffy, J.; Edwards, R.; Manga, N.; Worth, A.; Worgan, A. The importance of hydrophobicity and electrophilicity descriptors in mechanistically-based QSARs for toxicological endpoints. *SAR QSAR Environ.* **2002**, *13*, 167–176. [CrossRef]

94.	Ganta, S.; Amiji, M. Coadministration of Paclitaxel and Curcumin in Nanoemulsion Formulations To Overcome Multidrug Resistance in Tumor Cells. *Mol. Pharm.* **2009**, *6*, 928–939. [CrossRef]

95.	Jemal, A.; Bray, F.; Center, M.M.; Ferlay, J.; Ward, E.; Forman, D. Global cancer statistics. *CA Cancer J. Clin.* **2011**, *61*, 69–90. [CrossRef]

96.	Reya, T.; Morrison, S.J.; Clarke, M.F.; Weissman, I.L. Stem cells, cancer, and cancer stem cells. *Nat. Cell Boil.* **2001**, *414*, 105–111. [CrossRef]

97.	Murali, R.; Varghese, B.A.; Nair, R.; Konrad, C.V. The role of cancer stem cells in tumor heterogeneity and resistance to therapy. *Can. J. Physiol. Pharmacol.* **2017**, *95*, 1–15.

98.	Ahmad, G.; El Sadda, R.; Botchkina, G.; Ojima, I.; Egan, J.; Amiji, M. Nanoemulsion Formulation of a Novel Taxoid DHA-SBT-1214 Inhibits Prostate Cancer Stem Cell-Induced Tumor Growth. *Cancer Lett.* **2017**, *406*, 71–80. [CrossRef]

99.	Gillet, J.-P.; Calcagno, A.M.; Varma, S.; Marino, M.; Green, L.J.; Vora, M.I.; Patel, C.; Orina, J.N.; Eliseeva, T.A.; Singal, V.; et al. Redefining the relevance of established cancer cell lines to the study of mechanisms of clinical anti-cancer drug resistance. *Proc. Natl. Acad. Sci. USA* **2011**, *108*, 18708–18713. [CrossRef]

100. Botchkina, G.I.; Zuniga, E.S.; Das, M.; Wang, Y.; Wang, H.; Zhu, S.; Savitt, A.G.; Rowehl, R.A.; Leyfman, Y.; Ju, J.; et al. New-generation taxoid SB-T-1214 inhibits stem cell-related gene expression in 3D cancer spheroids induced by purified colon tumor-initiating cells. *Mol. Cancer* **2010**, *9*, 192. [CrossRef]

101. Jones, R.J.; Hawkins, R.E.; Eatock, M.M.; Ferry, D.R.; Eskens, F.A.; Wilke, H.; Evans, T.R. A phase II open-label study of DHA-paclitaxel (Taxoprexin) by 2-h intravenous infusion in previously untreated patients with locally advanced or metastatic gastric or oesophageal adenocarcinoma. *Cancer Chemother. Pharmacol.* **2008**, *61*, 435–441. [CrossRef]

102. Moura, J.A.; Valduga, C.J.; Tavares, E.R.; Kretzer, I.F.; Maria, D.A.; Maranhão, R.C. Novel formulation of a methotrexate derivative with a lipid nanoemulsion. *Int. J. Nanomed.* **2011**, *6*, 2285–2295.

103. Winter, E.; Pizzol, C.D.; Locatelli, C.; Silva, A.H.; Conte, A.; Chiaradia-Delatorre, L.D.; Nunes, R.J.; Yunes, R.A.; Creckzynski-Pasa, T.B. In vitro and in vivo Effects of Free and Chalcones-Loaded Nanoemulsions: Insights and Challenges in Targeted Cancer Chemotherapies. *Int. J. Environ. Res. Public Health* **2014**, *11*, 10016–10035. [CrossRef] [PubMed]

104. Periasamy, V.S.; Athinarayanan, J.; Alshatwi, A.A. Anticancer activity of an ultrasonic nanoemulsion formulation of Nigella sativa L. essential oil on human breast cancer cells. *Ultrason. Sonochem.* **2016**, *31*, 449–455. [CrossRef] [PubMed]

105. Migotto, A.; Carvalho, V.F.M.; Salata, G.C.; Da Silva, F.W.M.; Yan, C.Y.I.; Ishida, K.; Costa-Lotufo, L.V.; Steiner, A.A.; Lopes, L.B. Multifunctional nanoemulsions for intraductal delivery as a new platform for local treatment of breast cancer. *Drug Deliv.* **2018**, *25*, 654–667. [CrossRef] [PubMed]

106. Natesan, S.; Sugumaran, A.; Ponnusamy, C.; Thiagarajan, V.; Palanichamy, R.; Kandasamya, R. Chitosan stabilized camptothecin nanoemulsions: Development, evaluation and biodistribution in preclinical breast cancer animalmode. *Int. J. Biol. Macromol.* **2017**, *104*, 1846–1852. [CrossRef] [PubMed]

107. Kretzer, I.F.; Maria, D.A.; Guido, M.C.; Contente, T.C.; Maranhão, R.C.; Maria, D. Simvastatin increases the antineoplastic actions of paclitaxel carried in lipid nanoemulsions in melanoma-bearing mice. *Int. J. Nanomed.* **2016**, *11*, 885–904.

108. Favero, G.M.; Paz, J.L.; Otake, A.H.; Maria, D.A.; Caldini, E.G.; De Medeiros, R.S.; Deus, D.F.; Chammas, R.; Maranhão, R.C.; Bydlowski, S.P.; et al. Cell internalization of 7-ketocholesterol-containing nanoemulsion through LDL receptor reduces melanoma growth in vitro and in vivo: A preliminary report. *Oncotarget* **2018**, *9*, 14160–14174. [CrossRef]

109. Monge-Fuentes, V.; Muehlmann, L.A.; Figueiró Longo, J.P.; Rodrigues Silva, J.; Fascineli, M.L.; de Souza, P.; Faria, F.; Anatolievich Degterev, I.; Rodriguez, A.; Pirani Carneiro, F.; et al. Photodynamic therapymediated by acai oil (Euterpe oleracea Martius) in nanoemulsion: A potential treatment for melanoma. *J. Photochem. Photobiol. B Biol.* **2017**, *166*, 301–310. [CrossRef] [PubMed]

110. Journo-Gershfeld, G.; Kapp, D.; Shamay, Y.; Kopeček, J.; David, A. Hyaluronan Oligomers-HPMA Copolymer Conjugates for Targeting Paclitaxel to CD44-Overexpressing Ovarian Carcinoma. *Pharm. Res.* **2012**, *29*, 1121–1133. [CrossRef]

111. Liebmann, J.; Cook, J.; Mitchell, J. Cremophor EL, solvent for paclitaxel, and toxicity. *Lancet* **1993**, *342*, 1428. [CrossRef]

112. Matsubara, Y.; Katoh, S.; Taniguchi, H.; Oka, M.; Kadota, J.; Kohno, S. Expression of CD44 Variants in Lung Cancer and Its Relationship to Hyaluronan Binding. *J. Int. Med.* **2000**, *28*, 78–90. [CrossRef] [PubMed]

113. Kim, J.-E.; Park, Y.-J. Improved Antitumor Efficacy of Hyaluronic Acid-Complexed Paclitaxel Nanoemulsions in Treating Non-Small Cell Lung Cancer. *Biomol. Ther.* **2017**, *25*, 411–416. [CrossRef]

114. Chen, H.; Zhang, W.; Zhu, G.; Xie, J.; Chen, X. Rethinking cancer nanotheranostics. *Nat. Rev. Mater.* **2017**, *2*. [CrossRef]

115. Fernandes, D.A.; Fernandes, D.D.; Li, Y.; Zhang, Z.; Rousseau, D.; Gradinaru, C.C.; Kolios, M.C.; Wang, Y. Synthesis of Stable Multifunctional Perfluorocarbon Nanoemulsions for Cancer Therapy and Imaging. *Langmuir* **2016**, *32*, 10870–10880. [CrossRef] [PubMed]

116. Li, H.; Krstin, S.; Wink, M. Phytomedicine Modulation of multidrug resistant in cancer cells by EGCG, tannic acid and curcumin. *Phytomedicine* **2018**, *50*, 213–222. [CrossRef] [PubMed]

117. Loureiro, A.; Cavaco-Paulo, A. Size controlled protein nanoemulsions for cancer therapy. *Cent. Biol. Eng.* **2017**, *45*, 4710.

118. Hsu, H.J.; Huang, R.F.; Kao, T.H.; Inbaraj, B.S.; Chen, B.H. Preparation of carotenoid extracts and nanoemulsions from Lycium barbarum L. and their effects on growth of HT-29 colon cancer cells. *Nanotechnology* **2017**, *28*, 135103. [CrossRef]

119. Chen, G.; Wang, K.; Wu, P.; Wang, Y.; Zhou, Z.; Yin, L.; Sun, M.; Oupický, D. Development of fluorinated polyplex nanoemulsions for improved small interfering RNA delivery and cancer therapy. *Nano Res.* **2018**, *11*, 3746–3761. [CrossRef]

120. Guan, Y.-B.; Zhou, S.-Y.; Zhang, Y.-Q.; Wang, J.-L.; Tian, Y.-D.; Jia, Y.-Y.; Sun, Y.-J. Therapeutic effects of curcumin nanoemulsions on prostate cancer. *J. Huazhong Univ. Sci. Technol.* **2017**, *37*, 371–378. [CrossRef]

121. Tagami, T.; Ozeki, T. Recent Trends in Clinical Trials Related to Carrier-Based Drugs. *J. Pharm. Sci.* **2017**, *106*, 2219–2226. [CrossRef] [PubMed]

122. Superficial Basal Cell Cancer's Photodynamic Therapy: Comparing Three Photosensitizers: HAL and BF-200 ALA Versus MAL. Available online: https://ClinicalTrials.gov/show/NCT02367547 (accessed on 19 May 2019).

123. Benaouda, S.; Jones, A.; Martin, G.P.; Brown, M.B. Localized Epidermal Drug Delivery Induced by Supramolecular Solvent Structuring. *Mol. Pharm.* **2016**, *13*, 65–72. [CrossRef] [PubMed]

124. Luan, L.; Chi, Z.; Liu, C. Chinese White Wax Solid Lipid Nanoparticles as a Novel Nanocarrier of Curcumin for Inhibiting the Formation of Staphylococcus aureus Biofilms. *Nanomaterials* **2019**, *9*, 763. [CrossRef]

125. Curcumin in Reducing Joint Pain in Breast Cancer Survivors with Aromatase Inhibitor-Induced Joint Disease. Available online: https://ClinicalTrials.gov/show/NCT03865992 (accessed on 19 May 2019).

126. Servitja, S.; Martos, T.; Sanz, M.R.; Garcia-Giralt, N.; Prieto-Alhambra, D.; Garrigos, L.; Nogues, X.; Tusquets, I. Skeletal adverse effects with aromatase inhibitors in early breast cancer: Evidence to date and clinical guidance. *Ther. Adv. Med. Oncol.* **2015**, *7*, 291–296. [CrossRef]

127. Pilot Study of Curcumin for Women with Obesity and High Risk for Breast Cancer. Available online: https://ClinicalTrials.gov/show/NCT01975363 (accessed on 19 May 2019).

128. Senapati, S.; Mahanta, A.K.; Kumar, S.; Maiti, P. Controlled drug delivery vehicles for cancer treatment and their performance. *Signal Transduct. Target. Ther.* **2018**, *3*, 7. [CrossRef]

129. Neubi, G.M.N.; Opoku-Damoah, Y.; Gu, X.; Han, Y.; Zhou, J.; Ding, Y.; Damoah-Opoku, Y. Bio-inspired drug delivery systems: An emerging platform for targeted cancer therapy. *Biomater. Sci.* **2018**, *6*, 958–973. [CrossRef]

130. Soenen, S.J.; Rivera-Gil, P.; Montenegro, J.-M.; Parak, W.J.; De Smedt, S.C.; Braeckmans, K. Cellular toxicity of inorganic nanoparticles: Common aspects and guidelines for improved nanotoxicity evaluation. *Nano Today* **2011**, *6*, 446–465. [CrossRef]

131. Jiao, M.; Zhang, P.; Meng, J.; Li, Y.; Liu, C.; Luo, X.; Gao, M. Recent advancements in biocompatible inorganic nanoparticles towards biomedical applications. *Biomater. Sci.* **2018**, *6*, 726–745. [CrossRef]

132. Ma, Y.; Liu, D.; Wang, D.; Wang, Y.; Fu, Q.; Fallon, J.K.; Yang, X.; He, Z.; Liu, F. Combinational Delivery of Hydrophobic and Hydrophilic Anticancer Drugs in Single Nanoemulsions To Treat MDR in Cancer. *Mol. Pharm.* **2014**, *11*, 2623–2630. [CrossRef] [PubMed]

133. Mahato, R. Nanoemulsion as Targeted Drug Delivery System for Cancer Therapeutics. *J. Pharm. Sci. Pharmacol.* **2017**, *3*, 83–97. [CrossRef]

134. Chrastina, A.; Baron, V.T.; Abedinpour, P.; Rondeau, G.; Welsh, J.; Borgström, P. Plumbagin-Loaded Nanoemulsion Drug Delivery Formulation and Evaluation of Antiproliferative Effect on Prostate Cancer Cells. *BioMed Res. Int.* **2018**, *2018*, 9035452. [CrossRef] [PubMed]

135. Deli, G.; Hatziantoniou, S.; Nikas, Y.; Demetzos, C. Solid lipid nanoparticles and nanoemulsions containing ceramides: Preparation and physicochemical characterization. *J. Liposome* **2009**, *19*, 180–188. [CrossRef]

136. Verma, P.; Meher, J.G.; Asthana, S.; Pawar, V.K.; Chaurasia, M.; Chourasia, M.K. Perspectives of nanoemulsion assisted oral delivery of docetaxel for improved chemotherapy of cancer. *Drug Deliv.* **2014**, *23*, 1–10. [CrossRef]

137. Hörmann, K.; Zimmer, A. Drug delivery and drug targeting with parenteral lipid nanoemulsions—A review. *J. Control. Release* **2016**, *223*, 85–98. [CrossRef] [PubMed]

138. Clares, B.; Calpena, A.C.; Parra, A.; Abrego, G.; Alvarado, H.; Fangueiro, J.F.; Souto, E.B. Nanoemulsions (NEs), liposomes (LPs) and solid lipid nanoparticles (SLNs) for retinyl palmitate: Effect on skin permeation. *Int. J. Pharm.* **2014**, *473*, 591–598. [CrossRef] [PubMed]

139. Asmawi, A.A.; Salim, N.; Ngan, C.L.; Ahmad, H.; Abdulmalek, E.; Masarudin, M.J.; Rahman, M.B.A. Excipient selection and aerodynamic characterization of nebulized lipid-based nanoemulsion loaded with docetaxel for lung cancer treatment. *Drug. Deliv. Transl. Res.* **2019**, *9*, 543–554. [CrossRef] [PubMed]

MDPI

St. Alban-Anlage 66

4052 Basel

Switzerland

Tel. +41 61 683 77 34

Fax +41 61 302 89 18

www.mdpi.com

Nanomaterials Editorial Office

E-mail: nanomaterials@mdpi.com

www.mdpi.com/journal/nanomaterials